D1745113

PREFACE TO THE SIXTH EDITION

It is a matter of great pleasure to release the thoroughly revised, improved and adequately enlarged edition of "Animal Physiology". To make the new edition more useful, almost all the chapters have been partially rewritten and extensively revised. A number of new tables and illustrations have been added at appropriate places. Three new chapters, *viz.*, **Cell, Osmoregulation** and **Reproduction** have been added afresh. To review the subject matter for the students, revision questions have been given in the end of each chapter.

Finally the authors express their gratitude to the generous readers whose many suggestions, comments and criticism, have been very useful in the improvement of the book. Such suggestions, comments, and criticism for further improvement of the book will be thankfully acknowledged and incorporated in the future edition.

DEPARTMENT OF ZOOLOGY,
MEERUT COLLEGE, MEERUT

Authors

PREFACE TO THE FIRST EDITION

A number of Indian Universities have incorporated **"Animal Physiology"** in their syllabi of B.Sc., B.Sc. (Hons) and M.Sc. classes during the past few years to bring them more in line with current trends in the subject. There are certain books on physiology in the market but they do not fulfil the complete requirements of the teachers and students of B.Sc. and M.Sc. classes as they either lack in the subject matter or are quite elaborative. Keeping in view the requirements of the teachers and the students the present book entitled "Animal Physiology" has been written in which the authors have tried to make the subject as comprehensive, uptodate and concise as was possible. Authors feel that the present book will certainly meet the complete requirements of the teachers and the students. During the preparation of this book several standard latest books and research papers have been freely consulted and the authors do not claim any originality in the matter of this book because this is merely a compilation.

The book contains fourteen chapters, besides glossary, selected reading and index in the end of the book. Each chapter has been written in a very simple and lucid language and deals not only with the physiology of the system but related histology and biochemistry have also been discussed so that the readers may understand the subject easily. To make the matter more understandable many new and well imformative diagrams have been incorporated wherever they were required.

In this book more stress has been given on the mammalian physiology but other animal groups have also been discussed inorder to provide a comparative knowledge of the subject. The last chapter entitled "experimental physiology" deals with the various practical experiments which are helpful in providing the practical knowledge of the subject.

Inspite of our best efforts there might have left certain printing and other mistakes in the matter of the text and it is our request to the colleagues and the students that if they come across any type of mistake while going through the book, kindly point out the same to the authors so that these may be removed in the next edition, for which the authors will feel obliged. Other suggestions and criticisms for the improvement of the book shall also be highly appreciated and will be incorporated whole heartedly.

The authors are thankful to Dr. A. P. Tyagi, D. A. V. College, Muzaffarnagar; Dr. R.G. Sharma and Dr. (Mrs.) R. K. Sharma, Agra College, Agra; Prof. Kuldeepak Kumar, S.D. College, Muzaffarnagar; Prof. Smt. Rekha Agarwal N.R.E.C. College, Khurja; Prof. Nagendra Kumar, M.B. Degree College, Dadri; Dr. Mahesh Chandra Sharma, M.S. College, Saharanpur and Prof. Smt. Sadhana Kaushik, R. G. College, Meerut for rendering various types of help during the preparation of the manuscript of this book.

Finally the authors wish to express their thanks to Shri Shyam Lal Gupta "Padma Shri" and Shri T. N. Goel, M/s. S. Chand and Company Ltd., New Delhi for their sincere cooperation in the publication of this book.

Authors

DEPARTMENT OF ZOOLOGY,
MEERUT COLLEGE, MEERUT

ANIMAL PHYSIOLOGY

(For B. Sc., B. Sc. (Hons.) and M. Sc. Classes of All Indian Universities)

By

P. S. Verma
M Sc., Ph. D., F.E.S.I., F.A.Z.

B. S. Tyagi
M. Sc.

V. K. Agarwal
M. Sc.

DEPARTMENT OF ZOOLOGY
MEERUT COLLEGE,
MEERUT

S. CHAND
AN ISO 9001 : 2000 COMPANY

S. CHAND & COMPANY LTD

RAM NAGAR, NEW DELHI-110 055

S. CHAND & COMPANY LTD.
(An ISO 9001 : 2000 Company)
Head Office : 7361, RAM NAGAR, NEW DELHI - 110 055
Phones : 23672080-81-82; Fax : 91-11-23677446
Shop at: **schandgroup.com**
E-mail: **schand@vsnl.com**

Branches :
- 1st Floor, Heritage, Near Gujarat Vidhyapeeth, Ashram Road, **Ahmedabad**-380 014. Ph. 27541965, 27542369..
- No. 6, Ahuja Chambers, 1st Cross, Kumara Krupa Road, **Bangalore**-560 001. Ph : 22268048, 22354008
- 152, Anna Salai, **Chennai**-600 002. Ph : 28460026
- S.C.O. 6, 7 & 8, Sector 9D, **Chandigarh**-160017, Ph-2749376, 2749377
- 1st Floor, Bhartia Tower, Badambadi, **Cuttack**-753 009, Ph-2332580; 2332581
- 1st Floor, 52-A, Rajpur Road, **Dehradun**-248 011. Ph : 2740889, 2740861
- Pan Bazar, **Guwahati**-781 001. Ph : 2522155
- Sultan Bazar, **Hyderabad**-500 195. Ph : 24651135, 24744815
- Mai Hiran Gate, **Jalandhar** - 144008 . Ph. 2401630
- 613-7, M.G. Road, Ernakulam, **Kochi**-682 035. Ph : 2381740
- 285/J, Bipin Bihari Ganguli Street, **Kolkata**-700 012. Ph : 22367459, 22373914
- Mahabeer Market, 25 Gwynne Road, Aminabad, **Lucknow**-226 018. Ph : 2226801, 2284815
- Blackie House, 103/5, Walchand Hirachand Marg , Opp. G.P.O., **Mumbai**-400 001. Ph : 22690881, 22610885
- 3, Gandhi Sagar East, **Nagpur**-440 002. Ph : 2723901
- 104, Citicentre Ashok, Govind Mitra Road, **Patna**-800 004. Ph : 2671366, 2302100

Marketing Offices :
- 238-A M.P. Nagar, Zone 1, **Bhopal** - 462 011. Ph : 5274723
- A-14 Janta Store Shopping Complex, University Marg, Bapu Nagar, **Jaipur** - 302 015, Phone : 0141-2709153

First Edition 1979
Subsequent Editions and Reprints 1980, 83, 85, 88, 89, 91, 93, 94, 95, 97, 99, 2000, 2002
Reprints 2005

ISBN : 81-219-0351-3

PRINTED IN INDIA

By Rajendra Ravindra Printers (Pvt.) Ltd., Ram Nagar, New Delhi-110 055 and published by S. Chand & Company Ltd., 7361, Ram Nagar, New Delhi-110 055

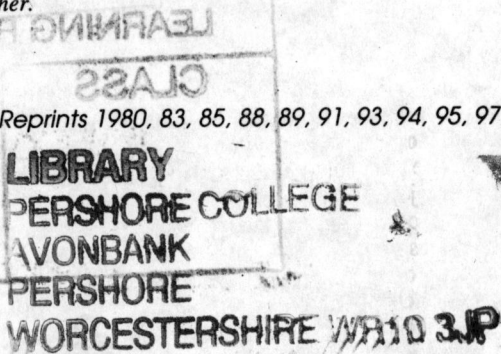

Contents

ANIMAL PHYSIOLOGY

CHAPTER 1

Introduction

Physiology (The term comes from two greek words : *physis*—nature, and *logos*—discourse) is the principal branch of biological sciences and stands above all. *It is the science of processes taking place in living organisms. It studies the functions of the organism, the activities of its different organs in their interaction with environment,* for example the work of muscles, heart, brain and spinal cord. The knowledge of morphology and anatomy is very essential for the science of physiology because the science of physiology is indissolubly associated with these sciences. This statement may be confirmed by the structures and functions of various organs of the body of the organism : the structure of lung is connected with the function of gas exchange, that of kidney with the formation of urine, that of stomach with the digestion of food. It follows that the structure of the organism and its various organs must be considered in relation to their function.

Physiology is very important subject in a medical curriculum as without its knowledge it is impossible to understand the changes caused by disease in the various organs and the organism as a whole. It follows that the knowledge of physiology is essential for the study of any medical subject. In addition, physiology reveals the common origin of man and other animals, and the material character of all the processes taking place in the body of living organism.

Physiology as an independent science was founded in the 17th century after the discovery of WILLIAM HARVEY (1578–1657), the English physician who discovered the circulation of blood, the first physiological activity of the body of that period which came into light.

Physics and chemistry have great significance for the advancement of physiology, since the fundamental task of physiology is to study the various processes or activities going on regularly in the body of living organisms, it is very essential to know the physical and chemical laws underlying each physiological function.

To study physiology the principle of the integrity of the organism and its relation with its external environment is of great importance because all organs of the living organism are interconnected, they continuously interact and make up a common complex system. The organism is also closely connected with its external environment.

BRIEF HISTORY OF PHYSIOLOGY

The basic idea for the development of physiology was the requirements of medicine. The ancients had considerable interest in physiology as they realized that it is impossible to cure a disease without knowing the structure of the body of the organism and the functions of its organs. The physicians of ancient Greece and Rome tried to obtain an idea about the activities or the functions of the different organs of the body, but their conclusions were neither systematic nor scientific as they were based entirely on speculative reasoning resulting mainly from a study of the rough structure of those organs.

Physiology in 17th century : The 17th century is regarded as the date of the foundation of physiology, it is because one of the most essential function of the organism was investigated for the first time. WILLIAM HARVEY was the first who in 1628, investigated the most important activity of the body—**the circulation of the blood**. In 1661, MALPIGHI discovered the capillaries. Due to lack of good apparatuses and techniques further investigations could not be made and thus physiology could not flourish more in the 17th century.

Physiology in 18th century : In the 18th century there were great advances in various fields of physiology which provided the foundation for its further development. ST. HALES was the first who in 1732 measured the blood pressure in the arteries. BERNOULLI in 1738 observed the movement of fluids in the body and laid down the foundation of further haemodynamics. BORELLI pointed out the movements of thorax during the process of respiration, particularly in higher animals. In 1753, LIND gave the idea of certain substances, now called hormones, which regulate the different activities of the body. In 1777, REAMUR and SPALLANZANI discovered the functions and properties of gastric juice secreted by the stomach cells or glands. P. ZAGORSKY studied the vascular system. WHYTT discovered the theory of reflex action and he showed that after the destruction of central nervous system, a frog no longer responds to external stimuli.

The most important discovery of the physiology in 18th century was the application of the laws of inanimate nature, the laws of chemistry to processes taking place in the body of organism. MIKHAIL LOMONOSOV investigated the law of constancy of matter and motion and formulated the kinetic theory of the structure of matter. He also proved that air is a mixture of various gases. LAVOISIER gave the idea of respiration as he showed that the oxidation of organic compounds of the body is carried out by the oxygen present in the air. Later LAVOISIER and LAPLACE demonstrated that oxidative processes yield same amount of heat both during the oxidation of "carbonic combustions" in the organism and during their combustion outside the organism. The concept of irritability was also originated in the 18th century. FRANCIS GLISSON was the man who gave irritability concept. JOHANNES, MULLER and MARSHALL proposed the theory of reflex action.

Thus in the end of 18th century physiology has accumulated certain facts concerning blood circulation and respiration. In that time it was already in a position to make use of scientific data on the microscopic structure of various organs.

Physiology in the 19th century : Great advances came in every field of physiology in the 19th century. The subsequent development of physiology was determined by new methods of scientific observations by the general development of science. Among the most distinguished 19th century scientists who worked in physiology are TAKAKI, EIJKMANN, WEBBER, PFLUGER, BICHAT, MAGENDIE, BELL, A. FILOMOFITSKY, PLESGAFT, V. BASOY and S. BOTKIN. Some of them made discovery in the physiology of endocrines, some in blood and blood-circulation, some in digestion and still others in respiration and nervous system.

Before the middle of 19th century it had taken as a full fledged branch of science. In the same century three great discoveries were made in the natural science, the law of conservation of energy, the cellular theory and the theory of evolution, which brought the swift progress in all biological sciences including physiology. HELMHOLTZ, LUDWIG, MAREY and others tried to find out the facts that how do muscles contract and how impulses are travelled along the nerve fibre. MEYER, PFLUGER, PASHUTIN and LIKHACHOV discovered the law of conservation and transformation of energy which made the physiology to arrive at a quantitative description of physiological processes from the aspect of energy-loss and energy-acumulation.

In the second half of the 19th century physiology attained great success as many great discoveries were made in that period. CLAUDE BERNARD as early as 1859 pointed out that complex organisms live in two environments–an external environment and an internal environment which surrounds the cells and tissues of the body, is characteristic of the animal species and remains relatively constant and serves as medium for the exchange of foods and wastes and for the distribution of chemical messengers. WEBBER (1860), PFLUGER (1900) tried to observe the mechanism of absorption. WEBBER, BEZOLD, CYON and PAVLOV investigated the influence of nervous system on the heart. BERNARD, LUDWIG, WALTHER and OVSYANNIKOV tried to observe the influence of nervous system on the blood vessels. MAGENDIE, HELMHOLTZ, WEDENSKY studied the influence of nervous system on the skeletal muscles, while PFLUGER, MISLAVSKY and LANGLY studied the influence of nervous system on smooth muscles of the alimentary canal.

In the end of 19th century PAVLOV investigated the most important concept of digestion. DUBOIS and REYMOND investigated the electrical phenomena in nerves and muscles, the work which was already started in the 18th century by GALVANI. HERMANN and WEDENSKY gave the concept of nerve impulse. NENTSKY and PAVLOV studied the transformation of products of metabolism in the liver.

Physiology in 20th century : In the 20th century physiology had reached at the peak of its development. With the development of sophisticated instruments and techniques the various problems of structure and functions of different organs of the organism are being elaborated by modern scientists. The characteristic feature of latest scientific research is that it is conducted on a molecular, sub-microscopic and cellular level. For this purpose very fine and complex experiments and methods are used. These experiments and methods make it possible to study processes occurring in the cell and in the various structures which form part of the cell. In the 20th century all the aspects of physiology have been studied by the modern scientists. BAYLISS and STARLING, later GABRIEL and FOGEL, in 1902, discovered the first hormone called **secretin** produced by the wall of the alimentary canal. S. GYORGYI, WELSH 1957, GORBMAN and BERN 1962, BARRINGTON 1963, VON EULER, HELLER 1963, also discovered certain neurosecretory substances and hormones. LANGLY 1921 discovered the concept of autonomic nervous system. HYMAN 1951, ANDREW 1959, MORTON 1960 gave the information about the movements of the gut contents in different animals. YONGE 1937, OWEN 1950, NICOL 1960, BARRINGTON 1962, discovered various types of feeding habits in animals. SCHILLING 1902, VAN HERWERDEN 1908, GREEN 1912, FRAZER 1940 and MELLANBY described the mechanism of absorption. BARD 1961, CARTER, WHITFORD and HUTCHINSON 1963, FOXON 1964 elaborated the knowledge of the process of respiration. RUDD 1954, FOX, VEVERS 1960, MANWELL 1960, GRATZER and ALLISON 1960, ANTHONY 1961 discovered the transportation of respiratory gases, *e. g.*, O_2 and CO_2. NEEDHAM 1931, SUMNER 1951, SMITH 1953, KREBS, COHEN and BROWN 1960, PROSSER and BROWN 1961, BALWIN 1963, discovered the various aspects of the process of excretion in different animals. VERWORN 1899, HEILBRUNN 1953, BERNARD and BAYLISS elaborated the various aspects of the concept of irritability which was already originated by FRANCIS GLISSON in the 17th century. SZENT GYORGYI 1944, INOUE 1959, HOFFMAN, BERLING 1960, SATIR 1961, HOYLE, ALLEN 1961, LEHNINGER 1962, PERUTZ 1964 and RINALDI discovered the various activities of the effector organs including muscles and nerves. In this way the various aspects have been studied by the modern scientists. Now physiology has made its independent existence amongst biological sciences.

FIELDS OF PHYSIOLOGY

For practical purposes the physiology can be divided into following categories :

1. **The general physiology** or **Cellular physiology** : The general physiology or cellular physiology deals with those basic characteristics which are extremely stable with respect to the environment and are common to most living organisms. Different organisms have more common structures at the cellular level because different cells resemble in their fundamental properties. Cellular specializations may brought some diversity in the cells.

2. **Physiology of special groups** : The physiology of special groups deals with the characteristics that are most common in particular group either of plants or animals, such as insect physiology, fish physiology and physiology of parasites.

3. Comparative physiology : Comparative physiology deals with the organ function in a wide range of groups of organisms or in other words it studies the physiological phenomena in their phylogenetic development in various species of invertebrates or vertebrates. It integrates and also co-ordinates functional relationship which transcent special groups of animals. It is concerned with the ways in which different organisms perform similar functions.

BRANCHES OF PHYSIOLOGY

The important branches of physiology, which have made their independent entity, are as follows :

1. Pathological physiology : This branch of physiology deals with disturbances that occur in various functions of the organism during certain diseases.

2. Clinical physiology : This branch of physiology deals with problems that come during the fundamental tasks of clinical practice.

3. Ecological physiology : This branch of physiology studies the peculiarities of physiological processes in various species of animals in relation to the various conditions of their existence.

Revision Questions

1. **What is physiology ? Describe brief history of physiology.**
2. **Describe different fields and branches of physiology.**

CHAPTER 2

Cell

Although the term **cell** was first used by ROBERT HOOKE, an English microscopist, in 1665 to describe a honey-comb like compartment seen in the structure of cork, the concept of the cell as the unit of structure and function of all living things was recognized about 1839 when the researches of MATHIAS JAKOB SCHLEIDEN ano THEODOR SCHWANN came into light as the **cell theory**. VIRCHOW, in 1859, confirmed the cellular hypothesis showing that all cells must necessarily be derived from pre-existing cells : *Omnis cellula e cellula.* Due to the employment of various improved micro-techniques such as micromanipulators, chromatography, electrophoresis, isotopic techniques, ultra-microtomes, electron microscopy, X-ray microscopy and various micro-methods of dissection, in the study of cells, the validity of the **cell theory** has become vague and the place of this theory has been taken by the **"modern cell theory"** which explicitly states that (1) all living matter is composed of cells ; (2) all cells arise from pre-existing (other) cells ; and (3) all the metabolic reactions of a living organism ; including all energy exchanges and all biosynthetic processes, take place within cells. It also states that cells contain the hereditary information of the organisms of which they are a part, and that this information is passed from parent cell to daughter cell.

In modern terms the cell may be defined as "*a dynamic, self-directed, and highly organised complex system of molecules and molecular aggregates which appropriates and utilizes the energy of its surrounding for the purpose of growth and reproduction*. The energy of the environment is available to most cells for their life activities either as energy stored in the chemical bonds of such substances as carbohydrates, fats, and proteins ; or as light energy, a form used exclusively by green plants and some bacteria. LOEWY and SIEKEVITZ (1965) defined the cell as "*a unit of biological activity delimited by a selectively permeable membrane and capable of self reproduction in a medium free of other living systems*".

The cell, as it fulfills the requirements of a living system in terms of structure and function, has been said to be the fundamental unit of life. The typical activities of it are an expression of the co-ordinated behaviour of the most highly integrated and organized group of components or subcellular parts present in the cell. So to understand the cell we must, therefore, understand its component parts as isolated structures ranging from the gross to the molecular, and as whole structures interacting within the integrity of the intact cell.

The living cell performs all the functions of life, such as intake of nutrients, metabolism, growth, respiration and reproduction, etc. To carry out these functions of life the cell has various cellular components and subcellular parts in its interior. Thus, in unicellular living forms the activities of the cell constitute the activities of the organism. In multicellular organisms the integrated activities of the various constituent cell types are responsible for the characteristic activities and behaviour of the whole organism. So to understand the essential physiology of organisms we must know the complete physiology of cell.

EUKARYOTIC CELLS

The cells which occur in animals (from Protozoa to Mammalia) and plants (from algae to angiosperms) and contain true nuclei are called **eukaryotic cells**. They are very much larger than prokaryotic cells (the cells which lack a distinct nucleus enclosed by a nuclear envelope, the nuclear material being scattered in the cytoplasm). The eukaryotic cells are essentially two **envelope systems** and secondary membranes envelope the nucleus and other cellular organelles and to a great extent they pervade the cytoplasm. These cells may vary in shape, size, and physiology but all the cells have a typical structure with little variations in number and location of cellular organelles.

The classical model of a generalized cell, based on the observations of the light microscope, shows that the cell possesses an outer limiting membrane, called the **plasma membrane** or **cell membrane** ; a mass of **cytoplasm** which seemed to possess a vague internal organization or cytoskeleton and which contains a membrane-bounded **nucleus** and various cellular organelles, such as the **mitochondria, Golgi complex, centrioles** and various other ill-defined structures of granular and vacuolar nature.

The introduction of electron microscope in 1939–40 made possible detailed examination of cellular organelles and other subcellular entities and it altered the model of the generalized cell. Cellular organelles such as the mitochondria, which in the light microscope appear as rod-shaped bodies, in the electron microscope are seen to be membranous structure of highly organized nature. The matrix of the cell, originally called the protoplasm by VON MOHL (1846), which in the light microscope appears as a homogeneous, clear, colloidal suspension, in the elelectron microscope is shown to consist of a multiphased complex system, containing numerous membranous elements. Before going into the details of eukaryotic cell and its various cellular components, it will be advisable to consider the general features of different types of eukaryotic cells and these are the following :

Shape : The animal and plant cells exhibit diverse forms and shapes. The typical shape of animal cell is spherical but irregular, triangular, tubular, cuboidal, polygonal, cylindrical, oval, rounded and elongated cells have also been reported in animals and plants. The shape of the cell may vary from animal to animal and from organ to organ. Even the cells of the same organ may exhibit variations in shape.

Variations in the shape of cells may be because of cell walls, or because of attachments to and pressure from other, neighbouring cells or surfaces. Moreover, external or internal environment may also cause variations in the shape of cells.

Size : The eukaryotic cells are usually microscopic but definitely they are larger in size than the prokaryotic cells. The size of cells may range from 1μ to 175000μ (175 mm). The ostrich egg cell is usually considered as largest cell about 175 mm diameter and certain nerve cells have been found to have the length of 3 or 3·5 feet.

A principal restriction on cell size seems to be the relationship between volume and surface area. A second restriction on cell size appears to involve the capacity of the nucleus, the cell's control centre to regulate the cellular activities of a large, metabolically active cell.

STRUCTURE

The eukaryotic cell comprises the following components :

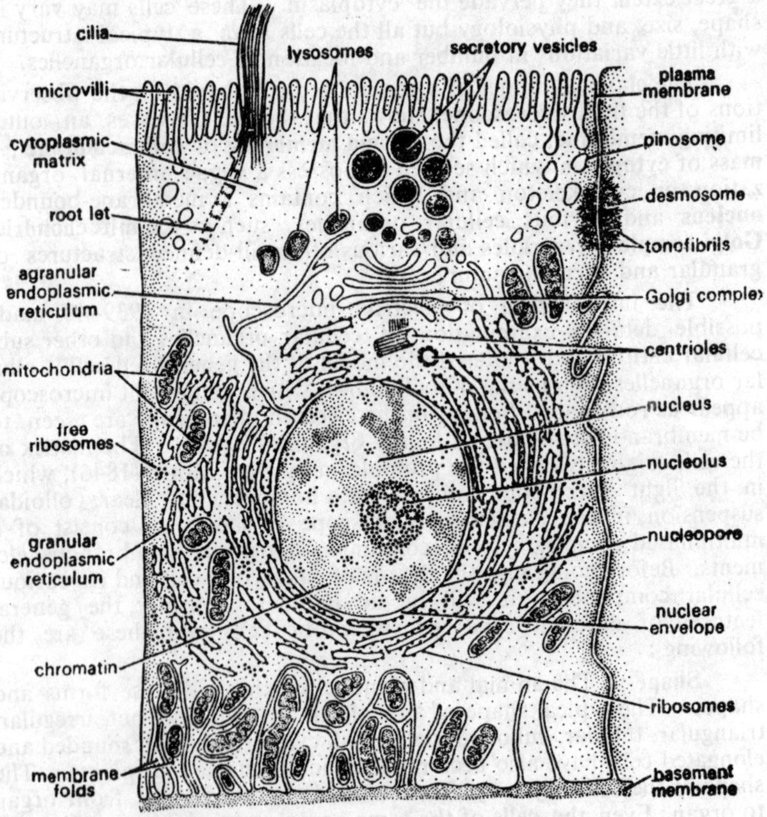

Fig. 2·1. Structure of a typical animal cell.

1. Cell wall and plasma membrane ; 2. Cytoplasm ; and 3. Nucleus.

CELL WALL AND PLASMA MEMBRANE

Cell wall : The cell wall is the outer most covering of the plant cells and which is absent in animals. It separates the protoplasm of plant cell from the external environment and provides protection to the cell as a whole. It is a thin, semi-rigid, flexible, laminated and external non-living covering. It is composed chiefly of **cellulose** and is secreted, molecule by molecule, by the cell as it grows. Between walls of adjacent cells is a thin glacy layer called the **middle lamella** which is composed of pectins and other non-cellulose polysaccharides and

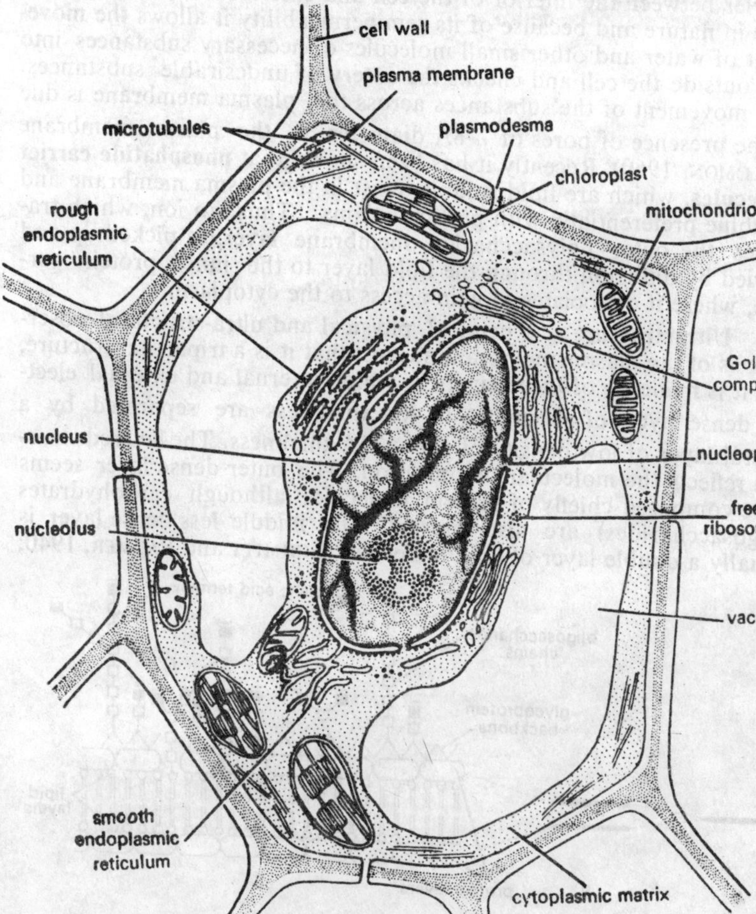

Fig. 2·2. Structure of a typical plant cell.

serves to hold the cells in place. Adjacent cells in the plant body remain interconnected by **plasmodesmata**, which are fine strands of cytoplasm crossing through cell walls and middle lamella.

Plasma membrane : Cells of all types are bounded at their outer surface by a specialized limiting membrane called the **plasma membrane** or **plasmalemma** or **cell membrane**. The plasma membrane is a living, ultra-thin, elastic, porous, semipermeable membranous covering. It is so thin (75 to 100Å (angstroms), one Å$=10^{-4}$ μm) that its actual structure can be observed only under electron microscope.

Functions : The plasma membrane provides an important barrier between the interior of the cell and exterior. It is semipermeable in nature and because of its semipermeability it allows the movement of water and other small molecules of necessary substances into and outside the cell and checks the entery of undesirable substances. The movement of the substances across the plasma membrane is due to the presence of pores of 7–8Å diameter in the plasma membrane (SOLOMON, 1960). Recently it has been shown that **phosphatide carrier molecules,** which are lipid soluble, exist in the plasma membrane and combine preferentially with particular ions. Thus, an ion, which traverses the outer polysaccharide membrane layer, is picked up and carried by diffusion through the lipid layer to the inner protein surface, where it dissociates to gain access to the cytoplasm.

Ultrastructure : Chemical, physical and ultra-structural observations of the plasma membrane suggest that it is a tripartite structure, *i.e.*, it is made up of three layers in which internal and external electron dense layers of about 20Å to 40Å thickness are separated by a central layer of low density of about 35Å thickness. The layered structure reflects the molecular organisation. The outer dense layer seems to be composed chiefly of protein molecules although carbohydrates (oligosaccharides) are also present. The middle less dense layer is actually a double layer of lipid molecules (SCHMITT and PALMER, 1940;

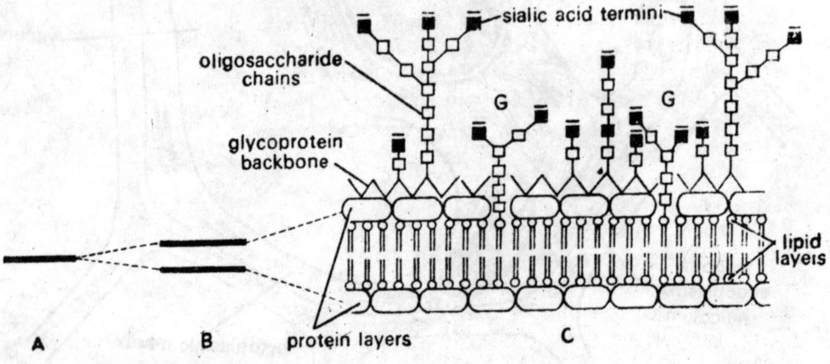

Fig. 2·3. The structure of plasma membrane as observed. A—at low magnification of compound microscope, B—at high magnification of compound microscope, C—ultrastructure in electron microscope.

MITCHISON, 1953). The lipid molecules occur in chain and two molecular chains of lipids remain parallel to each other forming a **bimolecular** or double-layered structure. The inner dense layer is also made up of protein molecules. There is a considerable evidence that some of the protein of both the outer and inner layers, protrudes into the middle layer.

The lipid layers of the plasma membrane are linked with each other by the inner ends of lipid molecules which are non-polar and hydrophobic (Gr., *hydra*—water, *phobe*—hate) in nature. Both these lipid layers are held together due to VANDER WAAL'S forces at these non-polar ends. The lipid molecules remain linked with the molecules of protein layers by their outer, polar and hydrophilic (Gr., *hydra*—water, *phil*—loving) ends. In the hydrogen bonds, ionic linkages or electrostatic forces bind the molecules of lipids and proteins together.

The tripartite structure of the plasma membrane was proposed by DANIELLI and DAVSON in 1935 and later PICKEN (1960) and ROBERTSON (1962, 1964) confirmed the DANIELLI and DAVSON'S observations about the plasma membrane. ROBERTSON in 1960 called the plasma membrane as the **unit membrane** but KORN (1966, 1969) criticized the unit membrane concept.

Recently GREEN and PERDUE proposed that biological membranes are built of lipoprotein macromolecular repeating units. FREY-WYSSLING (1957) ; SJOSTRAND (1963) and others have reported the presence of globular particles as repeating units in a variety of animal and plant cells.

Specialized structures of plasma membrane : Plasma membrane often exhibits certain specialized structures in some regions of its surface. According to SELBY (1959) these structures are of following five types :

(1) Microvilli : The plasma membrane at its surface has certain narrow elongated projections called **microvilli**. Their function is to increase its absorptive surface.

(2) Caveolae : These are the blind inpushings or cave-like invaginations of plasma membrane, first reported by YAMADA (1955). These also increase the absorptive surface of the cell.

(3) Desmosomes or **Maculae adhaerentes :** In certain cells the plasma membrane of adjacent cells becomes thicker in certain regions and from these thickened areas (or plaques) many fine filaments known as **tonofilaments** or **tonofibrils** are radiated towards the interior of the cell. Such thickened areas of the plasma membrane are called **desmosomes** or **Maculae adhaerentes**. The intercellular space between desmosomes contains a coating material which provides cellular adhesion to the cell.

(4) Plasmodesmata : These are fine strands of cytoplasm through which the adjacent cells remain interconnected in the plant body. These structures were first of all observed by TANGEL (1879) but finally by STRASBURGER (1882).

(5) Tonoplast : It is a vacuolar membrane found in plants. It is semipermeable in nature and perhaps arises from Golgi body or endoplasmic reticulum.

CYTOPLASM

The cytoplasm is a complex, heterogeneous substance which occupies the space between the plasma membrane and the nucleus. Basically it consists of a rather watery medium **"the ground substance"** in which is suspended, a variety of distinct structures. It includes all the functioning substances of the cell (the protoplasm) except the nucleus. It consists of various inorganic substances, such as water, salts of sodium, potassium and phosphates and other metals and organic compounds, *viz.*, enzymes, macromolecules, ATP, amino acids, nucleotides and electron carriers. Depending on the type of cell, it may also contain specialized materials, such as haemoglobin, starch granules, pigments, yolk granules, secretory granules, glycogen granules and oil droplets. It also includes the cellular organelles which perform various important functions and biosynthetic and metabolic activities. The most important cytoplasmic organelles are—**endoplasmic reticulum, ribosomes, Golgi complex, lysosomes, peroxisomes, mitochondria, plastids, cytoplasmic vacuoles, microfilaments, microtubules, cilia** and **flagella, centrioles** and **basal granules** or **kinetosomes.**

PORTER (1961) reported that the cytoplasm or hyaloplasm occurs between the canals of the endoplasmic reticulum. It appears structureless under the electron microscope except for the presence of scattered ribosomes and very fine microfilaments and microtubules (LEDBETTER and PORTER, 1963 ; SLAUTTERBACK, 1963).

Endoplasmic reticulum : The 'interior of the cell' or cytoplasm is interlaced by a highly complicated membranous system of vesicles, flattened sacs, cisternae and canals, originally called the **"ergastoplasm" by** GARNIER in 1897, and now known as the **endoplasmic reticulum**. Much of the surface of the endoplasmic reticulum is covered with small granules ($\sim$ 150Å in diameter). These are the RNA-containing ribosomes and such portions of the endoplasmic reticulum are termed **rough endoplasmic reticulum (RER)**. The RER is the site of protein synthesis in the cytoplasm. The portions of the endoplasmic reticulum which lack granules (ribosomes) are the **smooth endoplasmic reticulum (SER)**. Smooth endoplasmic reticulum has connections with outer unit membrane of the nuclear envelope and in some cases with the plasma membrane (ROBERTSON, 1959). It has the same basic structure as the cell membrane and the nuclear membranes. In some cells it also appears to connect with the Golgi complex and is also sometimes intimately associated with the mitochondria (BRANDES, 1965).

Functions : The endoplasmic reticulum divides the cell into separate compartments, making it possible for the cell's different chemical products and activities to be segregated from each other. Many of the enzymes that carry out these activities form a part of the lipoprotein structure of the membrane forming the endoplasmic reticulum. The tubules and vesicles have surfaces which may well play a role in enzymic reactions (SJOSTRAND, 1964).

The endoplasmic reticulum in addition to being the site of cytoplasmic protein synthesis, is also thought to function as a transport and storage system (ROBERTSON, 1959). It also plays important role in lipid metabolism (SIEKEVITZ, 1963, 1965). Glycogen synthesis may occur in the membranes of endoplasmic reticulum (COIMBRA and LEBLOND 1966 ; FREEMAN, 1966). The membranes of endoplasmic reticulum also possess an electron transport system, the details of which are under study (SIEKEVITZ, 1965, 1972).

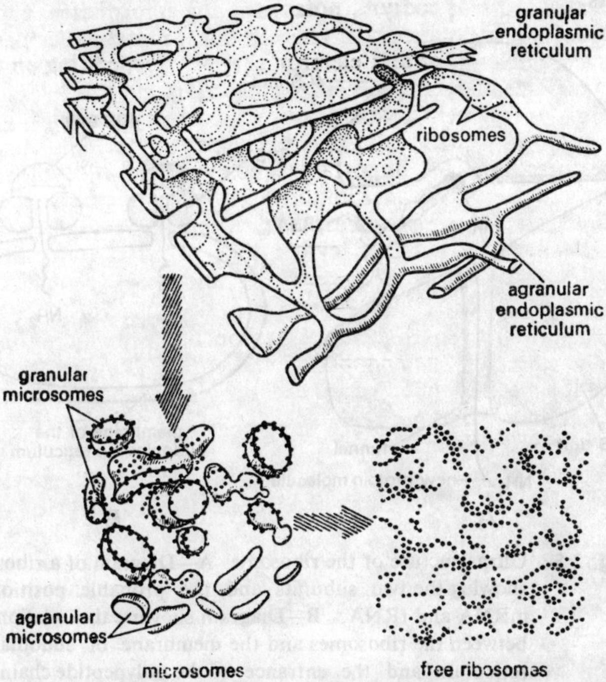

Fig. 2·4. Three dimensional structure of endoplasmic reticulum showing microsomes and ribosomes.

Origin : DALLMER *et al.*, 1966, reported that the endoplasmic reticulum is originated from the plasma membrane by invagination giving rise to a canalicular system. But DE ROBERTIS *et al.*, 1970, suggested that the endoplasmic reticulum originates from the nuclear envelope which it resembles physically and chemically.

Ribosomes : Both eukaryotic and prokaryotic cells contain ribosomes. These are the smallest structures of the cell which can be seen only under the electron microscope, first reported by PALADE (1953). These are roughly spherical bodies of about 150Å diameters and are not bounded by any membrane. They are found usually attached to outer surfaces of the ER and its vesicles (PALADE, 1953). Each ribosome consists of two structural **sub-units,** one slightly larger (**60 Svedberge** or **60S sub-unit**), than the other (**40S sub-unit**). The ribosomes remain attached with the membranes of ER by 60S subunits. The 40S sub-units occur on the larger sub-units and form cap-like structure.

Ribosomes are originated in nucleolus and are composed of relatively small molecules of RNA and of protein. PETERMANN, 1964, reported several kinds of RNA from ribosomes on the basis of analysis of purine and pyrimidine bases. These are ribosomal RNA's or rRNA. 5S, 18S and 28S rRNA's. The 28S and 5S rRNA's occur in large (60S) subunits, while 18S rRNA occurs in the smaller ribosomal subunits.

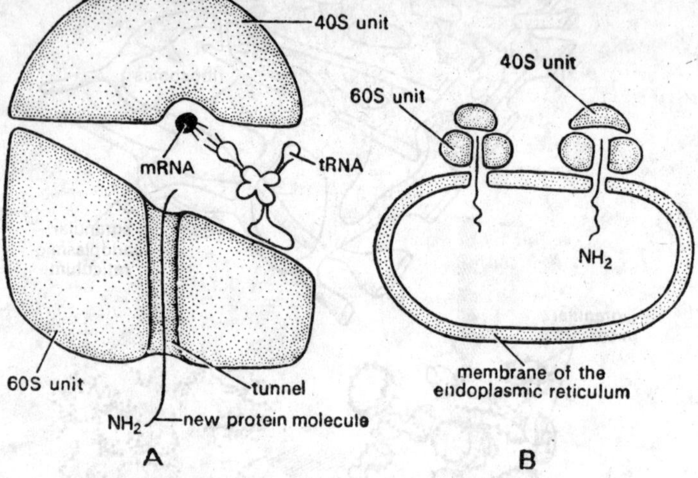

Fig. 2·5. Ultrastructure of the ribosome. A—Diagram of a ribosome showing the two subunits and the probable position of mRNA and tRNA, B—Diagram showing the relationship between the ribosomes and the membrane of endoplasmic reticulum and the entrance of the polypeptide chain into the centre of endoplasmic reticulum during the process of protein synthesis.

Ribosomes are the sites where all the proteins of the cell, including all enzymes, are synthesized. These literally carry out the instructions contained within the DNA code of the nucleus. In cells with rough endoplasmic reticulum, the membrane net-work forms a pathway by which substances are transported in and out of the cell.

Golgi apparatus : The Golgi apparatus (formerly complex) is found in the cytoplasm in almost all types of eukaryotic cells. Animal cells usually contain one; plant cells may have as many as several hundred. In glandular cells the Golgi apparatus is usually situated close to the nucleus and has characteristic polarity (GRASSE, 1957).

The Golgi apparatus, reported by **Golgi** in 1898, consists of a smooth surfaced membrane-limited structure. It has a system of outer flattened cisternae which appear as roughly parallel membranes in the form of an ellipse enclosing a space 60–90Å with a distance of about 200Å between them; clusters of vesicles around the ends and outer surfaces of the cisternae; and large, clear vacuoles which are associated with the inner cisternae particularly in glandular cells.

Functions : The Golgi apparatus is believed to play an important role in the synthesis and storage of secreted products (glucoproteins, mucopolysaccharides and enzymes) (CARO and PALADE, 1964; NEUTRA and LEBLOND, 1966, 1969). It is also considered as the source of primary lysosomes-membrane bounded particle containing hydrolytic enzymes. It also functions as packing centre of the cell. Evidence suggests that proteins synthesized for export, formed by ribosomes on the surface of the endoplasmic reticulum, are channeled into Golgi apparatus where they accumulate and are packaged in vesicles in which they travel to the outer cell membrane and are discharged to the exterior of the cell. The Golgi membrane then becomes part of the plasma membrane (NORTHCOTE, 1971).

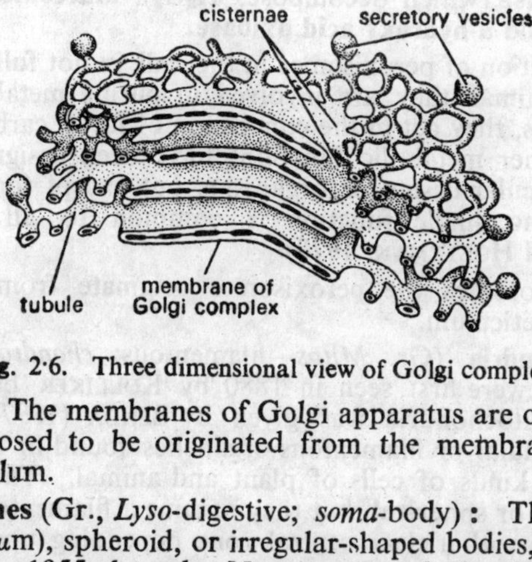

Fig. 2·6. Three dimensional view of Golgi complex.

Origin : The membranes of Golgi apparatus are of lipoproteins and are supposed to be originated from the membranes of endoplasmic reticulum.

Lysosomes (Gr., *Lyso*-digestive; *soma*-body) **:** These are tiny, (0·25 to 0·8 μm), spheroid, or irregular-shaped bodies, first reported by de DUVE in 1955, later by NOVIKOFF *et al.*, in 1956. They are distinguished from other types of particles in the cytoplasm mostly on the basis of their biochemical properties, namely, their content of specific enzymes, which are primarily hydrolytic in function. Like endoplasmic reticulum the lysosomes are covered by a single unit membrane.

Functions : The basic function of the lysosomes is to digest the food material which comes in the cell by pinacocytosis and phagocytosis, because they are characteristically rich in digestive enzymes with pH optima in the acid range. About 36 hydrolytic enzymes have been identified in lysosomes, including those which digest proteins, nucleic acids, polysaccharides, lipids, organic-linked sulfate and organic-linked phosphate though all do not occur in any one lysosome. Some important lysosomic enzymes are **acid ribonuclease, acid deoxyribonuclease, acid phosphatase, cathepsin, collagenase,**

β-glucuronidase, β-galactosidase, d-mannosidase, d-glucosidase, β-N-acetylglucosaminidase, and **aryl sulfatase**. There are many kinds and sizes of lysosomes (NOVIKOFF and HOLTZMAN, 1970).

Origin : The lysosomes originate either directly from the endoplasmic reticulum or from vesicles of the Golgi apparatus, or from both (DALTON, 1961, NOVIKOFF *et. al.*, 1964; de DUVE and WATTIAUX, 1966).

Peroxisomes : Many cells, both plant and animal, contain a different kind of membrane-bounded structures that somewhat resemble lysosomes. These structures first called **microbodies** but it is thought that they participate in metabolic oxidations involving hydrogen peroxide and they are, therefore, called **peroxisomes (glyoxysomes)** (BAUDHUIN, *et. al.*, 1965, de DUVE, 1965). They are smaller than mitochondria and lysosomes and contain a very dense crystalline material in a dark granular matrix.

Peroxisomes contain enzymes and other materials. They contain enzymes-**catalase** (which decomposes H_2O_2), **urateoxidase, D-amino acid oxidase** and **α-hydroxy acid oxidase**.

The function of peroxisomes in the cell is not fully understood but in some animals they are involved in purine metabolism and in some mammals, they act in the conversion of fat to carbohydrate. A number of other metabolic functions have been assigned to peroxisomes in particular tissues or organisms **but except** for the activities of **catalase,** no single function is common to all peroxisomes (NOVIKOFF and HOLTZMAN, 1970).

Like lysosomes the peroxisomes originate from the smooth endoplasmic reticulum.

Mitochondria (Gr., *Mitos*—filamentous; *chondros*—granules) : Mitochondria were first seen in 1880 by KOLLIKER but the present name, *i.e.*, mitochondria was given by BENDA (1897–98) to them. These are granular or filamentous organelles found in the cytoplasm of almost all kinds of cells of plant and animal. They are usually roughly ovoid or spherical, but they assume a filamentous shape or a vesicular form with a clear central zone depending upon the physiological conditions of the cells.

Mitochondria vary in size. They are usually 0·5 to 1·0 μm in diameter and in length upto a maximum of 7 μm. The number of them in a cell depends on its type and stage of development. In a liver cell as many as 2500 mitochondria may be present. They are numerous in brown fat cells while in plant cells their number is very small.

Mitochondria contain a limiting double unit membrane of lipoprotein. The outer membrane of the mitochondrial limiting membrane forms a bag-like structure around the inner unit membrane. The inner unit membrane has a complex structure consisting numerous parallel, flattened, interdigitating invaginations, the mitochondrial **cristae** (composed of membrane pairs). These cristae project from both sides into the interior as sheets or tubules.

The unit membrane of the mitochondria appears to be similar

to the cell membrane or plasma membrane. It is also made up of three layers, a middle lighter layer less opaque to electrons-sand wiched between two electron-opaque (darker) layers. According to ROBERTSON (1959) the lighter middle-inner layer is made up of two rows of lipid molecules with their non-polar groups in the centre, the two denser outer layers are composed of protein molecules.

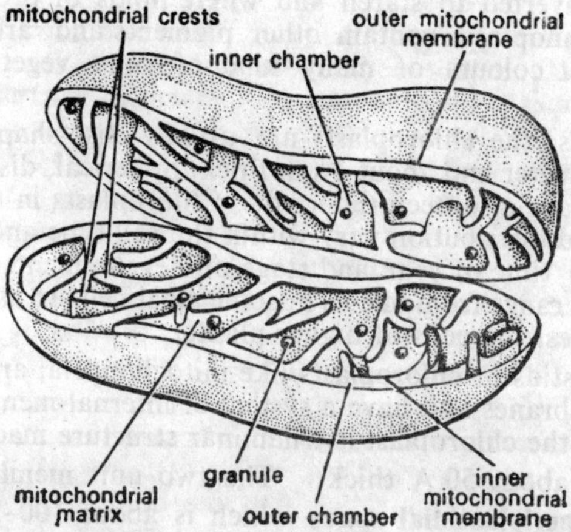

Fig. 2·7. Detailed structure of a typical mitochondrion.

Cristae are most common in the mitochondria of cells with a high metabolic activity. A finely granular, homogeneous **matrix** fills the lumen of the mitochondria. It also contains some very **dense granules** which are formed from the precipitation of insoluble salts of calcium and other ions accumulated by the mitochondria. Adhering to their surface are a very large number of different kinds of enzymes.

Mitochondria are frequently located to intracellular lipid droplets. Sometimes the lipid droplets appear to be fused to the mitochondria, particularly after a short of starvation or in rapidly metabolizing tissue such as cardiac muscles. They are also found located at the surface of cells particularly in those which are actively engaged in absorption or excretion, for example the intestinal epithelium or renal tubular cells.

Functions : The function of the mitochondria is quite clear. The membranes of cristae contain several enzymes and other molecules which carry out oxidation-reduction reactions and phosphorylation reactions, associated with energy metabolism (NOVIKOFF and HOLTZMAN, 1970). COPELAND and DALTON, 1959, suggested that they might serve to supply energy for protein synthesis.

Origin : There have been suggestions that mitochondria arise from various membrane systems in the cell, *viz.*, plasma membrane, endoplasmic reticulum (ROBERTSON, 1961), Golgi apparatus and nuclear envelope (PAPPAS and BRANDT, 1959., ROODYN and WILKIE, 1968).

Plastids : Plastids are microscopic organelles found in the cells of plants and some protozoans, such as *Euglena* and the cells of most photosynthetic microorganisms. They are of three general types : **leucocytes, chromoplasts** and **chloroplasts**. Of these chloroplasts which contain chlorophyll (green pigments) are biologically the most important because they are the organelles responsible for photosynthesis. Leucocytes are colourless and are the sites at which glucose is converted to starch and where lipids or proteins may be stored. Chromoplasts contain other pigments and are responsible for the bright colours of many kind of fruits, vegetables, flowers and leaves.

In plants, the chloroplasts are usually disc-shaped structure, 5–8 μm in diameter and about 1 μm thick. Spherical, discoid or ovoid chloroplasts have also been reported. Chloroplasts in their number, shape, size and distribution vary within the cell type and species, and often with the time of year and stage of development. The chloroplasts in a cell range from one-as in some algal cells-to as many as fifty in photosynthesizing cells of higher plants.

The plastids or chloroplasts like mitochondria, are bounded by two unit membranes and have a system of internal membranes. Each membrane of the chloroplast is trilaminar structure made up of lipoproteins and about 50 Å thick. The two unit membranes remain separated by **periplastidial space** which is about 100–300Å wide in *Nicotiana*. The inner periplastidial space of the chloroplasts is filled with a watery, proteinaceous and transparent substance known as the **matrix** or **stroma**. In the matrix the unit structures of the chloroplasts, the **grana** and **intergrana connecting membranes** are found embedded.

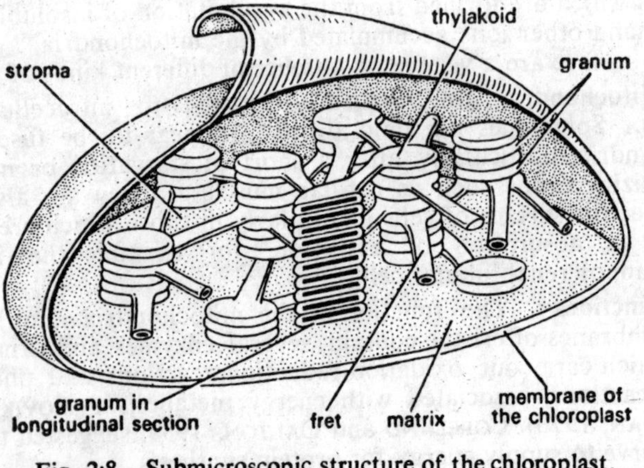

Fig. 2·8. Submicroscopic structure of the chloroplast.

The grana are lamellar or membranous, granular or chlorophyll bearing bodies. They vary in size from 0·3 to 1·7μm in various species; in spinach chloroplasts they are about 0·6 μm in diameter. In a given species grana may be present in chloroplasts in some locations and absent from those in others; for eaxmple, in maize the mesophyll chloroplasts have grana but sheath parenchyma chloroplasts do not (BISHOP, 1971).

Each granum is made up of 10–100 disc-like, super imposed stacks of membranous sacs called **thylakoids** or **thylacoids** (thylacoid-sac-like). Each thylacoid remains separated from the stroma or matrix of the chloroplast by its unit membranes. The grana of chloroplast are interconnected by a network of numerous tubules, the **stromal lamellae** or **frets**. The stromal lamellae or frets are extensions of membranes of certain thylacoids into the matrix. They are made up of many paracrystalline spheroid particles called **quantasomes** (PARK, 1966). They are beliveved to be unit of photosynthesis, each consisting 250 molecules of chlorophyll.

Origin : Chloroplasts arise from nunute sub-microscopic amoeboid **proplastids**. The proplastids have double membranes like the mitochondria. When the proplastid attains a diameter of 1 μm., its inner membrane invaginates to form lamellae which in light continue developing and differentiating to mature condition found in the chloroplast. Chloroplasts have also been seen to elongate and divide in the some plants (KAMEYA, 1972).

Cytoplasmic vacuoles : The cytoplasm of many plant and some animal cells (*i.e.*, ciliate protozoans) contains numerous small or large-sized hollow spaces filled with water and solutes. The spaces which are less than 100 nanometer in diameter are called **vesicles** while which are larger than 100 nanometer are known as **vacuoles.** Vacuoles are supposed to be greatly expanded endoplasmic reticulum or Golgi apparatus.

The vacuoles of animal cells are bounded by a lipoproteinous membrane and their function is the storage, transmission of materials and the maintenance of internal pressure of the cell. The vacuoles of the plant cells are bounded by a single semi-permeable membrane called **tonoplast**. Immature plant cells characteristically have many vacuoles and as the plant cells mature, the numerous smaller vacuoles coalesce into one large, central, fluid-filled vacuole that then becomes a major supporting element of the cell (KITCHING, 1956). In plant cells such vacuoles also may serve the functions of storing sucrose or of accumulating of waste products.

Microfilaments : Microfilaments as their name implies, are fine, thread-like fibrils some 40–60 Å in diameter. They are believed to be capable of contracting. Large numbers of them form masses or network at various places in the cell.

Microfilaments are believed to be made up of one of the major protein components (**actin**) of muscle cells. More recently, these microfilaments have been shown to be involved with the movements of the cytoplasm, called amoeboid movement, or cyclosis, or cell streaming depending on what type of cells they are observed in.

Microtubules : These are straight tubular structures and are found in many cells. They are usually about 250Å in diameter but **filaments** as thin as 150 Å have also been observed. They are extended for several microns or more and being stiff provide some rigidity to those parts of the cell in which they are located. They have a dense wall 45–70 Å thick surrounding a less dense central area. In addition, each microtubule appears to be surrounded by a slightly staining region about 100 Å thick from which organelles and particulate matter of the cytoplasm are excluded. The ultrastructure of microtubules was first described by LEDBETTER and PORTER, 1964.

Microtubules play an important role in cell division and are also used in the construction of centrioles, cilia and flagella. They also transport the water, ions and small molecules.

Cilia and flagella : Many cells have long and thin (about 0·2 micrometer) cytoplasmic structures extending from the surface of the cell, these are known as **cilia** or **flagella**. Cilia are short motile organelles, generally present in large numbers, covering the surface of some cells (*e.g.*, the ciliates). Flagella (characteristic of the flagellates, for example) are similar but larger and less numerous as well as more complex in their movement.

Cilia and flagella fundamentlly have the same origin and structure. In each case they grow out of a basal body. These have the same structure as centrioles and are formed by them. Ultrastructurally the cilia and flagella consist of nine outer fibrils around the two large central fibrils identical to microtubules in their construction. Each outer fibril consists of two microtubules. In both cilia and flagella, the entire assembly is sheathed by a unit membrane which is simply a part of the cell membrane.

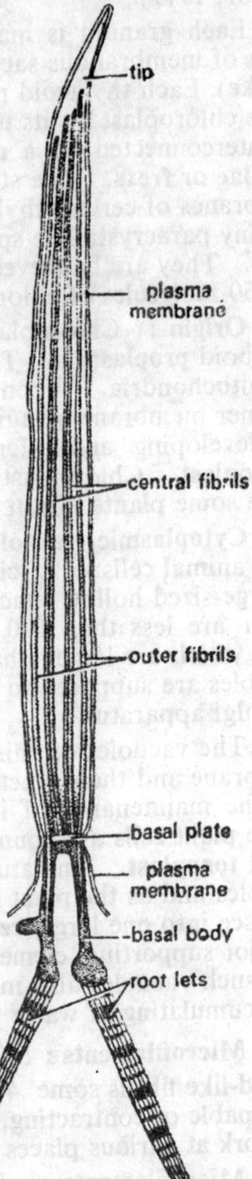

Fig. 2·9. Structure of a flagellum.

Cilia and flagella are generally used for locomotion. However, many animals have cells, the cilia of which serve simply to move materials past the cells.

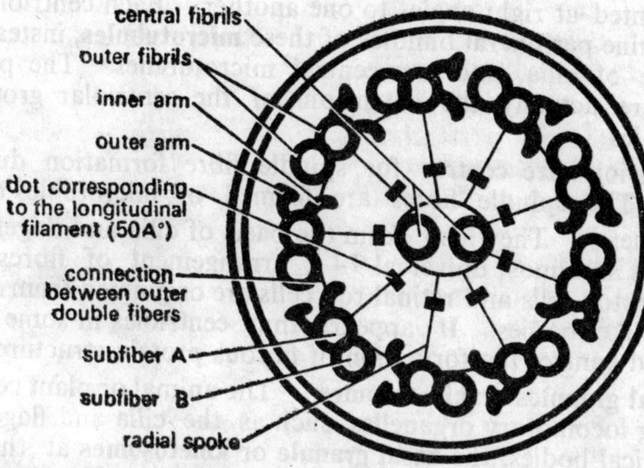

Fig. 2·10. Diagrammatic transverse section of a flagellum.

Centrioles : Animal cells and the cells of some microorganisms and lower plants contain two membranous and fibrous bodies near the nucleus in the cytoplasm. These are **centrioles**, each one is a structure about 0·2 μm in diameter. They were first reported by BENDON in 1887.

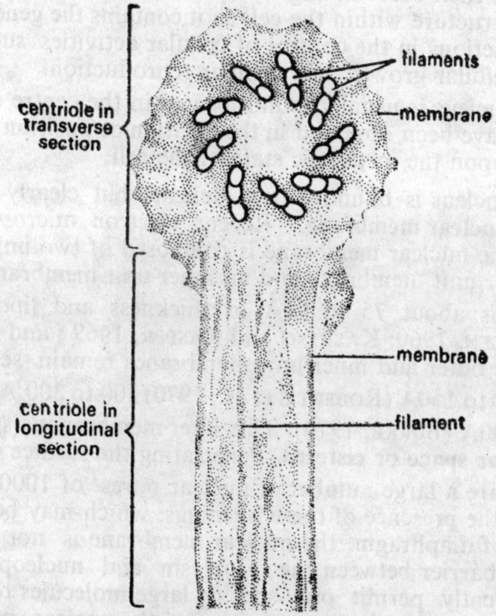

Fig. 2·11. A centriole in transverse and longitudinal section.

Ultrastructurally the centrioles are cylindrical elements having a structural plan essentially like that of basal granule and the cilium. Centrioles are generally paired, the two cylindrical elements always being oriented at right angles to one another. Each centriole is composed of nine peripheral bundles of these **microtubules**, instead of two as in case of cilia, and two central microtubules. The peripheral bundles are actually the extensions of the centriolar groupings of microtubules.

Centrioles are centres for spindle fibre formation during cell division. The spindle fibres are formed of microtubles of 230 to 270Å diameter. They also form the basis of cilia and flagella whose structure is an almost universal 9+2 arrangement of fibres. Many visual receptor cells and retinal rod cells are originated from centrioles and cilia-like bodies. It appears that centrioles in some way can initiate and control the formation of fibrous protein structures.

Basal granules or **kinetosomes :** The animal or plant cells which are having locomotory organelles, such as the cilia and flagella, contain spherical bodies, the **basal granule** or **kinetosomes** at the base of the cilia and flagella. They are found embedded in the ectoplasm and are composed of nine fibrils. Each fibril consists of three microtubules, out of which two enter in the cilia or flagella. The basal granule may contain both DNA and RNA.

NUCLEUS

All eukaryotic cells contain a large, pale, spherical or oval body in their surrounding cytoplasm, the **nucleus**. It is most prominent structure within the cell as it contains the genetical material and also functions in the control of cellular activities such as protein synthesis, cellular growth, and cellular reproduction.

The nucleus is usually found located in the centre of the cell but variations have been observed in the location or position of the nucleus depending upon the metabolic state of the cell.

The nucleus is bounded by a delicate but clearly defined membrane, the **nuclear membrane**. Recent electron microscopic studies show that the nuclear membrane is composed of two unit membranes, *viz.*, an outer unit membrane and an inner unit membrane. Each unit membrane is about 75 to 90Å in thickness and lipoproteinous in nature (PICKEN, 1960; KASHING and KASPER. 1969 ; and SMITH *et. al.*, 1969). The outer and inner unit membranes remain separated by a space of 100 to 150Å (ROBERTIS *et al.*, 1970) 100 to 300 Å (COHN 1970) or 400 to 700Å (BURKE, 1970). This inter membranous space is known as **perinuclear space** or **cisterna**. Penetrating the surface of the nuclear membrane are a large number of nuclear pores of 1000 Å diameter. Because of the presence of these openings, which may be closed with some type of diaphragm, the nuclear membrane is not considered as a complete barrier between the cytoplasm and nucleoplasm. These pores apparently permit only specific large molecules to go through, thus keeping the chemical composition of the nuclear material different from that of the cytoplasm.

It is believed that the nuclear membrane is composed of terminal parts of the endoplasmic reticulum or cisternae surrounding the nucleus (Watson, 1959, 1965 ; Grell, 1964) and that the pores which provide direct continuity between the nucleus and the cytoplasmic sap represent gaps between the **"end-feet"** of the endoplasmic reticulum. The nuclear membrane is of profound importance because during interphase the nucleus can pass information to the cytoplasm only through the nuclear membrane in controlling or in directing the synthetic activities occuring in the cytoplasm of the cell (Zbarskii, 1969).

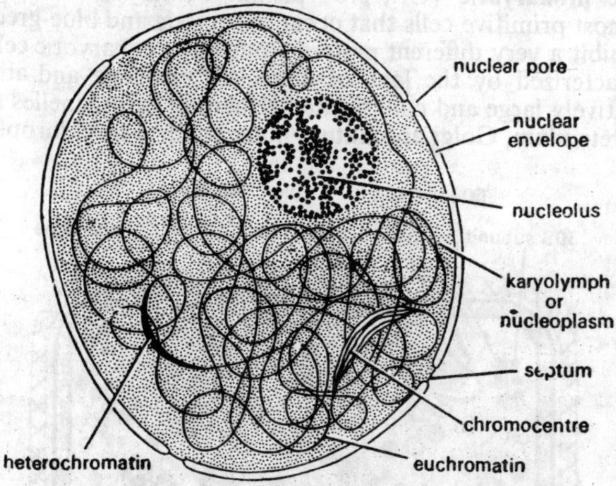

Fig. 2·12. Structure of a metaphase nucleus.

Within the nuclear membrane there is a semifluid medium, the **nuclear sap** or **nucleoplasm** or **karyolymph**. It is composed of mainly the nucleoproteins but it also contains other inorganic and organic substances including nucleic acids, proteins, enzymes, and minerals.

The nucleoplasm contains a tangle of fine coiled and elongated threads called the **chromatin**. Such chromatin fibres are observed when the cell is not dividing, *i.e.*, at interphase. During the cell division (mitosis or meiosis) chromatin fibres become thick ribbon-like structures which are known as **chromosomes**. Chemically the chromosomes are made up of DNA and histone proteins.

The most conspicuous body within the nucleus is the **nucleolus**. It is a large, spherical and acidophilic granule reported first by Fontona in 1781. The nucleolus, which is composed of DNA and protein, like the chromosomes, is formed in association with certain portions (called the **nucleolar organizers**) of certain chromosomes. It also contains long fibres, presumably composed of protein and perhaps serving as structural elements. It is the site at which a particular kind of RNA—the ribosomal RNA—is formed ; hence, the nucleolus is rich in RNA. The nucleolus is most prominent when the cell is synthesizing protein.

The nucleus in general performs two crucial functions for the cell. First, it carries the hereditary information for the cell ; second, it regulates the ongoing activities of the cell, ensuring that the complex molecules the cell requires are synthesized in the number and of the kind needed. These molecules are involved in carrying out the various activities of the cell and also in forming organelles and other structures.

PROKARYOTIC CELLS

The **prokaryotic** (Gr., *pro*—primitive ; *karyon*—nucleus) cells are the most primitive cells that occur in bacteria and blue-green algae. They exhibit a very different pattern from those eukaryotic cells. They are characterized by the lack of a nuclear envelope and absence of such relatively large and complicated structures or organelles as endoplasmic reticulum, Golgi apparatus, mitochondria or chloroplasts.

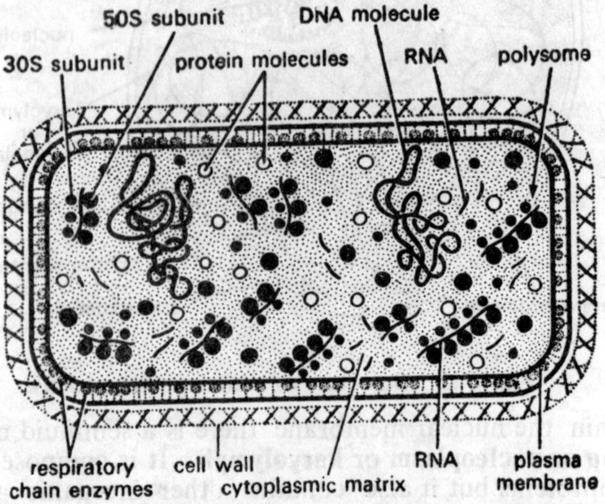

Fig. 2·13. A prokaryotic cell of *Escherchia coli.*

A prokaryotic cell is essentially a **one-envelope** system organized in depth. It consists of central nuclear components (*viz.*, DNA molecules, RNA molecules and nuclear proteins) surrounded by cytoplasmic ground substance, with the whole enveloped by a plasma membrane. The cytoplasm is filled with dense granules of 100-200Å diameter. Many of these granules are ribosomes and the granules do not appear to be associated with any membranous structures. In the photosynthetic bacteria there are vesicular infolding of the cell membrane in which pigments are located. These are known as chromatophores. Occasionally, the outer cell wall projects into an invagination of plasma membrane, forming a **septum**. This invagination may extend beyond the edge of the septum, forming a highly organized structure called a **mesosome** to which a **chromatin body** is frequently attached. The cell membrane is often reduplicated to form the **polar membrane.**

Revision Questions

1. Describe in brief a cell, either plant or animal cell as viewed under electron microscope.

2. Give an account of ultrastructure and functions of the following : endoplasmic reticulum, Golgi apparatus and mitochondria.

3. Discuss the structure, nature and functions of plasma membrane.

4. Describe different organelles of cytoplasm.

5. Write short notes on :

(i) Cilia and flagella,

(ii) Peroxisomes,

(iii) Lysosomes,

(iv) Ribosomes,

(v) Nucleus,

(vi) Nucleolus,

(vii) Centrioles,

(viii) Prokaryotic cell.

CHAPTER 3

Food

All living organisms require constant replenishment of energy giving and building substances for work, for maintenance of heat and for repair of continuously disintegrating cells of various tissues. Plants synthesize these substances from simple chemical substances present in the soil and air where as the animals by contrast get these substances for these purposes from their environment in the form of animal or vegetable food and water. **Food, thus, can be defined as any essential substance that, when absorbed into the body tissues yields materials for the production of energy, the growth and repair of tissue and the regulation of life processes, without harming the organisms.**

Sources of food : Wheat, rice, meat, butter, milk, fruits and vegetables are said to be the best sources of food. The ratio of these food may differ in different diet.

COMPOSITION OF FOOD

Whatever the food (meat, bread or milk, etc.) animals take that is composed of **essential nutritive substances** mainly of **proteins, carbohydrates** and **fats** and hence the foods are classified as protein, carbohydrate and fat foods. In addition to these nutritive substances each food stuff also contains **mineral-salts, water** and **vitamins** that form part of the protoplasm of all cellular structures. Mention should also be made of the **micro-elements** of which the negligible quantities are contained in the foods.

CLASSIFICATION OF NUTRITIVE SUBSTANCES OF FOOD

The nutritive substances of the animal food can be classified in different ways : 1. On the basis of their chemical constitution and 2. On the basis of their functions in the body.

1. On the basis of their chemical constitution the nutritive substances are classified as : (i) **Proteins,** (ii) **Carbohydrates,** (iii) **Fats,** (iv) **Mineral salts,** (v) **Water** and (vi) **Vitamins.** They may again be classified as **nitrogenous** and **non-nitrogenous** substances subject to the presence or absence of nitrogen element in their constitution. Protein is the only nitrogenous substance because it contains nitrogen element in its constitution, while rest of the substances are known as non-nitrogenous because they do not have nitrogen element

in their constitution. Similarly they may also be classified as **organic** and **inorganic** substances depending upon the presence or absence of carbon-element in their constitution. Thus, proteins, carbohydrates and fats are referred to as organic substances because they have carbon element in their constitution while mineral-salts and water are referred to as inorganic substances as they don't have any carbon element in their constitution.

2. On the basis of their functions in the body the nutritive substances fall under the following heads :

1. **Body builders :** Proteins, mineral salts and water are referred to as body-builders because they are the only substances which enter into the composition of the body.

2. **Energy producers :** Carbohydrates and fats on their oxidation yield certain amount of energy which is consumed by the animal in order to perform the various vital activities of the body. These substances, therefore, are referred to as "energy producing substances" or energy producers. Energy is always valued in terms of **calorie**.

Carbohydrates are said to be the best energy producers because they are completely oxidised easily. Fats are said to be the second best energy producer, while proteins are oxidised only in the deficiency of carbohydrates and fats.

3. **Regulators :** Mineral salts and vitamins are capable enough to regulate the different functions of the body, therefore, they are referred as regulating substances or regulators.

PROTEINS

Proteins are the most abundant chemical compounds of the living condition. They are versatile, complex, and fragile macromolecules with high molecular weights ranging from about 6500 to several million. They not only serve as fuel to yield energy but also play vital role in every aspect of the structural and functional characteristics of the living condition because they are essential constituents of the protoplasm and also form the physical basis of life. Therefore, a liberal supply of protein is necessary to the organisms and without it no life is possible. It must be constantly supplied to organisms for growth and repair. **Mulder** was the first man who introduced the word protein in 1840. Actually the term 'protein' comes from the Greek word '*protos*' meaning first.

Proteins are polymeric molecules in which the subunits are monomers or amino acids which are co-valently bonded by **peptide linkages** or **peptide bonds** formed between the carboxyl group of one amino acid and the amino group of the next. Proteins may be composed of a single polypeptide chain (a polypeptide may be defined as a chain of three or more amino acids joined together by peptide linkages) or of several polypeptide chains. The composition of the protein in terms of these amino acids as well as their actual arrangement in the molecule is of great importance not only to the structure of the polymer but also to its biologic activity. In other words, biologic activity is to be understood in terms of protein structure. The actual sequence of amino acids, called the **primary structure,** is determined genetically and resides in the language of the DNA molecules.

Proteins are responsible for most fundamental features of structure and function which underlie all living systems and are also the cause of the tremendous diversity of living systems. Enzymes are proteins which catalyze the different chemical reactions taking place in the living systems. Thus, proteins play the unique role of determining the pattern of chemical transformations in biological systems. As important components of muscle, tendon, bone, membrane, and skin, proteins are involved in structure and motility. Similarly there are specific proteins which transport and storage the small molecules and ions and give high tensile strength to skin and bones and capable for generation and transmission cf nerve impulses. Thus, their diverse physical, chemical, and biologic properties are dependent on the nature and arrangement of their constituents subunits, the **amino acids.**

Chemical composition of proteins : Proteins are the complex organic compounds of high molecular weight. They are mainly composed of carbon, hydrogen, oxygen, nitrogen and sulphur elements in their molecules. However, other elements like phosphorus, iron and copper may also be present.

Basic structural units of protein molecules : When protein molecules are carefully dismantled chemically, it is found that they consist of chains of much smaller units of the same fundamental chemical nature called **α-amino acids** of which more than twenty are known to occur naturally as components of proteins. These units have been so named, on account of having a-basic amino group (—NH_2) and carboxyl group (—COOH) in their constitution. Thus, proteins are polymers of amino-acids but the amount, the number and the arrangement of the amino-acids may differ in different protien molecules.

Amino acids : Amino acids are the basic structural units of proteins, therefore, EMIL FISCHER called them as building stones of the proteins. An amino acid consists of an amino group (—NH_2), a carboxyl group (—COOH), a hydrogen atom, and a distinctive R group bonded to a carbon atom, which is called the **α-carbon** (Fig. 3·1). An R group is referred to as a **side chain.**

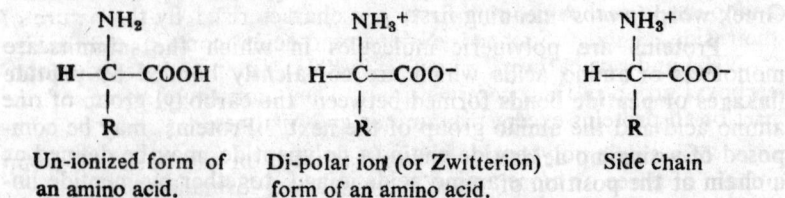

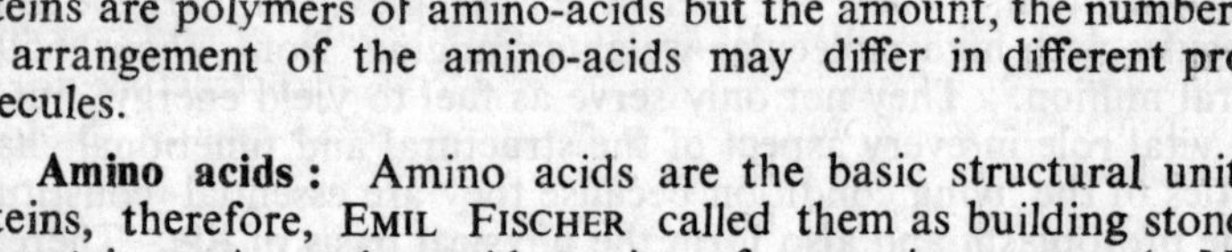

Un-ionized form of an amino acid. **Di-polar ion (or Zwitterion) form of an amino acid.** **Side chain**

Fig. 3·1. Structure of the un-ionized and Zwitterion forms of an amino acid.

Amino acids in solution at neutral pH are predominantly dipolar ions (or **Zwitterions**) rather than un-ionized molecules. In the dipolar form of an amino acid, the amino group is protonated ($—NH_3^+$) and the carboxyl group is dissociated ($—COO^-$). The ionization state of an amino acid varies with pH. In acid solution, the carboxyl group is un-ionized (—COOH) and the amino group is ionized ($—NH_3^+$). In alkaline solution (*e. g.*, pH 11), the carboxyl group is ionized ($—COO^-$) and the amino group is un-ionized ($—NH_2$).

More than twenty amino acids have been recorded so far which are used by the living organisms to form the proteins. It is not necessary that all amino acids should found in every protein. Some of the amino acids are referred as essential amino acids because they can not be synthesized by the organisms. These, therefore, must be included in the diet of the organisms from any external source. The remaining amino acids are referred to as dispensible or non-essential amino acids because they can be synthesized very well by the organisms themselves provided the other nutrient elements along with nitrogen and carbon compounds continue to be available. Threonine, phenylalanine, lysine, tryptophan, valine, methionine, leucine, isoleucine, arginine are the essential amino acids, while alanine, aspartic acid, cysteine, glutamic acid, glycine, proline, serine, tyrosine are the non-essential amino-acids of man.

All amino acids except glycine are optically active because of the asymmetry of the α-carbon atom, which is covalently bonded to four different groups. It is, therefore, possible to have a D-form and L-form of each amino acid. Only **L-amino acids** are constituents of proteins.

Twenty kinds of side chains varying in size, shape, charge, dehydrogen-bonding capacity, and chemical activity are commonly found in proteins. Indeed, all proteins in all species, from bacteria to man, are constructed from the same set of twenty amino acids. The remarkable range of functions mediated by proteins results from the diversity and versatility of these twenty kinds of building blocks because they create the intricate three-dimensional structure for proteins that enable proteins to participate in so many biological processes.

CLASSIFICATION OF AMINO ACIDS

Amino acids, on the basis of number of amino and carboxyl groups, are classified as follows :

1. **Neutral amino acids :** The neutral amino acids are mono-amino-mono-carboxylic acids and are characterized by the nature of their side chains. Such type of amino-acids contain one amino ($—NH_2$) and one carboxyl group (—COOH), *i.e.*, mono-amino mono-carboxyl group in their composition. They constitute most of the part of all proteins except protamines and histones.

These amino acids are further classified on the basis of group present at the position of 'R' of the emperical formula :

(i) **Aliphatic amino acids :** Such amino-acids contain aliphatic group at the position of 'R'. Examples—glycine, alanine, valine, leucine and isoleucine.

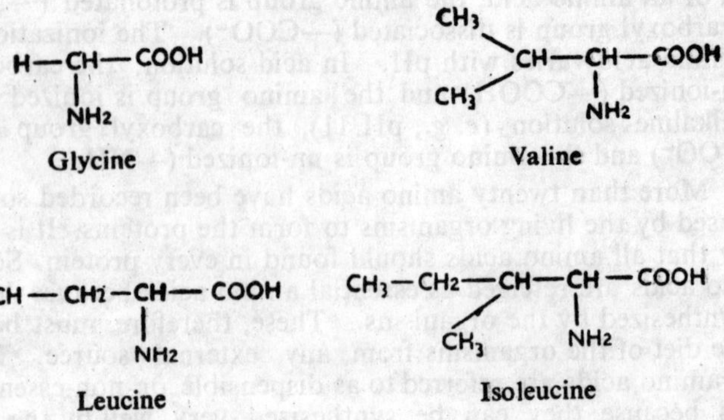

Fig. 3·2. Structure formulae of glycine, valine, leucine and isoleucine.

(ii) **Aromatic amino acids :** These amino-acids contain aromatic group at the position of R. Examples—phenylalanine, tyrosine and tryptophan.

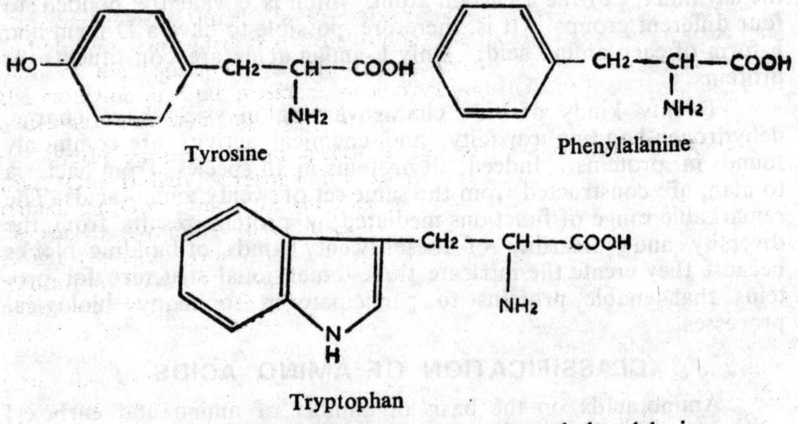

Fig. 3·3. Structure formulae of tyrosine, and phenylalanine tryptophan.

(iii) **Sulphur containing amino acids :** In these amino-acids R is represented by sulphur containing group. Examples—cysteine, cystine, methionine.

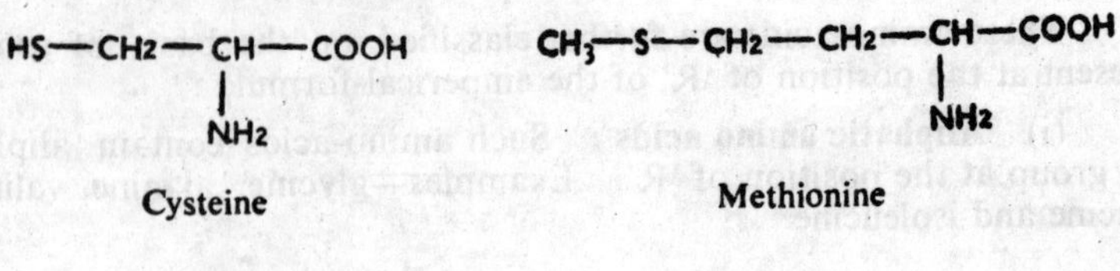

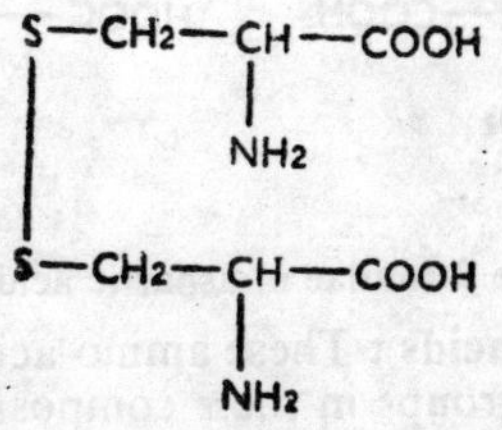

Cystine

Fig. 3·4. Structure formulae of cystein, methionine and cystine.

(iv) **Hydroxy amino acids :** These amino-acids contain hydroxyl group at the position of R. Examples—serine, threonine.

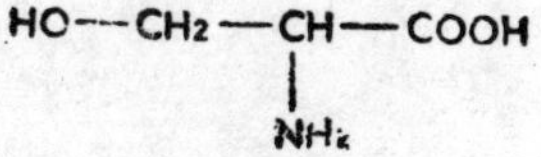

Serine

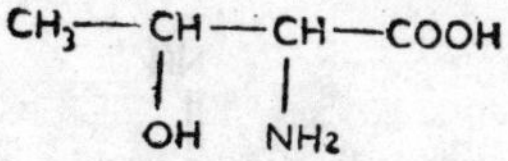

Threonine

Fig. 3·5. Structure formulae of serine and threonine.

(v) **Heterocyclic amino acids :** In these amino-acids such as histidine, proline and hydroxyproline, R is represented by heterocyclic group.

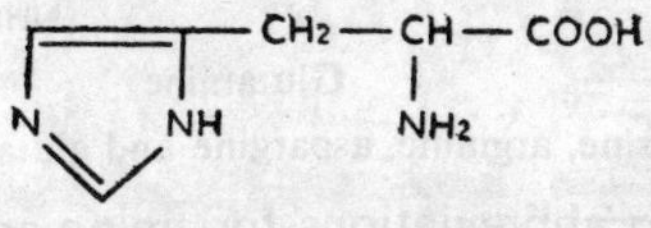

Histidine

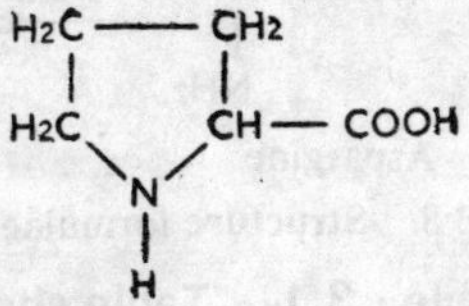

Proline

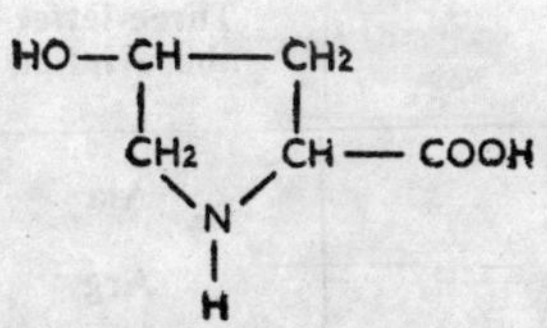

Hydroxyproline

Fig. 3·6. Structure formulae of histidine, proline and hydroxyproline.

Proline differs from other amino acids in the basic set of twenty in containing a secondary rather than a primary amino group. Strictly speaking, proline is an **imino-acid** rather than an amino-acid. The side chain of proline is bonded to both the amino group and the α-carbon, thereby forming a cyclic structure.

2. Acidic amino acids : Such type of amino-acids contain one amino group ($-NH_2$) and two carboxyl groups ($-COOH$) in their composition. Examples—asparatic acid, glutamic acid.

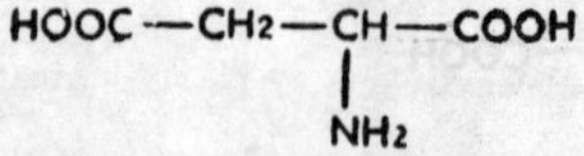

Aspartic acid

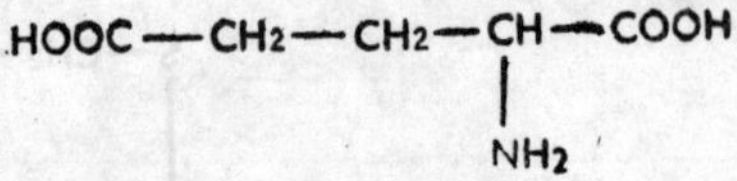

Glutamic acid

Fig. 3·7. Structure formulae of aspartic acid and glutamic acid.

3. **Basic amino acids** : These amino-acids contain one carboxyl group and two amino groups in their composition. Examples—lysine, arginine, glutamine and aspargine.

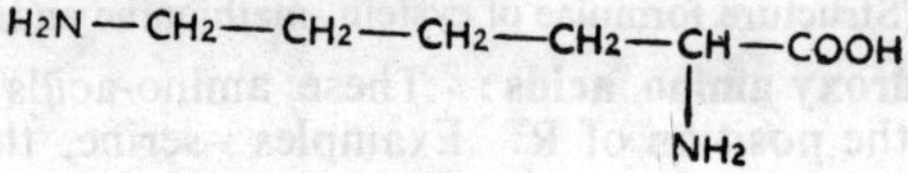

Lysine

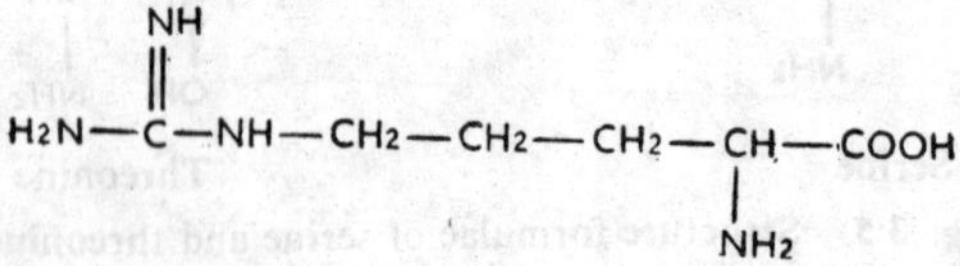

Arginine

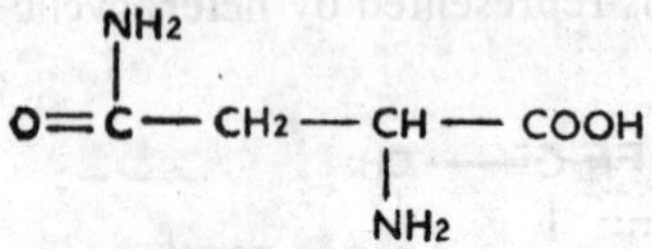

Aspargine

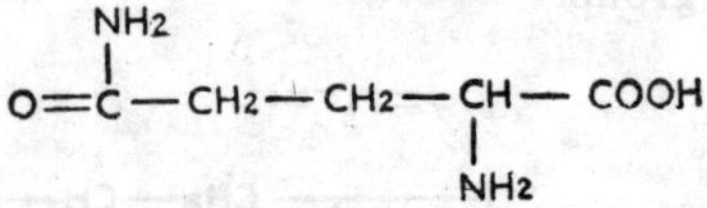

Glutamine

Fig. 3·8. Structure formulae of lysine, arginine, aspargine and glutamine.

Table : 3·1. Table showing abbreviations for amino acids.

Amino-acid	Three-letter abbreviation	One-letter symbol
Alanine	Ala	A
Arginine	Arg	R
Asparagine	Asn	N
Aspartic acid	Asp	D
Asparagine or aspartic acid	Asx	B
Cysteine	Cys	C
Glutamine	Gln	Q
Glutamic acid	Glu	E
Glutamine or glutamic acid	Glx	Z

Amino-acid	Three-letter abbreviation	One-letter symbol
Glycine	Gly	G
Histidine	His	H
Isoleucine	Ile	I
Leucine	Leu	L
Lysine	Lys	K
Methionine	Met	M
Phenylalanine	Phe	F
Proline	Pro	P
Serine	Ser	S
Threonine	Thr	T
Tryptophan	rp	W
Tyrosine	Tyr	Y
Valine	Val	V

HOW PROTEINS ARE FORMED FROM AMINO ACIDS

Since amino-acids contain both amino ($-NH_2$) and acidic or carboxyl group ($-COOH$) in their molecules, therefore, they may act as acids or alkali bases at a time. The molecules of such acidic and basic properties are known as **amphoteric molecules**. Due to amphoteric properties the amino acids unite with one another forming large and complex protein molecules. The amino-acids are linked together in the protein molecule in chains. So, when the amino-acids link together one after another during protein synthesis, then the terminal OH of carboxyl group ($-COOH$) of one amino-acid combines with one H of the amino group ($-NH_2$) of the next amino-acid to form water which is removed from there and the amino-acid link is joined. The $-OC.NH$ link so formed at the place where the water molecule has been removed is known as **peptide bond**.

When two amino-acids are united by peptide bond, the resultant product thus formed, is known as **dipeptide**. This sort of process of

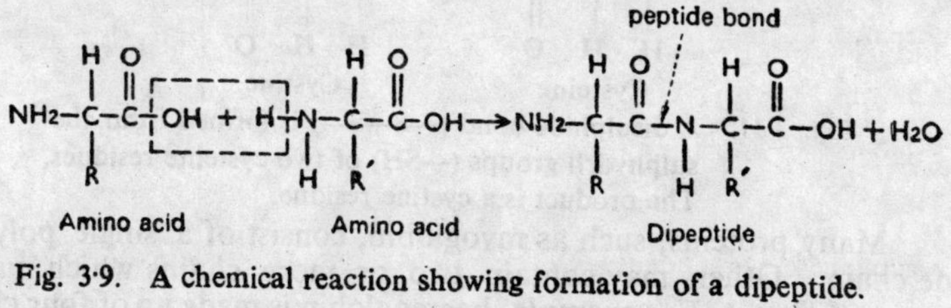

Fig. 3·9. A chemical reaction showing formation of a dipeptide.

linking can go on repeatedly to form **tripeptide** (with three amino-acid links) or **polypeptides** with larger number of amino-acid links. When a polypeptide contains fifty-or more amino-acid links it begins to show the characteristics of protein. Usually large peptides are sometimes also called **peptones**.

A polypeptide chain of amino-acids is an unbranched structure and an amino-acid unit in a polypeptide chain is called a **residue**. A polypeptide chain has direction because its building blocks have different ends-namely, the α-amino and the α-carboxyl groups. By convention, the amino end is taken to be the beginning of a polypeptide chain. The sequence of amino-acids in a polypeptide chain is written starting with the amino-terminal residue. Thus, in the tripeptide alanine-glycine-tryptophan, alanine is the amino-terminal residue and tryptophan is the carboxyl-terminal residue.

A polypeptide chain consists of a regularly repeating part, called the **main chain**, and a variable part, the distinctive **side chain**. The main chain is sometimes termed the **backbone**.

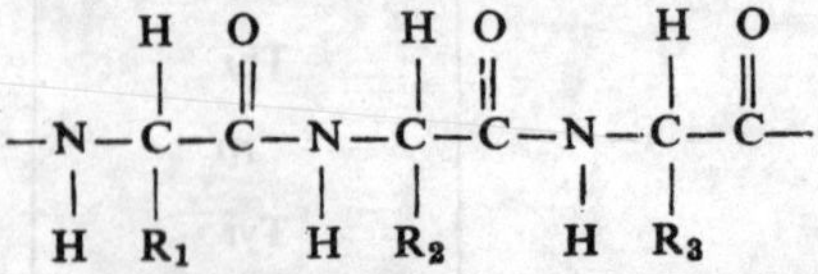

Fig. 3·10. A polypeptide chain made up of regularly repeating **back bone** and distinctive **side chains** (R_1, R_2, R_3).

In some proteins, a few side chains are crossed-linked by **disulphide bonds**. These cross-bonds are formed by the oxidation of cysteine residues. The resulting disulphide is called cystine. There are no other covalent cross-bonds in proteins.

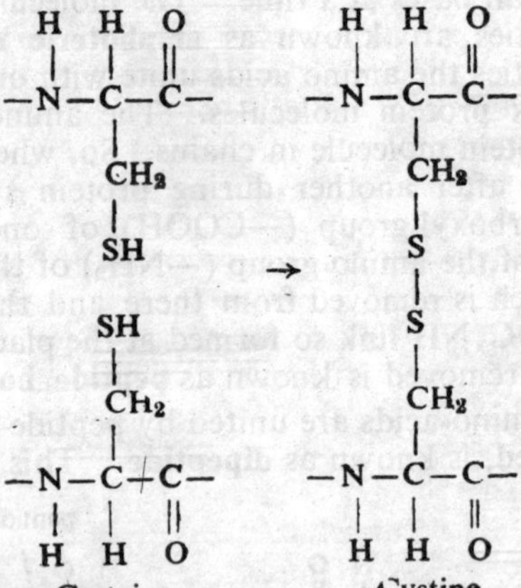

Fig. 3·11. A disulphide bond (—S—S—) is formed from the sulphydril groups (—SH) of two cysteine residues. The product is a cystine residue.

Many proteins, such as myoglobin, consist of a single polypeptide chain. Others may contain two or more chains which may be same or different. For example, haemoglobin is made up of four chains out of which two are of one kind and two of another kind. These four chains are held together by noncovalent forces. Alternatively, the polypeptide chains of some proteins are linked by disulphide bonds as in case of insulin.

Proteins have unique amino-acid sequences that are specified by genes, *i.e.*, the amino-acid sequences of proteins are genetically determined. The sequence of nucleotides in deoxyribonucleic acid, specifies a complementary sequence of nucleotides in a ribonucleic acid (RNA), which in turn specifies the amino-acid sequences of a protein.

STRUCTURE OF PROTEINS

Primary structure : Protein structure can be described at four levels of organization. The **primary** structure of a protein refers to its unique sequence of amino-acids linked by peptide bonds. The primary structure is thus a complete description of the covalent connection of protein. SANGER, *et al.*, 1955, established the sequence in insulin which contains fifty-one amino-acids. Later, the primary structure of ribonuclease, an enzyme containing 124 amino-acids, was elucidated by STEIN *et. al.*

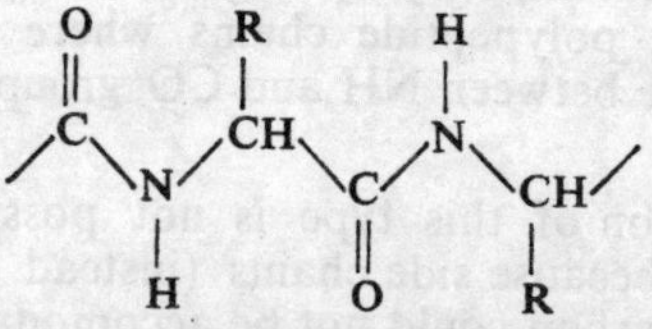

Fig. 3·12. Diagram showing a primary structure of protein.

Secondary structure : The secondary structure of proteins refers to the spatial arrangement of the polypeptide chains as a result of hydrogen bond formation. The secondary structure is a direct consequence of the sequential arrangement of the amino-acids in the polypeptide chain. The α-helix and the β-pleated sheet are best examples of secondary structure (PAULING, CORY *et al.*, 1951).

The α-helix is a rod-like structure. The tightly coiled polypeptide chain forms the inner part of the rod, and the side chains extend outward in a helical array. The α-helix is stabilized by hydrogen bonds between the NH and CO groups of the main chain. The CO group of each amino-acid is hydrogen bonded to the NH group of the amino-acid that is situated four residues ahead in the linear sequence. Thus, all the main-chain CO and NH groups are hydrogen-bonded. Each residue is related to the next one by a translation of 1·5Å along the helix axis and a rotation of 100°, which gives 3.6 amino-acid residues per turn of helix. Thus, amino-acids spaced three or four apart in the linear sequence are spatially quite close to one another in an α-helix. In contrast, amino-acids two apart in the linear sequence are situated on opposite sides of the helix and so are unlikely to make contact.

The pitch of the α-helix is 5·4Å, the product of the translation (1·5Å) and the number of residues per turn (3·6). The screw-sense of a helix can be right-handed or left-handed ; the α-helices found in proteins are right-handed.

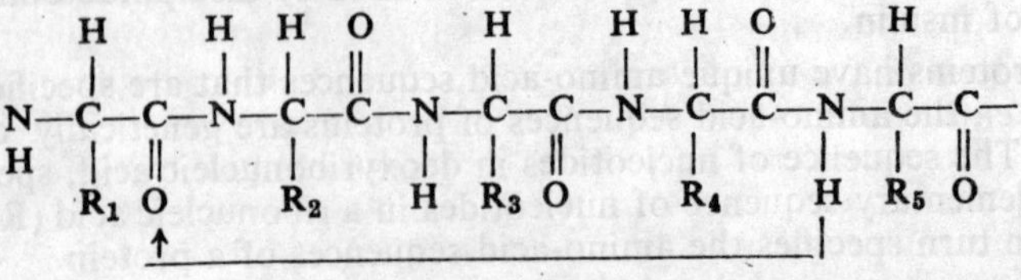

Fig. 3·13. In the α-helix, the NH group of residue *n* is hydrogen-bonded to the CO group of residue (*n*–4).

The **β-pleated sheet** structure of protein differs markedly from the α-helix in that it is a sheet rather than a rod. The polypeptide chain in the β-pleated sheet is almost fully extended rather than being tightly coiled as in the α-helix. Adjacent amino-acids in a polypeptide chain occur every 3·6Å, in contrast to 1·5Å for the α-helix, and every amino-acid residue is rotated 180° with respect to adjacent ones. The β-pleated sheet is stabilized by hydrogen bonds between NH and CO groups in different polypeptide chains where as in the α-helix the hydrogen bonds are between NH and CO groups in the same polypeptide chain.

A configuration of this type is not possible with amino-acids other than glycine because side chains (instead of hydrogen atoms) attached to the α-carbon could not be accomodated. Silk fibroin, a member of the class of β-keratins rich in alanine or serine, is an example of a protein with a related structure.

A somewhat more complex type of molecular organization is illustrated by the protein collagen. Collagen consists of three polypeptide chains with parallel orientation but with the glycine of every triplet of amino-acids displaced one residue compared to the register of the other two chains. The three chains are wound around in a coil so that hydrogen bonding occurs between the glycine residues in adjacent chains. While the α-carbons of the glycine residues tend to be oriented toward each other and the centre of the three-stranded unit, the other two-thirds of the amino-acid residues are oriented with the amino acid side chains facing outward. The pyrrolidine rings of proline and hydroxyproline tend to be closely packed, producing a helical ridge on the outer surface of the molecule. Adjacent amino-acids residue in a chain occur 2·86Å apart and are rotated 108Å.

Tertiary structure : When a long peptide chain, with or without a helix, is coiled and variously folded in itself, the resulting highly specific three-dimensional configuration of the protein is termed as the **tertiary structure**. This tertiary structure is found especially in the case of globular proteins.

Tertiary structure results from various weak molecular forces within the protein and from interactions between the protein and the solvent water. Disulphide covalent bonds and, rarely, interchain peptide bonds are responsible for the tertiary structure. Accumulating evidences suggest that hydrogen bonds and hydrophobic bonds are also involved in the tertiary structure of proteins.

Quaternary structure : A protein is said to have **quaternary structure** if it is composed of several polypeptide chains which are not covalently liked to one another. Each polypeptide chain in such a protein is called as **subunit**. The enzyme phosphorylase, for example, contains four subunits which are identical to each other but separately inactive catalytically. However, when they are joined together, the enzyme becomes active. This type of structure in which all the subunits are identical is termed as **homogeneous quaternary structure**. But, when the subunits are dissimilar, *e.g.*, the tobacco mosaic virus, the structure is said to be **heterogeneous quarternary structure**.

CLASSIFICATION OF PROTEINS

Proteins are classified in different ways :

1. **According to the structure of molecule the proteins are of two fundamental types :**

(i) **Fibrous proteins :** These proteins are composed of long and rod-shaped molecules of polypeptide chains which are arranged or twisted together, in a variety of ways so as to form strands. These are probably the simple proteins because these are stable on account of being relatively insoluble in water and also have the capacity for extension and contraction. Such type of proteins include the hard substances like collagen of cartilage and tendon, contractile substances like myosin of muscles, horny substances like keratin of hair, nails and feathers and fibrin of the coagulated blood, etc.

(ii) **Globular proteins :** These proteins are composed of globular or spherical or oval molecules. These are soluble in water and can form crystals. These do not possess contractility. Most of the proteins belong to this very category. Examples are—albumen of egg, haemoglobin, enzymes, venoms of snake and scorpion, wasp and bees.

2. **According to the chemical composition the proteins are of three principal types :**

(i) **Simple proteins :** These proteins are simply formed of amino-acids only, such as albumin, globulin histone, globins, prolamine, etc.

(ii) **Conjugated proteins :** These proteins in addition to amino acids, contain some other non-protein substances (the prosthetic groups) in their molecules. For example, the protein **globin** is combined with an iron containing porphyrin compound **haeme** to form **haemoglobin**. These are classified according to the nature of prosthetic group, as indicated below :

(a) **Lipoproteins :** These proteins are composed of amino acids as well as lipids, such as lipovitellin of egg yolk, serum, protein of brain, etc.

(b) **Nucleoproteins :** These proteins are made up of amino acids and nucleic acids. The chief nucleic acid found in the nucleus is deoxyribonucleic acid (DNA), it is united with protamines, histones and other proteins of the cell nucleus. The nucleoproteins of the cytoplasm are ribonucleoproteins found in the ribosomes.

(c) **Glyco—or mucoproteins :** These proteins, such as mucin of the saliva, chorionic gonadotropins and pituitary hormones such as follicile stimulating hormone (FSH) and luteinizing hormone, contain amino acids and carbohydrate group in their molecules. The carbohydrate in these is usually mucopolysaccharide.

(d) **Chromoproteins :** These proteins contain aminoacids and coloured pigments in their molecules. Examples are haemoglobin, haemocyanin, cytochrome and riboflavin, etc.

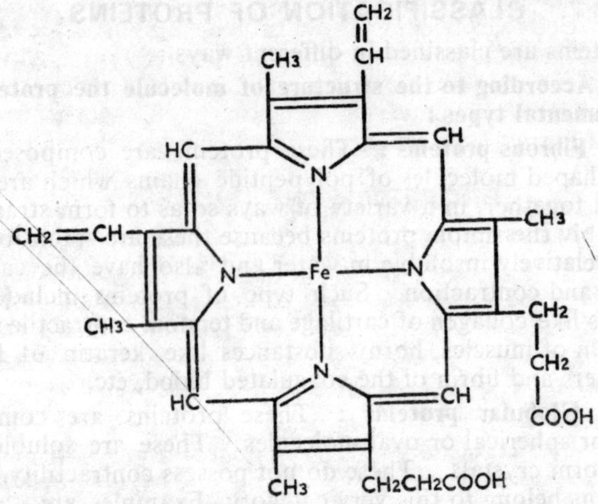

Fig. 3·14. Chemical formula of haeme portion of haemoglobin.

(e) **Phosphoproteins :** These proteins contain amino acids and phosphoric acid in their molecules. Examples are casein of milk and ovovitellin of the egg.

(f) **Metalloproteins :** These proteins contain metals in addition to amino acids in their molecules. For example, enzyme such as tyrosinase (having copper) and carbonic anhydrase (having zinc).

(iii) **Derived proteins :** These include the proteins which are derived from the original proteins by the action of heat, enzymes or chemical reagents. The best examples of such type of proteins are proteoses, peptones and metaproteins.

3. **According to their solubility the proteins are classified as follows :**

(i) **Albumins :** These are called simple proteins as these are soluble in water, dilute salts solutions and dilute acids and bases. Moreover these may coagulate on heating. Examples are egg albumin serum albumin and milk albumin.

(ii) **Globulins :** These are also simple proteins but insoluble in water. These can dissolve in dilute salt solutions such as 95%

sodium chloride solution. Examples are globulins of white egg and blood serum.

(iii) **Globins :** These are highly soluble proteins which can dissolve in ammonium hydroxide. Examples are haemoglobins.

(iv) **Protamines :** These are also highly soluble proteins which can be dissolved in ammonium-hydroxide. Examples are proteins of fish sperm.

(v) **Histones :** These proteins are soluble in water but insoluble in ammonium hydroxide. Moreover these can not coagulate on heating—such as nucleoproteins.

(vi) **Prolamines :** These proteins are insoluble in water but soluble in 70% ethyl alcohol. Examples are gliadin of wheat, zein of corn and hardein of barley, etc.

(vii) **Scleroproteins :** These proteins are insoluble in water as well as in all neutral solvents. They are partially or highly resistant to digestive enzymes and form important constituents of the connective tissue and external coverings of animals. Examples are keratin of hair and horny tissue, collagen of cartilage and elastin of tendons, etc.

PROPERTIES OF PROTEINS

The various properties of proteins can be studied under the following heads :

1. Physical properties : All proteins exist in a colloidal state. The protein molecules due to their big size either diffuse very slowly or are unable to pass membranes through which substances of smaller molecular diameter pass readily, as such they are free from diffusible contaminants.

2. Solubility : Proteins, due to colloids of big molecules, form turbid solutions in water. They are insoluble in alcohol and ether. They precipitate in certain concentration of acids. All proteins are soluble, when treated with strong acids or alkalies forming soluble acid and alkali-proteins.

3. Chemical reaction : All proteins are amphoteric as they behave as acid to alkalis and as alkalis to acids. They, thus, form salts with either.

4. Coagulation and precipitation : All proteins, on addition of alcohol, are coagulated. They are also coagulated on heating.

Most of the proteins are precipitated out of their solutions by salts of heavy metals such as, copper, mercury and lead. They may also be precipitated from slightly acid solutions like phosphotungistic, phosphomolybdic, tannic and picric acids.

5. Optical properties : All amino acids are optically active except glycine. Most of them are laevo-rotatory, while others such as haemoglobin and nucleoproteins are dextro-rotatory.

6. Colour reactions : Proteins, on being treated with certain chemical reagents, exhibit characteristic colour reactions which help in qualitative identification of proteins. Some important colour reactions of protein are as follows :

(i) *Biuret Reaction* : All proteins (excepting dipeptides) give biuret reaction when treated with strong alkali, such as, caustic soda and dilute copper sulphate solutions, giving a pinkish violet colour. **Biuret** ($NH_2CONHCONH_2$) also gives a pinkish violet colour with NaOH and $CuSO_4$ solution. Owing to this similarity in colour reaction it is assumed that there is a similar linkage (=CH—CO—NH—CH) in the protein molecules too.

(ii) *Millon's Reaction* : When a solution of protein or a solid protein is treated with **Millon's reagent** (a mixture of several nitrates of mercury in nitric acid solution) a brick red colour is obtained after heating. This is due to the presence of the hydroxyphenyl group (C_6H_4OH) in the protein molecule. **Tyrosine** is the only one of the amino acids containing this group. The proteins respond to this test due to the presence of this amino acid in their molecules.

(iii) *The Xanthoproteic Reaction* : When strong nitric acid is added either to solid proteins or solution of protein a yellow colour is obtained which changes to orange on the addition of alkali. The reaction is primarily due to the presence of the benzene ring or phenyl group ($—C_6H_5$) in the protein molecules. Tyrosine, tryptophan and possibly also phenylalanine contain this phenyl group.

(iv) *The Glyoxylic Acid Reaction (Hopkins-Cole Reaction)* : When a protein solution is treated with glyoxylic acid, a violet colour is produced on addition of sulphuric acid. This reaction is due to the presence of indol group which is only present in the tryptophan amino-acid.

(v) *Ninhydrin Reaction* : When ninhydrin reagent (triketo-hydrindine-hydrate) is added to protein solution, a deep blue or violet pink or red colour is produced. This reaction is given by proteins and by α amino-acids generally. It is more sensitive test than some of the other reactions.

(vi) *Diazo Reaction (Pauli's)* : When a protein solution is treated with diazo-benzene sulphonic acid in the presence of milk alkali, a red colour is obtained. This colour reaction is due to presence of histidine or tyrosine.

7. **Hydrolysis :** When proteins are boiled with dilute mineral acids, they undergo a series of chemical changes which is known as **hydrolysis**. Hydrolysis results in the breakdown of longer molecules into simpler ones. The first product is **proteose** which gives biuret reaction. It is soluble in water and precipitated by saturation with ammonium sulphate. Next product is **peptone** which is still more soluble in water. It gives clear solution and biuret test. It is not precipitated by ammonium sulphate but phosphotungistic acid or tannic acid would give a precipitate. The final products are amino acids which give no biuret test except histidine amino-acid.

8. **Biological properties :** There are certain proteins which have great biological significance.

(i) **Enzymes :** These act as biological catalysts in the reactions going on in the body.

(ii) **Hormones :** These control and regulate the various vital activities.

(iii) **Haemoglobin :** It helps in the transportation of oxygen and carbon dioxide gases in the body.

(iv) **Anti-bodies :** These combat the foreign agents and thus protect the organism from any infection.

(v) **Nucleoproteins :** These control the hereditary transmission in the organisms.

9. **Denaturation and reversal :** All proteins have a unique conformation in their structure because of which they are capable to perform different biologic activity. It has been established that under adverse environmental conditions the proteins loose the secondary and tertiary structure characteristic of the native conformation and as a result of which the proteins assume a more disordered and less compact conformation. This process is called **denaturation** and is accompanied by a loss of specificity and catalytic activity and by changes in physical properties such as solubility. The polypeptide back bone is not broken during denaturation, but intramolecular hydrogen bonds or other linkages and other weak interactions which hold the protein in its native conformation are disrupted. Agents which bring about denaturation are termed **denaturating agents.**

When the globular proteins, which have closely folded polypeptide chains, are denatured, they get unfolded and become insoluble protein of the fibrous nature. The commercial protein fibres are obtained by this manner.

There is great variation in the tendency for proteins to denature; some proteins denature remarkably easily while some others denature temporarily and after a short period assume the original form. For example, when pepsin is moderately heated, it loses its proteolytic (catalytic) property; but when it is cooled, the catalytic activity is restored. The phenomenon is known as **reversal.**

CARBOHYDRATES

Carbohydrates are compounds normally characterised by having carbon, hydrogen and oxygen elements in their molecules in which the ratio of the hydrogen and oxygen is the same as in water (2 : 1). There are exceptions to this, for example, pentose sugars of the nucleic acid DNA which have a formula $C_6H_{10}O_4$. There are also a very small number of carbohydrates that contain nitrogen, *e.g.*, chitin.

The basic units of the carbohydrate molecule are known as **monosaccharides** and glucose is the most important of these.

It has the crude formula $C_6H_{12}O_6$, and its normal structure is that of a 6-membered ring involving 5 of the carbon atoms and 1 oxygen, as shown in the diagram. In all the monosaccharides each carbon atom typically has 1H and 1OH group attached, and the individual compounds differ in the pattern of arrangement of these groups, as to whether they project above or below the plane of the ring.

The monosaccharides can be attached together by specific enzyme systems to form polymers. The process involves removal of the elements of water, H from one monosaccharide unit and OH from the other, to form a C—O bond, usually in the 1, 4′ or 1,6′ positions. The reaction is not a simple dehydration but involves phosphorylated intermediates. This synthesis is endergonic and requires coupling with an energy donating reaction.

Carbohydrates are very important widespread biological compounds as they are the chief source of energy and also structural constituents of the protoplasm. They are defined as aldehyde and ketone derivatives of the polyhydric alcohols or in other words the carbohydrates are polyhydroxy-aldehyde or polyhydroxyketone. They are represented by an emperical formula $C_x(H_2O)_y$.

In general, carbohydrates are white, solid, sparingly soluble in organic liquids but, except for certain polysaccharides, soluble in water.

CLASSIFICATION OF CARBOHYDRATES

Carbohydrates are classified into three main groups depending upon the number of saccharides contained :

1. **Monosaccharides :** These are simplest carbohydrates having only one sugar or saccharide molecule in their constitution. They can not be hydrolysed into small molecules. They are further classified, according to the number of carbon atoms they contain, into trioses, tetroses, pentose, hexoses and heptoses and so on.

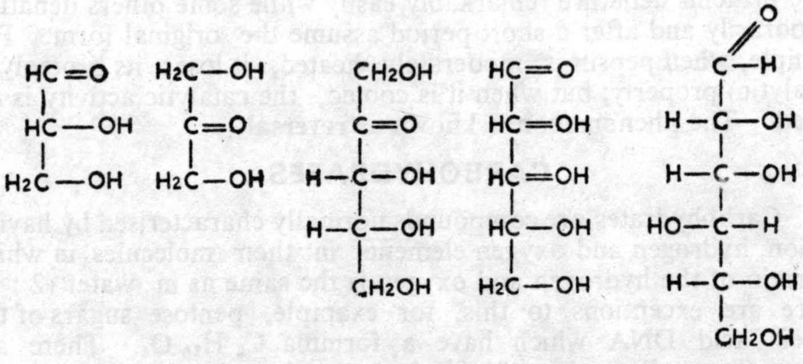

(Aldose) (Ketose) (Ribulose) (Ribose) (D. Glucose)

(Triose) (Pentose) (Hexose)

Fig. 3·15. Structure of some monosaccharides.

(i) **Trioses :** These are smallest molecules of carbohydrates which contain three carbon atoms in their molecules, *e.g.*, glyceraldehyde and dihydroxyacetone. The emperical formula is $C_3H_6O_3$.

(ii) **Tetroses :** Monosaccharides containing four carbon atoms are called tetroses, *e.g.*, erythrose and threose. The emperical formula is $C_4H_8O_4$.

(iii) **Pentoses :** Monosaccharides containing five carbon atoms in their molecules are termed as pentoses, *e.g.*, xylose, ribose and arabinose. The emperical formula is $C_5H_{10}O_5$.

(iv) **Hexoses :** These contain six carbon atoms in their molecules, *e.g.*, glucose, fructose and galactose.

(v) **Heptoses :** Heptoses contain seven carbon atoms in their molecules.

Monosaccharides are the most active carbohydrates as they form the main respiratory substrate of all living cells. The common monosaccharides are the pentoses and hexoses. They play fundamental role in cellular nutrition. The important hexoses are glucose, dextrose, fructose and galactose. These are sweet and neutral in nature. They can be crystallized and are soluble in water. They can also undergo alcoholic fermentation.

The important pentoses are ribose, deoxyribose and ribulose. Ribose is the important constituent of the ribonucleic acid (RNA) and of certain coenzymes as nicotinamide adenine dinucleotide (NAD), NAD phosphate (NADP), adenosine triphosphate (ATP) and coenzyme A (CoA). Deoxyribose is one of the constituents of deoxyribonucleic acid (DNA). Ribulose is a pentose sugar which is essential for photosynthetic mechanism.

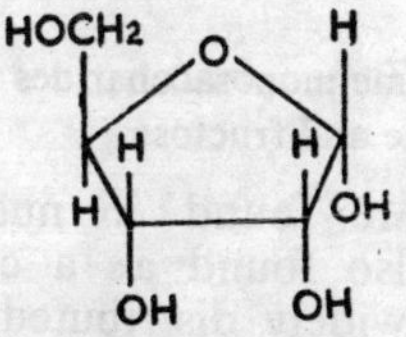

Fig. 3·16. Chemical formula of ribose sugar.

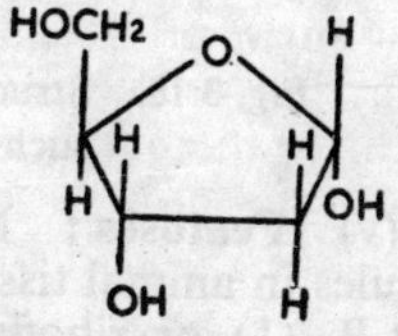

Fig. 3·17. Chemical formula of deoxyribose sugar.

(i) **Glucose :** It occurs in nature in fruits, such as grapes and in honey along with fructose. It is found in the blood, in the intestines during digestion and in the urine of the diabetics. It occurs in the body in combination with other substances, *e.g.*, as esters of glucose and hexose phosphates and as lactose in milk in combination with galactose. It is a white crystalline solid readily soluble in water and sweet in taste. With yeast it is converted into alcohol with liberation of CO_2 in bubbles.

$$\underset{\text{Glucose}}{C_6H_{12}O_6} = \underset{\text{Ethyl alcohol}}{2C_2H_5OH} + 2CO_2$$

(ii) **Mannose :** It occurs in various plant seeds. It is a colourless substance readily soluble in water but slightly in alcohol. It is sweet in taste.

(iii) **Fructose :** It is found in sweet fruits and in honey. Although it is termed *d*-fructose but it is strongly laevo-rotatory and hence called **evulose**. Chemically it is a **ketose**. It is readily converted into glycogen in the animal body. It is absorbed slowly in the intestine than other monosaccharides. It's excess, in the body, may be converted into fat, starch and glycogen and may pass out in the urine as levulose (levulosuria).

(iv) **Galactose :** Galactose occurs in nature in combined form as a constituent of lactose, and in certain complex proteins and lipids. It is rarely found in plants but in mammals mammary glands prepare it from the blood glucose. It with excess of glucose forms lactose or milk sugar. It is generally used in the body in the same manner as glucose. It is absorbed rapidly in the body.

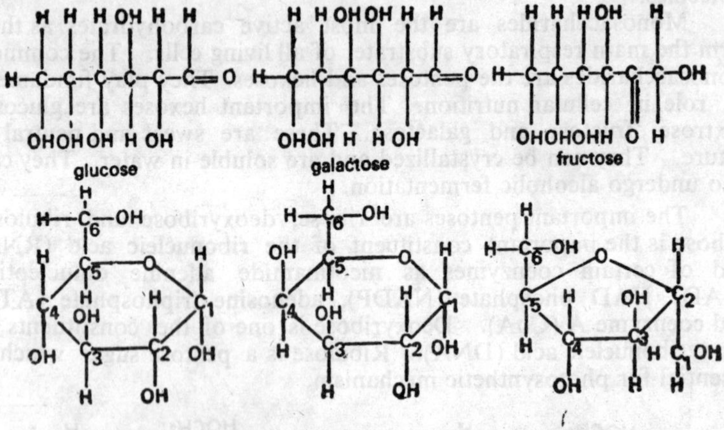

Fig. 3·18. Structure formulae of some monosaccharides such as glucose, galactose and fructose.

(v) **Pentoses :** Normally they are found in nucleic acid molecules in animal tissues. They are also found as a component of Vit B_2 (1) or riboflavine. They are widely distributed in plants as pentosome (complex polysaccharides) yielding pentoses on hydrolysis. They are all reducing sugars and do not ferment with pure yeast. They yield acids on oxidation and alcohols on reduction.

Chemical properties of monosaccharides : The various chemical properties the monosaccharides exhibit can be studied under the following heads :

1. **Reactions in acid solution :** When monosaccharides (pentose) are treated with a strong mineral acid, the product furfural is obtained. This is the characteristic reaction of sugars and is known

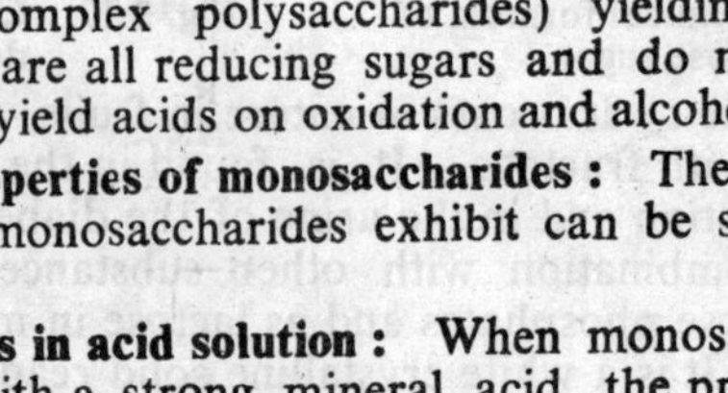

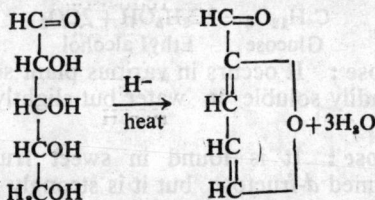

as **dehydration**. With hexoses, the analogous reaction leads to formation of 5-hydroxymethyl furfural, which on further heating is transformed to levulinic acid.

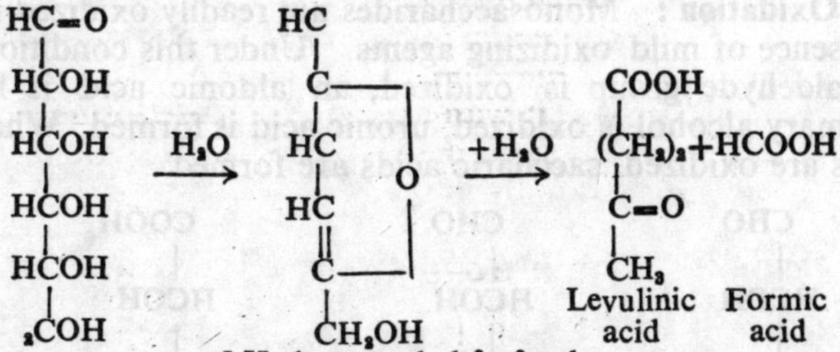

5 Hydroxymethyl furfural

If the furfurals are treated with variety of phenolic compounds, coloured products are produced. These reactions are important for qualitative and quantitative tests of carbohydrates.

(i) Molish Test : Take 5 ml sugar solution. Add 2 ml of Molisch reagent (5% α-naphthol in alcohol) and 3 ml of concentrated sulphuric acid to sugar solution. A reddish-violet zone appears between the acid and sugar layer.

2. **Reaction in alkaline solution :** When the sugars are treated with dilute alkaline solution, they change to cyclic α-and β-forms (*i.e.*, mutarotation).

If the above mixture is heated at 37°C, the acidity increases as a result of which a series of **enols** called **'endiols'** are formed. Sugars with strong alkali solution produce decomposition products and yellow and brown pigments develop.

3. **Reduction :** When monosaccharides are treated with H_2 gas under pressure in the presence of a metal catalyst, or with an active metal, such as Ca, in water, the carbonyl group is reduced to an alcoholic hydroxyl group, yielding a polyhydric alcohol.

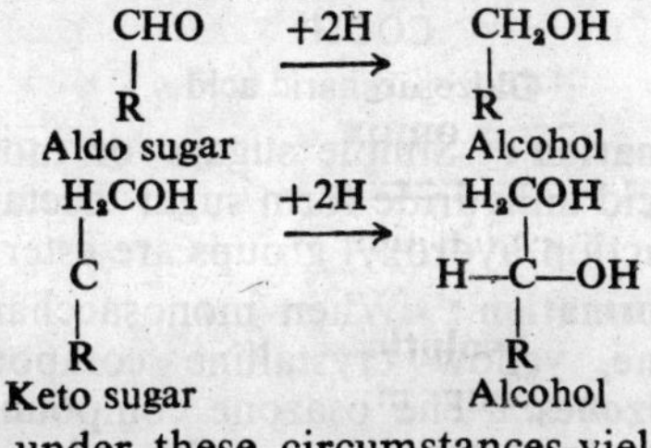

D-Glucose, under these circumstances-yields sorbitol (D-glucitol), which is also a product of the reduction of L-sorbose.

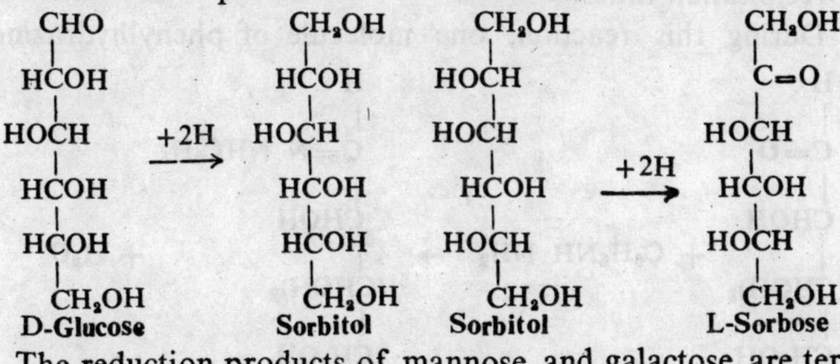

The reduction products of mannose and galactose are termed, respectively, mannitol and dulcitol.

4. **Oxidation :** Monosaccharides are readily oxidized in acids in the presence of mild oxidizing agents. Under this condition when only the aldehyde group is oxidized, an aldonic acid is formed. When primary alcohol is oxidized, uronic acid is formed. When both the groups are oxidized, saccharic acids are formed.

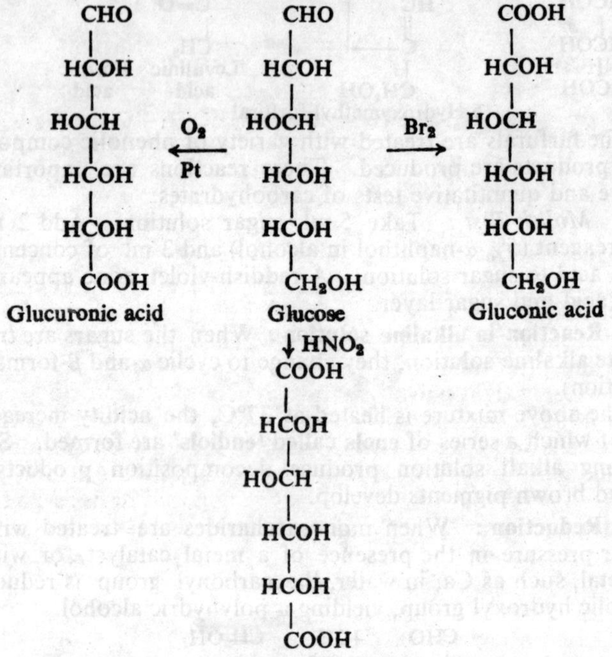

Glucosaccharic acid

5. **Ester formation :** Simple sugars or monosaccharides on treatment with an acid anhydride form sugar acetate and benzoates, etc. During this reaction hydroxyl groups are esterified.

6. **Osazone formation :** When monosaccharides are treated with phenylhydrazine, yellow crystalline compounds are formed which are called **osazones.** The osazone compounds of sugars are coloured and possess characteristic crystalline forms, melting points and precipitation times.

During this reaction, one molecule of phenylhydrazine first

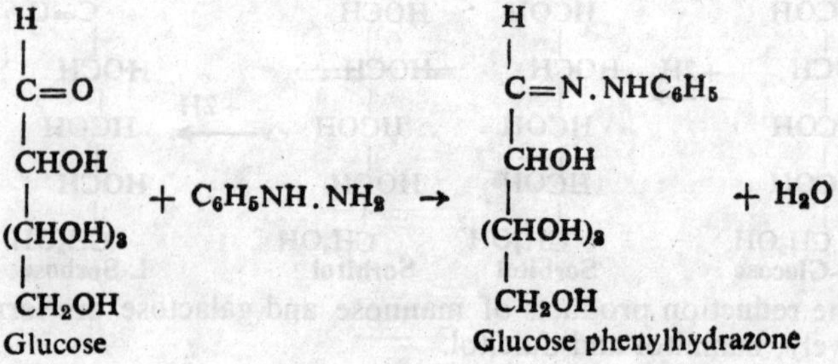

reacts with one molecule of an aldose or ketose sugar to form a hydrazone.

The hydrazone, thus formed, is oxidized in the presence of excess phenylhydrazine to form glucosazone and phenylhydrazine is reduced to form aniline and ammonia.

$$\begin{array}{lcl} H & & H \\ | & & | \\ C{=}N.NHC_6H_5 & & C{=}N.NHC_6H_5 \\ | & & | \\ CHOH & +2C_6H_5NH.NH_2 \rightarrow & C{=}N.NHC_6H_5 + C_6H_5NH_2 + NH_3 \\ | & & | \qquad\qquad\qquad \text{Aniline} \quad \text{Ammonia} \\ (CHOH)_3 & & (CHOH)_3 \\ | & & | \\ CH_2OH & & CH_2OH \\ \text{Glucose phenylhydrazone} & & \text{Glucosazone} \end{array}$$

2. **Oligosaccharides :** The oligosaccharides consist of 2 to 10 monosaccharides (monomeres) in their molecules. In a oligosaccharide sugar the monomeres or monosaccharide units are linked with each other by the glycosidic linkages. The distinction between oligo-and high-molecular weight polysaccharides is arbitrary since the properties of higher oligosaccharides merge with those of the lower members of the polysaccharide class. A large number of oligosaccharides have been described, many do not occur as such in nature but have been obtained, by partial hydrolysis of larger oligosaccharides and polysaccharides. Some important oligosacci.arides are as follows :

I. **Disaccharides :** These are most abundantly naturally occuring oligosaccharides. These are water soluble sugars formed by the union of two molecules of monosaccharides, one molecule of water being removed in their combination. In this process of dehydrolysis or condensation the —OH group of one monosaccharide is joined to the 'H' of the OH group of another monosaccharide to form one water, and remaining 'O' forms a **glycoside bond** between the two sugar molecules, The common disaccharides are maltose, (malt sugar) sucrose (cane sugar) and lactose (milk sugar).

(i) **Maltose :** Maltose does not occur free in nature. It is formed by condensation of two molecules of glucopyranose at the α-position. This sugar is not of importance in itself but form carbohydrate. It is hydrolysed back to two glucose molecules under the influence of maltase enzyme.

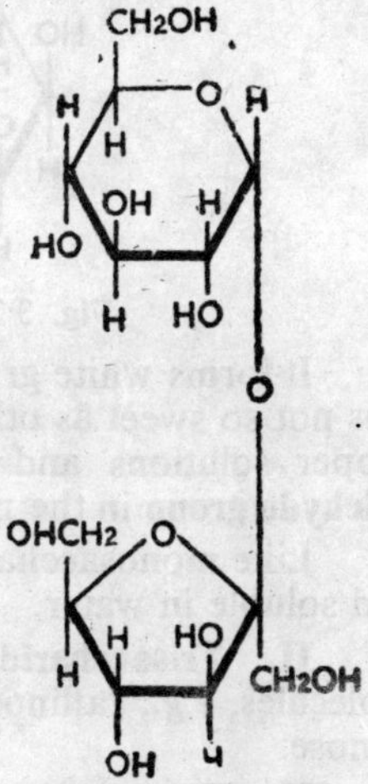

Fig. 3·19. Structure formula of sucrose showing condensation of two molecules of monosaccharides.

It is not so sweet as sucrose. It reduces copper solutions and forms osazone. It, in the presence of yeast, ferments readily.

(ii) **Sucrose :** It is found in greatest abundance in certain plants such as, sugar cane, beets, carrots and some sweet fruits. It is very sweet, crystalline and freely soluble in water. It is neither an **aldehyde** nor a **ketone** as it does not reduce copper sulphate solutions nor does form osazone.

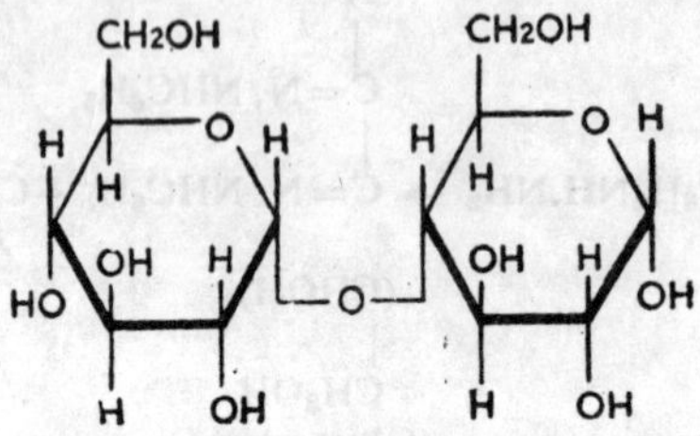

Fig. 3·20. Chemical formula of maltose.

It is formed by the condensation of fructofuranose and α-glucopyranose. It is very important sugar and the mainway in which carbohydrates are transported in plants. It is dextrorotary and hydrolysed by the action of **sucrase** (sometimes called invertase).

(iii) **Lactose :** It is the sugar found in milk. It is formed by the condensation of glucose and galactose. It on hydrolysis yields one molecule of glucose and one molecule of galactose. It is hydrolysed by the enzyme called **lactase**.

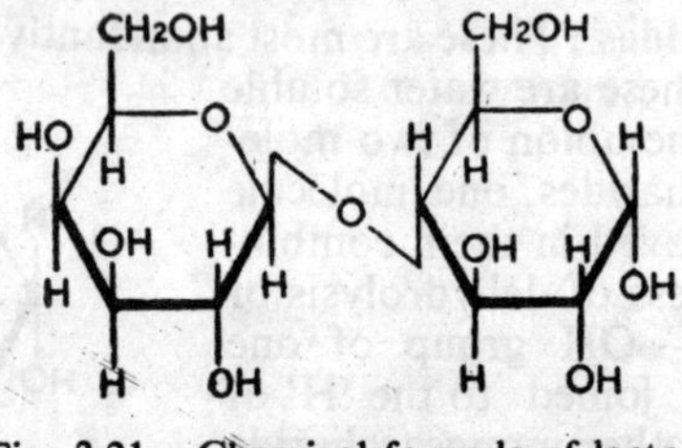

Fig. 3·21. Chemical formula of lactose.

It forms white gritty crystals on concentration of the milkwhey. It is not so sweet as others. It is not found in plants. It also reduces copper solutions and forms osazone, indicating the presence of an **aldehyde group** in the molecule.

Like monosaccharides these sugars are also sweet, crystallizable and soluble in water.

II. Trisaccharides : These contain three monomeres in their molecules, *e.g.*, raffinose, mannotriose, rabinose, rhaminose and gentianose.

III. Tetrasaccharides : These contain four monomeres in their molecules. *e.g.*, stachyose and scordose.

IV. Pentasaccharides : These contain five monomeres in their molecules, *e.g.*, verbascose.

3. Polysaccharides : Polysaccharides are made up of ten to many thousands monosaccharide monomeres in their macromolecules.

Unlike monosaccharides and disaccharides, polysaccharides are often insoluble. They constitute a storage form for sugar. When there is a surplus sugar polysaccharides are formed, when there is shortage, the polysaccharides are hydrolyzed (broken apart again).

The various polysaccharides differ from one another not only in constituent polysaccharide composition but also in molecular weight and other structural features. Thus, some polysaccharides are linear polymers and others are highly branched. In all cases the linkage that unites the monosaccharide units is the glycosidic bond. This may be α or β and may join the respective units through linkages that are 1, 2 ; 1, 3 ; 1, 4 ; or 1. 6 in the linear sequence or between those units that are at **"branch points"** in the polymers. An enormous number of variants is, therefore, possible, and, indeed, a great many polysaccharides have been described.

Polysaccharides are represented by the formula $(C_6H_{12}O_6)n$ where the value of n may be 200 or even more. Being of large size, the, polysaccharides can assume the size of a colloidal solution or even a solid state and cannot pass through the natural membranes.

Chemically they are inactive and thus do not ionise ; consequently they are most suitable as energy stores. The complex polysaccharides are found both in plants and animals and serve as a food material. Starch, insulin, dextrin, glycogen, cellulose and vegetable mucilages are the common polysaccharides.

They do not form clear solutions in water due to large size of their molecules but some of them, on boiling form opalescent solutions. They neither reduce **Fehling's solution** nor form osazone as other sugars. They are optically active and are not sweet in taste.

Polysaccharides in combination with proteins form **mucopolysaccharides** which are generally found in mucus in order to lubricate the food during eating and its passage through the gut.

There are certain polysaccharides which are important from the **immunological** point of view as they are responsible for the specific immunity produced by the individual species of bacteria.

The polysaccharides may be of two types, *viz.*, homopolysaccharides and heteropolysaccharides.

Homopolysaccharides : The homosaccharides contain same type of monomeres in their molecules, *i.e.*, they yield, on hydrolysis a single monosaccharide constituents. Some of the important homopolysaccharides are the following :

(i) **Starch** : It is a complex substance which is formed by the condensation of **amylose** and **amylopectin** which is a branched polysaccharide with shorter chains. It is found in abundance in plants, seeds, fruits and tubers. It is an amorphous white powder which forms opalescent solutions in water on boiling. With iodine starch forms an adsorption compound of blue iodide of starch, the colour of which disappears on heating but appears on cooling.

Amylases or **diastases** hydrolyse starch but these enzymes are themselves complexes with members specific for the various linkages within the starch. The first product of hydrolysis is maltose and the final product is glucose.

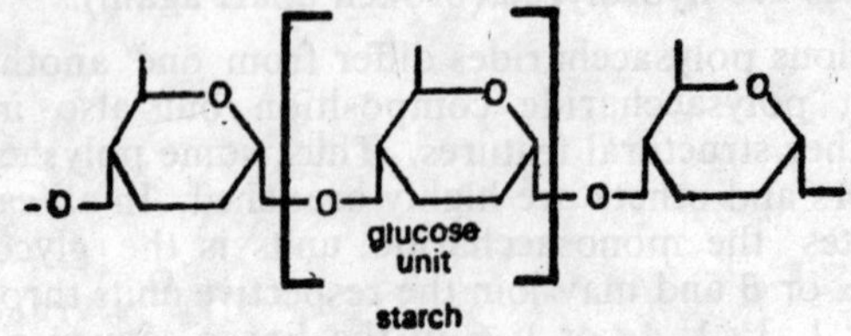

Fig. 3·22. Structure formula of starch.

(ii) **Insulin :** It is a storage polysaccharide found in the tubers of dahlia, artichoke, etc. It is an inert substance when injected into the body and is composed of large numbers of fructofuranose molecules condensed together in chains.

(iii) **Dextrin :** It forms hazy solution in water which when concentrated becomes thick and adhesive. It gives a port-wine colour with iodine. It is hydrolysed with the help of dilute acids. The first product of hydrolysis is maltose and the final product is glucose.

(iv) **Glycogen :** It is main carbohydrate storage substance of animals and fungi (**Haworth**). It is made up of molecules rather like amylopectin but with more numerous side chains. Physically and chemically it resembles with dextrin. It forms opalescent solutions in water as dextrin and gives brown colour on treating with iodine. It is also hydrolysed by the dilute acids into glucose.

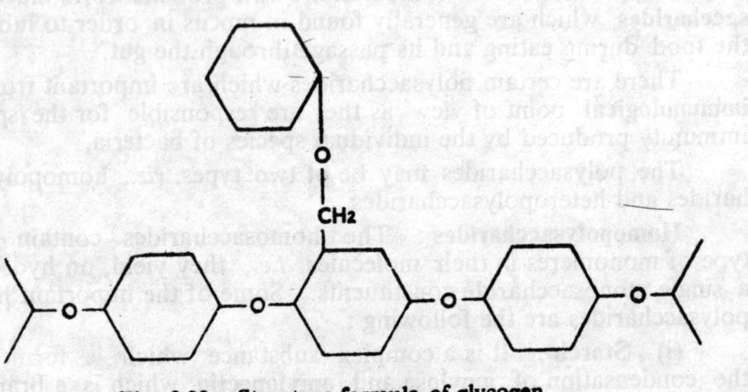

Fig. 3·23. Structure formula of glycogen

(v) **Cellulose :** It makes the cell wall of plants and the external covers of all grains also. It is made up of long chains of β-glycosides as in cellobiose and it may contain many thousands of monosaccharides to a single molecule. These long chains may be further strengthened by hydrogen bonding from one chain to another.

It is absolutely insoluble in water and resistant to the action of dilute acids or alkalis. Strong acids would hydrolyse it into glucose. It is also hydrolysed by **cellulose** enzyme into cellobiose and eventually to glucose.

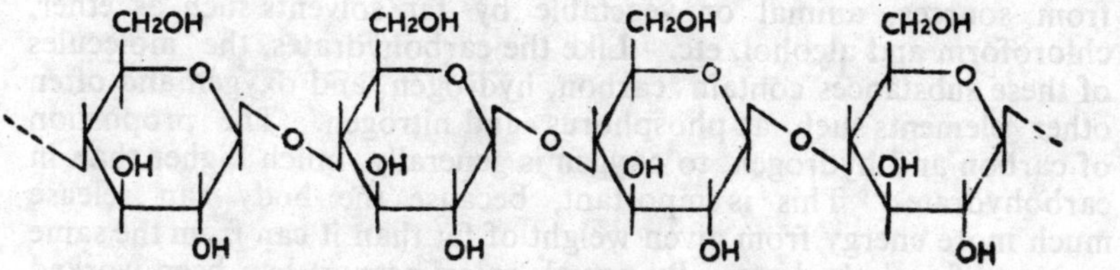

Fig. 3·24. Structure formula of cellulose.

(v) **Lignin :** It is a structural chemical forming the wood of plants. Unlike cellulose and other carbohydrates it is made up of aromatic units which form long chains with cross linkages. It is thought that the original units themselves synthesised from monosaccharides. Lignin is very important in the support of woody plants.

Heteropolysaccharides : Polysa harides that, on hydrolysis, yield mixtures of monosaccharides and derived products are numerous in both plants and animals. The simplest heteropolysaccharides are those constructed by repetitive use of a mixed disaccharide. The most important heteropolysaccharides are the following :

(i) **Neutral heteropolysaccharides :** These contain monosaccharides and the acetylated amino-nitrogen in their molecules and are known as **acetyl glucosamines**. The most important acetyl glucosamine is **chitin**.

Chitin : It is also a structural chemical forming the cuticle of the insects. It also forms the cell wall of fungi. The chitin molecule is also made up of long chains of monosaccharide units but in this case they have the amino group, NH_2, *e.g.*, attached. It is very tough and durable.

(ii) **Acidic heteropolysaccharides :** These contain different kinds of monosaccharides and sulphuric or other acids in their molecules, *e.g.*, hyaluronic acid, chondroitin sulphate and heparin.

The **hyaluronic acid** forms the cementing material of the connective tissues. It occurs in the skin, connective tissues and synovial fluid of the joints. The **chondroitin sulphate** occurs in the cells of the cartilage, skin, and cornea and serves as matrix for the bone formation. The **heparin** is a blood coagulant found in the liver, lung, thymus, spleen and blood.

(iii) **Mucoproteins** or **glycoproteins :** These contain monosaccharides and proteins in their molecules, *e.g.*, blood group polysacchardes. They occur in blood corpuscles, saliva, **gastric** mucin, serum and albumins.

LIPIDS

The lipids are organic compounds which are characterized by solubility in non-polar solvents like ether, alcohol and chloroform. They include not only the true fats but also some others, such as oils and waxes. They are insoluble in water and could be extracted from sources, animal or vegetable by fat solvents such as ether, chloroform and alcohol, etc. Like the carbohydrates, the molecules of these substances contain carbon, hydrogen and oxygen and often other elements such as phosphorus and nitrogen. The proportion of carbon and hydrogen to oxygen is generally much higher than in carbohydrates. This is important, because the body can release much more energy from given weight of fat than it can from the same amount of carbohydrate. By actual calorimetry it has been worked out that, 1 gm of carbohydrate, on being oxidised, would produce 4·1 calories, while 1 gm of fat, on being oxidised, produces 9·3 calories.

CLASSIFICATION OF LIPIDS

The lipids on the basis of their occurrence and constitution are classified into three groups :

1. Simple lipids,
2. Complex lipids,
3. Derived lipids.

1. Simple lipids : The simple lipids are the esters of fatty acids with trihydric alcohol, **glycerol**. Ester is a combination of an alcohol with an acid. Simple lipids include the true **fats, oils** and **waxes**.

(i) **Fats** : Fats are the most important lipids which also include the oils occurring in liquid state. They are used as fuel in the biological oxidation. The chief sources of fats are butter, ghee, liver oils, groundnut and almonds. There are certain seeds, which contain oil, these are castor oil seeds, soyabeans, linseeds (flax seeds), cotton seeds and palm kernels. They occur as storage forms of energy in the vertebrates, insects and some other animals.

The true fats, sometimes called neutral fats, are the simple lipids. They are known as the triglycerides of the fatty acids because the fat molecules are made up of one molecule of glycerol (glycerine) and three molecules of fatty acids. Glycerol molecules contain a chain of three carbon atoms and have the formula $C_3H_5(OH)_3$. Fatty acids vary in their chemical make up but those found most often in common fats contain sixteen to eighteen carbon atoms. The structural formula of a fat may be represented as follows :

$$\begin{array}{c} H_2{=}C{-}O{-}R \\ | \\ H{-}C{=}O{-}R \\ | \\ H{-}C{-}O{-}R \end{array}$$

Fig. 3·25. **The generalized formula for a fat where R represents a fatty acid.**

During the formation of fats, each of the three fatty acid molecules is joined to the glycerol molecule by a **dehydration synthesis** reaction, in which a molecule of water is released for each

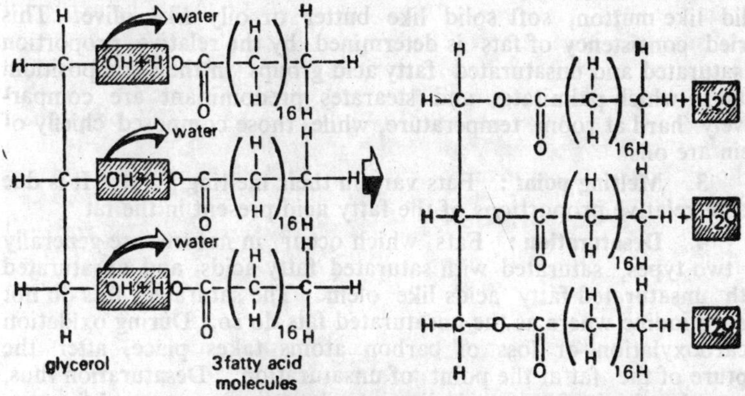

Fig. 3·26. **The dehydration synthesis of a molecule of fat. Three fatty acid molecules combine with a molecule of glycerol, and yield the fat molecule and there molecules of water.**

fatty acid molecule joined. In fat digestion water is combined with fat molecules in a process known as hydrolysis that results in the breaking of the fat molecule into three fatty acids and glycerol again. In the body of the animals fats are found mostly as saturated fats in the several depots of adipose tissues, over the kidneys, in the great omentum, etc.

(a) **Fatty acids :** Fatty acids are the compounds of straight chain of carbon atoms with one carboxyl group at the end. Although in some cases these chains are very long and may have side chains also. The fatty acids, on which the consistency of fat depends and which take part in the formation of fats and oils, are of two types ;

(i) Saturated fatty acids,

(ii) Unsaturated fatty acids.

(i) **Saturated fatty acids :** Myristic, palmitic and stearic acids are said to be the most common saturated fatty acids as all the bonds of carbon atom are saturated. These are found in abundance in all naturally occurring fats.

(ii) **Unsaturated fatty acids :** Oleic, linoleic, linolenic and arachidonic acids are the most common unsaturated fatty acids as their double bonds are not satisfied. These fatty acid have special physiological value and are easily oxidised or may even oxidize spontaneously on exposing to sunlight or ultraviolet rays. They may change into saturated fatty acids on the addition of reducing agents. It is all due to the presence of unsatisfied double bonds.

Properties of fats : The variable but important properties of fats can be studied under the following heads :

1. **Solubility :** Fats are tasteless, colourless, odourless and semisolid substances which are insoluble in water but readily soluble in non-polar solvents like ether, alcohol and chloroform.

2. **Consistency :** The fats occur either in the form of semi-solid like mutton, soft solid like butter or oily like olive. This varied consistency of fats is determined by the relative proportion of saturated and unsaturated fatty acid groups in their composition. Fats in which palmitates and stearates predominant are comparatively hard at room temperature, while those composed chiefly of olein are oils.

3. **Melting point :** Fats vary in their melting points. It is due to the relative proportions of the fatty acid present in the fat.

4. **Desaturation :** Fats which occur in nature are generally of two types, saturated with saturated fatty acids and unsaturated with unsaturated fatty acids like olein. The saturated fats do not absorb iodine where as the unsaturated fats do so. During oxidation decarboxylation or loss of carbon atoms takes place, after the rupture of the fat at the point of unsaturation. Desaturation thus, means breaking of longer chains into shorter ones, needed for easy transport, esterisation and combustion.

5. **Saponification :** Fats, when boiled with an aqueous or alcoholic solutions of alkalis (NaOH and KOH), undergo saponification forming soaps. From a dilute solution the soap could be thrown down after adding the neutral salts.

$$\underset{\text{Tristearin}}{C_3H_5(COOC_{17}H_{35})_3} + \underset{\text{Caustic soda}}{3NaOH} = \underset{\text{Sodium steorate}}{3C_{17}H_{35}COONa} + \underset{\text{Glycerol.}}{C_3H_5(OH)_3}$$

6. **Emulsification :** When a fat is shaken with water, it breaks down and separates out into definite layers of oil on water to form an emulsion. Mucilage, alkalis and bile salts are known more emulsifying agents than water.

7. **Hydrolysis :** Natural fats are hydrolysed under prolong heating and readily break up into their components, fatty acids and glycerol.

(ii) **Waxes: :** Waxes are esters of fatty acids and complex monohydric alcohol instead of glycerol of fats. They are fairly hard and their melting points are sufficiently high. They form protective and water resistant layers. They do not occur in human body. The fat spliting enzyme, **lipase** has no action on them and as such they are not suitable as food.

2. **Complex lipids :** Lipids with additional radicals are called complex lipids. Some of the important complex lipids are as follows :

(i) **Phospholipids** or **phosphatides :** Such type of lipids contain fatty acids, glycerol, phosphoric acid and an organic nitrogenous base in their molecules. Lecithins, cephalins and sphingomyelin are the most important phospholipids.

These lipids probably occur in every living cell and play an important part in the structure of the cell particularly the cell membrane. These lipids can also react with protein to form lipoproteins. They are found in abundance in the brain and nervous tissues. They possess both hydrophilic and hydrophobic proporties.

(ii) **Glycolipids** or **glucolipids :** These lipids contain fatty acids, carbohydrate radicals and complex amino alcohols like sphinogosine in their molecules. These were long known as the **cerebrosides**, according to THUDICHUM as they are found in abundance in nervous tissue. The two cerebrosides which have been distinguished so far, are **phrenosin** and **cerasin**. These on hydrolysis yield fatty acids, sphinogosine and galactose.

(iii) **Chromolipids :** These lipids, besides fatty acids and glycerol contain carotenoids and other related pigments in their composition, example—carotene.

(iv) **Aminolipids** and **sulpholipids** : These lipids, in addition to fatty acid and glycerol, contain amino acids and sulphur compounds in their composition respectively.

3. **Derived lipids :** These lipids are obtained by the hydrolysis of simple and compound lipids. Although these are the products of hydrolysis of lipids, but even then they have some properties of lipid. Examples—cholesterol and ergosterol.

MINERAL SALTS

Mineral salts are essential components of the cell which by their small amount regulate the different metabolic activities of the animals. Since there is a steady loss of these salts from the body of the animal, therefore, they are obtained from the food material of the diet and from drinking water from time to time in order to make up the deficiency of these mineral salts. For the normal functioning of the body the following mineral salts are important :

1. **Calcium :** It along with phosphorus helps in the formation of many skeletons, in clotting of blood, growth, contraction of muscles and excitability of nerves. Its deficiency in the body leads to the poor development of skeletons and manifestation of rickets in childrens.

The calcium is found in abundance in milk, cheese, vegetables, eggs, butter, orange, and carrot. One should take 1·0 gm of calcium per day in his diet.

2. **Phosphorus :** It is also needed for the formation of skeleton, teeth, muscles and blood. It activates the activity of the enzymes. It plays great role in the formation of phospholipids and nucleic acid and also controls the metabolism of the fat. Its deficiency in body leads to the poor development of the skeleton and the retardation of growth. The important sources of phosphorus are milk, egg yolk, chees, meat, fish and certain cereals and vegetables. According to SHERMANN one should take 0·88 gm of phosphorus in his diet in a day.

3. **Potassium :** It is very essential for growth and normal functioning of the body. It also controls the osmotic pressure in the

body. Its deficiency in the body leads nervous disorder, irregular heart beat and also poor muscular control.

4. **Sodium** : It is very important salt which regulates the osmotic pressure of the body. Its deficiency leads to the nervous disorder. The most abundant source of sodinm is sodium-chloride, so called common salt.

5. **Magnesium** : It is known as an activator of many enzymes. It also activates oxidative reactions and controls the actions of intestinal **amino peptidase** and of bone **phosphatase**. Its deficiency leads to malformation of the skeleton, nervousness, poor growth and irregular heart beats.

6. **Iron** : It is essential for the formation of haemoglobin (a respiratory pigment) and chromatins. It is also present in several oxidative enzymes including cytochromes. Its deficiency leads to anaemia. Meat, green vegetables and certain fruits such as raisins, are the important sources of iron.

7. **Sulphur** : It is the essential constitnent of the protein, hence it is found in the amino acids, such as cystine and methionine. It is also found in bile salt, insulin and thiamine. It is generally used in the formation of feather and eggs of the birds.

8. **Copper** : It is a essential consituent of haemocyanin present in the blood of most arthropods and molluscs. It is also present in a number of vertebrate enzymes. In the fowl and several mammals it is important for the formation of blood. Its deficiency causes nutritional anemia particularly in the cattles.

9. **Cobalt** : It is a consituent of Vitamin B_{12} and also helps in the production of blood. Most vertebrates like cattle and sheep get it from their symbiotic organisms which have themselves probably synthesized it from inorganic cobalt.

10. **Zinc** : It is a constituent of carbonic anhydrase and other enzymes present in the R. B. C. It helps in the conversion of carbonic acid into carbon dioxide and water.

11. **Iodine** : It is essential for the formation of **thyroxine** hormone of the thyroid gland. Its deficiency causes the simple goiter. Drinking water and sea foods are the chief source of iodine.

12. **Chlorine** : It is found in the body as chloride ion in combination with sodium. It is highly concentrated in the cerebro-spinal fluid. It maintains body, acid, base and water balance. The chief sources of it is sodium chloride.

Minerals as they are vital to the body in many ways, each of them, however must be in a compound form before it can be used by the body. Eating chemically pure elements such as sodium or chloride would be fatal. When these are in compound forms such as, sodium chloride they are harmless and in fact essential to the body.

WATER

It is an inorganic substance which does not yield any energy to the tissues. However, it is so vital in the maintenance of life that a person deprived of it dies sooner than he would if deprived of other types of food.

It forms 90% of the protoplasm, which is a physical basis of life. It is obtained in large quantity, from the food, fruits, vegetables and milk, etc. It is also produced during the oxidation of food materials inside the body.

The water requirements of the body are met in three ways :

(i) Some water is present in the food which the animals take.
(ii) Some is a by product of oxidation reaction.
(iii) Some is consumed as drinking water.

It is important due to following points :

1. It is the chemical source of hydrogen and some of the necessary oxygen.

2. It is the medium in which the materials of the protoplasm are dispersed.

3. It is the medium of transport of foods, minerals, and other vital substances in the body of the organisms.

4. It is the medium in which soluble materials are absorbed from the environment.

5. It provides the environment to aquatic organisms.

6. It regulates the body temperature.

7. There are certain reactions in the body which are ionic in nature and the ionisation takes place in the presence of water.

VITAMINS

There are certain substances which are present in very small amounts in food and are essential for normal growth, body activity and in the prevention of certain diseases like **deficiency diseases.** Dr. Funk called them **vitamins**. Vitamins are organic substances that are indispensable for life but not required as a source of energy. They act as catalysts similar to digestive enzymes. The chief sources of vitamins are fruits and vegetables.

Table 3·2. Food Substances.

Substance	Kind of Substance	Essential for	Source
Proteins	Organic nutrients	Growth, maintenance and repair of protoplasm	Milk, wheat, meats, eggs, peas, cheese and beans.
Carbo-hydrates	Organic nutrients	Energy (stored as fat or glycogen in the body) bulk in diet.	Bread, pastries, cer eals, fruits, vegetables.
Fats	Organic nutrients	Energy (stored as fat or glycogen in the body)	Butter, cream, cheese, lard, oils, nuts, meats and oleomagarine.

Substance	Kind of Substance	Essential for	Source
Sodium compound	Mineral salts	Blood and other body tissues	Vegetables, table salt.
Calcium compound	Mineral salts	Formation of bone, teeth heart and nerve action clotting of blood	Milk, meat, vegetables, whole grain cereals.
Phosphorus compounds	Mineral salts	Formation of bones, teeth, ATP and nucleic acids	Milk, vegetables and meats.
Magnesium	Mineral salts	Muscle and nerve action	Vegetables.
Potassium compounds	Mineral salts	Blood and cell activities	Vegetables.
Iron compounds	Mineral salts	Formation of R.B.C.	Meats, leafy vegetables liver, raisins, prunes.
Iodine	Mineral salts	Secretion of thyroxine	Water, sea foods, iodized salts.
Water	Inorganic compound	Composition of protoplasm and blood dissolving substances	All foods (released during oxidation).
Vitamins	Organic compounds	Normal growth, body activity and the prevention of certain disease	Fruits, vegetables, milk.

Revision Questions

1. What is food ? Describe different components of food.
2. What are proteins ? Discuss the various properties of proteins.
3. What is amino acid ? Justify that the amino acids are the building stones of proteins.
4. What are carbohydrates ? Classify carbohydrates.
5. Describe the structure, functions and properties of fats.

CHAPTER 4

Vitamins

In the late nineteenth century it was found that young rats, fed-on synthetic milk made from appropriate concentration of protein, fat, carbohydrate and mineral salts in water, died quickly MAGENDIE (1816), LUNIN (1880) but when they were provided with natural milk the health of the rats was maintained and they began to grow again PEKELHARING (1905), HOPKINS (1912). This paved the way for the discovery that a further class of organic compounds, now called **vitamins,** was essential for the healthy maintenance of the body.

The studies suggesting the presence of indispensable substances in natural foods were brought into focus by LIND (1753), LUNIN (1881), TAKAKI (1887), EIJKMANN (1897) and HOPKINS (1912) who called them **"accessory food factors"**. The work of OSBORNE and MENDAL and of MCCOLLUM (1913) established beyond doubt the presence of these factors in milk. FUNK (1911) introduced the term **Vitamine** to these substances in the belief that he had isolated one of these factors and established that it possessed the properties of an amine. The last 'e' was dropped when it was shown that these substances as a class are not related to amines.

Vitamins are organic compounds of varying complexity that are indispensable for the organism. These organic compounds do not furnish energy, but they are responsible for the transformation of energy and for the regulation of the body metabolism. In the body, a vitamin may become part of a complex enzyme system and it is these enzyme systems that catalyze many of the important metabolic reaction of the body (FLORKIN, 1971). Thus, these are popular as **"biological catalysts"** in the organisms. In most instances, these substances do not seem to be synthesized in the body in significant amounts and as such have to be supplied in the diet according to the requirement to ensure proper growth and maintenance of health and life. Sometimes a little amount of a particular vitamin may be present in the body but that cannot suffice the optimum requirement of the body.

Although, these substances are required normally in very small amount but absence of any one in the diet for a long period may cause **deficiency diseases** (GRIJNS, 1901). Thus, their absence (**avitaminosis**) or deficiency in diet (**hypovitaminosis**) provokes more or less severe metabolic disturbances charaterized by specific symptoms. The **avitaminosis** may be **primary** or **direct**, due to inadequate intake of vitamin or its precursor over a prolonged period of time. The **avitaminosis** may be **secondary**, a **"conditioned deficiency"** due to other factors such as gastrointestinal disorders, poor teeth, allergies, malabsorption, increased excretion and imbalance. Avitaminosis from any cause, if prolonged, leads to (1) a gradual decrease in tissue levels of the vitamin or vitamins deficient, (2) a biochemical lesion, and, in time, (3) an anatomic lesion and finally cellular pathology and disease. This sequence is shown schematically in the given figure (Fig. 4·1).

Deficient intake (From dietary history) ↘	↙ **Secondary conditioning factors** (From clinical history)
Gradual decrease in tissue level	(evaluated by blood, urine or tissue analysis)
↓ Biochemical lesion	(reduced enzyme levels, altered metabolites, etc.)
↓ Anatomic lesion	(clinical evaluation)
↓ Pathology-disease	(clinical symptoms)

Fig. 4·1. Schematically representation of the sequence of events occurring in a typical avitaminosis.

Some vitamins can be stored in the body, while others must be supplied constantly as they are needed to neutralize and to convert certain toxic substances produced by the breakdown of the complex organic substances present in food. Excess of vitamins (**hypervitaminosis**), however, can also be unfavourable.

It is not a hard and fast rule that vitamins which are essential for some organisms, can also be important for others. For example, the deficiency of vitamin 'C' is the cause of drastic metabolic disorders in the man, monkeys and guineapigs but not in rats and a number of other mammals. Since vitamin 'C' is synthesized in their bodies, they do not need it in their food.

Chemical nature of vitamins : Vitamins vary greatly in their chemical nature and often have nothing in common. Some are protein, some alcohol, some sterol and some quinone in their chemical nature.

Nomenclature of vitamins : When these substances came into knowledge their chemical nature was unknown, therefore, letters of the alphabet were used to designate the vitamins in the order in which they were identified. This led to the use of the terms vitamin A, vitamin B, vitamin C, vitamin D and so on. As the knowledge in the vitamin field increased, certain difficulties arose. For example, it was found that certain vitamins thought to be simple were made up of many different components such as vitamin B-complex, it includes vitamin B_1, B_6, B_{12}. Then such names B_1, B_2, B_6, B_{12} and so on further were adopted. Today most of the vitamins have their names that indicate their chemical composition.

CLASSIFICATION OF VITAMINS

Vitamins, on the basis of their physiochemical properties particularly in their solubility, can be classified into two groups :

1. Fat-soluble vitamins, and
2. Water-soluble vitamins.

Fat-soluble vitamins are A, D, E and K, while water-soluble vitamins are B (B-complex), C and P.

The distinction between water-soluble and fat-soluble vitamins is of no fundamental significance, but is important in assessing the likely adequacy of a given diet since one that is deficient in fat will also be deficient in the vitamins that are soluble in it.

FAT-SOLUBLE VITAMINS

1. **Vitamin A :** Vitamin A, because of its action, also called **vitamin of growth** or **antixerophthalmic vitamin** or **anti-infective vitamin**. It influences growth of the body. It is also necessary for the maintenance of normal epithelial tissues of the organs including cornea and conjunctiva. It serves as a carrier of specific proteins to places for special processes such as in the formation of visual purple which in the light is converted into visual yellow. It is also essential for specific synthesis of the tissue of the nerve cells and fibres and for maintaining the integrity of the structure and functions of the reproductive organs.

Vitamin A is a yellow viscous oil. Chemically, it is related to **carotene,** a vegetable pigment. **Carotene** $C_{40}H_{56}$ is a common pigment found in abundance in carrots. Ordinary carotene was shown to consist of three isomers : α-carotene (15%), m.p. 184°c β-carotene (85%), m.p. 184°c, and γ-carotene (trace), m. p. 177°c. By the addition of two molecules of water β-carotene forms two molecules of vitamin A as it is the most active of the three carotenes, while α-carotene and γ-carotene can yield only one vitamin A molecule. The conversion of β-carotene into vitamin A occurs partly in the liver but mainly in the wall of the small intestine. For this conversion a normally functioning thyroid gland is necessary. **Carotene is,** thus, called as **provitamin A.**

Vitamin A is no longer a hydrocarbon like carotene but a mono-atomic alcohol generally represented by the following formula :

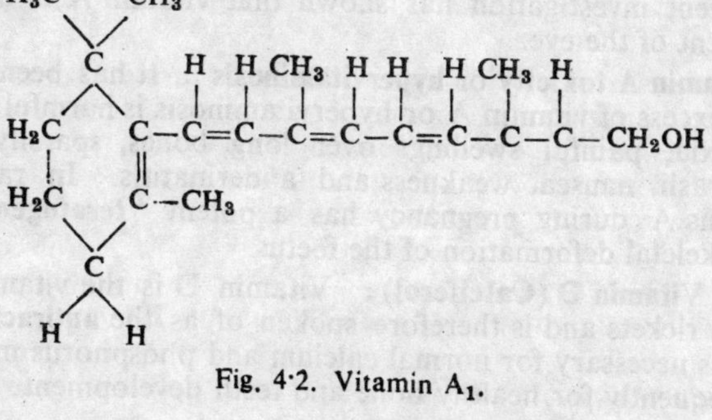

Fig. 4·2. Vitamin A_1.

It is fairly thermostable in the absence of oxygen but it is oxidized and loses its biological activity when it is heated in the presence of oxygen. In a vaccum, it can be distilled.

Vitamin A occurs in two forms, *e.g.*, A_1 and A_2 which are very closely related. Each form is derived from a separate carotene. Vitamin A_1 is formed from α-carotene, while A_2 from β-carotene. The two carotenes differ in spectrum and colour tests. Vitamin A_2 has one more conjugated double bond as shown below :

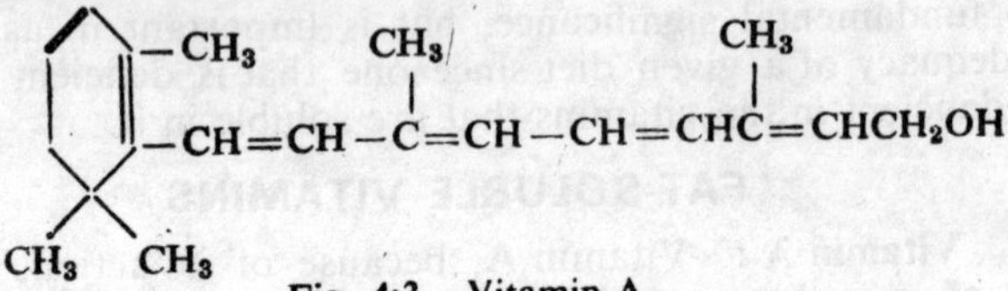

Fig. 4·3. Vitamin A_2.

Occurrence and availability : Vitamin A and provitamin A, the so called carotene, are found in considerable amounts in butter, liver, egg yolk, milk and especially in fish liver oil and in certain vegetables such as carrots, tomatoes and spinach, etc.

Daily requirement : The minimum daily requirement of vitamin A recommended by Food and Nutrition Board, National Academy of Sciences, National Research Council, is about 5000 units for men and women. For pregnant and nursing women, from 6000 to 8000 units daily are recommended. Infants and growing children require 1500 to 5000 units daily, depending on age. The one unit represents 0.000344 mg or 0.344 μg.

Vitamin A deficiency or **avitaminosis :** It is belived that a vitamin A deficiency is associated with a loss of weight and inhibition of growth in young animals. Its deficiency also causes morbid changes in the epithelium of the cornea, the respiratory and digestive tracts, and of other organs. A dryness and keratinization of the skin and an increase in pigmentation are observed. **Xerophthalmia** is a very characteristic eye disease also-caused by vitamin A deficiency. In this disease, the eyes become haemorrhagic, encrusted and infected. The earliest sign of vitamin A deficiency in man is **nyctalopia** (night blindness). In such patients vision is abnormally poor or fails completely when it is dark.

Recent investigation has shown that vitamin A is involved in the pigment of the eye.

Vitamin A toxicity or **hypervitaminosis :** It has been observed that the excess of vitamin A or hypervitaminosis is harmful as it leads to anorexia, painful swellings over long bones, sparsity of hair, pruritic rash, nausea, weakness and a dermatitis. In rats, hyper vitaminosis A during pregnancy has a potent **"teratogenic effect'**, causing skeletal deformation of the foetus.

2. **Vitamin D (Calciferol) :** Vitamin D is the vitamin that is related to rickets and is therefore spoken of as the **antirachitic vitamin**. It is necessary for normal calcium and phosphorus metabolism and consequently for healthy bone and teeth development.

It is colourless, odourless, crystalline alcohol possessing steroid-like structure, that means, it is related to sterol chemically. It is soluble in fats and fat-like solvents and is insoluble in water. A decrease in potency is noted when it is exposed to increased temperatures, sunlight or oxygen. It is formed by the action of ultraviolet rays on sterols (steroids with 8 to 10 carbon atoms in the side chain at position 17 and an alcoholic hydroxyl group at position 3 are closed as sterols) (STEENBOCK and HESS 1924). The ultraviolet rays cause rearrangement of atoms in the sterol molecule that results in the formation of calciferol or vitamin D. Cholesterol ($C_{27}H_{45}OH$) and ergosterol ($C_{28}H_{43}OH$) are important sterols which on irradiation give rise to vitamin D and as such are called provitamin D. Besides these other provitamins of vitamin D have also been noticed such as 22, 23 dehydrocholesterol, 7 dehydrocholesterol, etc. The 7 dehydrocholesterol is a natural provitamin which is always found in mammals. It changes to vitamin D_3 under the action of sunlight or ultraviolet rays.

Vitamin D (or D_1) occurs in two forms D_2 and D_3. Vitamin D_2 and D_3 are closely related. In 1932, WINDANS by the ultra-violet irradiation of ergosterol, isolated the first pure active substance, vitamin D_2. Ergosterol is a substance found only in lower plants such as yeast, and fungi (*e.g.*, ergot, from which it is named). Now vitamin D_2 has been named as **ergocalciferol**, of vegetable origin. Vitamin D_3 was obtained later by the irradiation of 7 dehydrocholesterol and now it has been named as **cholecalciferol**, of animal origin.

Vitamin D_2 contains four double bonds, three of which are conjugated, while vitamin D_3 contains three double bonds, all of which are conjugated.

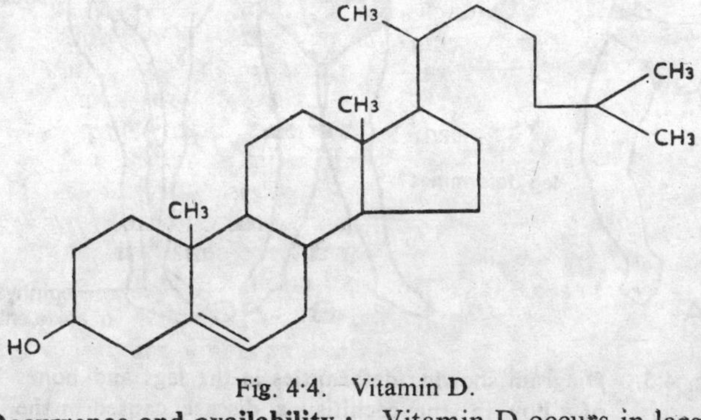

Fig. 4·4. Vitamin D.

Occurrence and availability : Vitamin D occurs in large quantities in fish-liver oils (*e.g.*, cod- or halibut-liver oil) and in yolk of eggs; it is present in a small degree in animal fats (*e.g.*, beef suet, milk) and is practically absent from vegetable oils (*e.g.*, olive oil). Milk is generally fortified with vitamin D and thus becomes an ideal food for growing children. It also contains appreciable quantities of calcium and phosphorus.

Daily requirement : It is difficult to estimate human requirements, but it is suggested that the daily requirement of vitamin D of a child is 400 International Units and that of an adult about 400 units. The daily requirement for other animals varies to a great extent with the species, age, sex and time of year. One International Unit (IU) is equivalent to the biologic activity of 0·025 μg of pure crystalline vitamin D_3.

Vitamin D activity and deficiency : When vitamin D is present in insufficient amounts or essentially absent in the diet, no metabolism of calcium and phosphorus salts takes place as a result of which no calcium is deposited in the bones that results in the development of a disease called **rachitis** or **rickets.** In **rickets** the bones do not calcify properly. In children, rickets is associated with knock-knees, bowlegs and distorted joints. In general, in a rachitic patient the ends of the bones show incomplete calcification. The sutures of the skull and the fontanelles do not close, the bones of the limbs become too flexible and curve under the weight of the body that one feels great difficulty in movement.

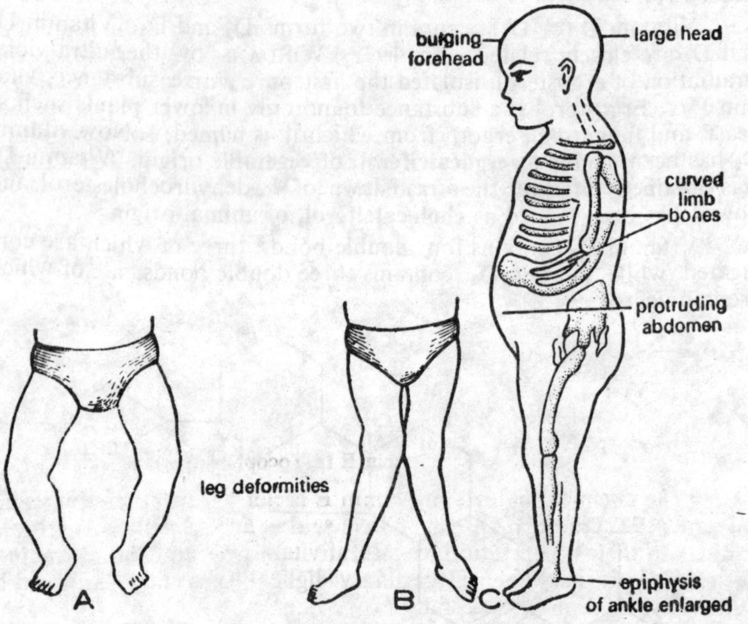

Fig. 4·5. Diagram showing deformities in the legs and bones of a boy in the rachitis, a disease caused in the deficiency of Vitamin D.

In adults its deficiency leads to **osteomalacia**, a condition characterized by the softening of the bones specially the pelvic girdles, ribs and femora, leading to deformities.

Vitamin D toxicity or **hypervitaminosis :** Although one should strive to furnish the body with an adequate supply of vitamin D, care should be taken that the body will not become overdosed with vitamin D as overdoses of vitamin D are harmful causing anorexia, nausea, vomiting, headache, drowsiness, polyuria, and polydipsia. If both serum Ca and serum P are markedly increased, resulting in metastatic calcification of many soft tissues and formation of renal calculi. The latter may block the renal tubules sufficiently to cause a secondary hydronephrosis. The signs of hypertension may also develop.

3. **Vitamin E (Tocopherol) :** Vitamin E investigated by EVANS and SURE in 1922, as it maintains the normal state of tne sex-organs and ensures the continuation of pregnancy and normal child birth, has been named **tocopherol** (Gk. *tocos*, child birth; *phero*, to bear; *ol*, an alcohol).

The physiological effects of vitamin E are caused by a series of seven naturally occurring, closely related compounds what are known as **tocopherols.** The tocopherols are characteristically soluble in fat solvents and insoluble in water. They are colourless, non-crystallizing oils, stable to heat and acids.

The most important isomers of tocopherol are alpha (α), 5, 7, 8-trimethyltocol; beta (β), 5, 8-dimethyltocol; gamma (γ), 7, 8-dimethyltocol ; and delta, (δ) 8-methyltocol. These differ in their biological activity. β and γ tocopherols differ from the α-form in the positions of the methyl groups attached to the benzene ring. For example, β-tocopherol has methyl groups only on carbons 5 and 8. The structure formula for α-tocopherol is as follows :

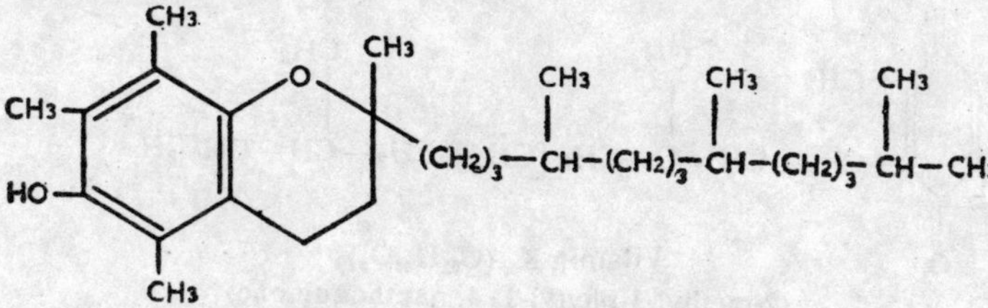

Fig. 4 6. Vitamin E (α-Tocopherol).

The chemical nature of vitamin E is not yet definite. EVANS and BISHOP (1922) thought that the above said α, β, γ tocopherols are very similar in their composition to natural vitamin E and the formula is $C_{29}H_{50}O_2$. It is stable in ordinary light but readily destroyed by ultraviolet light and rancid fat.

Occurrence and availability : Vitamin E is generally found in some of the natural oils such as wheat germ oil, cotton seed oil, rice germ oil and in the green leaves of lettuce and alfalfa. It is also found in meat, egg yolk, fish liver oil but in poor amount.

Daily requirement : The recommended daily requirement of vitamin E for adults is 30 International Units for men and 25 I. U. for women. The amount for women is increased to 30 I. U. during pregnancy and lactation. For infants and children, 1 to 1·25 I. U. per kilogram body weight are recommended.

One International Unit (I. U.) of α-tocopherol is defined as the biologic activity of 1·1 mg of the pure compound (or 0·67 mg α-tocopherol).

Vitamin E deficiency or **avitaminosis :** In the deficiency of this vitamin the germinal epithelium of the testes is destroyed and testes fail to produce spermatozoa. In the female, ovulation and fertilization take place, but the foetus dies and is resorbed within a short time. Nursing animals develop lactation disorders. Lack of this vitamin also causes a muscle ailment (muscular dystrophy).

Vitamin E toxicity or **hypervitaminosis :** Its overdoses lead certain metabolic disorders and disturbances which occur mainly in the gonads and neuromuscular systems and to a minor extent in the pituitary and thyroid glands.

4. Vitamin K : Vitamin K is a **antihaemorrhagic** vitamin. It is necessary for the formation of prothrombin (in the liver) which participates in blood clotting. DAM (1934) named this vitamin K from German and Scandinavian term for coagulation : koagulation. It occurs in two forms, K_1 and K_2. They are formed from naphthoquinones. Vitamin K_1 is 2-methyl-3-phytyl-1, 4-naphthoquinone. Vitamin K_2 is 2-methyl-3-difarnesyl-1, 4-naphthoquinone, called **farnoquinone.** Both these natural types have the same general activity. Recently it has been reported that **menadione,** *i. e.,* 2-methyl, 4-naphthoquinone, also has the same structure as vitamin K_1 or K_2 contains and is as effective as vitamin K_1 on a molar basis. The exact structures of K_1 K_2 and menadione are as follows :

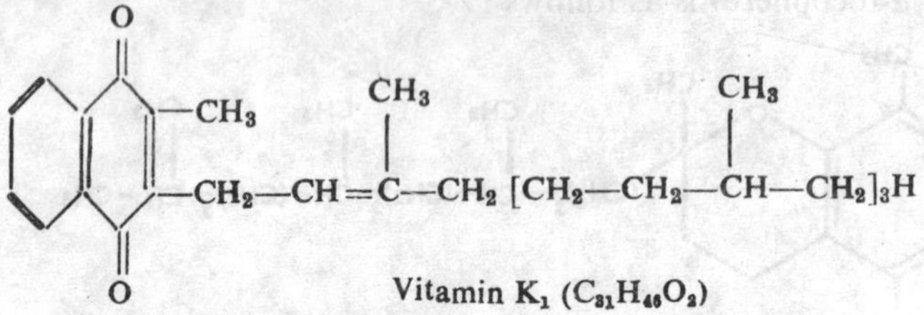

Vitamin K_1 ($C_{31}H_{46}O_2$)
(2-methyl-3-phytyl-1, 4, napthoquinone)

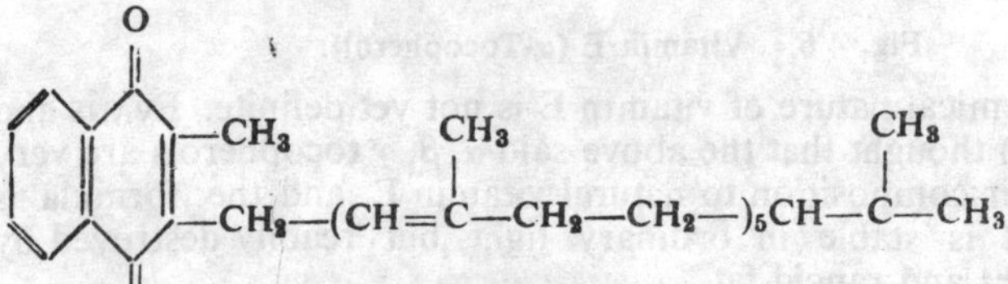

Vitamin K_2 (farnoquinone)
(2-methyl-3-difarnesyl-1, 4-naphthoquinone)

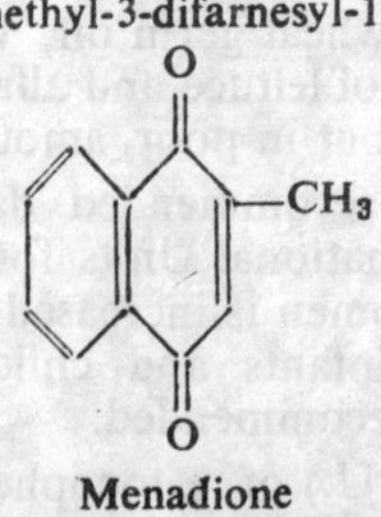

Menadione
(2-methyl-1, 4-naphthoquinone)

Fig. 4·7. Vitamin K_1, K_2 and menadione.

Vitamin K is very much soluble in fat solvents and insoluble in water. It is quite stable at increased temperature but is very sensitive to alkali and various light sources such as ordinary and ultra-violet rays.

Vitamin K_1 is a yellow oil at ordinary temperature and its melting point is 20°C, while vitamin K_2 is a yellow crystalline solid of which the melting point is 55°C.

Occurrence and availability : In general, vitamin K_1 occurs in plant materials such as green leafy tissues, alfalfa, spinach, kale, cabbage, chestnut. A rich source of vitamin K_2 is purified fish meal. Milk and eggs also contain vitamin K_1 and K_2 in small amounts. Microorganisms such as microbacteria, molds and yeast are said to be the best sources of vitamin K_2.

Daily requirement : It is difficult to estimate the daily requirement of vitamin K because the intestinal flora provide the vitamins in the amounts required.

Vitamin K deficiency or **avitaminosis :** A deficiency of vitamin K may cause a deficiency of prothrombin in the circulating blood, a condition known as **hypoprothrombinemia**. In the absence of prothrombin blood fails to clot and results in profuse bleeding (for example, from the gums) and haemorrhages (into joints and retina, etc.). Recent studies indicated that vitamin K has a genetic action in including RNA formation for the synthesis of blood-clotting proteins (OLSEN and PHILLIPS).

Vitamin K toxicity or **hypervitaminosis :** Over doses of vitamin K are toxic, due to which malfunction of any body organ may occur.

Coenzymes-Q : The coenzymes-Q group has been classed as vitamins because these compounds are capable to cure or protect against vitamin E deficiency in several species of animals. Some of the coenzymes are also active in electron transport and for oxidative phosphorylation.

Antistiffness factor : Stigmasterol has been called as antistiffness factor. It is a plant sterol with a formula quite similar to that of ergosterol. Its absence causes stiffness of the wrists and elbows of guinea pigs. The muscles atrophy and become streaked with bundles of fine white lines of calcium deposits.

Stigmasterol occurs in fresh kale or alfalfa and in fresh cream.

WATER-SOLUBLE VITAMINS

5. **Vitamin C (Ascorbic acid) :** Vitamin C ($C_6H_8O_6$) is known as ascorbic acid (scurvy preventing) and also as cevitamic acid. It is also known as **antiscorbutic vitamin**. According to **Wolbach**, this vitamin is necessary for formation of all intercellular substances. It is also essential for the immunity or body defence mechanism. It is an activator for the immunity or body defence mechanism. It is an activator for the growth of the tissues. It plays important role in blood formation and maintenance of physiological level of erythrocytes, the so called red blood corpuscles. It also activates certain intracellular enzymes such as **proteolytic enzymes, cathepsin arginase** and **amylase**. It takes part in the metabolism of tyrosine.

Vitamim C is a colourless, odourless, crystalline material. Its melting point varies from 190 to 192°C. It is very much soluble in water and quite insoluble in the fat solvents such as ether, alcohols sterols, etc. Its pH value ranges from 2–3, depending upon the concentrations. It is sensitive to light and air. It is reasonably stable in mild acid solutions while in alkaline solutions there is a considerable loss of activity. It is destroyed by excessive boiling or prolonged cooking. Drying, storing and aging of food may also destroy the vitamin C activity.

Vitamin C is closely related to the hexoses in structure and is, in fact, conveniently synthesized from glucose. It is not a typical organic acid in that it has no free carboxyl group (—COOH) ; actually, a lactone structure is present (a lactone is an inner ester of an alcohol and an acid group in the same molecule).

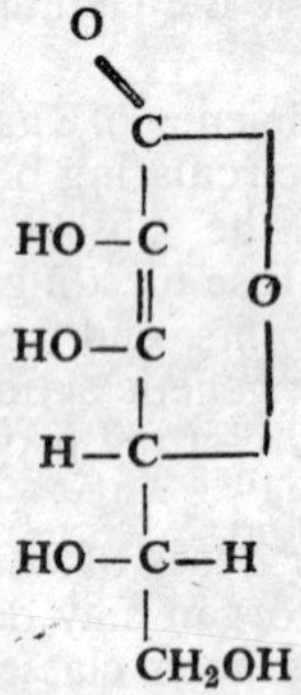
I

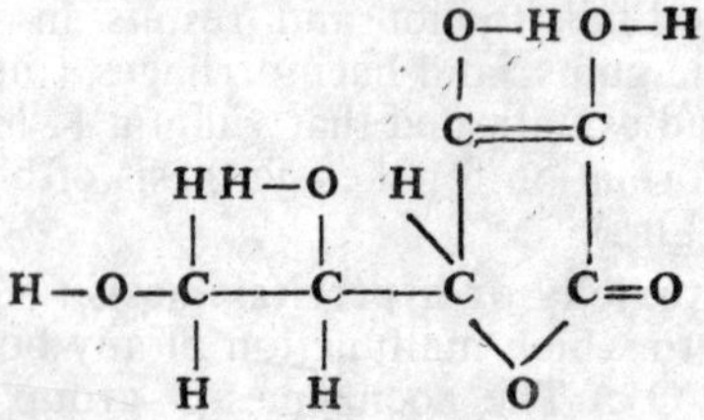
II

Fig. 4·8. Vitamin C (Ascorbic acid).

(The structure of ascorbic acid may be written as I but structure II is more indicative of the stereochemical relationships).

Occurrence and availability : Vitamin C occurs abundantly in certain fresh fruits such as, citrus fruits, lemons, limes, and grape fruits are excellent sources. Other fruits like raspberries, currants, gooseberries and strawberries also contain vitamin C. It is also found in plenty in vegetables such as tomatoes, kale, spinach, broccoli, cabbage.

Daily requirement : The exact daily requirement is uncertain, but for men and women (during pregnancy and lactation) the recommended dietary requirement is 60 mg. Infants should have 35 mg per day, and, as the child gets older, a gradually increasing amount is required until a maximum need is reached at adolescence, 50 to 60 mg daily for boys and 50 to 55 mg for girls.

Vitamin C deficiency or **avitaminosis :** Its deficiency in the body leads to **scurvy**, a disease characterized by bleeding gums, loosening and falling out of teeth and subcutaneous and intramuscular haemorrhages. At the first sign of scurvy, there is a loss of weight and appetite and pains develop in the joints and muscles. As the disease progresses, the gums are particularly affected, showing swelling, tenderness, gingivities, redness, ulceration and even gangrene. Weakness and emaciation are seen. There are definite defects in skeleton calcification without much disturbance in mineral metabolism. This leads to tenderness in legs.

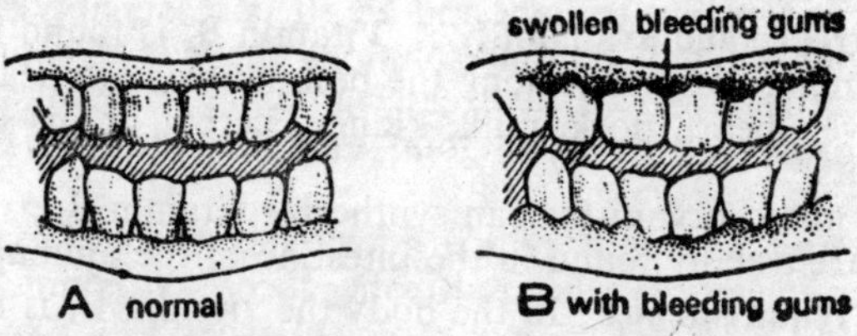

Fig. 4·9. Diagram showing the symptoms of scurvy, a disease caused due to deficiency of vitamin C.

It is believed that vitamin C plays an important role in biological oxidation and reduction as it is easily oxidized and reduced again. However, the exact enzyme systems and chemical reactions involving ascorbic acid are still not too clear. In the absence of this vitamin the collagen and the connective tissue proteins are not synthesized properly.

6. **Vitamin B** or **B-complex :** Historically, the term **"vitamin B"** was first applied to a water soluble substance separated from protein free milk, wheat germ and yeast and found to be necessary for the nutrition of the infants. At that time, vitamins C and A were recognized ; hence the term **"vitamin B"** was used to distinguish it from these two.

When it was later realized that many substances are included in this "vitamin B" subscripts such as B_1, B_2, B_3, B_4, B_5, B_6, and so on were used to distinguish the various vitamins of this group.

Now vitamin B or B-complex represents a group of eleven vitamins. The common feature of this group is that they all contain nitrogen. Each of these vitamins differs from the rest in chemical structure and in the effects it produces in the organisms.

It is now known that the members of this group are universally distributed in all living cells and are essential constituents of all living matter. The enzymes of this group form enzyme systems which are important in catalyzing many of the metabolic reactions in the body. We will discuss these B vitamins separately :

(i) **Vitamin B_1 (Thiamine hydrochloride) :** This vitamin B_1 is commonly known as **antineuritic vitamin** or **aneurin** or **antiberiberi factor**. It is a white, crystalline solid readily soluble in water but insoluble in ether and chloroform. Its melting point is 248°C. It has the odour and flavour characteristic of yeast and salty taste. Its pH value ranges from 3–4, depending upon the concentration.

Chemically, this vitamin B_1 is composed of two different heterocyclic nuclei : the pyrimidine and the thiazole rings. These two rings are connected by a methylene group-CH_2—. The pyrimidine ring contains a methyl group-CH_3 and an amino group-NH_2. The thiazole ring contains a methyl group and a hydroxyethyl group, $-CH_2CH_2OH$.

The structure of thiamine, which was first isolated in crystalline form by JANSEN in Holland and WINDAUS in Germany, was established by R. R. WILLIAMS *et. al.*

Occurrence and availability : Vitamin B_1 is found in almost all plant and animal tissues. Among the best sources are peas, dry yeast, egg yolk, liver, heart, pork, milk, kidney, cereal grains (soy, wheat, etc.) and nuts.

Some micro-organisms can synthesize vitamin B_1 ; these micro-organisms are usually found in the intestine of the animals.

Daily requirement : In the body the vitamin B_1 is not stored in sufficient amounts to last more than a brief period. Therefore, the body should receive a daily amount adequate for its need. The recommended daily vitamin B_1 requirement for an adult is about 1·2 to 1·4 mg for men and 1·0 mg for women, increasing to 1·1 to 1·5 mg during pregnancy and lactation. The thiamine requirement of the infants is between 0·2 to 0·5 mg. daily.

Vitamin B_1 deficiency or **avitaminosis :** The deficiency of vitamin B_1 causes **anorexia** (lack of appetite) **arrested growth**, and **polyneuritis** in animals and **beriberi** in man. Beriberi is a common disease in countries where rice is the principal food. In the adult it is characterized by polyneuritis, with muscular atrophy, cardiovascular changes and oedema. At first there is weakness and fatigue, followed by headache, dizziness, gastro-intestinal, symptoms and tachycardia. Later the major symptoms may follow any one of the following patterns :

(1) Nervous symptoms (**dry beriberi**) ; (2) Symptoms associated with oedema and serous effusions (**wet beriberi**) ; (3) Symptoms of heat involvement (**acute pernicious beriberi**). Often the symptoms are characteristic of more than one of these three classes and are called **mixed beriberi.**

Early symptoms of vitamin B_1 deficiency are lack of appetite, fatigue, and heaviness of the legs. As the disease progresses, an increasing neuritis manifests itself, causing the knee and reflexes to be lost. Pain and tenderness develop in the muscles particularly of legs. In the final stages, the calf muscles are degenerated. The heart becomes enlarged and painful and the pulse weakens.

Biochemical relationship : Vitamin B_1 and two molecules of phosphoric acid form an ester or coenzyme what is known as **cocarboxylase** or **carboxylase.** Chemically which is thiamine pyrophosphate (TPP) or diphosphothiamine. This coenzyme participates in all oxidative decarboxylations of α-keto acids, including pyruvic and α-ketoglutaric.

In the deficiency of vitamin B_1 the coenzyme, **cocarboxylase,** is not synthesized in sufficient quantities, owing to which the **oxidase** of pyruvic acid in the animal proves insufficiently active.

(ii) **Vitamin B_2 (Riboflavin) :** This vitamin is also known as **lactoflavin** or **ovoflavin** or vitamin G. It participates in the metabolism of carbohydrates and other substances : It also influences respiration, **haematopoiesis** and the activity of the nervous system. It also takes part in the synthesis of visual purple.

Riboflavin $C_{17}H_{26}N_4O_6$ is an orange-yellow crystalline substance which is sparingly soluble in water. Its melting point is 280°C. It is essentially a odourless but it has a bitter taste. The water solution of vitamin B_2 shows a yellow-green fluorescene. It is quite stable to heat, air and oxygen but it is very much sensitive to light.

Chemically, the vitamin B_2 contains three ring system (iso-alloxazine) combined with an alcohol derived from ribose sugar (ribitol). Two of the rings are nitrogen heterocycles, each containing two nitrogen atoms. The basic ring structure is isoalloxazine, and the sugar part is directly related to D-ribitol. The structure formula of vitamin B_2 is as follows :

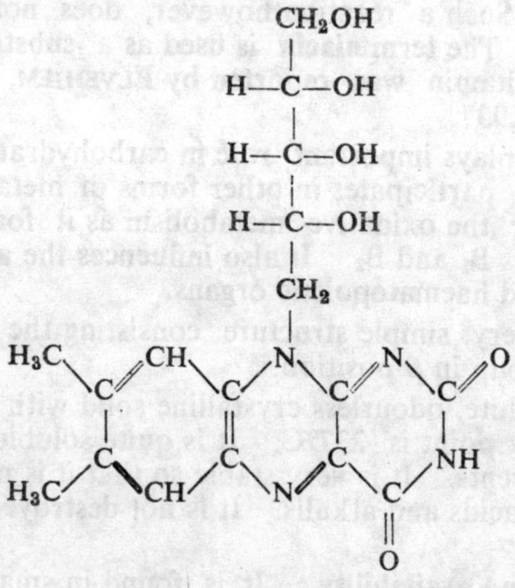

Fig. 4·10. Vitamin B_2 (Riboflavin).

Occurrence and availability : Riboflavin occurs widely in nature. Milk is an important source of it. Other excellent sources are meats, especially liver and kidney, fish and eggs. Leafy vegetables and fruits are also good sources of riboflavin.

Daily requirement : The recommended dietary requirement of riboflavin per day varies from 0·6 to 1·7 mg for children and adults while pregnant and lactating women may need 2·5 mg daily.

Vitamin B_2 deficiency or **avitaminosis :** The continuous vitamin B_2 deficiency in diet leads morbid changes in the skin **cheilosis** (a disease characterized by inflammation of the lips, fissures at the corners of the mouth, scaliness, greasiness, and fissures in the folds of the ears and nose), **glossitis**, (a disease characterized by inflammation of the tongue), **seborrheic dermatitis** (a disease characterized by waxy accumulations in the skin), and **keratitis** (a disease characterized by the roughening of the skin at the mouth and nose). Some initial trauma or infection is likely to be followed by a skin lesion if a riboflavin deficiency is present. There may be ocular disturbances like inflammation of the cornea, blood shot eyes, photophobia, dimness of vision, and itching, burning, and dryness of the eyes with redness of the conjunctiva.

Biochemical relationships : Riboflavin with proteins forms flavoproteins which are enzymes that function in tissue respiration as components of the electron-transport system and in a number of enzymes including **Warburg's** "yellow enzyme", L-and D-amino acid oxidase, xanthine oxidase, cytochrome-c reductase, and certain dehydrogenases. The flavoproteins thus form a varied and important group of intracellular enzymes involved in oxidation reduction reactions.

(iii) **Nicotinic acid** or **niacin (the pellagra preventing factor) : The term nicotinic** acid comes from **nicotine**, a tobacco constituent, because one can isolate acid when nicotine is oxidized by vigrous chemical means. Such a reaction however, does not occur when smoke is inhaled. The term **niacin** is used as a substitute for nicotinic acid. This vitamin was reported by ELVEHJEM, WOOLEY and their associates in 1937.

This vitamin plays important role in carbohydrate metabolism in tissues and also participates in other forms of metabolism. It is very important for the oxidative metabolism as it forms enzymatic system like vitamins B_1 and B_2. It also influences the activity of the degestive glands and haematopoietic organs.

Niacin has very simple structure consisting the pyridine ring with a carboxyl group in β position.

Niacin is a white, odourless crystalline solid with an acidic tart taste. Its melting point is 237°C. It is quite soluble in water and insoluble in fat-solvents. It is very stable so that it is not inactivated by air, light, heat, acids and alkalis. It is not destroyed in ordinary cooking processes.

Occurrence and availability : It is found in small amounts in various plants and animal tissues. Excellent sources are liver, yeast, wheat germ, lean meat, whole cereals and pulses, etc. Common fruits and vegetables contain very little of it.

Daily requirement : The recommended dietary requirement of niacin for children is from 8 to 15 mg per day and for adults, from 13 to 18 mg. For women in pregnancy and lactation, the requirement may be increased to some extent.

Niacin deficiency or **avitaminosis :** The deficiency of niacin causes **pellagra**, a disease common in the southern states where the people can not afford a proper diet or where corn is a main staple. The specific manifestations of this disease include patches of dermatitis, soreness and inflammation of the tongue and mouth, alimentary disorders (achlorhydria and diarrhoea) and pigmentation and thickening of the skin. There is usually a rash that appears symmetrical on the sides of the body and backs of the hands and arms. Nervous disorders and mental disturbances have also been observed. The mucosa in the oral cavity especially on to the tongue, swells and is frequently ulcerated. In dog its deficiency causes **black-tongue disease**, this is characterized by a sudden refusal to eat the deficient diet, apathy, and lesions in the mouth. The inner surfaces of the lips and cheeks become covered with pustules, and the mucous lining comes away in shreds. Intense salivation and bloody diarrhoea are additional symptoms.

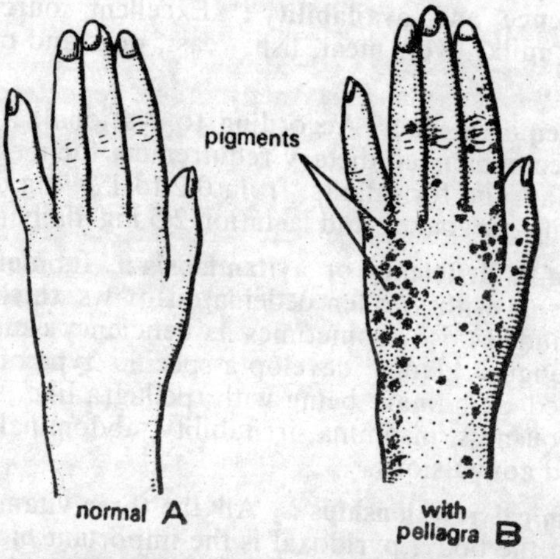

Fig. 4·11. Diagram showing the symptons of pellagra, a disease caused due to deficiency of niacin.

Biochemical relationships : The amide of nicotinic acid is called as nicotinamide. Nicotinamide forms part of two coenzymes, NAD^+ and $NADP^+$, also called DPN and TPN and coenzyme I and II, respectively. They are composed of phosphoric acid, D ribose and adenine in different proportions. Both of these coenzymes are hydrogen acceptors and take part in oxidation and reduction reactions of metabolic nature in the body.

Nicotinic acid can be synthesized biologically from tryptophan amino acid. This has been established by the fact that people who obtain sufficient amount of tryptophan, instead of nicotinic acid, do not exhibit pellagra.

(iv) **Vitamin B_6 (Pyridoxine) :** This vitamin plays active role in the metabolism of proteins, fats and carbohydrates. It controls growth in rats and essential for the life of many lower organisms such as yeast, mosquito larvae and some micro-organisms. It also influences formation of RBC and synthesis of haemoglobin. It is also involved in the conversion of protein to fat.

Although pyridoxine was at first thought to be the vitamin B_6, it has since been shown that there are three naturally occurring substances of which pyridoxine can be converted to either of the other two, but neither of them can be changed to pyridoxine in the body. They are pyridoxal and pyridoxamine and referred to as vitamin B_6. These are simple substituted pyridine derivatives which differ only in group attached in the γ position of the ring. The group may be either a primary alcohol, an aldehyde or a primary amine.

Pyridoxine is a white crystalline solid. The melting point

ranges from 204 to 206°C. Pyridoxal and pyridoxamine are also white crystalline solids but they melt at about 165°C and 227°C respectively. They are quite stable in heat, acids and alkaline but not stable in ultraviolet rays.

Occurrence and availability : Excellent sources of vitamin B_6 are egg yolk, milk, liver, meat, fish, yeast, peas and other leguminous plants.

Daily requirement : According to National Research council (1968) the recommended dietary requirement for men and women is 2·0 mg per day and for infants from 0·2 to 1·2 mg daily, depending on age. During pregnancy and lactation 2·5 mg daily is recommended.

Vitamin B_6 deficiency or **avitaminosis :** Its deficiency in rats causes **acrodynia** with swollen oedematous paws, thickened encrusted ears, and denuded nose. Sometimes its deficiency causes ulceration of mouth and tongue. Dogs develop a specific type of anaemia which is also noticed in human being with pellagra like symptoms, *e.g.*, extreme nervousness, insomnia, irritability, abdominal pain vomiting, diarrhoea and convulsions.

Biochemical relationships : All the three vitamin B_6 are inter-convertible in the body, pyridoxal is the important biologically active form. The coenzyme containing vitamin B_6 is pyridoxal phosphate, it is the coenzyme for the decarboxylases that act on a number of amino acids and for two enzyme systems involved in the metabolism of sulphur-containing amino acids. The pyridoxine serves as coenzyme for the transaminase and it is believed to be involved in the metabolism of the unsaturated fatty acids.

The definite requirements of vitamin B_6 are not known but reactions catalyzed by the enzyme system containing pyridoxal phosphate are well known. These reactions are simple decarboxylations, transamination and condensations, in these reactions either the carbon or carbon bonds are created or ruptured.

It has been definitely established that the vitamins B_6 participates in the metabolism of tryptophan amino acid. In the deficiency of this vitamin, humans do not metabolize tryptophan properly and a marked increase in xanthurenic acid is noted in the urine.

(v) **Pantothenic acid :** Pantothenic acid (B_3) was at first called the **filtrate factor** and was given its present name by its discoverer, R.J. WILLIAMS. This vitamin has been known as growth factor for small organisms such as bacteria and protozoans. It is also essential for cellular metabolism as it forms part of an enzyme system.

The term pantothenic acid is derived from the Greek word *pantothen* meaning from everywhere. The vitamin had been given many names before its chemistry and identity were completely established.

The pure acid is a pale yellow viscous oil, readily soluble in water but insoluble in alcohol and other fat solvents. It is unstable and for this reason it is solid as a calcium salt.

Chemically pantothenic acid $C_9H_{17}NO_5$ is an aliphatic compound containing an amide linkage, a free acid group, and two free hydroxyl groups, one primary and the other secondary. It is represented as follows :

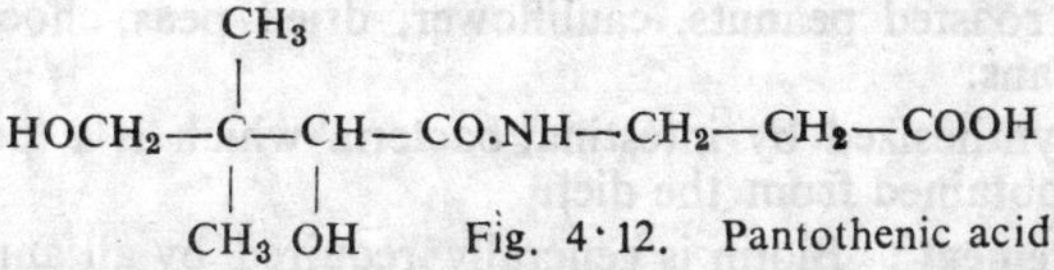

Fig. 4·12. Pantothenic acid.

Occurrence and availability : Pantothenic acid is richly found in liver and kidneys. It is also found in the heart, brain and tongue. Yeast, egg yolks, crude crane molasses, and cereal brans are other good sources. In animal tissue, it is found in enzyme systems and does not occur in free state.

Daily requirement : The daily requirement of pantothenic acid for man has not been established and deficiency disease has also not been recognized. Sheep and cattle require pantothenic acid and they obtain it from the bacterial action on food.

Pantothenic acid deficiency or **avitaminosis :** Pantothenic acid deficiency in rats is associated with poor growth, early aging, a graying of the hair and a dermatitis (inflammation of the skin). The lack of this vitamin causes deficient feathering in chicks, degeneration of nervous system in pigs and haemorrhages in the adrenal gland of human beings. In humans there occurs "burning feet syndrome". Fatigue, cardiovascular disturbances, gastrointestinal disturbances, numbness and tingling of the extremities, and a number of other distressing conditions have also been reported.

Biochemical relationships : Coenzyme A contains about 27% pantothenic acid which is essential for various reactions going on in the body, an important one being acetylation. Coenzyme also participates in carbohydrate metabolism in the Krebs cycle. It is a key enzyme in the interconversion of carbohydrates and fats.

(vi) **Biotin :** This vitamin is said to be an important factor in human nutrition and plays important role in the lipid metabolism as it forms prosthetic group of some of the enzymes such as **carbo-xylase**.

Chemically biotin is a rather complex substance containing a sulphur heterocycle ring, and a nitrogen heterocyclic ring, fused together. Its structure is as follows :

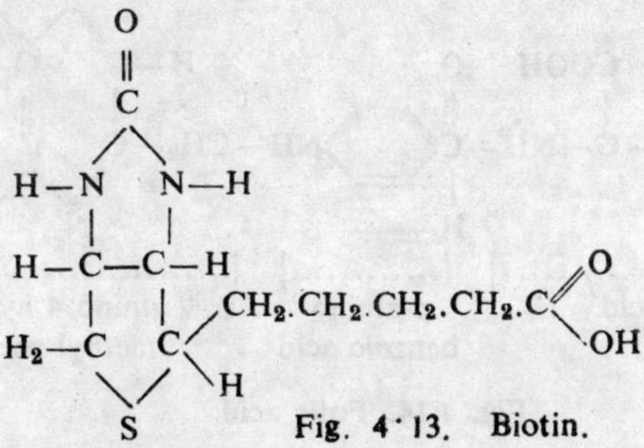

Fig. 4·13. Biotin.

Biotin is very much soluble in water but insoluble in fat solvents. It is reasonably stable towards dilute acids, alkali and heat.

Occurrence and availability : It is found in abundance in egg-yolks, liver, kidney, milk, yeast, wheat and other seeds. Its other good sources are roasted peanuts, cauliflower, dried peas, chocolate and dried lima beans.

It may be synthesized by intestinal bacteria which is a potent source than that obtained from the diet.

Daily requirement : Biotin is generally required by all animals investigated so far. It is essential for rats, pigs, rabbits, monkeys, dogs and men and chicks. The daily requirement of a man is 0·5 mg.

Biotin deficiency or **avitaminosis :** The deficiency of biotin in rats causes special dermatitis, marked by sealiness and desquamation (shedding of scales), the spectacled eye condition is also apt to appear. In addition, there is a loss of weight leading to extreme emaciation and death. It can also prevent the birth of young. Deficiency of this vitamin in hens results in high embryonic mortality and skeletal deformities in chicks.

In man, although the deficiency of biotin has not been noticed but experimental deficiency leads to polar desquamation, noticeable susceptibility to fatigue and muscular pains, and heart distresses.

Biochemical relationships : The role of biotin is primarily of carbon dioxide fixation, or carboxylation, as occurs in the conversion of pyruvic to oxaloacetic acid. It also takes part in other carboxylation reactions, including the conversion of acetyl-CoA to malonyl-CoA, in the biosynthesis of fatty acids and of propionyl-CoA to methylmalonyl-CoA.

(vii) **Folic acid :** This vitamin serves as a essential growth factor not only for bacteria but for higher organism as well. It has been called variedly such as vitamin M because it can prevent a nutritional anemia in monkeys, vitamin Bc because it can prevent nutritional anemia in chicks, factor U, when found essential for growth of chicks and **lactobacillus casei factor**. We shall use the generally accepted term folic acid.

Folic acid is also known as pteroylglutamic acid. It is composed of an L-glutamic acid group, a p-aminobenzoic group and a substituted pterin. The structure formula of folic acid is as follows :

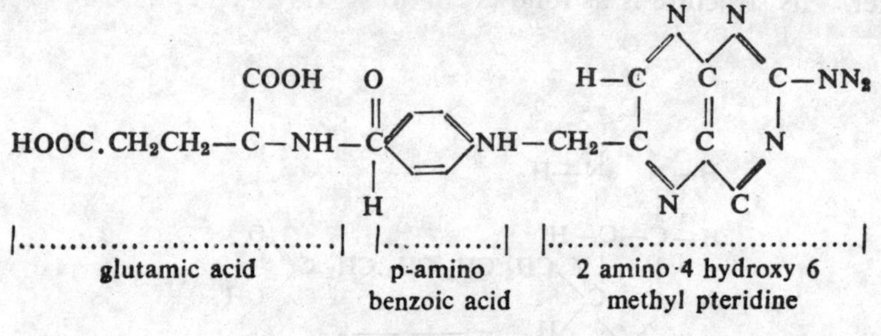

Fig. 4·14. Folic acid.

This vitamin is tasteless and only slightly soluble in water. It is sensitive compound which is inactivated by acid, base. sunlight, heat and oxidation and reduction.

Occurrence and availability : The most potent sources of this vitamin are liver, kidneys, yeast and mushrooms. It is also richly found in wheat germ, dried lima beans, spinach, chicken, peanuts and whole wheat, etc.

Daily requirement : The daily requirement of this vitamin for an adult is about 0·4 mg. The requirement for infants and children varies with age and body weight.

Folic acid deficiency or **avitaminosis :** The deficiency of this vitamin in man is associated with certain ailments such as macrocytic anemia, characterized by increase in average size of red blood corpuscles, and sprue (a deficiency disease).

Many animals including man, generally do not exhibit a folic acid deficiency, because this vitamin is produced in sufficient amounts by the intestinal bacteria.

(viii) **Vitamin B_{12} :** Vitamin B_{12} is a growth factor in some bacteria, in chickens, rats and mice. It is also a nutritional factor necessary for normal metabolic functions. It influences the haematopoietic process by participating in the synthesis of proteins required for the maturation of the erythrocytes.

Vitamin B_{12} is a red crystalline solid with a high complex structure. It contains a dimethyl-benzimidazole group along with other heterocyclic rings.

Occurrence and availability : The best sources of this vitamin are liver, eggs, meat, beef, pork, milk and milk products.

Daily requirement : The daily requirement of this vitamin has not been estimated but it should be taken in sufficient amount.

Vitamin B_{12} deficiency or **avitaminosis :** The vitamin B_{12} deficiency is characterized by disturbances in haematopoiesis. This disease develops not only as a result of vitamin B_{12} deficiency, but also as a morbid states of the stomach and small intestine, when absorption of this vitamin from the food is impaired.

The deficiency also causes pernicious anaemia, typical sore tongue and several neurological involvements of the spinal cord.

(ix) **Inositol :** This vitamin plays vital role in fat metabolism. Chemically it is hexa-hydroxycyclohexane (*i.e.*, a cyclohexane ring containing one—OH group on each carbon atom). Its structure formula is as follows :

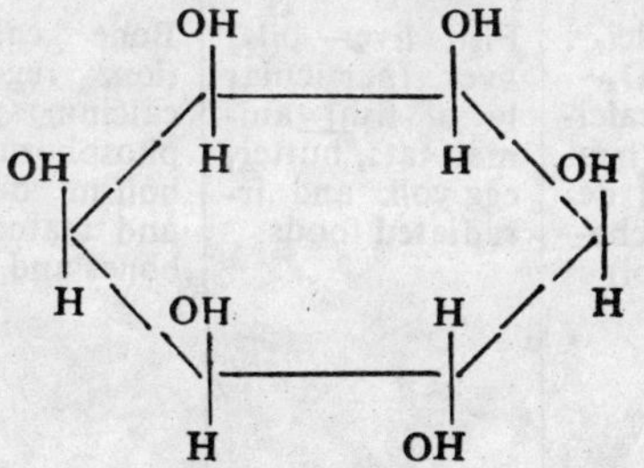

Fig. 4·15. Inositol.

Occurrence and availability : It is richly found in both animal and plant tissues. Good sources of this vitamin are liver, meat, lemons oranges, cereal grains, yeast, molds, bacteria, water germ and beef heart.

Daily requirement : The optimum requirement of this vitamin has not been estimated but one should take it in adequate amount.

Deficiency of inositol or **avitaminosis :** Inositol deficiency in man is little understood as it is rapidly synthesized by micro-organisms in the intestine. Its deficiency in rat produces a special type of fatty liver containing much cholesterol. It also leads to alopecia (loss of hair) with ball spots and dermatitis, often accompanied by spectacled eye. Its deficiency also causes haemorrhagic degeneration of the adernal gland.

(x) **Choline :** This vitamin is generally found in egg yolk, liver and kidney. Although its deficiency has not been noticed in man but it is believed that it plays important role in nutrition and fat metabolism. Its deficiency in most of the animals is associated with **haemorrhagic** changes in kidney and **cirrhosis** in liver.

Table 4·1. Vitamins with their functions and important sources.

Vitamin	**Chemical nature**	**Foods most rich in the vitamin**	**Essential for**	**Characteristic deficiency symptoms**
Fat soluble vitamins 1. Vitamin A (A_1, A_2, A_3)	**Carotenoid**	Fish, liver oils, liver (especially of fish), kidney egg yolk, milk, butter, cheese green and yellow vegetables, yellow fruits as orange, tomatoes etc. It is formed in the human body from the carotene, a vegetable pigment contained in carrots, spinach, etc.	Growth, health of the eyes, structure and functions of the cells of skin and mucous membrane.	Dryness and keratosis of epithelium and of the cornea. Night blindness, increased susceptibility to infections, defective tooth formation, retarded growth.
2. Vitamin D (antirachitic factor) D_2, D_3	D_2-calciferol D_3-cholecalciferol (irradiated dehydrocholesterol)	Fish liver oils, liver (particularly of fish) animal fats, butter, egg yolk and irradiated foods.	Bone calcification, regulating calcium and phosphorus metabolism building and maintaining bones and teeth.	Soft bones, poor development of teeth, dental decay, rickets in children. In elderly people a surplus vitamin D causes excessive bone fragility and spasmophilia.

Vitamin	Chemical nature	Food most rich in the vitamin	Essential for	Characteristic deficiency symptoms
3. Vitamin E	Tocopherol	Wheat, germ oil, leafy vegetables, egg yolk, butter and milk.	Normal functioning of sex-glands and neuro-muscular system, reduction of vitamin A activity.	Sterility in male rats death of young in uterus effect not proved in man.
4. Vitamin K (coagulation or anti-haemorrhagic factor)	Naphthoquinone K_1-Phylloquinone K_2-Farnoquinone	Green vegetables rowan berries, tomatoes, soyabeen oil. It is synthesized in the human gut by certain bacteria.	Normal clotting of blood, normal liver function.	Haemorrhages.
Water	soluble	vitamins.		
5. Vitamin B_1	Thiamine	Sea food, yeast, cereals, liver, kidney, egg yolk, milk, wheat bran green vegetables spinach, cabbage carrot, etc.	Growth, carbohydrate metabolism, normal functioning of heart, nerves and muscles, normal appetite	Beriberi and loss of appetite, retarted growth, heart trouble and disorder in digestive tract.
6. Vitamin B_2	Riboflavin	Liver, cheese, milk, eggs, meat, soyabean, green vegetables.	Growth, health of skin and mouth, carbohydrate metabolism, normal functioning of the eyes.	Retarded growth, loss of hair, soreness of the tongue and lips, premature againg intolerance to light.
7. Nicotinic acid or Niacin (anti-pellagra)	Nicotinic acid	Meat, milk, liver, kidney, egg, peanut, butter, potatoes, whole grains, tomatoes and leafy vegetables.	Growth, co-enzyme in tissue reactions, functioning of the stomach, intestine and nervous system.	Redness of the skin, diarrhoea mental disturbances (pellagra), skin-eruptions, smoothness, of the tongue.
8. Vitamin B_6	Pyridoxine	Yeast, cereals milk, eggs, liver, meat and wheat germ.	Growth co-enzyme in the metabolism of tyrosine and other amino acids.	Inflammatary skin diseases. anaemia, mental disorders, loss of weight in man, retarded growth in rats.
9. Pantothenic acid	Pantothenic acid	Eggs, meat, milk, liver, yeast, kidney, wheat, riabram and peanuts, etc.	Nerves and skin health, part of coenzyme.	Dermatitis, spectacle eye, greying of hair deficiency in man has not been noticed.

Vitamin	Chemical nature	Foods most rich in vitamin	Essential for	Characteristic deficiency symptoms
10. Biotin	Biotin	Egg, liver, kidney, milk, yeast, fresh-vegetables and fruits.	Growth in insects carbohydrate metabolism.	Dermatitis in rats, pig, fowl. Paralysis and muscle pains.
11. Folic acid	Folic acid	Green leaves, vegetables, yeast, beef, wheat and soyabean, etc.	Growth, blood-formation co-enzyme of nucleic acid metabolism.	Anaemia in man, slow growth and anaemia in chicks and rats.
12. Vitamin B_{12}	Cyanocobalamine	Liver, eggs, milk, meat, fish, bacteria and fruits.	Growth, blood formation.	Pernicious anaemia in man, slow growth in small organisms, disturbances in blood-formation.
13. Vitamin C	Ascorbic acid	Citrus fruits such as lemons, orange, grapes and other fruits tomatoes, apples, green vegetables.	Growth, maintaining strength of blood vessels, development of teeth, gum health.	Loss of weight, fatigue, disturbance of protein-metabolism, increased susceptibility to infection, dental disease.
14. Inositol	Hexahydroxy-cyclohexane	Liver, meats, lemons, oranges, cereal grains, yeast, molds bacteria, water germ and beef heart.	Growth, fat metabolism and other body activities.	Development of fatty liver in rats, alopecia, (loss of hair) dermatitis, spectacled eye.
15. Choline	Choline	Egg yolk, liver, kidney, fruits and vegetables.	Nutrition and fat metabolism.	Haemorrhagic changes in kidney and cirrhosis in liver in rats, other mammals but in man its deficiency has not been noticed.

Revision Questions

1. What is a vitamin ? Discuss the roles the vitamins play in the nutrition.
2. Vitamins are essential constituents of the body. Justify this statement with suitable examples.
3. Describe in brief the water-soluble and fat-soluble vitamins.
4. Explain some deficiency diseases caused by various vitamins.

CHAPTER 5

Nutrition

Nutrition is that physiological activity in which organisms make or obtain from their environment those materials in the form of food which are essential for life, not only to supply the energy for all other metabolic activites but also for tissue repair and growth.

Nutritive requirements : Animals in general require in their food the following :

1. The elementary components of cells and tissues, these include carbon, hydrogen, nitrogen, oxygen, phosphorus and sulphur element and the mineral constituents such as copper, iron, magnesium, calcium, sodium and potassium. These serve as components of enzymes or enzyme systems or activators.

2. A sufficient amount of organic compounds such as carbohydrates, organic acids, alcohols and amino acids. These are required partly because of a need for energy and partly because of the inability of the animals to synthesize certain important molecules from simpler substances as they are needed for life.

3. An optimum supply of major organic tissue constituents which the organisms fail to synthesize from other dietary constituents. These include amino acids, in some species smaller amounts of lipids, such as linoleic acid, cholesterol and nitrogenous compounds like purines, pyrimidines and choline.

4. An adequate supply of vitamins which can not be synthesized by the animals. Vitamins being components of enzyme systems, play important role in metabolic activities of the body.

TYPES OF NUTRITION FOUND AMONG ORGANISMS

Among organisms nutrition is of two types :

1. Autotrophic or Holophytic nutrition, and
2. Heterotrophic nutrition.

1. Autotrophic or **Holophytic nutrition :** This type of nutrition is characteristic of green plants which are able to synthesize all essential organic compounds such as carbohydrates, fats and proteins, from inorganic mineral sources. Plants, as being themselves capable enough to synthesize organic compounds, are called **autotrophs**. They nourish themselves independently on existing organic compounds, therefore, this mode of feeding is known as **autotrophic** or **holophytic nutrition.**

This type of nutrition is not only restricted to plants but it is also found in some animals, particularly which contain chlorophyll in their body, such as *Euglena, Volvox*. It is also found in some bacteria which contain **bacterio-chlorin** in their body.

There are two kinds of autotrophs :

(i) **Chemosynthetic autotrophs :** These use the energy from oxidation of substances already present on earth to 'fix' carbon to organic compounds, such as sulphur bacteria.

(ii) **Photosynthetic autotrophs :** These use the energy of light to fix carbon to organic compounds, such as green plants.

2. **Heterotrophic nutrition :** This type of nutrition is characteristic of animals and some non-green plants. Here the animals unable to use free energy to synthesize organic compounds necessary for life but depend upon organic source of carbon, they are, thus, dependent upon plants and are called **heterotrophs**.

There are three kinds of heterotrophs.

(i) **Holozoic heterotrophs :** These are those which, like animals, feed exclusively on solid material which they take in through mouth.

(ii) **Saprophytic heterotrophs :** These include organisms such as the fungi which take in dissolved organic material often all over the body.

(iii) **Parasitic heterotrophs :** These ("parasitic" means also a way of life as well as a method of nutrition) are those which feed in or on another organism, the host and cause it harm, *e.g.*, tapeworm, and potato blight.

FEEDING AND FEEDING MECHANISMS

Simply the intake of food from the environment into the body is known as feeding which is basic to animal life. Animals feed in different manners, therefore, it appears that the feeding mechanisms of animals generally exhibit some relation to the type of food and in most of the cases the structure and function of the digestive tract also depend upon the type of food and feeding mechanism.

Several attempts have been made to classify the different feeding mechanisms of the animals, but none appears entirely satisfactory. Difficulties in establishing a natural systematics of feeding types arise not only from the different feeding mechanisms, but also from the fact that often one organism is able to obtain food in more than one way and may thus be referred to more than one feeding type. Finally our knowledge about the feeding mechanisms in many animals is still incomplete. This is especially true for the invertebrates, which have the greatest number of variants and specialization.

We shall adopt here, somewhat modified, the classification of C. M. YONGE (1928), which is one of the most widely used classification. It was originally meant for invertebrates, but as fundamentally new feeding types have not appeared during vertebrates evolution, the classification is entirely adapted to encompass the entire animal kingdom (NICOL, 1960 a).

C.M. YONGE grouped feeding mechanisms into three major categories according to the type of food utilized.

1. Mechanisms for dealing with small particles,
2. Mechanisms for dealing with large particles or masses,
3. Mechanisms for taking in fluids or soft tissues.

YONGE'S classification provides a detailed and well-illustrated description of the feeding mechanisms of the animals.

MECHANISMS FOR DEALING WITH SMALL PARTICLES

Such mechanisms are characteristics of sedentary or sluggish aquatic animals which feed continuously on small pieces of food, detritus, diatoms, planktons and bacteria. These animals collect their food pieces by any of the following methods :

1. **Pseudopodial feeding :** This type of feeding is characteristic feature of rhizopods, such as *Amoeba*, which engulf their food with the help of pseudopodia, which extend around the food particle and capture it with a small amount of water. Radiolarians and foraminiferans capture the food with their reticulate pseudopodia.

2. **Ciliary and flagellar feeding :** This method of feeding is commonly used by a large number of animals, such as ciliate protozoans, sponges, many tubicolar annelids, echiuroids, rotifers, entoprocts, phoronids, brachiopods, some gastropods, most lamellibranchs and many tunicates and cephalochordates.

The cilia and flagella are not only the organs of locomotion but also useful in getting the food particles. A sort of water-current is established by the movement of cilia. Water-currents carry the food particles towards the mouth.

In **ciliate protozoans** the food particles are carried towards the cytoplasmic mouth by special ciliary currents of water. In **porifers** the water current is established by the movement of cilia or flagella of choanocytes. In **lamellibranchs**, the water current is established by the movement of cilia found in complicated arrangement on gills, palps and mantle. Ammocoete larva and tadpoles are also ciliary feeders.

3. **Mucoid feeding :** In some animals, such as *Chaetopterus* (a tube dwelling gastropod), *Vermetus* (gastropod) and larvae of *Chironomus plumosus* (Diptera), the mucus secretion often helps in feeding. These animals secrete threads or nets of mucus that float in the water. These threads or nets of mucus, after brief periods, are withdrawn and ingested together with the food particles which they have trapped.

4. **Tentacular** or **setose feeding** : In some animals such as polychaetes, and hydrozoans, their freely movable ciliated tentacles are used in collecting the food particles. The prey is often attached to the tentacles by a sticky secretion. On the contraction of tentacles the prey is brought close to the mouth. In some crustaceans the appendic setae (hair-like chitinous structures) are used in creating water current and directing the food particles to reach the mouth.

5. **Muscular feeding :** In some animals even muscular contractions help in feeding. In scyphozoan jelly fishes the rhythmic muscular contractions of the bell direct the water ladden food particles towards the mouth, while in *Chaetopterus*, the parapodial movements produce water currents, which bring in food particles which are collected by the mucus bags secreted by the animal itself, they are later swallowed by the animals.

MECHANISMS FOR DEALING WITH LARGE PARTICLES OR MASSES (MACROPHAGUS FEEDERS)

1. **For swallowing inactive food :** Burrowing animals such as *Lumbricus* and *Arenicola*, generally feed on the organic matters present in the mud. They use their eversible pharynx for swallowing mud whereas crustaceans such as *Upogebia* and *Callianassa*, on the other hand use their mouth appendages for swallowing mud. The organic matters of the mud are utilized and the residue is ejected as faeces.

2. **For scraping and boring :** Some animals possess structures for scraping and boring of the food stuffs. Examples are ARISTOTLE'S lantern of echinoids, radular apparatus of some molluscs, boring valve of *Teredo* and heavy mouth parts of termites.

3. **For seizing the prey :** Some animals possess specialized organs by which they are able to seize the prey. YONGE, as these organs are sometimes masticatory and digestive in function, classified these specialized organs into three following heads :

(a) **For seizing only :** Impaling proboscis of *Didinium*, nematocysts of coelenterate tentacles, the turbellarian pharynx, jawed pharynx of many polychaetes, teeth and radulas of some gastropods are referred to be the best seizing organs of the animals.

(b) **For seizing and masticating :** Toothed jaws of vertebrates, jaws and radulas of molluscs, the jaws and other mouth parts of crustaceans, insects, arachnids and myriapods are the organs which are used for seizing the prey and are also masticatory in function.

(c) **For seizing followed by external digestion :** This type of feeding mechanism is found in starfish which uses its tube feet to force open its shell and simultaneously its stomach is everted for digesting the muscular part of the prey. Some carnivorous gastropods, cephalopods and spiders seize prey with the help of cephalic appendages and also carry out some external digestion.

MECHANISMS FOR TAKING IN FLUIDS OR SOFT TISSUES

There are some animals which feed on already liquid food. These animals have some means of piercing the skin or outer covering of the body of the prey and sucking out blood or other fluids. The following three mechanisms have been recognized among these animals :

1. **For piercing and sucking :** The structures meant for piercing and sucking purposes are found in hookworms, leeches, ticks, cyclostomes, bugs, mammals and mosquitoes.

2. For sucking only : Some protozoans, nematodes and mammals possess only sucking organs.

3. For absorption through the body surface : There are some animals, such as sporozoans, cestcdes and *Sacculina*, which absorb food through the whole body surface as they have no digestive organs.

DIGESTION

The word **digestion** comes from two Latin words meaning "to earry" *(gerere)* and "apart" or "asuuder" *(dis)*. It is an essential physiological activity in all animals, whether they feed on minute food particles (**microphagus**) or on large plants and animals (**macrophagus**). Some of the internal body parasites, such as the tapeworms, can dispense with a digestive system and absorb food predigested by their hosts.

Whatsoever the food animals take that cannot be used directly by the cells for the metabolism cf their body. It is because of two reasons ; first, many substances of the focd are insoluble in water and could not enter the plasma membranes cf the cells even if they reached them. Second, these foods are too complex chemically for growth and repair by protein synthesis. Digestion brings about changes in both these conditions, with the result that substances of the focd are converted into a state where they can be used directly in the cells of the body for the resynthesis of new cell constituents or as substrate in the machinery that converts the energy in the focd into available energy for the organism. Thus, in digestion, complex focd molecules are broken down both mechanically and chemically, into smaller molecules of water soluble substances that can be used by the cells.

The first part of the change that occurs during digestion is mechanical, it involves the **chewing** of the focd in the mouth (particularly in higher animals like man) and the constant **churning** and mixing actions brought about by the muscular movements of the walls of the digestive organs. The second part of the change includes the breakdown of the large food molecules into smaller molecules or the component parts and the thorough mixing with various digestive juices secreted by different digestive organs. Digestive juices contain many enzymes which help in digesting the various food substances. Digestion, in higher animals, occurs in digestive tract.

The process of digestion can be summarized as follows : Focd molecules are too big to pass through the digestive tract wall and must be broken down to smaller ones so that they may pass through the wall of digestive tract into the body. After the molecules are broken down, they are absorbed. Generally, the absorbed material is resynthesized into large molecules in the body. The digestive tract wall serves as a typical semipermeable membrane and is a barrier that prevents large molecules from coming into the circulatory fluids cf the bcdy and the large molecules in the circulatory fluids from leaving the body. Small molecules, such as glucose, are in the simplest form and can be assi. milated without any digestive action. Digestion, in its narrow sense, can be regarded as a process that renders food particles suitable for absorption.

Agents of digestion : **Enzymes** are said to be the agents of digestion. They are secreted on to the food at various points along the length of the digestive tract or alimentary canal. They act as biological catalysts, accelerating the rate at which food will react with water to form a large number of smaller soluble molecules, suitable for absorption into the body. Enzymes are very sensitive to the acidity or alkalinity of the surrounding medium, and different enzymes work best under different conditions. We shall see that the conditions vary along the length of the alimentary canal to suit the enzymes present at any particular point.

KINDS OF DIGESTION

The process of digestion is essentially of following two types, depending upon both the phylogenetic position of the animal and the character of the food.

1. Intracellular digestion,
2. Extracellular digestion.

1. Intracellular digestion : When digestion of the food materials occurs within the cells, then this sort of digestion is known as intracellular digestion. This type of digestion is generally found in lower animals such as protozoans, sponges and few other complex animals like coelenterates where in the digestion is partially extracellular but largely intracellular (see BARRINGTON, 1962 ; BEATON and MCHENRY, 1964–1966 ; MEGLITSCH, 1967).

2. Extracellular digestion : When digestion of the food-materials takes place outside the cells in a lumen of the tube known as digestive tube or alimentary canal, then this type of digestion is known as extracellular digestion. The digestive tube or alimentary canal includes those structures that do not actually receive undigested food but that act on food in the digestive tube by means of secretions delivered to the lumen by various ducts. Extracellular digestion is the characteristic of higher animals.

DIGESTIVE TUBE OR ALIMENTARY CANAL

General organisation of the alimentary canal : The alimentary canal of higher animals is a long muscular tube with dilations and constrictions at places. It is lined throughout by mucous membrane beginning from the mouth cavity and ending at the anus. It consists of two sets of muscles, an inner circular and outer longitudinal muscles. These muscles provide the peristaltic movement for the passage of the food. Inside the muscle layers are the mucosa and submucosa which enclose the lumen of the alimentary canal. The mucosa is glandular and has a very large surface area. Between the mucosa and submucosa, there is a muscularis mucosa.

The alimentary canal all along its length consists of numerous glands, some of which are of special type. All have their ducts opening into the mucous membrane and pour their secretions into the lumen. Gastric and intestinal glands are situated within the mucosa-membrane but the salivary glands (submaxillary, sublingual and parotid), the pancreas and the liver lie separately. These glands pour their secretions by their respective ducts into the lumen of alimentary canal.

Functional regions of the alimentary canal : The alimentary canals of the animals show striking adaptations to the nature of foods and the feeding habits of themselves. According to C. M. YONGE the alimentary canal in animals has the following important functional regions : 1. the region of reception 2. the region of conduction and storage 3. the region of internal trituration and digestion 4. the region of final digestion and absorption 5. the region of faeces formation.

1. The region of reception : This region includes the mouth and its associated appendages and cavity. This region is primarily meant for food selection and mechanical break-down of the food. Food selection is made by taste, smell and texture.

The mouth cavity often contains salivary glands which secrete enzymes in few mammals such as man but in rest cases secrete copious mucus which is by its nature known as lubricating fluid. Blood sucking animals such as mosquitoes, usually have an anticoagulant in their salivary secretion which prevent the blood from clotting, while carnivores feeding on live prey may secrete paralyzing toxins. Many insects and vertebrates secrete carbohydrate splitting enzymes, while some carnivores and cephalopods secrete protein-splitting enzymes, in addition to poison and mucus.

2. The region of conduction and storage : This includes the oesophagus and the crop of some animals. They are often muscular and are meant for storing up the food as in case of leeches and conduction as well.

3. The region of internal trituration and digestion : This region includes the gizzard and the stomach of some animals. They are modified in different animals in order to carry out the internal trituration and digestion of the ingested food.

Digestive enzymes may be secreted into the stomach by unicellular gastric glands situated in the lining or by glandular diverticula or caeca which help in digesting the food contents.

4. The region of final digestion and absorption : This includes the small intestine where the final digestion and absorption of the food materials take place. Digestion of the food materials takes place with the help of enzymes which are either liberated in the preceeding region as in most insects or in the same region as in most vertebrates.

5. The region of faeces formation : It is the last region of the alimentary canal where the absorption of water from the undigested food takes place and subsequently undigested food materials are twisted together with mucus into faeces. This region is very conspicuous in terrestrial animal as in insect hindgut and vertebrate colon.

MOVEMENT OF FOOD ALONG THE ALIMENTARY CANAL

The food is propelled along the alimentary canal at appropriate speed by ciliary action or muscular activity or by combination of both. The movement of food by ciliary action is the characteristic of small animals like ectoprocts, endoprocts and pelecypod molluscs whereas the muscular activity is involved to a greater or lesser extent in food propulsion in a wide variety of animals including the vertebrates. The muscular activity not only propells the food but also serves to triturate and mix the food with digestive juices.

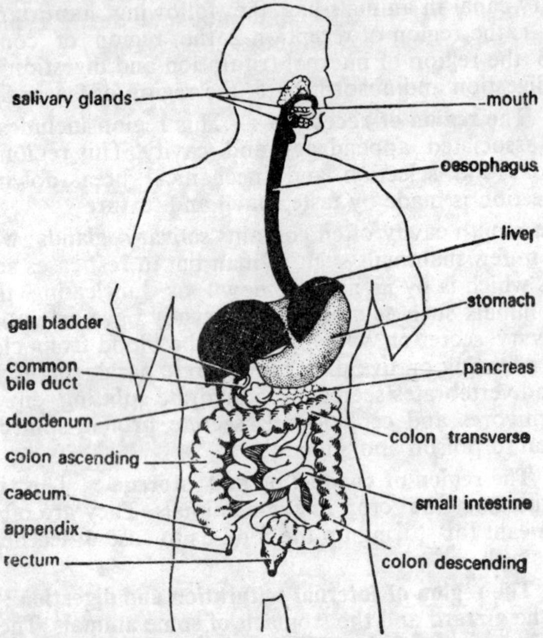

Fig. 5·1. The organs of digestion in the human body.

DIGESTION IN VERTEBRATES (MAMMALS)

All vertebrates possess alimentary canal as an organ for the digestion and absorption of the food materials. Digestion does not occur in a particular region of the alimentary canal but takes place in different regions of alimentary canal so that the digestion of the food materials may be completed. Hence the process of digestion in the alimentary canal can be classified into the following heads :

1. BUCCAL DIGESTION

Mouth—The first digestive structure : Mouth prepares the food for digestion. In higher animals, particularly in man, the digestion of nutritive substances of food stuffs begins in the buccal-cavity. As soon as the food is entered the mouth, the process of mastication starts with the help of teeth. Teeth break the food into smaller particles. During the process of mastication the food is thoroughly mixed with the fluid **saliva** secreted by the **salivary glands.**

Salivary glands : In the buccal cavity of most vertebrates four pairs of salivary glands are found. These are the infra-orbital, parotid, sublingual and submaxillary glands. In humans there are only three pairs, there being no infra-orbitals.

Salivary glands may contain either **mucous cells or serous cells. Mucous cells** contain large transluscent granules (consisting of a precursor of mucin) and such cells appear pale or translucent in histological sections. **Serous cells** contain opaque small zymogen granules (consisting of a precursor of amlase or ptyalin).

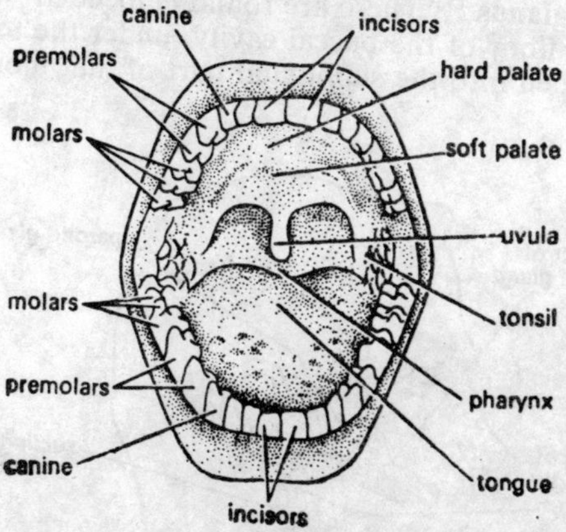

Fig. 5·2. Digestion begins in the mouth, the first organ of the alimentary canal.

Parotid glands : These are the largest of the salivary glands, found lying one on each side of the face below and in front of the ears. The secretion of each parotid gland passes via **Stensen's duct**

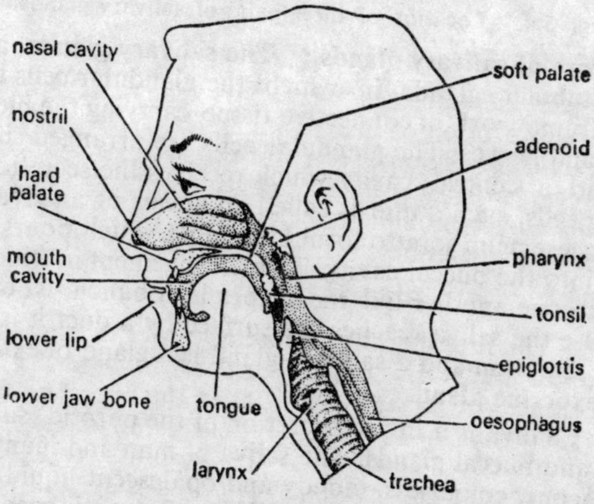

Fig. 5·3. A cross sectional view of the structure of the mouth and throat.

which opens into the mouth opposite the site of the second upper molar tooth. The disease called **mumps** is an infection of the parotid glands that causes swelling and irritation.

The parotid glands are purely serous and so they secrete zymogen granules.

Submaxillary glands : These glands are found within the angles of the lower jaws. These glands contain both types of cell, but predominantly serous. Their secretion passes via **Wharton's duct** into the floor of the mouth at the side of the frenulum linguae.

Sublingual glands : These are found embedded in the mucous membrane in the floor of the buccal cavity, under the tongue. Ducts of these glands open into the sublingual part of the mouth under the tongue.

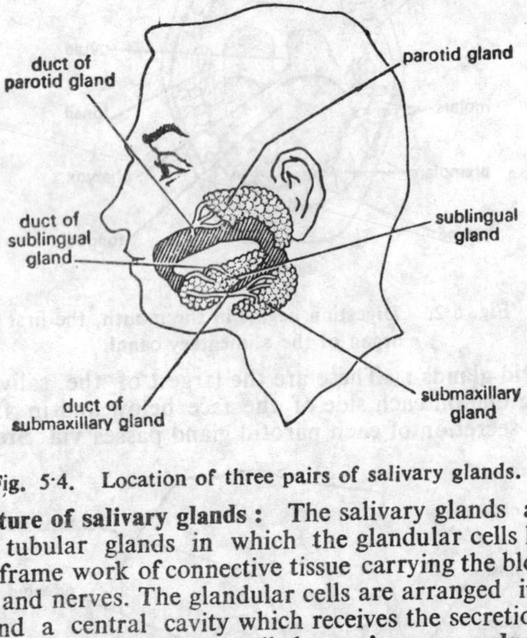

Fig. 5·4. Location of three pairs of salivary glands.

Structure of salivary glands : The salivary glands are typical compound tubular glands in which the glandular cells lie within a supporting frame work of connective tissue carrying the blood vessels, lymphatics and nerves. The glandular cells are arranged in a single layer around a central cavity which receives the secretion from the surrounding cells, such a unit is called an **acinus** or an **alvelous.** Small ducts from these acini join to form large duct which pours the secretion finally into the buccal cavity. This arrangement in which the acini are related to one another like the grapes in a bunch, is called **recemose.** Because the saliva reaches the surface by a duct it is said to be an external secretion and a salivary gland is a gland of external secretion or an **exocrine gland.**

Saliva : Saliva is a mixed secretion of the parotid, submaxillary, sublingual, and buccal glands. The saliva of man and many other animals is a viscous, colourless, cloudy and opalescent liquid. The average pH value is 6·8 with a range of 5·6 to 7·6 and specific gravity is 1·002 to 1·008. It is constantly secreted in small quantities to keep the buccal cavity moist, but when food is present the rate of secretion is increased because the saliva not only moistens the food, lubricating its subsequent passage through the alimentary canal but also begins the process of digestion.

It contains 98·5 to 99% water and 1 to 1·5% of a dense residue which includes the following :

(i) **Cellular components** : The cellular components of saliva are desquamated epithelial cells of the oral mucosa, leukocytes and numerous bacteria. Yeast cells and sometimes protozoans like *Amoeba salivaricus* are also found.

(ii) **Inorganic components** : Inorganic components of saliva include chlorides, sulphates, carbonates of sodium, potassium, calcium and magnesium and traces of ammonia as well. The saliva of man and some animals also has potassium sulphocyanate about 0·01%.

(iii) **Organic components** : Organic components of the saliva are chiefly **mucin** (which gives saliva its viscosity and lubricating properties) and **enzymes** which digest the food. Some other organic substances, such as urea, cholesterol, amino-acids, citrates and vitamins, have also been found in saliva. Glucose is practically absent.

Enzymes of saliva : Saliva contains a large number of enzymes such as **amylase, lysosome acid phosphotase, aldolase, cholesterase, lysozyme, maltase, catalase, lipase, urease** and **protease** of these probably only two namely **amylase** and **lysozyme** are of physiological importance.

Amylase : The saliva of cat, dog and some other animals lack salivary **amylase**, while the saliva of other animals such as man, pig and rats has strong **amylase activity.**

Salivary amylase was once named **ptyalin,** to distinguish it from pancreatic amylase (diastase), but these two enzymes are now known to be identical. It hydrolyses starch and glycogen to maltose, isomaltose, dextrin and some glucose.

Lysozyme : This enzyme is **polysaccharidase** which hydrolyses certain complex polysaccharides present in the cell wall of many different species of bacteria *(e.g., Micrococcus lysodeikticus)*, thereby killing and dissolving them. Its action is partially retarted by the presence of mucin..

Functions of saliva : The saliva subserves a number of functions :

1. It moistens dry food and facilitates swallowing by a lubricating action. Since water evaporates slowly from saliva it prevents desiccation of the oral mucosa.

2. It provides an enzyme called **salivary amylase** or **ptyalin** for the digestion of starch.

3. It keeps the mouth and teeth clean (according to PIGMAN and REID).

4. It dissolves the soluble substances such as sugar, and salts.

5. It makes the food delicious to taste.

6. Certain substances such as lead, mercury and iodides are

excreted in the saliva but they cannot be important since they are reabsorbed when the saliva is swallowed.

7. By facilitating movements of the tongue and lips it makes rapid articulation possible.

8. It subserves the sense of taste by acting as a solvent. The taste buds can be stimulated only when the sapid substances are actually present in solution.

9. It contains three buffering systems, bicarbonate, phosphate and mucin of which the first is most important. The concentration of bicarbonate and the buffering systems rises when salivary flow increases, particularly at the time of eating.

CONTROL OF SALIVA SECRETION

The secretion of saliva is generally coordinated by the intake of food. Olfactory and gustatory stimuli normally initiate the nervous reflex that results in the stimulation of salivary secretion. Secretion from the salivary gland cells is predominantly controlled by parasympathetic nerve fibres (Hilton and Lewis, 1957).

SALIVARY DIGESTION

Mastication or chewing mixes the food with the saliva and brings **ptyalin** into intimate contact with the starch of the food. Due to the action of enzyme the starch-molecules are hydrolysed to dextrins and then to maltose. Hydrolysis is finally arrested by a fall in pH and destruction of **ptyalin** as the acid gastric juice gradually penetrates through the food mass in the body of the stomach.

MECHANICAL PROCESS IN THE BUCCAL CAVITY DURING EATING AND DIGESTION

In addition to being chemically changed in the buccal cavity, the food is ground up and saturated with saliva due to mechanical process.

Mastication : It is a voluntary mechanism which involves the play of many muscles of the mouth, tongue and cheeks, in a perfectly co-ordinated manner. It grinds the food into small particles and aids in moistening it with saliva and forming a **bolus**.

PASSAGE OF FOOD FROM THE MOUTH TO THE STOMACH (DEGLUTITION)

After mastication the food undergoes the process of deglutition which involves all such motor reactions by which the food is moved from the buccal cavity through the oesophagus to the stomach. The process of deglutition can be divided under the following three stages :

1. Movement of food from the buccal cavity to the pharynx.
2. Movement of food from the pharynx to the oesophagus.
3. Movement of food from the oesophagus to the stomach.

1. MOVEMENT OF FOOD FROM THE BUCCAL CAVITY TO THE PHARYNX

This stage in under voluntary control during which the mouth remains closed and movements of tongue push the food (bolus) to the tongue's dorsum where a groove is formed (by contractions of the tongue muscles). Then, due to the contractions mainly of mylohyoid and styloglossus muscles of tongue the bolus is pressed against the hard palate and passed through the anterior pillars of the fauces (oropharyngeal isthmus). At the same time the soft palate is raised and approaches the posterior pharyngeal wall which is brought forward to close off the nasopharynx. At this time respiration is inhibited. The larynx begins to rise as bolus passes over the tongue and reaches the pharynx. From this moment the process of deglutition is involuntary.

2. MOVEMENT OF FOOD FROM THE PHARYNX TO THE OESOPHAGUS

Since the pharynx is communicated with both oesophagus and trachea, the movements occurring during this stage are simply designed to allow the bolus to enter the oesophagus while avoiding the air passage. The tongue moves back like a piston towards the posterior pharyngeal wall and forces the bolus back against the epiglottis. It then becomes folded to form a cowl-like hood over the laryngeal orifice. The enterance to the larynx is closed by the sphincteric action of the muscles surrounding it. The food passes over the lateral edges of the epiglottis in two streams into the part of the pharynx immediately posterior to the larynx and then on into the oesophagus. At this moment the cricopharyngeus (the lower fibres of the inferior constrictors) forming a sphincter at the mouth of the oesophagus, relaxes for about one second when the bolus is safely past the cricopharyngeus, the larynx reaches its original position, the vocal folds open and the epiglottis quickly resumes its initial position. The cricopharyngeus muscles then close firmly in order to prevent the air from entering the oesophagus during inspiration. This is the end of second phase.

3. MOVEMENT OF FOOD FROM THE OESOPHAGUS TO THE STOMACH

Oesophagus is a tubular structure that connects the bucco-pharyngeal region to the stomach. From the oesophagus, the bolus, (liquid or semiliquid) may pass direct to the stomach due to mylo-hyoid contractions. If the bolus is large and solid, it is carried down by the **peristaltic movements** of the oesophagus to the stomach.

Peristalsis (Peristaltic movements) : Peristalsis is an essential activity brought about by the successive contractions of rings of muscles arranged circularly around the tubular part of the alimentary canal, and is responsible for the progress of the food through the system. BAYLISS and STARLING defined true peristalsis as a **co-ordinated reflex, in which a wave of contraction preceded by a wave of relaxation, passes down a hollow viscus.**

In the oesophagus each successive part relaxes to receive the bolus and then contracts to drive it on. Stimulation of the vagus causes a tetanic contraction of the oesophagus.

Peristalsis is usually initiated by swallowing of food but may also arise in response to local stimulation at various levels in the oesophagus. There is a nerve plexus on the surface of the oesophagus called **plexus gulae**, it is mainly responsible for the propagation of the peristaltic wave along the oesophagus.

CARDIAC SPHINCTER

This is the posterior most part of the oesophagus which is more muscular and remains in apposition and does not open unless peristaltic force relaxes it. After a meal it is closed to prevent the backward movement of the food (*e.g.*, **regurgitation**). CANNON thought that the closing of the cardiac sphincter is due to the acidity of the stomach, while CARSON believed that it is due to the rise of the intra-gastric pressure after a meal.

DIGESTION IN THE STOMACH (GASTRIC DIGESTION)

After the food has been swallowed, it passes through the oesophagus to the stomach to undergo further mechanical disintegration and chemical changes, primarily in the protein constituents. The fundus is the large main part of the stomach and is attached to the oesophagus. The part of the stomach attached to the duodenum is called the pyloric portion and is a narrow constricted region.

The muscular walls of the stomach can expand when the stomach has received the food and contract in the emptying process of passing the partially digested food to the duodenum.

Digestion in the stomach is accomplished by **gastric juice**. SPALLANZANI (1783) was the first who pointed out about the gastric digestion. According to him the food is digested in the stomach with the help of liquid, now called gastric juice secreted by the cells or glands of stomach wall. He also noted that the juice was acidic but its acidic nature and active agents were not demonstrated until later.

A great advance in our knowledge of gastric digestion, particularly in man, was made through the observations of BEAUMONT on his patient, ALEXIS ST MARTIN who in 1822, following a gun shot wound, was left with an opening from the stomach through the abdominal wall to the exterior. Through this fistula BEAUMONT found it possible to follow the course of gastric digestion of different food under varying conditions of health and obtained pure gastric juice for digestion experiments outside the body.

PAVLOV also extended our knowledge of gastric digestion after creating, in dogs, a small stomach pouch separate from the main stomach. He studied the secretion of small pouch without interfering the proper stomach.

Secretion of gastric juice : The wall of the stomach has numerous deeply pitted **gastric glands** which secrete **gastric juice**. About

2 to 3 liters of gastric juice are secreted daily by a normal adult. Apparently there is a constant gastric juice flow, because there always appears to be some liquid in the stomach.

Each gastric gland has the form of a somewhat crooked tube with fine bore opening into the stomach cavity and thick walls composed of secreting cells. These glands always secrete a little amount of gastric juice into the lumen of stomach. As a result there is almost always present in the stomach even before the food about 50 ml of secretion which is called the **residium**.

The gastric glands, on the basis of their distribution in the stomach, can be classified under following heads :

1. Cardiac glands,
2. Fundal glands,
3. Pyloric glands.

1. Cardiac glands : These glands are found scattered in the cardiac part of the stomach. They are single or tubulo-racemose glands lined by small columnar granular cells. These glands probably secrete **mucus**, a substance which makes the food soft and slipery for onward movement in the system.

2. Fundal glands : These glands are compound tubular glands lined with columnar epithelium. The cells are polyhedral, coarsely granular and are called **central** or **Chief cells.** There are another group of spherical cells between the basement membrane and the layers of the central cells, these are known as **parietal or oxyntic** cells which secrete hydrochloric acid to make the appropriate medium for the gastric enzymes.

3. Pyloric glands : These are compound convoluted tubulo-alveolar glands of which the ducts are lined with columnar cells and the tubules by finely granular cubical cells.

The gastric glands open by single mouth at the bottom of tiny pits, **gastric foveola**. Sometimes a single **foveolus** receives secretions of two or more glands, and are throughout scattered over the gastric mucosa, the lining epithelium is columnar and continuous on both the surface as also in the foveola and gland tubules.

The central cells and pyloric gland cells secrete an enzyme called **pepsin**. The milk curdling enzyme is also secreted by these cells which is commonly known as **rennin**. But according to LIM the pyloric gland cells secrete a viscid alkaline juice without any pepsin.

Character and composition of gastric juice : Pure gastric juice of man and other animals (such as dog, cat) is a light coloured, thin and transparent fluid of an acid reaction. It has a low specific gravity ranging from 1·002 to 1·006. It contains about 0·5 per cent of solid matter including sodium chloride with traces of potassiam chloride and phosphates, mucin and the enzymes **pepsin, rennin** and **gastric lipase**. Its acidity is due to the presence of free hydrochloric acid in it. It also contains a small amount of mucus.

The gastric juice is a composite secretion from atleast three different types of cells in the ducts of the gastric mucosa, these are

(i) **parietal cells**, (ii) the **Chief cells** and (iii) the **mucous cells**. There is a good evidence that the parietal cells furnish the hydrochloric acid of gastric juice, the Chief cells supply pepsin and possibly other enzymes like rennin and gastric lipase and the mucous cells secrete mucin.

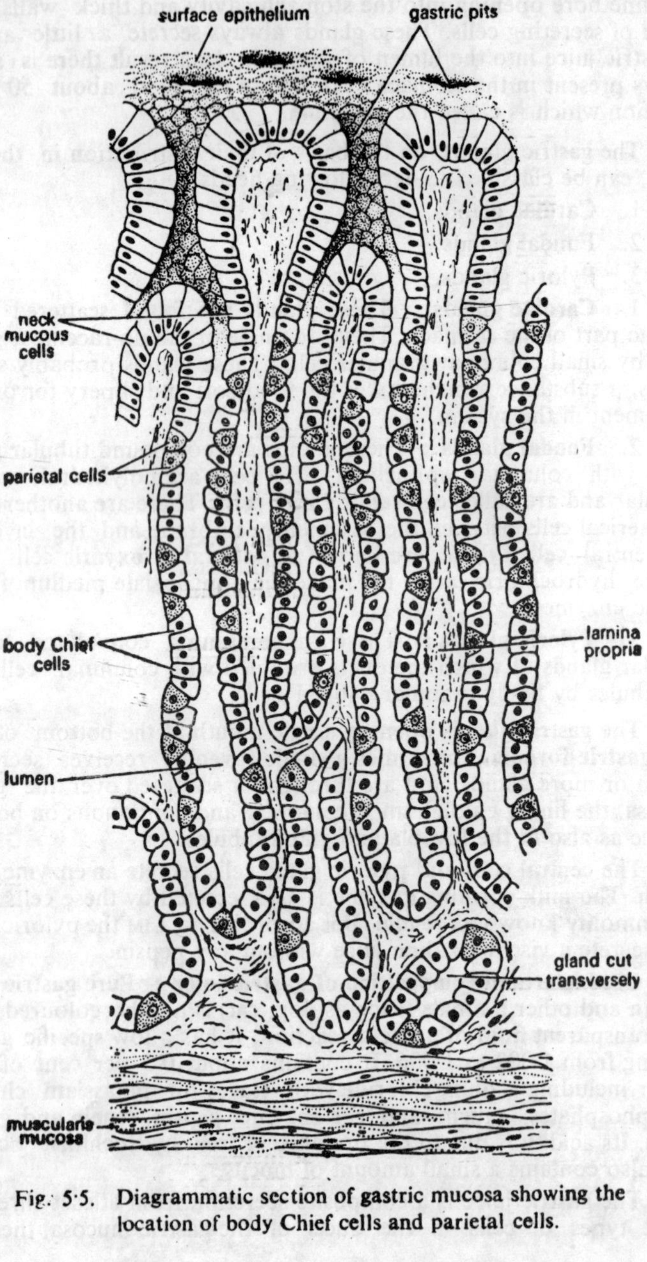

Fig. 5·5. **Diagrammatic section of gastric mucosa showing the location of body Chief cells and parietal cells.**

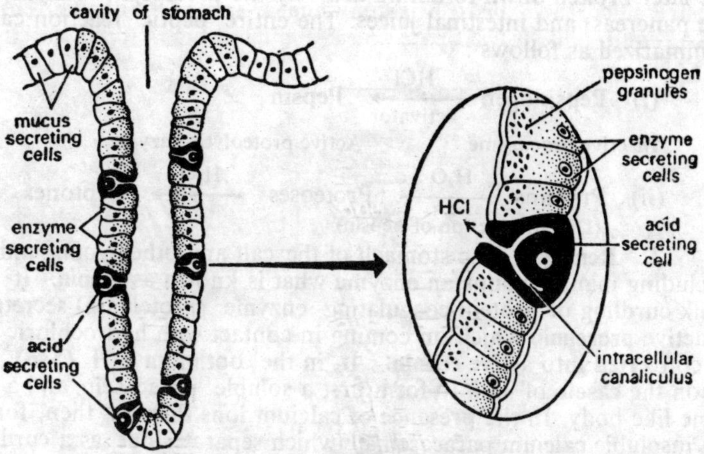

Fig. 5·6. Diagram of a gastric gland showing the distribution and differences of mucus, enzyme and acid secreting cells.

HYDROCHLORIC ACID OF THE GASTRIC JUICE

The parietal cells of stomach wall produce hydrochloric acid at a concentration of 0·16N (0·6%) which has a slightly higher osmotic pressure than the blood but probably the same as the parietal cell cytoplasm. It functions in several ways :

1. It dissolves the living material between the cells in order to make them readily available for enzyme-action.
2. It provides favourable medium in which the enzymes of gastric juice can do their work rapidly.
3. It kills the detrimental bacteria which have been taken in along with food.

ENZYMES OF GASTRIC JUICE AND THEIR ACTIONS

1. **Pepsin :** It is the proteolytic enzyme of gastric juice secreted by the peptic cells (Chief cells) of the fundus part of stomach in the form of an inactive precursor or zymogen what is known as **pepsinogen**, which, on coming into contact with hydrochloric acid, is converted into active **pepsin**. It is an usual enzyme in that its optimum pH is in the range 1·6 to 2·4 and it, therefore, meets conditions most suited to its action in the acid contents of the stomach.

A second proteolytic enzyme with an optimum pH of 3·3 to 4·0 has recently been found.

Although pepsin is a powerful proteolytic enzyme, it does not digest proteins to α-amino-acids, but instead it breaks down proteins to the stage of proteose or peptones, that is to say, it breaks the protein chain into shorter fragments but it does not liberate free amino-acids. The products of peptic digestion (*e.g.*, proteoses and peptones).

are later broken down to amino-acids by the proteolytic enzymes of the pancreas and intestinal juices. The entire peptic reaction can be summarized as follows :

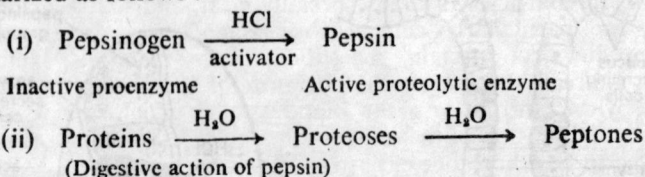

(i) Pepsinogen $\xrightarrow[\text{activator}]{HCl}$ Pepsin

Inactive proenzyme **Active proteolytic enzyme**

(ii) Proteins $\xrightarrow{H_2O}$ Proteoses $\xrightarrow{H_2O}$ Peptones

(Digestive action of pepsin)

2. Rennin : The stomach of the calf and other young animals including man, contains an enzyme what is known as **rennin**. It is a milk curdling or protein-coagulating enzyme (proteinase) secreted as inactive **prorennin** which on coming in contact with hydrochloric acid is converted into active **rennin**. It, in the optimum pH (5–6), acts upon the **casein** of milk to form first a soluble paracasein and a peptone like body. In the presence of calcium ions there is then formed an insoluble calcium paracaseinate which separates out as a curd that can be digested by pepsin. The entire reaction can be summarized as follows :

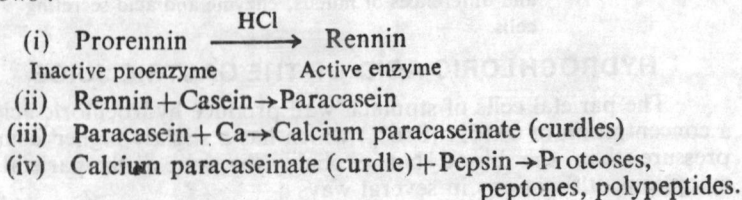

(i) Prorennin $\xrightarrow{HCl}$ Rennin

Inactive proenzyme **Active enzyme**

(ii) Rennin + Casein → Paracasein

(iii) Paracasein + Ca → Calcium paracaseinate (curdles)

(iv) Calcium paracaseinate (curdle) + Pepsin → Proteoses, peptones, polypeptides.

3. Gastric lipase : It is a weak fat-splitting enzyme which acts in an acid medium, unlike other lipolytic enzymes. It is probably secreted by Chief cells of the fundic glands and is destroyed by alkalies.

The digestion of fat in the stomach is, however, of slight importance as it is completed in the intestine through the action of the lipase of pancreatic juice.

CASTLE'S INTRINSIC FACTOR

It reacts with substances present in certain food and forms the **haematinic-principle**, essential for the proper formation of R.B.C. and necessary for the absorption of vitamin B_{12} or cyanocobalamine from the small intestine.

CONTROL OF GASTRIC SECRETION

Gastric secretion is governed by many factors. Atleast three phases are recognized :

1. Caphalic phase : Stimulation by the sight, taste, odour and more thought of food results in a psychic secretion of gastric juice even before the food enters the stomach. This cephalic phase of gastric secretion is of considerable importance in the dog but may not be so important in man. It is partly under the control of the nervous system by way of the secretory fibres in the vagus nerves.

2. **Gastric phase :** The presence of food in the stomach is itself capable of provoking the continued secretion of gastric juice ; this is, so called, the gastric phase of gastric secretion.

Certain articles of food especially meat, and soups made from meat, are particularly active as secretagogues, as are also some of the products of protein digestion. These materials exert their action indirectly by causing the liberation of **gastrin** from the mucosa of the pyloric antrum into the blood stream which carries it to the glands. Distension of the antrum also releases gastrin. **Gastrin** is believed to be a hormone of gastric-secretion.

Histamines are also powerful stimulant of gastric secretion in addition to their well known effects on blood pressure.

3. **Intestinal phase :** When the products of gastric digestion leave the stomach and enter the duodenum, they have a stimulating effect on gastric secretion. The mechanism of this action is not yet clear but it is probably due to substances present in the foods. These are absorbed and, perhaps, stimulate nervous structures.

Gastric emptying : The stomach does not retain its contents until gastric digestion is completed. As the food has reached the stomach, some contents are ejected into the duodenum as a result of which intestinal digestion and absorption begin. There are a number of factors that influence gastric emptying. One of the factors is **enterogastrone** which is secreted from the duodenal and intestinal mucosa on stimulation by the presence of fatty acids. It acts on the stomach to inhibit the release of gastrin and also of HCl. It is important, however, to appreciate the complex interplay of stimuli and checks and balances that serve to activate and inactivate digestive enzymes and to adjust the chemical environment.

SIGNIFICANCE OF STOMACH DURING THE PROCESS OF DIGESTION

The significance of the stomach seems to be first that it serves as a storage site because undigested food may remain there for several hours. From the stomach partly digested food is released into the intestine. The second role of stomach is mechanical. The walls of stomach are provided with typical musculature which reduces the food, often taken in as large particles, to a semi-fluid suspension called **chyme**. The third role of stomach is chemical. The stomach walls secrete gastric juice, the enzymes of which digest the different components of ingested food as discussed above.

PASSAGE OF CHYME THROUGH PYLORUS

During gastric digestion the food and gastric juice are held in the stomach by the **pyloric sphincter** or **valve**, a ring of circular muscle making the boundary between the stomach and intestine. When the stomach contents reach a semi-fluid state (chyme) this valve begins to open as each peristaltic wave reaches it, and the stomach contents are thus periodically ejected into the intestine. Thus the action of pyloric sphincter regulates the passage of gastric contents into the intestine. The material that passes through it called the **chyme**.

SECRETION OF BRUNNER'S GLANDS

In the submucosa of upper division of duodenum, there are located numerous glands called BRUNNER's **glands**. As soon as the food in the form of chyme, passes along this region the secretion of these glands get mixed with chyme and digest the contained proteins, fats and starch.

The secretion of BRUNNER's glands is colourless syrupy mass of an alkaline reaction with an admixture of mucus. It contains **pepsin** which acts in a mild acidic medium. Its digestive power is comparatively weaker than of gastric juice.

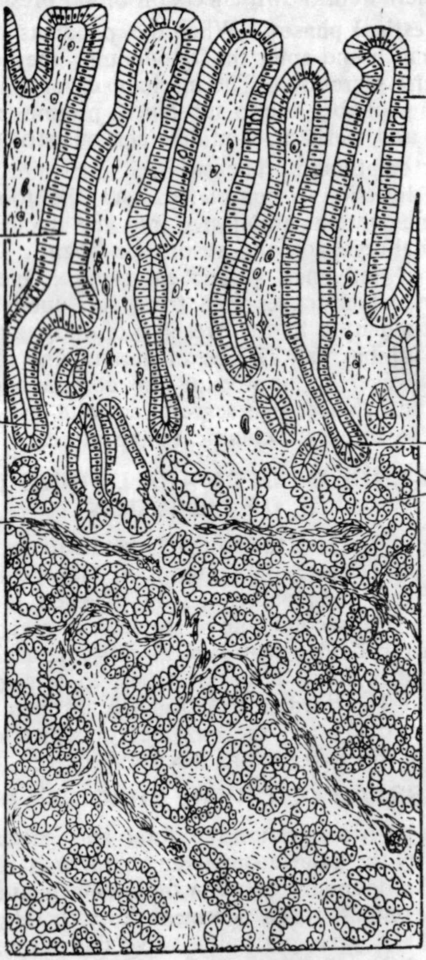

Fig. 5·7. Diagrammatic section of the wall of duodenum showing villi and duodenal (Brunner's) glands.

According to GROSSMAN, 1958, the principal function of the juice secreted by BRUNNER'S glands is to protect the mucosa of the first part of the duodenum against damage by acid chyme from the stomach.

DIGESTION IN THE INTESTINES (INTESTINAL DIGESTION)

The anterior part of the small intestine is called duodenum which is truely an important organ, because so much digestion occurs in this relatively small region. The duodenal secretions are also involved in the elimination of waste products, particularly when the kidneys are impaired.

As the chyme enters the duodenum in small amount by means of the pyloric sphincter, it is quickly mixed with the flow of BRUNNER'S gland secretion, pancreatic, bile and intestinal juices, because the pancreatic and bile ducts are only a few inches away from the pylorus. These juices are alkaline and neutralize the acidity of chyme; thus, they make the intestine alkaline.

The combined effects of pancreatic and bile juices cause the chief chemical changes of the intestinal digestion. Intestinal digestion is also dependent upon the intestinal juice, called the **succus-entericus**. This juice is secreted by several types of glands which are located abundantly in the mucous lining of the intestine.

Because the pancreatic, bile, and intestinal juices are very important although not completely independent of one another, we shall discuss them separately, one at a time.

PANCREAS

The pancreas is one of the most important digestive glands. The cells of which secrete **pancreatic juice**. This gland is located in the loop of duodenum. It is emptied by its two ducts, (in man and dogs), and sometimes several additional ducts, into the duodenum. It is remarkable in producing both an external and internal secretion. The bulk of the gland has a racemose structure resembling that of the salivary glands.

The acinar cells show a peripheral clear zone and a central zone containing zymogen granules which discharge into the lumen during the period of secretory activity. The main duct of the gland communicates with the bile duct at the ampulla of the bile duct (**ampulla of VATER**) and discharges the secretion into the duodenum through the **sphincter of ODDI**.

Embedded within the pancreatic acinar tissue there is a relatively small amount of cellular tissue, quite different in appearance, which occurs in little groups the **interalveolar cell-islets** or the **islands of LANGERHANS**. This tissue produces the internal secretion called **insulin**.

The pancreatic juice secreted by pancreatic gland reaches the intestine through the duct of WIRSUNG which usually joins with the common bile duct before opening into the first part of the duodenum.

As soon as the food enters the duodenum, a series of neurogenic and hormonal mechanisms come into play and maintain a secretion of these fluids into the duodenum, where they mix with the acid chyme from the stomach. The most important feature of the pancreatic juice is its high concentration of bicarbonate ions and enzymes capable of hydrolyzing virtually any type of food constituents.

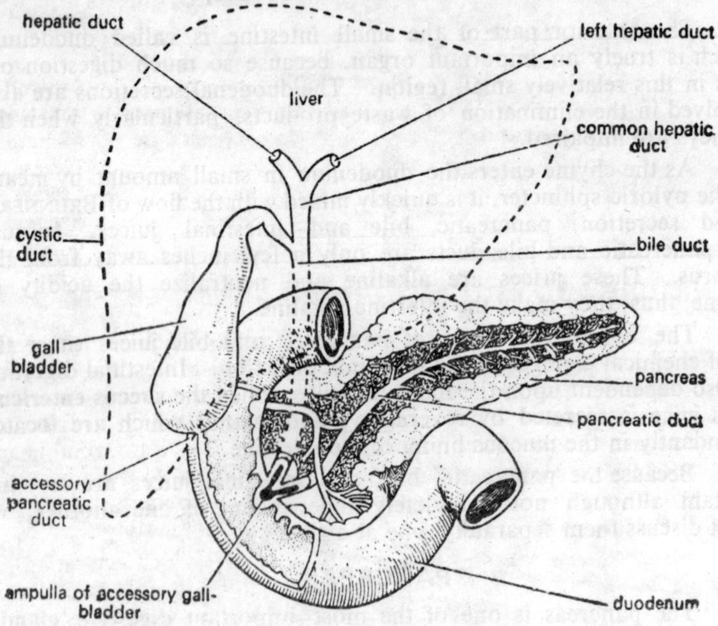

Fig. 5·8. Connections of the liver, gall-bladder and pancreas with duodenum.

PANCREATIC JUICE

Pancreatic juice is probably one of the most potent digestive juices available in the body. In man some 500 to 800 ml of it are secreted daily. The pancreatic juice is composed of some 98% of water and 2% solid substances. The solid substances include some potentially powerful enzymes which act on proteins, fats and carbohydrates. It also has high proportion of sodium bicarbonate ($NaHCO_3$) and di-sodium hydrogen phosphate (Na_2HPO_4). These basic salts make the juice quite alkaline, within the 7·5–8·0 pH range. Its specific gravity ranges from 1·008 to 1·030.

Activated pancreatic juice contains **trypsin**, **chymotrypsin**, **carboxypeptidase**, **pancreopeptidase E** or **elastase**, **amylopsin** (amylase) and **steapsin** (pancreatic lipase). Other enzymes such as **sucrase lactase** and **maltase**, may be present in trace quantities. In addition, certain enzymes, not yet characterized and capable enough to hydrolyze certain proteoses, peptones and polypeptides have also been reported. **Ribonuclease and deoxyribonuclease have also been reported.**

Table : 5·1. Composition of pancreatic juice (pH 7·1 to 8·3).

Inorganic constituents 1·0%	Na^+ K^+ Ca^{2+} Mg^{2+}	Cl^- SO_4^{2-} HPO_4^{2-} HCO_3^- (0·5 to 0·7%)
Organic constituents 1 to 2%	Trypsinogen Chymotrypsinogen Carboxypeptidase Nuclease Albumin and Globulin	Lipase Amylase Maltase Sucrase Lactase Esterase

NATURE AND ACTIONS OF PANCREATIC ENZYMES

1. **Trypsin :** This enzyme is secreted as the inactive proenzyme **trypsinogen**. This is rapidly activated into **trypsin** by the enzyme **enteropeptidase** (formerly called **enterokinase**) secreted into the intestinal lumen by the duodenal mucosa. **Enteropeptidase** in its function can be regarded as a proteolytic enzyme. It liberates a terminal hexapeptide of the **trypsinogen** and thereby converting it to the enzymatically active **trypsin**. Trypsinogen can also be activated in the same way as **trypsin**. This type of activation is called **autocatalytic**.

It is the most effective proteolytic enzyme which acts upon all proteins that are attacked by pepsin. It acts more rapidly and completely if the proteins have been partly hydrolyzed by **pepsin**.

Trypsin acting on native proteins causes them to pass through several intermediate stages such as globulins, alkaline metaproteins, deuteroproteoses, peptones, polypeptides and finally converting them into mono-amino acids (as leucine, aspartic acid) and diamino acids (as lysine and arginine, etc.).

2. **Chymotrypsin :** It is another proteolytic enzyme secreted as inactive proenzyme called **chymotrypsinogen**. It is activated by trypsin to form the active enzyme, **chymotrypsin**.

It acts in very much the same way as trypsin but in addition it possesses the power of clotting milk. It completes the protein digestion at an optimum pH of 7—8. It behaves like gastric **rennin** and digests casein of milk and gelatin.

3. **Carboxypeptidase :** It is probably another proteolytic enzyme of pancreatic juice and is activated **by trypsin** and **enterokinase**. It hydrolyses the amino acids, one by one from the end of a polypeptide chain which has a free carboxyl group. It does not generally attack dipeptides.

Two different carboxypeptidases of pancreatic juice named A and B have been described. The two differ in specificity depending on whether the free α carboxyl group is on a terminal aromatic or basic amino acid.

4. **Pancreopeptidase E** or **Elastase :** It is capable of solubilizing elastin by hydrolysis of peptic bonds, essentially those adjacent to neutral amino acid residue.

5. **Amylopsin** or **Amylase :** It is the most important starch splitting enzyme which converts starch, dextrin and glycogen into maltose. It requires the presence of chloride and phosphate as coenzyme and acts best at pH 7·0.

6. **Lipase** or **Steapsin :** This is most important fat-splitting enzyme of pancreatic juice.

It is secreted into the pancreatic juice in its active form. It has not yet been obtained in crystalline form but highly purified fractions have recently been prepared.

It rapidly acts on mixture of glycerides and fatty acids at a pH of 7—8. The fatty acids combine with alkali to form soap. It is essential for absorption.

7. **Sucrase :** It acts best at a pH of 5—7 and digests the sucrose into glucose and fructose.

$$\text{Sucrose} + H_2O \xrightarrow{\text{Sucrase}} \text{Glucose} + \text{Fructose}$$

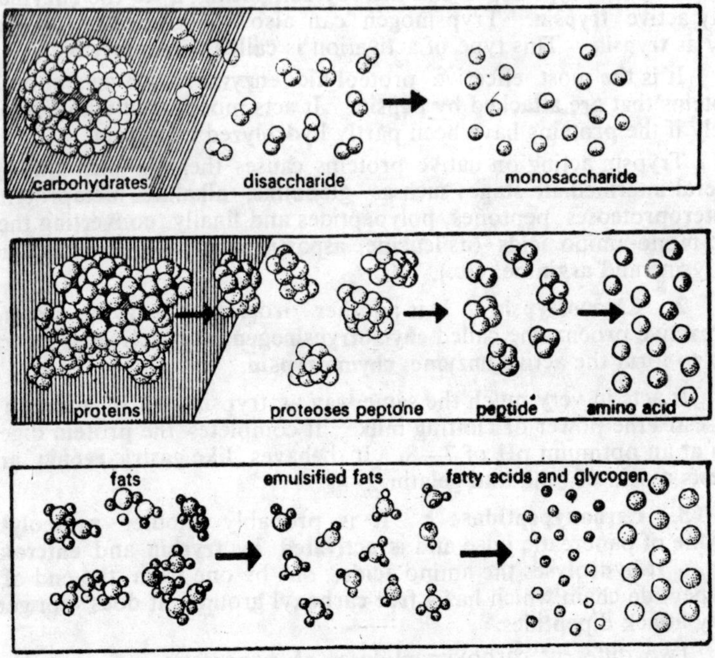

Fig. 5·9. Phases of digestion of carbohydrates, proteins and fats.

8. Maltase : It converts maltose into glucose at optimum pH 7·0.

$$\text{Maltose} + H_2O \xrightarrow{\text{Maltase}} \text{Glucose}$$

9. Lactase : It reacts with lactose converting it into glucose and galactose. It acts best at, pH of 5·4—6·0.

$$\text{Lactose} + H_2O \xrightarrow{\text{Lactase}} \text{Glucose} + \text{Galactose}$$

10. Ribonuclease and **deoxyribonuclease :** They split nucleic acids (of ribose and deoxyribose type respectively) into simple nucleotides.

CONTROL OF PANCREATIC SECRETION

Pancreatic secretion is under both nervous and hormonal control. When food (chyme) enters the intestine, the wall of duodenum produces an inactive substance called **prosecretin** which on coming in contact with HCl is converted into its active form **secretin** which is according to BAYLISS and STARLING (1902) a hormone. It on reaching into the pancreas through blood stream, causes it to produce pancreatic juice.

Besides secretin there is one more hormone called **pancreozymin** which simply controls the enzyme producing function of the pancrease (according to HARPER and RAPER, 1943). Thus secretin controls the volume of pancreatic juice where as pancreozymin controls the amount of enzymes which it contains.

Secretin producing cells in the epithelium of duodenum release the hormone in response to the presence of acids in the gut content. **Pancreozymin** release and the ensuing secretion of pancreatic enzymes into the duodenum are caused by the digestive products in the chyme that enter the duodenum from the pylorus, especially amino acids and long chain fatty acids.

RELATION OF BILE TO THE PANCREATIC DIGESTION

Pancreatic juice contains several enzymes which can act upon food constituents only in the alkaline medium. Food which enters the intestine remains acidic in its nature due to the presence of HCl. Bile juice simply creates the alkaline medium in the intestine so that the pancreatic enzymes could digest the food materials. In this way it is clear that bile juice has great affinity with pancreatic digestion.

Bile juice : Bile juice is produced by the **parenchymal cells** of the liver. This is an alkaline liquid, bitter in taste and varies in consistency and colour. There appears to be a continuous flow of bile about 500 to 1000 ml per twenty four hours, which varies according to the diet. It is secreted through villi at their borders into the **bile canaliculi** which are small slits between the liver cells. The canaliculi empty into small vessels called **ductules** or **cholangioles.** These unite to form bile ducts which in turn empty into the hepatic duct carrying the bile away from liver. In animals which do not have gall bladder, the bile passes directly from the hepatic duct through the **common bile duct** to the duodenum. In other animals including man it passes from the hepatic duct via the cystic duct into the gall bladder from which it passes intermittently into the intestine in response to the arrival of food in the duodenum.

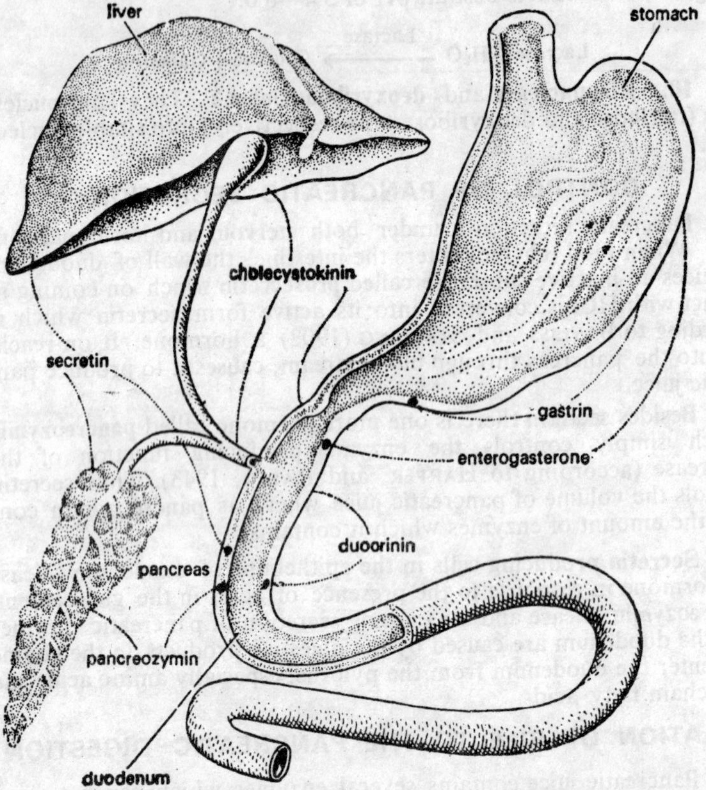

Fig. 5·10. Gastro-intestinal hormones. The places of origin and actions of each are entirely within the digestive tract but they are carried from one place to another by systematic circulation.

Bile has a variable composition. The variations depend not only upon the condition of the liver but also upon the length of time during which the bile is stored in the gall bladder.

Composition of bile : The bile, collected directly from the liver, contains over 98% of water but the bile collected from the gall bladder of recently killed animals, contains as little as 89% of water. In this way the bladder bile is less watery in comparison to liver bile. Bile whether, hepatic or gall bladder is always **isosmotic** with respect of blood plasma. The chief solid constituents of the bile are proteins, inorganic salts, **bile-pigments**, bile salts, cholesterol, lecithin and other lipid constituents. The bile also contains a little amount of bile acids including glycocholic acid and laurocholic acid. The composition of human bile, as secreted by the liver and as obtained from the gall bladder, is shown in the following table :

Table : 5·2. Composition of human bile.

Constituents of bile	Liver bile	Gall bladder bile
Water	98·00	89
Total solids	2 to 4	11·00
Bile salts	0·2 to 2	6·00
Bile pigments	0·02 to ·07	2·5
Cholesterol	0·1 to 0·3	0·4
Phospholipids (lecithin)	0·2 to 0·8	—
Inorganic salts	1·0	0·8
pH	8·0 to 8·6	7·0 to 7·6

From THURENBORN (1962)

Bile salts : Bile salts are chiefly the sodium salts of glycocholic and taurocholic acids, some potassium salts may also occur to a limited extent. These are soluble in water and alcohol, their solutions are distinctly alkaline and control the pH of the bile. They emulsify fats after lower the surface tension of water. They may be regarded as as they play important role in the digestive process. Bile salts are valuable for the following reasons :

(i) Fat soluble vitamins such as A, D, E and K could not be properly absorbed in the body if the bile salts did not emulsify the fat particles. Thus, we may regard the bile salts as instrumental in the assimilation of the fat soluble vitamins.

(ii) The fatty acid salts produced during the saponification reaction are not too soluble in water, it is because of law solubility. The bile salts react with the fatty acid salts forming more and readily soluble substances which can now be properly absorbed. Once absorbed the bile salts get separated from the fatty acids and are returned to the liver to be used again.

(iii) These salts promote the secretory power of liver cells.

(iv) These salts along with bile create an optimum medium so that the **steapsin** can properly function and hydrolyze the lipid particles to glycerol and fatty acid salts.

(v) They activate the inactive lipase and make it more effective.

2. **Bile pigments :** Bile pigments are **bilirubin** and **biliverdin.** They give distinctive colour to bile. Bilirubin is reddish pigment while biliverdin is greenish pigment. Bilirubin is converted into biliverdin by mild oxidation while biliverdin is converted into bilirubin by reduction.

These pigments can combine with bases because they are having two carboxyl groups in their molecules. They are found probably in the form of sodium salts in the bile.

3. **Cholesterol :** The cholesterol in bile may be considered a product to be excreted. The body is an efficient product of it and is synthesized from acetic acid. In addition, much cholesterol enters the body in the diet. Its great concentration in bladder bile causes its frequent crystallization forming one variety of **gall stone** which causes disturbances in the liver functioning.

4. **Lecithin :** Lecithin is a typical substance found in bile juice. It is accompanied by traces of soap. No significant value of it has been reported so far.

5. **Inorganic salts :** Inorganic salts are chlorides and bicarbonates, the latter being responsible for the slightly alkaline reaction of the bile.

Small amount of mucin, nucleoproteins, triglycerides and free acids are also present in the bile.

Alkaline phosphatase : It is a only bile enzyme probably derived from liver cells but its role in the process of digestion in gut is not yet clear. When the flow of bile into the intestine is prevented (**obstructive jaundice**), this phosphatase is found in the blood in increased concentration, thus providing an important means of distinguishing obstructive jaundice from other forms of jaundice.

Role of bile : The bile performs the following important functions :

1. Bile creates alkaline medium in the intestine in order to activate the pancreatic **lipase**.

2. The acids of bile augment the action of proteolytic and aminolytic enzymes.

3. It is capable enough to dissolve a large amount of fatty acids.

4. It helps in maintaining the fats in an emulsified state, it is due to reduced surface tension.

The emulsification of fats simply makes the fat particles accessible to the action of **lipase**.

5. It is essential for the absorption of the products of fat digestion.

6. It helps in neutralizing the acid chyme which enters the intestine from the stomach, in connection with which the action of pepsin ceases. It, thus, prevents the destruction of **trypsin** by **pepsin**.

CONTROL OF BILE SECRETION

There are several statements about the control of bile secretion, the important are as follows :

1. Stimulation of vagus nerve causes the secretion of bile juice.
2. Bile salts have also been stated to be responsible for the secretion of bile.
3. When food in the form of chyme enters the intestine, its mucosa secretes a hormone called **cholecystokinin** which circulates along with the blood and causes gall bladder to contract and deliver its bile.
4. **Secretin** also provokes bile production.

DIGESTION IN SMALL INTESTINE

The food, already digested in the stomach and duodenum, is pushed into small intestine (ileum) by peristalsis where it undergoes further hydrolysis so that it may be absorbed readily in the body.

The inner surface of the small intestine contains two distinct types of glands :

1. The duodenal glands, and
2. The intestinal glands.

1. The duodenal glands (BRUNNER'S **glands**) **:** These glands are mainly confined in the submucosa of the duodenum. These glands secrete an alkaline secretion which contains **mucin** and a weak **proteolytic enzyme**. Mucin makes the food slipery while proteolytic enzyme completes the protein digestion.

2. The intestinal glands : These glands are also called **crypts of** LIBERKUHN. These glands are found throughout the small intestine. They are simple tubular glands which open into the spaces between neighbouring villi. They contain a number of different types of cells that secrete mucus and a large number of enzymes. The secretion of these glands is called **succus entericus.**

Succus entericus : It has a pH about 7·6 and contains some 1·6% of solid matters out of which about 1% are inorganic and rest are organic. The detailed table of its composition has been given below :

Table : 5·3. Composition of intestinal juice (Succus entericus).

Organic constituents (0·6%)	(i) *Erepsin* (containing *aminopeptidases* and *Dipeptidases*)
	(ii) *Lipase*
	(iii) *Enteropeptidase* (*Enterokinase*)
	(iv) *Maltase*

	(v) *Lactase*
	(vi) *Invertase*
	(vii) *Sucrase*
	(viii) Nuclease
	(ix) Nucleotidase
	(x) *Esterase*
	(xi) *Phosphatase*
	(xii) Mucin
Inorganic constituent 1·0%	Na^+, Ca^{2+}, Mg^{2+}, Cl^-, HCO_3^-, K^+, PO_4^{3-}

ENZYMES OF SUCCUS ENTERICUS AND THEIR ACTIONS

1. **Erepsin :** It is a complex proteolytic enzyme which converts albumoses and peptones into amino-acids. This enzyme ends the conversion of food proteins begun by **pepsin** and **trypsin**.

2. **Lipase :** It is not a strong enzyme but is very stable. It completes the fat digestion.

3. **Enteropeptidase (enterokinase) :** This enzyme simply activates the **trypsinogen** to form active **trypsin**.

4. **Maltase, Lactase, Invertase** or **Sucrase :** These enzymes act on maltose, lactose and cane sugars respectively and form monosacharides.

5. **Nuclease and Nucleotidase :** These enzymes act on nucleic acid components of the nucleoproteins, finally splitting them into corresponding purine or pyrimidine nucleotides.

There are other enzymes such as **esterase, phosphatase, arginase, catalase, aminopeptidase, dipeptidase, proteinase,** and **tyrosinase** which act mostly on the broken down food stuffs during the passage through the epithelial cells of the small intestine on their way to the blood vessels.

CONTROL OF SUCCUS ENTERICUS SECRETION

The secretion of succus entericus from the intestinal mucosa is at least partly under the control of the nervous system. Mechanical stimuli reflexly cause a flow of this fluid. Secretin probably exerts a hormonal effect for this secretion. **Enterocrinin** secreted by the intestinal mucosa is said to control the production of this secretion.

Table : 5·4. Table showing the gastrointestinal hormones.

Hormone	Source	Stimulus for production	Action
Gastrin	Stomach	Food in stomach	Stimulates secretion of gastric juices.
Enterogastrone	Small intestine	Fatty acids in intestine	Inhibits secretion of HCl and gastric juice.
Secretin	Small intestine	HCl in duodenum	Stimulates secretion of pancreatic fluids containing bicarbonate; stimulates production of bile by liver.
Cholecystokinin (pancreozymin)	Small intestine	Food in duodenum	Stimulates release of pancreatic enzymes; stimulates release of bile from gall bladder.
Enterocrinin	Small intestine	Food in duodenum	Stimulates release of succus entericus.

Cellulose digestion : Cellulose is not digested in the human alimentary canal as it lacks the appropriate enzyme. In herbivores such as rabbit, ox and goat, however, cellulose is digested by a variety of micro-organisms with the production of fatty acids. Micro-organisms actually produce **cellulase**, an enzyme which digests the cellulose.

DIGESTION IN THE LARGE INTESTINE

The process of digestion does not take place in the large intestine as the wall of this region does not have any digestive gland. However, bacterial digestion of undigested carbohydrates by a **cytase**, unabsorbed or unutilized proteins and rarely fats may take place. Large intestine simply secretes mucus from the goblet cells in the epithelium which serves as lubricant for easy passage of the undigested food.

DIGESTION IN OTHER VERTEBRATES

The process of digestion in other vertebrates (excluding mammals) follows the same lines as in mammals because the general plan of the alimentary canal is practically the same.

The marked differences found in the process of digestion of other vertebrates are as follows :

1. Salivary digestion or buccal digestion, which is characteristic of mammals, is rarely found in other vertebrates but the saliva of frog, toad and of the fowl contains an enzyme called **amylase.**

2. The **pepsin** of mammals differs in specificity from that of salmon.

3. In gold fish, lizard and some passerine birds there is found an enzyme called chitinase which is completely lacking in mammals. This enzyme helps in the digestion of chitinous materials contained in food.

4. The birds, such as honey guides (**indicator**), like mammals, do not produce **esterase** enzyme but are dependent on their microflora.

5. In mammals the medium of stomach is acidic, it is due to the secretion of hydrochloric acid but in some rays and bony fishes the medium of stomach is alkaline even though the **pepsin** is present. Some sharks, by contrast, have been reported to have an acidity twice that of man.

6. There are several vertebrates such as holocephali, dipnoi, teleosts and protochordates which donot possess stomach and, thus, pepsin is not secreted.

7. In fowl preliminary digestion takes place in the crop by **autolysis** and bacteria.

8. In herbivorous animals special modifications are met in the alimentary canal that help in cellulose digestion. In these animals **cellulase** is found which digests the cellulose. **Cellulase** is not secreted by the animals but is produced by the bacteria present in the alimentary canal.

DIGESTION IN INVERTEBRATES

1. **Digestion in Protozoa :** In all protozoans except parasitic protozoans such as *Trypanosoma*, *Monocystis,* digestion takes place in the food vacuoles. Digestion is carried out with the help of certain digestive enzymes. It is still not clear just how the digestive enzymes enter the food vacuoles (BARRINGTON, 1962). According to DE DUVE 1963, the enzymes are hydrolytic in nature and are contained in special packages called **lysosomes**. It is believed that during the process of digestion the lysosomes are discharged in some manner into the food vacuoles and thus contained food is broken down into its constituents which then pass into the cytoplasm and be assimilated.

The parasitic protozoans either absorb the predigested food of their host or digest their food with the help of hydrolytic enzymes outside the body.

2. **Digestion in Porifera :** In porifers or sponges the digestion is intracellular, therefore, it occurs in food vacuoles of collar cells much as in protozoans (BARRINGTON, 1962), but these animals with their multicellular organisation require a distributing system. Apparently this function is performed by **amoebocytes** which transfer the food materials from the collar cells to other cells of the body.

3. **Digestion in Coelenterata :** In coelenterates both intracellular and extracellular digestions are met. During intracellular digestion the food or prey is engulfed and digested in a food vacuole

with the help of digestive enzymes as in protozoans. During extracellular digestion the food is digested in the **gastrovascular** cavity, the wall of which contains numerous glandular cells which secrete digestive enzymes. These enzymes help in the digestion of food materials.

4. **Digestion in parasitic worms :** In most turbellarians the digestion is extracellular but in some it is carried out in the meshes of a temporary syncytium formed by processes from amoeboid cells

In some of the Rhabdocoelida digestion is mainly or perhaps entirely extracellular. The pharyngeal glands of the turbellarians produce mucus while in some species of Rhabdocoelida their secretions contain digestive enzymes which are used in the digestion of food materials outside the body.

In liver-fluke, the so called *Fasciola hepatica*, the digestion is extracellular and it feeds on bile, blood, lymph, and cell debris. It can also absorb glucose, fructose, galactose and maltose but not lactose or sucrose, through the general body surface. It is believed that an enzyme called **proteinase** is secreted in the digestive tract which digests the protein of the solid tissues.

In parasitic tapeworms such as *Taenia solium* the digestion is supposed to be intracellular as they absorb digested food in the form of nitrogenous substances from the mucous membrance of the host. They also absorb glucose and galactose but they donot secrete any **carbohydrases**.

In some nematodes including *Ascaris* the digestion is extracellular and it is believed that **amylase, maltase, protease, peptidase** and **lipase** enzymes are secreted in the alimentary canal which digest the starch, maltose, proteins and fats respectively.

5. **Digestion in Annelida :** In most of the annelids the digestion is extracellular. Annelids take all types of food and their intestinal walls produce several enzymes such as proteolytic enzyme, **diastase,** glycogen hydrolyzing enzyme, lipase and amylase, these enzymes digest the .food materials in the intestine. In earthworm, the salivary glands of pharynx produce **mucin** and **proteolytic** enzyme, the former lubricates the food and passes it down and the latter starts the digestion of protein.

6. **Digestion in Arthropoda :** In Crustacea the digestion takes place outside the body (extracellular). In decapods the food is broken down into its constituents in the gastric mill and at the sametime attacked by a **proteinase** which is sent forward from the digestive diverticula. At the entrance to the latter there is a complicated filter which permits only fine food particles to pass. The food particles are taken into the diverticula by the contraction of longitudinal muscles and expelled by the contraction of circular muscles.

In some insects the digestion is extracellular. The labial glands of these insects secrete an enzyme called **amylase in** cockroach, **sucrase** in honeybees and moths. These enzymes digest the starch and sucrase respectively. The labial glands open immediately into the **crop** where digestion may take place by enzymes. The crop is the chief site of digestion of these insects. Yeast and bacteria may also carry out the digestion of food materials.

The insects which feed on solid food, the bolus on leaving the foregut is enclosed in a thin sac of chitin called the **peritrophic membrane** which is permeable to both enzymes and digested food. Absorption of digested food materials also takes place through it. This is completely lacking in fluid-feeders such as bugs, fleas, lice, etc.

Many adult butterflies and moths have no digestive enzymes except sucrase which digests the sucrose. In these insects the digestion is extracellular.

7. **Digestion in Mollusca :** In molluscs both types of digestion, *e.g.*, extracellular and intracellular, are found. Two specializations are of particular interest. The first of these is the digestive gland made up of branched glandular follicles communicating with the stomach by a system of ciliated ducts.

In some groups the epithelial cells are phagocytic, and digestion is intracellular. In nuclidae the digestion is entirely extracellular and is carried out with the help of digestive enzymes (OWEN, 1956).

In many forms there is a combination of the two processes (extracellular and intracellular). A curious process in connection with the intracellular digestion taking place in the epithelial cells is a fragmentation of the outer wall of the cell to form spherical bodies containing the food vacuoles together with waste products and enzymes. These bodies than pass into the stomach and may be sources of some of the enzymes there (OWEN, 1955, 56).

The second of these special features is the crystalline style which is generally found in lamellibranchs but rarely in the more advanced herbivorous gastropods. The gradual development of the style in connection with ciliary feeding is discussed by MORTON (1958, 1960). It is a thick gelatinous rod loaded with enzymes. It is rotated by strong cilia which force it gradually into the stomach where it rubs against the horny gastric shield to release its enzymes and to stir up the stomach contents. It is believed that **amylases** are the most abundant enzymes of the crystalline style, however, lipases seem also to be present.

In lamellibranchs and in some echinoderms their intracellular digestion is partially dependent on an extensive phagocytosis as the amoebocytes or phagocytes procure food particles, digest them and pass their products to the other cells of the body (YONGE, 1937, 1960 a).

In most gastropods the digestion is extracellular and digestion and absorption take place in the digestive diverticula. In these animals radula is an important feeding organ which is lubricated by the secretion of buccal glands. In cephalopods the digestion is entirely extracellular. The digestive diverticula of *Sepia* and *Octopus* but not of *Loligo* are absorptive as well as secretory in function.

8. **Digestion in Echinodermata :** In echinoderms the digestion is extracellular as well as intracellular. In starfish, digestion takes place outside the body with the help of digestive enzymes. The pyloric

caeca contain many phagocytes which probably take up small food particles and migrate through the gutwall. It is believed that in sea

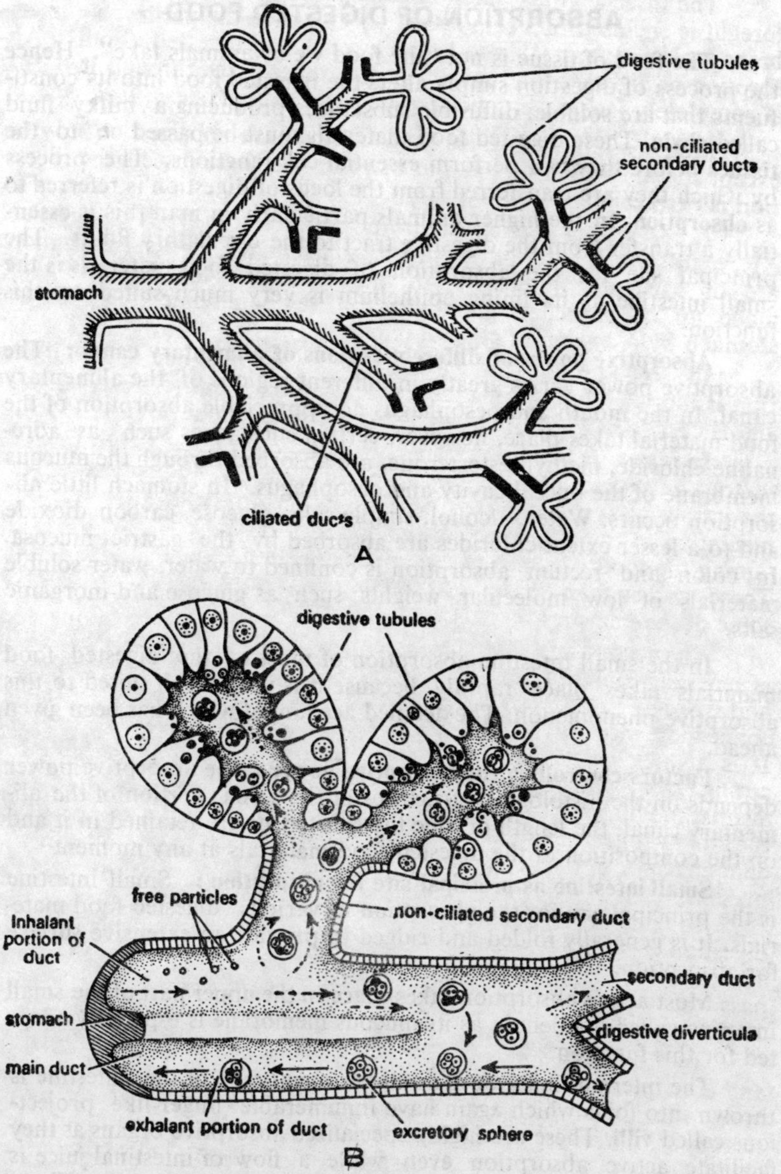

Fig. 5·11. The digestive diverticula of the Anisomyaria and Eulamellibranchia.
A—The arrangement of ducts and tubules ;
B—The probable circulation of particles.

urchin, brittle star and holothuria the digestion occurs partially in the gut and partially in the phagocytes.

ABSORPTION OF DIGESTED FOOD

The food of tissue is not "the food we or animals take". Hence the process of digestion simply alters the ingested food into its constituents that are soluble, diffusible substances producing a milky fluid called **chyle**. These digested food materials must be passed on to the tissues before they can perform essential cell functions. The process by which they are transferred from the locus of digestion is referred to as **absorption**. In the higher animals particularly in man this is essentially a transfer from the digestive tract to the circulatory fluids. The principal site for the absorption of digested food materials is the small intestine as its lining epithelium is very much suited to this function.

Absorptixe power of different regions of alimentary canal : The absorptive power varies greatly in different regions of the alimentary canal. In the mouth and oesophagus no appreciable absorption of the food material takes place, however, certain substances such as adrenaline chloride, methyl testosterone, are absorbed through the mucous membrane of the buccal cavity and oesophagus. In stomach little absorption occurs. Water, alcohol, simple salts, glucose, carbon dioxide and to a lesser extent chlorides are absorbed by the gastric mucosa. In colon and rectum absorption is confined to water, water soluble materials of low molecular weights such as glucose and inorganic salts.

In the small intestine absorption of most of the digested food materials takes place rapidly because it is very much suited to this absorptive phenomenon. The detailed account of which has been given ahead.

Factors controlling the absorptive power : The absorptive power depends on the histological structures of the given portion of the alimentary canal, the length of time the substances are retained in it and on the composition of the digested food materials at any moment.

Small intestine as principal site for absorption : Small intestine is the principal site for the absorption of various digested food materials. It is generally folded and ridged to provide an extensive surface for absorption.

Most active absorption takes place in the upper part of the small intestine called duodenum as its mucous membrane is especially adapted for this function.

The internal lining or the mucous membrane of the intestine is thrown into folds which again have innumerable finger-like projections called **villi**. These are highly specialized absorptive organs as they facilitate active absorption even while a flow of intestinal juice is simultaneously coming out of the glands of intestinal mucosa. These are adequately exposed to the liquid food in the intestine because intestinal contractions keep the liquid food constituents constantly in motion.

The length and diameter of villi vary (in man) from 0·2 to 1 mm. The entire surface of the villi is covered with epithelium, the cells of which contain fibrils, mitochondrial apparatus that changes during absorption. Each villus contains a network of capillaries derived from blood vessels in the wall of the alimentary canal and a central

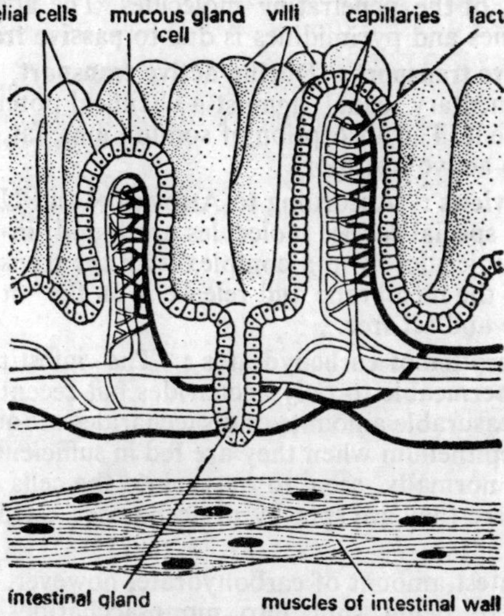

Fig. 5·12. The villi which increase the absorptive surface of small intestine.

lymph capillary or **lacteal** which begins blindly under the epithelium at the tip of the villus and drains, after forming a large plexus and a submucosa plexus, into the main lymphatic channels of the gut wall. In the submucosa plexus the lacteals possess valves owing to which the lymph flows only in one direction. The consensus is that lipids pass primarily into the lacteals, while the sugars and amino-acids are absorbed directly in the capillary blood. Both the villus and the intestinal fold contain longitudinally arranged smooth muscle fibres, the rhythmic contractions of which bring the villi into contact with the intestinal contents and thus maintain the circulation in the lacteals, lymphatics and small blood vessels.

PHYSIO-CHEMICAL MECHANISM OF ABSORPTION

The exact physio-chemical mechanism by which the digested food materials are absorbed is not known so far. It is believed that phenomena such as **passive transport**, and **active transport** (diffusion) are involved in the absorption of digested food materials. The movements of villi also cause some absorption. There are some evidences that the leucocytes (white blood corpuscles) found in the intestinal mucosa, actually pass through the intestinal wall and pick up the food particles and carry them back into the blood and lymph and thus help in absorption.

1. **Passive transport :** In **passive transport** materials move through cells, membranes and intercellular spaces because of differences in the concentration gradient. The substances are in aqueous solutian and they move either due to random motion of solute molecules, or due to osmotic properties of the cell. The extent of **passive transport** varies with the type of cell, its physiological condition and the properties of the penetrating molecules. The absorption of most vitamins, purines and pyrimidines is due to **passive transport.**

2. **Active transport :** During **active transport**, molecules move from the alimentary canal against a chemical potential gradient or diffusion gradient. The absorption of sodium ions, fat, glucose, amino-acids is due to active transport.

Current views : According to ANDERSON and USSING, 1963 and HOLTER, 1961 special carrier molecules are found in the alimentary canal. These molecules form a complex with the transported material on one side of the membrane and release it on the other side where they dissociate and set free.

1. **Absorption of carbohydrates :** The intestinal epithelium is practically impermeable to polysaccharides but recent findings have shown that measurable amounts of disaccharides are absorbed through the intestinal epithelium when they are fed in sufficient amounts, some disaccharides normally seen to pass into the cells of the intestinal epithelium where they are split by disaccharases such as **maltase** and **lactase**.

The greatest amount of carbohydrate, however, seems not to be absorbed before being split into monosaccharides. During passive diffusion monosaccharides such as fructose, mannose and most pentoses are absorbed, where as hexoses, glucose and galactose are absorbed as a result of active diffusion.

All the carbohydrates are absorbed in the form of monosaccharides directly into the blood stream. Blood carries these sugars into the liver where the excess of sugar is converted into glycogen in which form they are stored.

2. **Absorption of proteins :** Proteins, as a result of digestion, are hydrolyzed into α-amino-acids. Phosphoric acid and nucleic acids may also be formed but the α-amino-acids are the primary units liberated.

The early workers thought that all amino-acids are absorbed by diffusion (active) but later GIBSON and WEISMAN, 1951, proved that the absorption of amino-acid is a selective chemical process as the natural α-amino-acids are more rapidly absorbed than their optical mates, D-amino-acids. Some of the α-amino-acids are absorbed against a concentration gradient, *e.g.*, actively absorbed while others are not. In few cases, as with glutamic acid and aspartic acid considerable transamination occurs in the intestinal mucosa but others are absorbed without transformation.

3. **Absorption of fats :** In the process of digestion the fats are hydrolyzed into fatty acids. Fat absorption is an active process. The end products of fat digestion are absorbed by the lacteals of the in-

testinal villi. From the lacteals they pass into the lymph vessels and finally reach into the blood stream via thoracic duct.

Various theories have been formulated about the nature of fat absorption. The most important are as follows :

(i) **Old views :** Prior to 1900 it was believed that fats are absorbed by the intestinal wall only after they had been hydrolyzed into fatty acids and glycerol. It was further believed that fats could be absorbed as finely emulsified particles.

(ii) **PFLUGER's views :** According to PFLUGER no emulsified fats could be absorbed by the intestinal cells and all the fats must be hydrolyzed before absorption into the body.

(iii) **Lipolytic hypothesis :** According to this hypothesis the fats are absorbed in the form of soaps. The natural fats are hydrolyzed into fatty acids, and glycerol by the action of pancreatic **lipase**. The fatty acids, thus formed, are made water soluble by saponification or by hydrotropic action of bile salts. These saponified fatty acids along with glycerine are absorbed by the cells of intestinal epithelium where they are said to combine to yield first phospholipids and then triglycerides what are known as complex fats. These complex fats pass to lacteals for transport via thoracic duct to systemic blood.

(iv) **Partition hypothesis :** FRAZER formulated this hypothesis, according to which the fats are absorbed both in the hydrolyzed as well as in the emulsified form. FRAZER believed that the tiny droplets of fats enter the intestinal cells through the pores present in their outer boundaries.

(v) **Recent views :** According to recent findings the complex fats, that are found in the lacteals, are formed after absorption.

4. Absorption of water : The absorption of water begins in the stomach, but, since it passes rapidly into the intestine it is mainly absorbed in it. Enormous quantity of water can be absorbed by the intestine (in man 15 to 20 litres per day). Process of osmosis form the principal mechanism of water absorption since the osmotic pressure of the food is usually higher than that of the chyme.

5. Absorption of mineral salts : In addition to the organic food materials certain inorganic salts are also absorbed by the cells of the intestinal epithelium. The salts of alkaline metals are absorbed into the blood through cells of the intestinal epithelium and not through the intercellular spaces. The salts of haloid acids are absorbed better than sulphates and carbonates. Na, K and Cl ions are absorbed rapidly but Mg and SO_4 ions are absorbed with difficulty.

Absorption in large intestine : In the large intestine practically no absorption of food materials takes place because the chyle contains no absorbable substance by the time it reaches the large intestine. In the large intestine simply the absorption of water takes place.

PRODUCTION AND COMPOSITION OF FAECES

All the end products of digestion, except fats taken up by lacteals, pass from the absorbing cells into the finer blood capillaries of the intestinal wall and so to the portal vein, and the general circulation.

All the remaining contents of the small intestine are moved on by its peristaltic contractions into the colon or large intestine. Due to enormous absorption of water in the large intestine the contents take the form of semi-solid residues which include the following :

1. The undigested materials, those that have escaped from the action of digestive enzymes as well as those for which specific enzymes are lacking.
2. Bile pigments.
3. Other compounds that may have entered the intestine with secretions that flow into it, or from the blood.

In the colon further decomposition of the contents takes place through the activities of the bacteria found in it. The material remaining after the absorption of water, constitutes the faeces, the consistency of which depends on the compositon of food.

Vegetable food forms large amount of faeces than does animal food. Milk yields a relatively large amount of faeces, it is because of the large quantity of unabsorbable salts contained in it. Animal proteins are utilized upto 98 to 99%, while vegetable proteins are assimilated to a lesser extent.

DEFAECATION

The faecal mass collects in the distal end of the large intestine before reaching to the rectum. The removal of faeces from the body or alimentary canal is called **egestion** or **defaecation**. Due to accumulation of large quantity of faeces the lower segment of the large intestine (rectum) contracts that results in the elimination of faeces outside the body.

Table : 5·5. Summary of digestion.

Place of digestion	Glands	Secretion	Enzymes	Medium for enzyme activity	Digestive activity
Mouth	Salivary	Saliva	*Ptyalin*	Neutral	Changes starch to maltose.
	Mucous	Mucus	—	—	Lubricates the food.
Oesophagus	Mucous	Mucus	—	Neutral	Lubricates as well as makes the food soft.
Stomach	Gastric	Gastric juice	*Pepsin*	Acidic	Changes proteins into peptones and proteoses.
			Rennin	Acidic	Changes milk protein into paracasein.
	Parietal cells	Hydrochloric acid	—	Acidic	Activates pepsin, dissolves minerals, kills bacteria.

Place of digestion	Glands	Secretion	Enzymes	Medium for enzyme activity	Digestive activity
	Mucous	Mucus	—	Acidic	Lubricates the food.
Small intestine	Liver	Bile juice	—		Emulsifies fats.
	Pancreas	Pancreatic juice	*Trypsin*	Alkaline	Changes proteins, peptones and proteoses into peptides, activates *chymotrypsin.*
			Chymotrypsin		Changes proteins, peptones and proteoses into peptides. Clots milk protein.
			Carboxypeptidase		Changes amino acids into peptides.
			Amylase		Changes starch to maltose.
			Lipase or steapsin		Changes fats into fatty acids and glycerine.
	Intestinal glands	Intestinal fluid succus-entericus	*Erepsin*	Alkaline	Changes peptides to amino acids.
			Maltase		Changes maltose into glucose.
			Lactase		Changes lactose into glucose and galactose.
			Sucrase		Changes sucrose to glucose and fructose.
Large intestine	Mucous	Mucus	—	Alkaline	Lubricates the faeces.

Revision Questions

1. What is nutrition ? Describe various feeding mechanisms found in organisms.
2. What is digestion ? Describe the process of digestion in any mammal studied by you.
3. Describe the physiology of digestion in different regions of the alimentary canal of a vertebrate.
4. What role the pancreas and the liver play in the physiology of digestion ?
5. Describe the physiology of digestion in different invertebrates.
6. Describe in brief the mode of absorption of different components of digested food in the alimentary canal.

CHAPTER 6

Enzymes

Enzymes, commonly as known **biocatalysts,** are unique and highly specific proteinaceous substances (globular proteins). They are produced by the cell and have enarmous ability to catalyze virtually all of the chemical reactions or activities of living system in a highly and very effective manner. They are involved in such biological processes as digestion, respiration, the stepwise breaking down, building up and interconversion of carbohydrates, fats, proteins and nucleic acids, and the release and utilization of energy by living cells.

Enzymes greatly **accelerate chemical reactions** without themselves undergoing any apparent change in the process of work, Another distinguished feature of them is **enormity of reactive power.** *e.g.*, mere trace of particular enzyme may induce quite a tremendous amount of chemical reaction. Enzymes are universely present in living organisms, and the occurrence of metabolic reactions common to all cells reflects the specificity of the responsible enzymes. The great stereochemical specificity of enzymes results from complementarity of the conformation at the active side of the protein molecule with the geometry of the substrate.

Enzymes are produced within the cells but capable of action outside the cells. Each cell probably contains hundred of different enzymes. Each enzyme usually acts on a single substance (substrate) or on a group of closely related substances. Enzymes which act on a particular substrate are said to be highly specific in their action.

The catalytic efficiency of enzymes is extremely high. Pure enzymes may catalyze the transformations of as many as 10,000 to 1,000,000 moles of substrate per min per mole of enzyme.

A considerable number of evidences suggest that enzyme and substrate combine transiently to form an **enzyme-substrate complex.** The enzyme and substrate are probably not covalently linked but reather are held together by VAN DER WAALS interactions. The enzyme-substate complex dissociates into enzyme and products of the reactions.

Many enzyme-catalyzed chemical reactions require the presence of low molecular weight organic compounds called coenzymes. These coenzymes act as donors or recipients of hydrogen atoms or other groups, and like enzymes, function over and over again. Many enzymes require metal ions for their activation. In some instances these activators function in combination with the protein, in others, the metal ions forms a compound with the substrate, and it is the metal-substrate complex that reacts with the enzyme. The ions of Ca, Co, Cu, Mg, Mn, Mo, Na, K and Zn are known to participate in enzymic reactions.

Enzymes are organized both functionally and structurally into groups which carryout sequences of chemical reactions. The sequences of chemical reactions constitute the **metabolic path ways.** All of the enzymes in a metabolic pathway tend to be found in one type of intracellular organelle or in solution in the cell sap. The destructive or catabolic pathways usually are downhill reactions, while the constructive or biosynthetic or anabolic pathways usually are uphill reactions which must be coupled to energy-releasing processes.

Definition of enzymes : Originally enzymes were defined as **"catalytic substances produced by living cells".** PAVLOV defined them as "activators and keys to life" being involved in the processes of metabolism. CHITTENDEN in 1894, expressed the following to the enzymes : (1) **Enzymes are proteins.** (2) **Catalytic activity is in someway related to protein structure.** (3) **Enzymes are not passive catalysts but function by forming an intermediary complex with the snbstrate.**

Modern definition states that **"enzymes are catalysts of biological origin which accelerate the various cellular reactions, without themselves undergoing any apparent change during the course of action and that are not dependent on the intact cell for their activity**

Nomenclature : A systematic method of naming individual enzyme is very essential because a large number of enzymes have been recognized so far. In the modern system for naming a particular enzyme the suffix **"ase"** is added to the root word referring to the substrate of the enzyme. By substrate is meant the substance that is acted upon by the enzyme. Under this system the name **sucrase** obviously refers to an enzyme that causes the breakdown of the sugar, **sucrose.** Another system for naming enzymes is that the suffix **"lytic"** is added to the substrate such as proteolytic (protein-splitting) and amylolytic (starch-splitting) enzymes.

The enzymes may also be named according to the type of reactions they cause, for example, oxidation (oxidase), dehydrogenation (dehydrogenase), hydrolysis (hydrolase). etc. Some enzymes named before any formal system was adopted retained their old fashioned names, such as pepsin, rennin, trypsin and ptyalin.

Chemical nature of enzymes : Knowledge of the role of enzymes as catalysts has grown with knowledge of catalysis in general. BERZELIUS was first who defined and recognized the nature of catalysis and proposed in 1837 that **"ferments"** were catalysts produced by living cells. Nevertheless, little was known about the chemical nature of enzymes until the beginning of this century, when there was a growing conviction that enzymes were probably protein in nature. The crystallisation of **urease** from jack been seeds in 1926 by SUMNER, pepsin from the gastric juice of the cow in 1930 by NORTHROP and **trypsin** from the pancreas of the cow in 1931 by NORTHROP and KUNITZ showed that enzymes are infact proteins. Over 1000 enzymes of all classes are now identified and about 150 have been crystallized and purified (NORTHROP, 1948., FLORKIN and STOLZ, 1964; LEHNINGER, 1970).

So all enzymes isolated so far are proteins, out of which many are conjugted proteins in their chemical nature. They have the same properties and characteristics as proteins have in general. The clear evidences for the protein-like characters of enzymes are the colloidal nature of these biochemical catalysts, colour reactions, ultraviolet absorption spectra, inactivation by heat and radiations, and sensitivity to pH and salts, and so forth. Enzymes form colloidal solution in water and are not soluble in high concentration of alcohol and alkaline reagents by which they are precipitated. Enzymes like proteins do not pass through a dialyzing membrane. They migrate under the influence of an electrical current which indicates that they are protected by electrical charges. They can be destroyed by heat, which indicates a solvent protection. Enzymes contain protein molecules which are attached with other non-protein groups such as metals, vitamins and carbohydrates depending upon the particular enzyme. These non-protein groups are called **prosthetic groups**.

Catalytic abilities of enzymes : The catalytic abilities of enzymes are derived from their particular protein structure. A small portion of the surface of the enzyme protein constitutes the **active site** where the specific chemical reaction catalyzed by the enzyme occurs. At this site certain chemical reactions characteristics of each enzyme are facilitated by the physical and chemical interactions that occur there. Each protein molecule has a complicated three dimensional shape because of the particular folding adopted by the large molecule. The type of folding is dependent on its particular sequence of amino-acid residues, which form these long protein chains.

At pH value nearly neutrality the side chains of particular enzyme protein, *e.g.*, free $-NH_2$ and $-COOH$ are changed to $-NH_3^+$ and $-COO^-$ respectively and even the imidazole ring (of histidine) may also be positively charged. It is these charged groups which exert a powerful electrostatic attraction on other ions of the solution so that they may able to bind with the enzyme protein.

PHYSICAL PROPERTIES OF ENZYMES

In general enzymes behave as colloids or as substances of high moleculer weight. They diffuse either slowly or do not pass through colloidion or similar membranes. Like proteins they are also salted out. Their activities are destroyed by heat but could recover their activities if cooled again according to KUNITZ and NORTHROP, (1934).

CHEMICAL PROPERTIES

1. **Enzymes as catalysts :** Enzymes in their property act catalytically. Catalysts are substances which alter the speed of chemical reactions without themselves undergoing any permanent change during the mode of action.

This statement states that catalysts do not initiate chemical reactions but only speed up reactions already proceeding at a slow rate. For example, hydrogen peroxide gas undergoes a very slow spontaneous decomposition at room temperature with the formation of water and oxygen. The addition of a little part of platinum or of an enzyme called **"catalase"** enormously increases the rate of this decomposition. It, thus, may be assumed that the reaction is actually proceeding but at an immeasurably slow rate. From practical stand point, however, the presence of a catalyst or enzyme in this reaction leads to the formation of products which would not be found in its absence.

Specificity of enzymes : Enzymes are amazingly specific in their action. The degree of specificity varies with respect to the types of substrates and chemical reactions which the enzymes catalyze. Some enzymes are so highly specific that they catalyze only one chemical reaction involving a particular reactant or substrate and therefore, exhibit a **absolute specificity**; others appear to catalyze a number of related reactions involving a wider range of reactants or substrates and exhibit **relative specificity**. Proteolytic enzymes hydrolyze the peptide bonds of polypeptided and, therefore, exhibit relative specificity. Likewise some esterases such as lipases act upon the esters of different fatty acids with a variety of alcohols and split the ester bonds. Nevertheless, the esterases are specific in their esterase action; they do not catalyze other hydrolytic reactions, nor do they function as oxidases, decarboxylases, etc. Examples of absolute specificity are more restricted : urease acts only on urea; carbonic anhydrase acts only on carbonic acid; and fumarase acts only on fumeric acid. In addition, many enzymes exhibit a high degree of stereochemical specificity, *e.g.*, arginase acts only on L-arginine, not on the D-isomer. Likewise glucose oxidase acts only on the β-D anomer of glucose, not on the α-D-anomer.

The specific combination that occurs between enzyme and substrate to form the enzyme-substrate complex is believed to reflect a particularly suitable spatial relationship between the substrate and certain active sites on the enzymes analogous to a lock and key or jigsaw arrangement. It states that the structural configuration of the enzyme can only accomodate a particular type of substrate, as a key fits into its lock.

According to **induced fit** hypothseis proposed by KASHLAND that some enzymes have **"flexible"** active sites because of which the protein may not have a proper proximity of reactive groups unit the substrate binds to the enzyme. Only the appropriate substrate can cause the precise aligment of calalytic groups needed for enzyme action.

GENERAL OR OTHER PROPERTIES

Other properties of enzymes are as follows :

1. All enzymes are more powerful than the inorganic catalysts.

2. They simply initiate and accelerate the chemical reactions.

3. They may induce even reversible reactions.

4. They are very unstable compounds mostly soluble in water, dilute glycerol, sodium-chloride solution and in dilute alcohol as well.

5. They act actively at optimum temperature. At low temperature or with a fall of temperature enzyme activity decreases sharply but on heating to high temperature they become inactive.

6. Besides breaking down, enzymes are capable of rebuilding specific elements out of the final products either by reversible synthesis or else by specific synthesis by specific enzymes.

7. The activity of enzymes depends upon the acidity of the medium. Each enzyme is most active at a definite acidity which is optimal for it.

MECHANISM OF ENZYME-ACTION

Our knowledge about the mechanisms involved in the enzyme catalysis is very little. However, the view that has dominated all attempts to explain mechanisms of enzyme-action is that the enzyme forms an intermediate complex with the substrate or substrates. Before proceeding to a more detailed examination of this concept, it is desirable to summarize briefly some of the early experiments that led to this view.

1. In 1880, WURTZ observed that, after addition of the soluble proteinase **papain** to the insoluble protein, fibrin, repeated washing of the fibrin did not stop the proteolysis. He then concluded that the papain had formed a substance or complex with the fibrin.

2. In 1890, O'SULLIVAN and THOMPSON observed that the enzyme invertase could withstand higher temperatures in the presence of the substrate, sucrose, than in its absence. They also concluded the WURTZ'S observation that the invertase had formed a complex with its substrate, sucrose.

3. In 1890, EMIL FISCHER, after conducting so many experiments on different enzymes, quoted the following remark for the enzymes that "*inasmuch as the enzymes are in all probability proteins,... it is probable that their molecules also have an asymmetrical structure, and one whose asymmetry is, on the whole, comparable to that of the hexoses. Only if enzyme and fermentable substance have a similar geometrical shape can the two molecules approach each other close enough for the production of a chemical reaction. Metaphorically, we may say that enzyme and glucoside must fit into each other like lock and key.*

Keeping all these observations in knowledge, FISCHER, in 1894 suggested the **"lock and key"** hypothesis to explain that how do enzymes perform their activity. This hypothesis is still acceptable but in a modified way as **"induced fit"** hypothesis. These theories suppose the existence of **active-sites** on the surface of the enzyme molecules.

According to this hypothesis the substrate molecules are thought to fit into the active sites located on the surface of the enzyme molecules just as one particular kind of key fits into one particular kind of lock. This results in the rapid formation of intermediate compounds the so called **enzyme-substrate complexes** by reversible reaction **enzyme + substrate ⇆ enzyme-substrate complex.** These intermediate compounds are believed to be much less stable than the original substrate and so they break down spontaneously, the enzyme being again liberated.

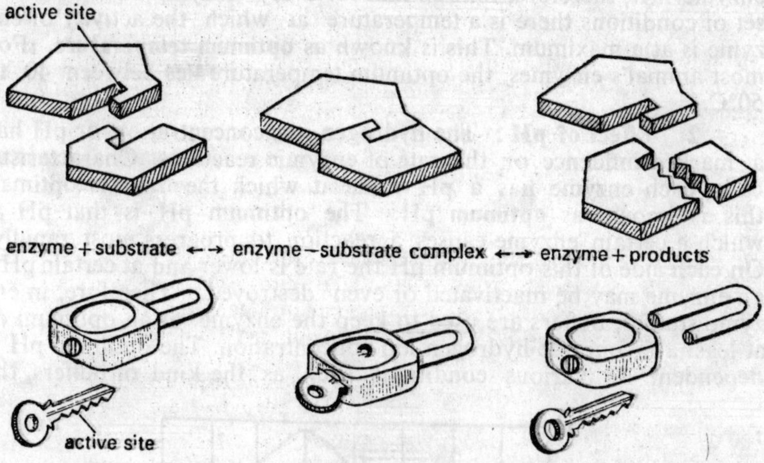

Fig. 6·1. A lock and key model for enzyme activity and specificity.

Recently in the year of 1958 FRUTON and SIMMONDE have also demonstrated that the **enzyme substrate** complex is an essential first step in enzyme action.

According to some workers the amino-acid residues comprising the active site of the enzyme are remained to be in direct contact with substrate molecule. Therefore, the size, shape and charge of the possible substrate are defined by the size, shape and charge of the active site, and it is these factors which determine the specificity of the enzyme towards certain substances only. Moreover there are three groups or constellations of amino-acids (perhaps more in some instances) on the enzyme which are actually involved and cause chemical changes. One group is responsible chiefly for binding the substrate to the enzyme, the second combining with another part of the substrate so as to reduce its stability and makes it more susceptible to reaction, while the third is responsible for specificity. This view has been developed particularly in connection with **peptidases, esterases** and other transferases.

EFFECT OF VARIOUS CONDITIONS ON ENZYME ACTIVITY

There are many factors which influence the enzyme activity. The important factors are the following :

1. Influence of temperature : Temperature affects enzyme cata-

lyzed reactions in the same way that it affects ordinary chemical reactions. As the temperature rises, the rate of a chemical reaction increases owing to an increase in the number of activated molecules. But when the temperature rises above a certain limit the enzymes looses its activity after destroying its tertiary structure. Similarly low temperatures, such as freezing temperatures, generally inactivate the enzyme. It, therefore, follows that for every enzyme under a given set of conditions there is a temperature at which the activity of enzyme is at a maximum. This is known as **optimum temperature.** For most animal's enzymes, the optimum temperature lies between 40 to 50°C.

2. **Effect of pH :** The hydrogen ion concentration or pH has a marked influence on the rate of enzymic reactions. Characteristically, each enzyme has a pH value at which the rate is optimal, this is known as **optimum** pH. The optimum pH is that pH at which a certain enzyme causes a reaction to progress most rapidly. On each side of this optimum pH the rate is lower and at certain pH's an enzyme may be inactivated or even destroyed. Therefore, in enzyme studies, buffers are used to keep the enzyme at an optimum or at least a favourable hydrogen ion concentration. The optimum pH is dependent on various conditions such as the kind of buffers, the

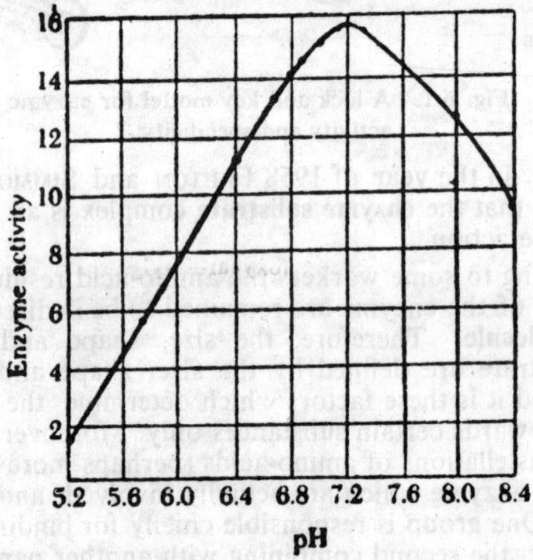

Fig. 6·2. **The effect of pH on enzyme activity. These data were obtained by determining the number of milligrams of reducing sugar formed in reaction mixtures in which the amounts of enzyme (pancreatic amylase) and substrate (soluble starch) were constant. The pH was adjusted by using phosphate buffers (KH_2PO_4/Na_2HPO_4) in which the different H^+ concentrations were attained by using different ratios of the acid and alkalies. (From MYERS, V.C,, and FREE, A.H., AMER, J. CLIN. PATH, 13 : 42, 1943).**

particular substrate, and the source of the enzyme may also have an influence. Table 6·1 shows the optimal pH value for some representative enzymes.

The influence of pH on enzymic reactions may involve several different types of effect. Enzymes, like other proteins, are ampholytes and possess many ionic groups. If the enzymic function depends on certain special groupings, these may have to be present in some instances in the un-ionized state and, in others, as ions. The ionic strength of the solution affects the velocity of enzyme reactions.

In some cases, the substrate are electrolytes, and the reaction may depend on a particular ionic or non-ionic form of the substrate.

Many enzymes are complex proteins having a non-protein part loosely bounded to the protein part. Since both moieties are essential for activity, conditions that influence the conjugation will destroy the enzymic activity, although this dissociation is usually reversible, *e.g.* peroxidase is split into its two components which are inactive at acid pH values ; readjustment of the pH solution to pH-7 restores the activity.

Table : 6·1. Optimal pH values for some enzymes.

Enzyme	Substrate	Optimum pH
Pepsin	Egg albumin	1·5
Pepsin	Casein	1·8
Pepsin	Haemoglobin	2·2
Pepsin	Benzyloxycarbonylglutamyltyrosine	4·0
Sucrase	Sucrose	6·2
Amylase	Starch	5·6—7·2
α-Glucosidase	α-Methylglucoside	5·4
α-Glucosidase	Maltose	7·0
Lipase (esterase)	Ethyl butyrate	7·0
Dipeptidase	Glycylleucine	7·3—8·1
Trypsin	Proteins	7·8
Pancreatic-α-amylase	Starch	6·7 to 7·2
Malt β-amylase	Starch	4·5

Enzyme	Substrate	Optimum pH
Carboxypeptidase	Various substrates	7·5
Urease	Urea	6·4 6 9
Phosphatase	Glycerophosphate	8·4
Catalase	Hydrogen peroxide	6·3—9·5
Arginase	Arginine	9·5—9·9
Xanthine oxidase	Xanthine	5·5– 8·5
Succinic dehydrogenase	Succinate	9·0

3. Concentration of enzyme : Because the enzyme itself enters into the reaction by combining temporarily with the substrate, the relative concentration of enzyme influences the velocity of the

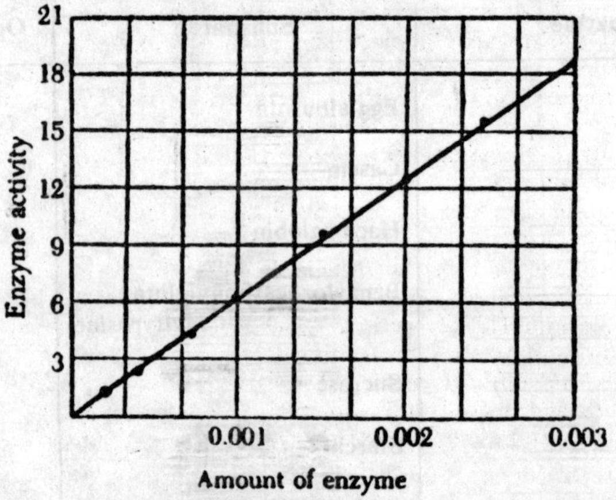

Fig. 6·3. Effect of enzyme concentration on the rate of a reaction when an excess of substrate is present. These data were obtained by determining the number of milligrams of reducing sugar formed in digestion mixtures containing different amounts of pancreatic amylase. The pancreatic amylase was supplied as duodenal contents, and the values of the abscissa indicate the number of milliliters of duodenal contents present in the digestion mixtures. The digestion mixture was buffered at an optimal pH and contained optimal amounts of chloride and substrate (From MYERS, V.C., and FREE, A.H., AMER J. CLIN. PATH. 13 : 42, 1943).

reaction. This means that the velocity of an enzyme reaction is directly proportional to the concentration of the enzyme, provided that the substrate is present in excess and is, therefore, not rate limting. This is true at the beginning of the reaction, but it may not hold true as the reaction continues, especially as the substrate is used up (Fig. 6·3).

4. Concentration of substrate : For a given quantity of enzyme, the velocity of the enzyme-catalyzed reaction increase as the concentration of substrate is increased, such a reaction is known as **first order**. At first, this relationship is almost linear but if the concentration of substrate is increased to a great extent than the rate of the reaction may prove independent of the substrate concentration. Such a reaction is said to be in a **zero order**. This means the enzyme becomes saturated with substrate or nearly so and further increase in substrate concentration has a negligible effect.

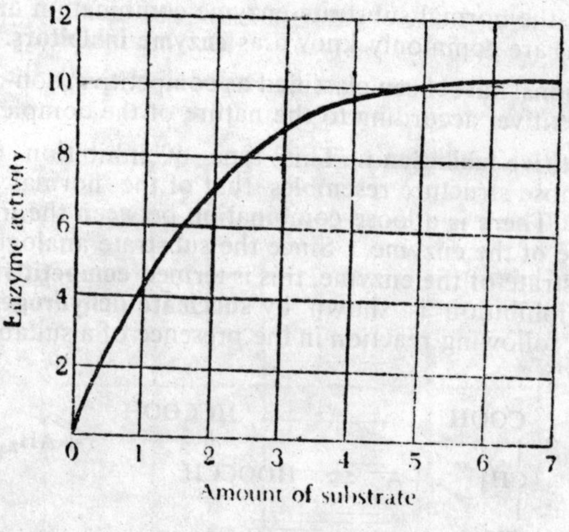

Fig. 6·4. Effect of substrate concentration on the rate of reaction when a limited amount of enzyme is present. Note that the enzyme becomes saturated with substrate so that further increase in concentration of substrate has no effect. These data were collected by determining the number of milligrams of reducing sugar formed in reaction mixtures in which the amount of enzyme (pancreatic amylase) was constant. The amount of substrate indicates the number of milliliters of 1% soluble starch present in the reaction mixtures, all of which had the same total volume (From MYERS, V C. and FREE, A.H., AMER, J. CLIN, PATH 13 : 42, 1943).

5. Other factors : In general enzymes are inactivated in strong light. Ultraviolet light is especially effective in destroying enzyme activity. X-rays may or may not destroy enzymes, depending on the circumstances. The presence of salts may influence the enzyme activity by participating with the enzyme due to its protein nature.

INHIBITION OF ENZYMES

The concept of enzyme action is that an enzyme reacts with its substrate to form an intermediate complex, *i.e.*, enzyme-substrate complex. Although the entire molecule of enzyme is required for its catalytic behaviour, there must be a small but definite locus on the surface of the enzyme molecule where the substrate can combine with the enzyme. In complex enzymes, the non-amino acid portion, or prosthetic group, is one place where the substrate may combine or react. In simple enzymes the places of combination or reactions are the **"active sites"** or **"active centres."**

Now it has been established that in many cases analogs of metabolic intermediates (antimetabolites) may inhibit enzyme activity. The compounds that can combine with an enzyme in a such a manner as to prevent the normal substrate-enzyme combination and the catalytic reaction, are commonly known as **enzyme inhibitors**.

Inhibitions have been classified as competitive, non-competitive, and uncompetitive, according to the nature of the complexes formed.

Competitive inhibition : This type of inhibition results from molecules whose structure resembles that of the normal substrate of the enzyme. There is a loose combination between the inhibitor and the active site of the enzyme. Since the substrate analogue competes with the substrate of the enzyme, this is termed **competitive inhibition**. This type of inhibition is shown by **succinate dehydrogenase,** which catalyzes the following reaction in the presence of a suitable hydrogen acceptor (A) :

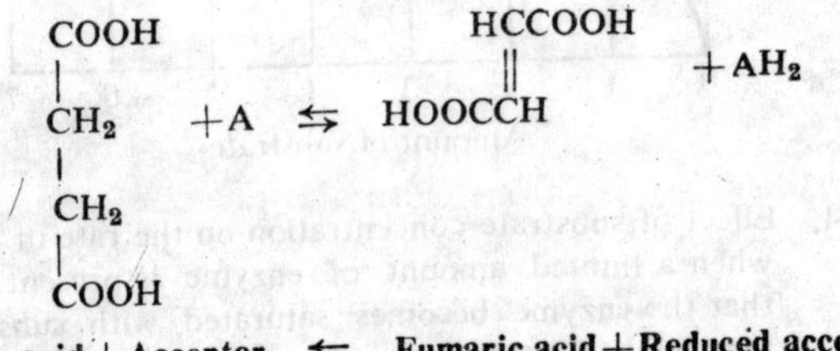

Succinic acid + Acceptor ⇆ Fumaric acid + Reduced acceptor

Many compounds that are structurally similar to succinic acid but which are not dehydrogenated can combine with the enzyme and inhibit catalysis of the normal reaction by blocking the active centres. Several compounds that inhibit succinate dehydrogenase are the following : Malonic acid, Oxalic acid, Glutaric acid, Phenylpropionic acid.

Competitive inhibition is reversible, that is, enzyme activity is recovered when the inhibitor molecule is removed from the enzyme. The actual rate of the catalyzed reaction is then strictly dependent on inhibitor concentration, substrate concentration, and relative affinities of inhibitor and substrate for the enzyme.

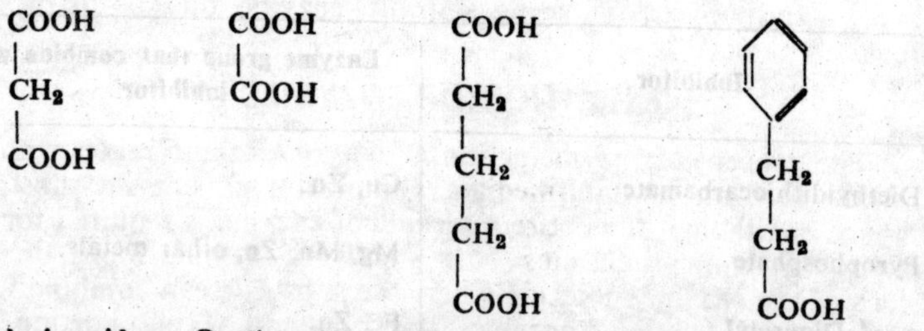

Malonic acid Oxalic acid Glutaric acid Phenylpropionic acid

Noncompetitive inhibition : In this instance, there is no relationship between the degree of inhibition and the concentration of substrate. This inhibition simply results from the combination of inhibitor molecule with a site on the enzyme other than the active centre. The velocity of the reaction depends only on inhibitor concentration and its affinity for the enzyme.

Examples of noncompetitive inhibition are the inhibition of many enzymes by heavy metal ions such as Ag^{+}, Hg^{2+}, Pb^{2+}, etc. Urease is extremely sensitive to traces of these ions.

Uncompetitive inhibition : In such a inhibition a complex is formed between enzyme substrate and the inhibitor molecule ; no enzyme-inhibitor complex is formed.

In many instances the product of the last of a chain of enzyme reactions is an inhibitor of an earlier step, and a feed back system is established that controls the rate of the whole sequence. Such feed back system is sometimes called **"end product inhibition"** (UMBARGER, 1964, ATKINSON, 1965, BURLAND *et al.*, 1965).

ALLOSTERIC EFFECT

The allosteric effect refers to alterations in enzyme activity caused by conformational changes brought about by activators or inhibitors. The activators or inhibitors are capable to change the ability of substrate to combine with the proteins (KOSHLAND, 1962, VAN ROSSUM, 1963, GOODWIN *et al.*, 1964).

Table : 6 2. Some commonly used enzyme inhibitors.

Inhibitor	Enzyme group that combines with inhibitor
Cyanide	Fe^{3+}, Cu, Zn, certain other metals.
Sulphide	various metals.
Fluoride	Ma, Ca, other metals.
Oxalate	Ca, Mg.
Carbon monoxide	Fe^{2+}, Cu.

Inhibitor	Enzyme group that combine with inhibitor
Diethyldithiocarbamate	Cu, Zn.
Pyrophosphate	Mg, Mn, **Zn,** othar metals.
α, α′-Dipyridyl	Fe, Zn.
Azide	Fe^{3+}-protoporphyrin enzymes.
Heavy metals (Ag^{+}, Hg^{3+} Pb^{2+}, etc.)	Sulphydryl ; may also cause non-specific protein precipitation.
Iodoacetate	Sulphydryl, imidazole, carboxyl, thiol ether.
Various arsenicals	Sulphydryl.

COENZYMES

Originally the term **coenzyme** was used to designate any substance like Ca, Co, Cu, Na, K and Zn ions whose presence in small amount is responsible for the action of a given enzyme. These simple substances are not the coenzymes of modern Enzymology but they are **metalloenzymes.** BERSIN simply called them **"complements".** In some enzymes the metal ions may be tightly bound to the proteins while in others they may be quite easily dissociated.

Many enzymes, but not all, possess in addition to their protein moiety, a non-proteinaceous organic component of relatively smaller molecular weight (about 500) called **coenzyme.** It is essential for enzymatic activity. The protein moiety constitutes the **apoenzyme.** The apoenzyme and coenzyme are collectively known as **holoenzyme.**

The coenzymes generally act as acceptor or donors of hydrogen atoms or other functional groups that are removed from or contributed to the substrate. Since these organic cofactors are frequently readily dissociable from the enzyme protein, they may properly be regarded as cosubstrates. A specific coenzyme is associated with each enzyme catalyzed reaction as a nonprotein **prosthetic** group. Alone, coenzymes can not perform any catalytic function, nor is the enzyme alone active. Thus, though the coenzyme is involved in the actual transfer of a functional group or a hydrogen atom, the protein component of enzyme is required for catalytic activity. The enzyme determines the proper alignment of all the reactants and thereby orients the coenzyme for specific reaction.

Coenzymes are present in all cells in minute amounts and like enzymes, they function over and over again. Hydrogen atoms picked up in one reaction are passed on to another molecule in another reaction, generating the coenzyme. Plants are capable to synthesize the required coenzymes while the higher animal can not.

Analysis of various coenzymes has shown that they all either are vitamins or contain vitamins as component of their structure. The important coenzymes are listed in table 6·3.

Table : 6·3. Some organic coenzymes and prosthetic groups of enzymes.

Coenzyme or Prosthetic group	Enzymic and other functions	Essential nutritional factor or vitamin
Adenosine triphosphate (ATP)	Transphosphorylation	None
Ascorbic acid	Hydroxylation reactions	Ascorbic acid
Biotin	CO_2 transfer	Biotin
Cobamide	Group transfer	Cobalamine
Coenzyme A (CoA)	Acetyl or other acyl group transfer : fatty acid synthesis and oxidation	Pantothenic acid
Flavin adenine dinucleotide (FAD)	Hydrogen carrier	Riboflavin
Flavin mononucleotide (FMN)	Hydrogen carrier	Riboflavin
Iron-sulphide chromophore	Electron transport	None
Nicotinamide adenine dinucleotide (NAD)	Hydrogen carrier	Nicotinic acid
Nicotinamide adenine dinucleotide phosphate (NADP)	Hydrogen carrier	Nicotinic acid
Pyridoxalphosphate	Aminotransferases, amino acid decarboxylases racemáses, etc.	Pyridoxine
Lipoic acid	Oxidative decarboxyla-	Required by some micro-

Coenzyme or Prosthetic group	Enzymic and other functions	Essential nutritional factor or vitamin
	tion ; as hydrogen and acyl acceptor	organism
Tetrahydrofolic acid	One-carbon transfer	Folic acid
Thiamine pyrophosphate	Oxidative decarboxylation, active aldehyde carrier	Thiamine
Ubiquinone	Hydrogen carrier	None

PROENZYME

When cells secrete enzymes, these are often first, produced in an inactive form. These inactive forms of enzymes are said to be the **"proenzymes"**. For example, the gastric glands of stomach secrete inactive **pepsinogen** which is later converted into active pepsin, upon the addition of hydrochloric acid. The reaction is **autocatalytic** and is due to primarily to the action of pepsin of itself. Pepsinogen, as it produces pepsin is called **"proenzyme"**. These, in the older literature were called **"zymogens"**. The change from inactive proenzyme to enzyme apparently involves some alternations in the protein molecules.

ANTIENZYMES

According to Bersin **"anti-enzymes"** are those proteins or protein-like substances which are produced in the body of living organism, following the injection of enzymes or proenzymes, and inhibit the catalytic action of the enzyme molecules. The antienzymes are thought to produce inhibition as a result of some surface reaction with the enzyme molecules. These substances mainly include antipepsin, antitrypsin and antiamylase, etc.

ISOENZYMES

There are certain enzymes which exist in tissues in two or more components, with the same catalytic activity by chemically, immunologically and electrophoretically distinct. These are known as **"isoenzymes"** or **"isozymes"**. For example human blood-plasma contains at least five isoenzym of lactatees dehydrogenase which can readily be separated by **electrophoresis**. The isoenzymes have relation with the heredity (Latner and Skillen, 1969, and Weyer, 1868).

CLASSIFICATION OF ENZYMES

By the early 1960s more than seven hundred enzymes had been

recognised and classified and more are being added to this total each year. While there have been various attempts to standardise the nomenclature and classification of enzymes, the position is still rather flexible.

For many years it was the custom to classify enzymes solely by the reactions which they catalyze but now they are classified not merely in terms of what they do but also in terms of their true chemical composition. According to one classification the enzymes, that are indispensable in living organisms are classified into the follow ing classes :

1. Hydrolases
2. Oxidoreductases
3. Transferases
4. Isomerases
5. Hydrases
6. Coagulative enzymes
7. Lysases and
8. Ligases or synthetases.

1. HYDROLASES

These enzymes bring about the hydrolysis or condensation of substrates by the addition or removal of water.

$$AB + H.OH \underset{\text{Condensation}}{\overset{\text{Hydrolysis}}{\rightleftarrows}} AH + B.OH$$

These are more familiar as the intestinal juices of animals. They also occur in seeds and other parts of plants. Hydrolases are further sub-divided, according to the class of substrates on which they act, as follows :

(i) *Proteases* or *Proteolytic enzymes* : These enzymes hydrolyse proteins into proteoses and peptones and some still further into polypeptides and amino acids. During their course of action they simply catalyze the breaking of the peptide linkages which join amino acids, —CO.NH to —COOH and NH_2. Those proteases such as **pepsin** and **trypsin** which act on peptide linkages within the body of the protein are termed **endopeptidases** while o hers, the **exopeptidases**, hydrolyze terminal peptide links. The latter group consists of **amino-peptidases** which attack peptide links next to the NH_2 end of the molecule and **carboxypeptidases** which hydrolyse the links at the —COOH end. These exopeptidases tend to act on smaller units such as the **tri** and **dipeptides** which themselves result from endopeptidase activity on proteins. Some important proteases or proteolytic enzymes are as follows :

(a) **Pepsin :** It is the enzyme of gastric juice which hydrolyses the proteins into proteoses and peptones.

(b) **Trypsinogen** and **Chymotrypsinogen :**—These are inactive enzymes of the pancreatic juice which after activation are converted into trypsin and chymotrypsin respectively. These split up proteins into the stage of polypeptides.

(c) **Erepsin :** It is the enzyme of intestinal juice which splits up polypeptides into amino acids.

(d) **Papain :** It is a vegetable enzyme found in *Carciapapaya* ordinarily hydrolyses proteins to the peptone stage but in the presence of traces of HCN, the action may go even upto polypeptide stage.

(ii) *Amylolytic enzymes (amylum,* **starch**-*lysis to* **breakdown***)* : These enzymes hydrolyse the starch into dextrin and finally into maltose. Some known amylolytic enzymes are as follows :

(a) **Diastase :** It converts starch into maltose.

(b) **Ptyalin :** It is the enzyme of salivary juice which converts the starch into maltose.

(c) **Amylopsin** or **Pancreatic amylase :** It hydrolyses boiled or unboiled starch into maltose.

(d) **Glucogenase :** It in the liver and muscles converts glucose into glycogen and very possibly the same enzyme by reversible reaction converts glycogen into glucose.

(e) **Phosphorylase :** It helps the first stage of breakdown of glycogen and also to synthesize a polysaccharide undistinguishable from glycogen, so its action is also reversible.

Glycogen or starch + inorganic phosphate ⇆ glucose-1-phosphate (coriester)

(f) **Nucleases :** These hydrolyse nucleic acids into nucleotides which again are hydrolysed by **nucleotidases** into nucleosides and the latter once again by **nucleosidases** into pentose sugars, bases and phosphates out of which they are made.

(iii) *Amidases :* These enzymes hydrolyze amides and include **urease, arginase** and **purinamidases.** These act on urea, arginine and purine bodies respectively to carry further dissociation of these products.

(iv) *Kathepsins :* These serve like proteinases in the cells of animal tissues.

(v) *Lipolytic enzymes :* They breakdown neutral fats into glycerol and fatty acids, some known lipolytic enzymes are as follows :

(a) **Gastric lipase :** It splits up small amount of fat emulsified by mucin in the stomach. It also splits milk fat.

(b) **Steapsin** or **Pancreatic lipase :** This enzyme splits emulsified fat into glycerol and fatty acids.

(c) **Intestinal lipase :** This enzyme splits phosphatides, as lecithins, into glycerol, phosphoric acid, fatty acids and choline.

(d) **Esterase** and **phosphatases :** These enzymes act on simple esters and phosphoric esters respectively in ossification of bone and many other processes.

(vi) *Invertase :* It is known as **inverting enzyme** which transforms disaccharides into monosaccharides.

Maltase and **succus-entericus** also break up maltose into molecules of glucose in the same way as invertase transforms disaccharides into monosaccharides.

2. OXIREDUCTASES

These enzymes allow oxidation or reduction reactions to take place. They are essential for both aerobic and anaerobic respiration. They are of following types :

1. **Dehydrogenases :** These enzymes remove the hydrogen from substrate, but only in the presence of suitable hydrogen acceptor; some of them called **reductases.** An example is the action of **ethanol dehydrogenase**, in which nicotinamide adenine dinucleotide (NAD) is the hydrogen acceptor :

$$\text{Ethanol} + NAD^+ \leftrightarrows \text{acetaldehyde} + NADH + H^+$$

2. **Oxidases :** These have O_2 as hydrogen acceptor, *e. g.*, aldehyde oxidase :

$$\text{Aldehyde} + H_2O + O_2 \leftrightarrows \text{Acid} + H_2O_2$$

3. TRANSFERASES

These enzymes catalyze the transfer of specific group from one substance to another. They may be of following types :

1. Those transfering 1-carbon groups, *e. g.*, **methyl transferases.**
2. Those transferring aldehyde or ketonic groups, *e. g.*, **transketolase.**
3. Those transferring acyl groups, *e. g.*, **amino acyl transferase.**
4. Those transferring sugar groups, *e.g.*, **sucrose-1 fructosyl transferase.**
5. Others—Those transferring alkyl, nitrogenous, phosphorus containing and sulphur containing groups.

4. ISOMERASES

These enzymes catalyze intramolecular rearrangement, *e. g.*, the interconversion of aldose and ketose sugars. For example, **phosphohexose isomerase** catalyzes the following inter-conversion :

$$\text{Glucose-6-phosphate} \leftrightarrows \text{Fructose-6-phosphate}$$

5. HYDRASES

These enzymes catalyze the addition to or removal of water from their specific substances. **Fumarase** catalyzes the interconversion of malic and fumaric acids :

$$\underset{\text{Malic acid}}{HOOC{-}CH_2{-}CHOH{-}COOH} \leftrightarrows \underset{\text{Fumaric acid}}{HOOC{-}CH{=}CH{-}COOH} + H_2O$$

6. COAGULATIVE ENZYMES

Rennin and **Rennet** are the best known co-agulative enzymes which convert soluble calcium-caseinogen into insoluble calcium caseinate.

7. LYSASES

These enzymes split groups from their substrate (not by hydrolysis), leaving double bonds, or conversely, add groups to double bonds. These act on C—C, C—O, C—N, C—S and C—halide bonds.

8. LIGASES OR SYNTHETASES

These catalyze union of two molecules, coupled with breakdown of a pyrophosphate bond in ATP or similar triphosphate. They form C—O, C—S, C—N, or C–C bonds, *e.g*, pyruvate carboxylase, which carboxylates pyruvate (using ATP) to form oxaloacetate.

Revision Questions

1. What are enzymes ? Describe in brief the physical and chemical properties of them.
2. Classify the enzymes and also give suitable examples for its support.
3. What is enzyme specificity. Give several factors that influence the enzyme activity.
4. Describe the mechanism of enzyme action.
5. Write short notes on :
   - (i) Allosteric effect,
   - (ii) Inhibition of enzymes,
   - (iii) Coenzymes,
   - (iv) Proenzymes,
   - (v) Isoenzymes,
   - (vi) Antienzymes,

CHAPTER 7

METABOLISM

The term **metabolism** which literally means *"change"* is used to refer to all the enzyme catalyzed chemical reactions taking place within an organism or its individual cells, whether they are involved in growth, reproduction, tissue repair or energy production. It is one of the principal vital functions of the organism. The reactions or changes which are concerned with production of heat to maintain the body temperature and supplying of energy for other vital activities constitute **"energy metabolism"** in contrast to **"intermediary metabolism"** which is concerned with specific chemical reactions or changes through which the end products of digestion are passed before they could be fully utilized by the body.

The chemical reactions or changes which undergo during the process of metabolism can be divided under two categories. Those reactions concerned with the synthesis of cell constituents and cell products from simpler substances are referred to collectively as **anabolism**. In general. these do not supply the energy needed by the cell. Indeed, they are usually, endergonic reactions requiring an input of energy. Energy production is the function of second category of metabolic reactions, collectively called **catabolism**. In catabolism complex molecules are broken down to simpler ones by exergonic processes that ultimately involve energy liberating oxidative reactions.

In metabolism the two categories, *e.g.*, **anabolism** and **catabolism** are insolubly connected as there is continuous building up and breaking down of organic compounds. Moreover in nature anabolism is always in excess when compared with catabolism, it is only because of this, the growth in animals and plants is effected.

Complex substances which become part of the cells and intercellular structures are continuously being formed in the living organisms. At the same time complex organic substances are being broken down into simpler ones. The final products of metabolism which can not be transformed by the organism are eliminated through the specific organs called excretory organs.

Metabolism involves proteins, carbohydrates, fats, water and mineral salts. In addition to these substances the organism requires vitamins.

Metabolism is subject to humoral regulation and regulation by the nervous system. The fact that metabolism in the organs and tissues is controlled by the nervous system was established by IVAN PAVLOV and confirmed by BYKOV and others. The influence of nervous system is called trophic influence. The humoral influence on metabolism is exerted through blood by hormones secreted by the endocrine glands.

METABOLIC RATE

The metabolic rate is usually expressed in **Calories** per kilogram weight of the body per hour. It may be determined either from the quantity of carbon dioxide eliminated or the quantity of oxygen consumed over a definite period of time.

The metabolic rate is affected by muscular activity, low environmental temperatures and ingestion of food. It may be lowered in hypothyroidism, malnutrition. pituitary disorder and deficiences of adernal cortex.

All metabolic reactions take place because of favourable factors such as temperatures, pH, enzymes and essential co-enzymes and co-factors. Nearly 600 enzymes have been identified so far which take part in metabolic reactions.

GENERAL METABOLISM

General metabolism includes the total energy exchanges of a cell or organism as a whole.

BASAL METABOLISM

Basal metabolism refers to the minimum energy required to maintain the normal activities of the body during fasting or at complete rest in a warm atmosphere 12–18 hours after the intake of food. The basal metabolism of an adult ranges from 1000 to 2000 Calories a day for man and from 1000 to 1700 Calories for woman. In addition to sex differences, basal metabolism also depends upon the person's weight, height and age.

The rate of basal metabolism can not be determined easily. It can be determined only by two factors : (i) the inherent rates of chemical reactions of the cells which relatively remain constant in different animals and (ii) the activity of thyroid hormone on the cells. The rate of basal metabolism is influenced by age, sex, body weight health and internal secretions of the body.

ENERGY METABOLISM

In organisms energy is continuously required in order to maintain the various activities of the body, such as muscular work secretory activity of glands and nervous activity, etc. It is simultaneously and continuously formed. The organism derives it from the breakdown of food which organism consumes from the external environment. Under the action of tissue enzymes the food is broken down and transformed into simpler substances. During this transformation certain amount of energy is liberated in the form of chemical energy which is consumed by the organism. The various activities involving the expenditure of internal energy is accompanied by the liberation of heat. So, all these reactions which are concerned with energy and heat productions constitute **energy metabolism**. The intensity of energy metabolism can only be measured by the amount of heat formed in the body of organism.

SPECIAL METABOLISM

Special metabolism includes changes undergone by a particular substance or group of substances in the body of the animal. For. example, the metabolism of proteins, carbohydrates and of fats.

PROTEIN METABOLISM

Proteins are the essential constituents of the protoplasm of cells and also of intercellular substances. The huge variety of different proteins imparts to each species its own specific characteristics, and each organ within an animal differs in the nature of its protein content. All the enzymes without which the metabolic processes can not go on, are protein bodies. Proteins are also present in the form of hormones which play vital roles in intracellular chemical processes. The phenomena of muscular contraction are also connected with proteins, myosin and actin. The blood contains oxygen carriers, they are also protein in nature. In the red blood cells, the conjugated protein haemoglobin is the means of transporting oxygen to the cells and helps in regulating acid-base balance in the body. Proteins of blood plasma serve a number of purposes including regulation of water, electrolyte balance, providing antibodies. The proteins of the body, like carbohydrates and lipids, serve as a source of energy.

Proteins are highly specific. Each organism and each tissue have proteins which differ from those that form part of other organisms and other tissues.

The main significance of proteins is that they provide the material from which the cells and intercellular substances are built and substances which take part in the regulation of physiological functions, are synthesized. However, proteins are in some measure used along with carbohydrates and fats to cover the energy output. This shows that proteins are essential compounds from the biological point of view.

During the process of digestion proteins contained in the food are hydrolyzed to α-amino acids under the influence of proteolytic enzymes found in the digestive tract. Phosphoric acid and nucleic acid (subject to further break down) may also be formed, but the α-amino acids are the primary units liberated. The requirement for proteins is fundamentally the requirement for amino acids because proteins are rebuilt from the amino acids. It should be possible to express the protein's requirement of an animal in terms of the amount and kind of amino-acids rather than the protein itself.

The α-amino acids derived from the digestion of protein are soluble in water in the ionic form. They are normally absorbed through the lumen of the small intestine and carried directly to the liver via the portal blood stream where the necessary proteins can be synthesized from the α-amino acids and sent, via the blood stream, as plasma protein to whichever part of the body needs them. There is also evidence that α-amino acids circulate as such in the blood stream and may go in this simple form to the various tissues. By a process of rigid selection, the various parts of the body take or synthesize the proteins of particular shape and composition that they need and incorporate them as tissue.

Amino acids like carbohydrates and fats are not stored in the body. However, after a rich protein meal at the end of a fast, temporary storage of amino-acids may take place in the liver and in other tissues and is followed by a limited increase in protein in the tissues which may be regarded as restoration of protein lost during fasting.

Amino acids, which are not required for growth, tissue repair or the construction of important secretions and enzymes pass into the fuel system so that they may be metabolized to liberate energy. Their fate depends upon the amount of protein in the diet, the age of the animal and the other sources of fuel available to it. In emergencies amino acids may be withdrawn from the protoplasm of the cell and thus during starvation the cellular protein content declines.

Different tissues synthesize their respective proteins from the amino acids available in blood stream. A certain proportions is utilized towards the renewal of wornout protoplasm where as the surplus amino acids are simplified into excretable nitrogenous wastes after undergoing deamination and transamination processes. They are broken down to (i) the non-nitrogenous part and (ii) the nitrogenous part. The non-nitrogenous part enters the KREBS **cycle** and is oxidized to CO_2 and H_2O, thus furnishing energy to the body or is converted into body fat where as the nitrogenous part goes to form urea or uric acid or ammonia which is excreted in the form of urine in animals.

During the metabolism of amino acids there is an increase in the production of heat by the body which RUBNER referred to as **specific dynamic action**. It is apparently due to reactions which the amino acids undergo in the liver, or to the stimulation of other reactions in the cells by products formed in the liver. The phenomenon of specific dynamic action does not occur after removal of the liver.

Nitrogen balance and equilibrium : The relation between the amount of nitrogen entering the body from the diet in the form of amino acids, and the amount of nitrogen excreted from the body in the form of metabolic end products (chiefly as urea, but to some extent as uric acid, creatine, etc.) is known as the **nitrogen balance**. The nitrogen atoms entering the body in the food are not the same nitrogen atoms leaving the body because the nitrogen atoms entering the body become incorporated in the body for a certain interval of time. The proteins already present in the body break down and their nitrogen is converted into urea, etc., and is excreted.

If the amount of nitrogen entering the body is greater than the amount of the nitrogen leaving the body, this is what is known as **positive nitrogen balance**. This means that some nitrogen must be staying and accumulating in the body, presumably to build new tissue. This is a common occurrence in the growing child. The body in this case is building up tissue faster than breaking it down.

If the amout of nitrogen entering the body is less than the amount of nitrogen leaving the body, this is known as **negative nitrogen balance**. This means that the body is tearing itself down faster than body tissues are being built up. This condition would prevail during a wasting disease, starvation or under conditions whereby the proper protein diet is lacking.

If the amount of nitrogen entering the body and the amount of nitrogen leaving the body are essentially equal, **nitrogen equilibrium** results.

Dynamic equilibrium of body proteins : Researches on the stability of proteins have shown that the tissues of the body are constantly broken down and the nitrogen eliminated, and the new tissues are synthesized from the proteins already present in the plasma (dynamic equilibrium). The saying mean is that amino acids are constantly liberated into the blood from tissue proteins which are brokendown during normal metabolism, and to a much greater extent, in starvation, in chronic wasting diseases, in fever and in other states of increased metabolism. Such amino acids together with those derived from the food form a common stock or "**metabolic pool**" in the blood and tissues. In this metabolic pool it is not possible to distinguish between amino acids derived from dietary sources and those derived from protein catabolism. The pool simply serves as a source of amino acids some of which may be used whatever their origin, to build nitrogenous substances such as proteins of tissues or certain hormones, while the vast majority are degraded by a process to be described later, their nitrogen being excreted as urea. This means that every part of the body undergoes a complete protein overhaul every few months.

The fact, that an animal kept on a protein-free-diet will continue to eliminate nitrogen in urine and faeces, indicates that this breakdown process may go on even though no new nitrogen compounds enter the body. This shows that the body will continue to have plasma protein despite a protein free diet. The plasma protein must come from easily expendible tissue, the location of which is unknown. This tissue is of a reserve nature, but it is not like glycogen or fat depots. The reserve supply is not inexhaustible, but the body will continue to form plasma protein for a rather long period.

GENERAL TRANSFORMATION OF AMINO ACIDS

The general metabolic reactions which the α-amino acids undergo during their mode of metabolism are as follows :

1. **Deamination :** Deamination is the initial process in the catabolism of amino acids. It usually takes place in the liver and kidney tissues. During this process the amino group is removed from the α-amino acid to form **α-keto acid**. The disposition of α-keto acid is almost an individual process, differing in a manner dependent on to amino acid from which it was derived. For the most part the amino group is converted to urea in man and excreted. It may also be excreted as ammonia or as other nitrogenous waste products. The entire process is caused by several enzymes which are either of oxidizing or of reducing type.

2. Oxidative deamination : In some cases, the deamination process may be accompanied by an oxidation reaction such a combination is referred to as **"oxidative deamination"**. This process, although, prominent in tissues, is by no means the only way in which amino groups are removed from the donor amino acids. A series of non-oxidative deamination enzymes are associated with certain specific amino acids such as the hydroxy amino acids (dehydrase) the sulphur containing amino acids (desulphydrases), histidine (histidase), tryptophan (tryptophanase) and others.

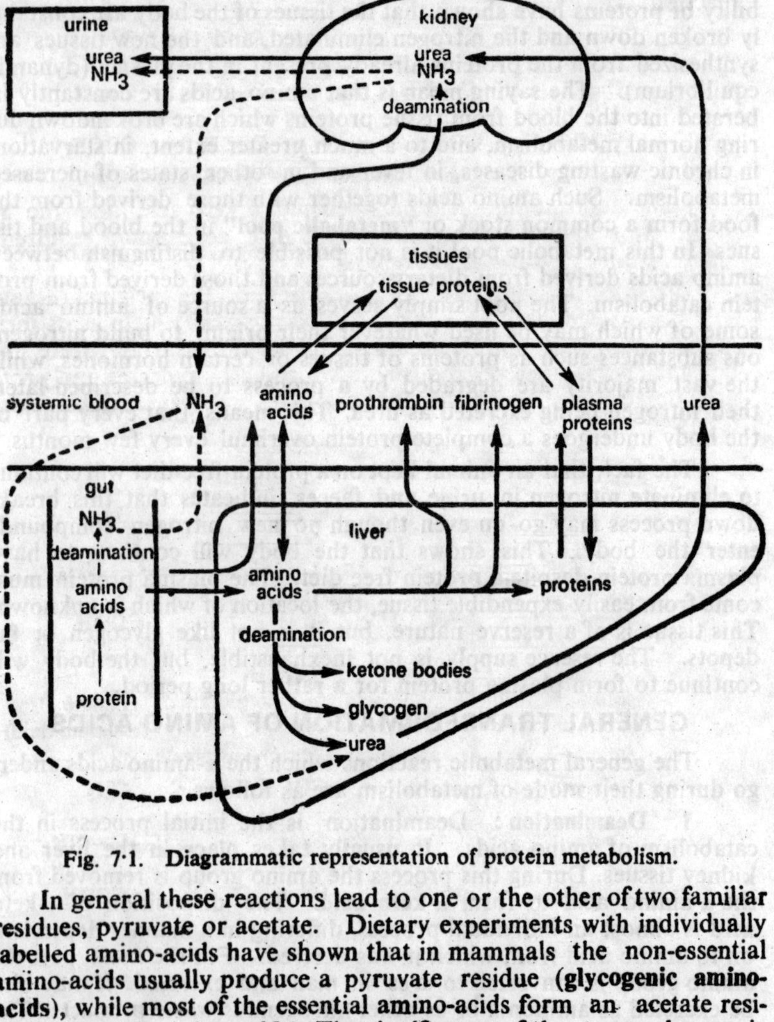

Fig. 7·1. **Diagrammatic representation of protein metabolism.**

In general these reactions lead to one or the other of two familiar residues, pyruvate or acetate. Dietary experiments with individually labelled amino-acids have shown that in mammals the non-essential amino-acids usually produce a pyruvate residue (**glycogenic amino-acids**), while most of the essential amino-acids form an acetate residue (**ketogenic amino-acids**). The significance of the terms glycogenic and ketogenic is obvious. The first group of amino-acids in large quantities leads to the production and storage of carbohydrate, while fat metabolism is emphasized in the case of the ketogenic group. Through deamination at least half of the common amino-acids are reduced to these two familiar carbon residues, and from this point, as acetyl CoA, they can. enter the energy-yielding reactions of the citric acid cycle.

3. **Transamination :** Transamination is a process which brings about the transfer of the amino group from the donor amino-acid to a recipient keto acid, under the influence of a **transaminase** or **amino transferase** enzyme. This is represented by the following chemical reaction.

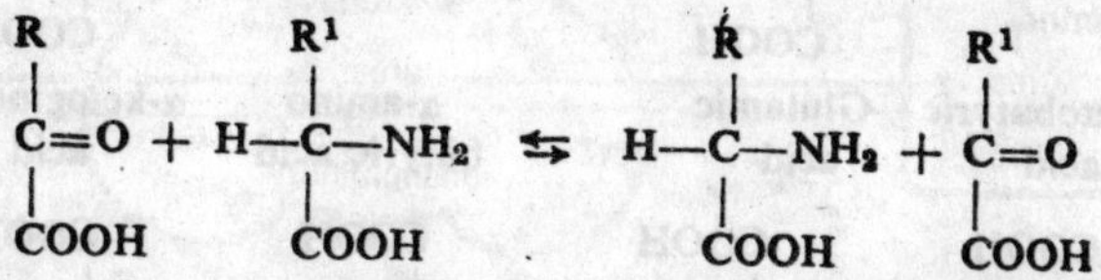

Fig. 7·2. A chemical reaction showing transamination.

The donor amino-acid thus becomes a keto acid and the recipient keto acid becomes an amino-acid. The co-enzyme required for this reaction is **pyridoxal phosphate.**

In both transamination and deamination the original α-amino-acid is converted into an α-keto acid. Most of the α-keto acids formed in these conversions are compounds involved in carbohydrate metabolism. This is precisely the way in which carbohydrates and proteins are interrelated. Some specific examples of transamination are as follows :

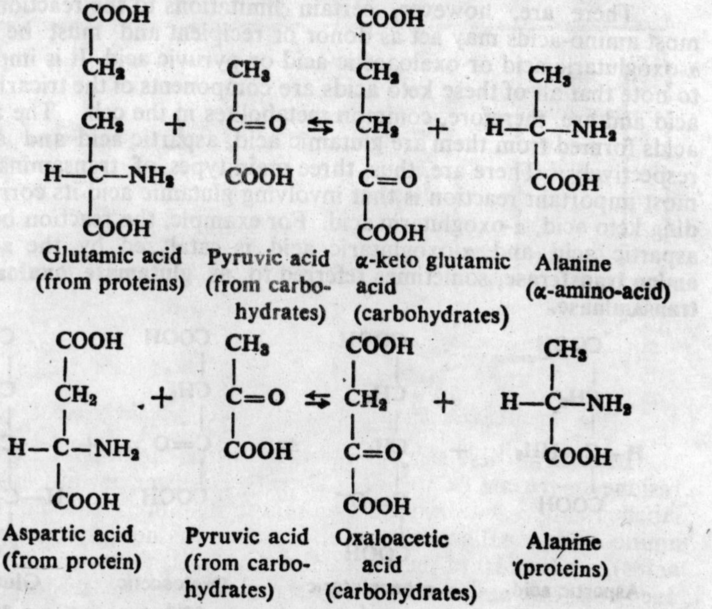

Fig. 7·3. Chemical reactions showing the process of transamination.

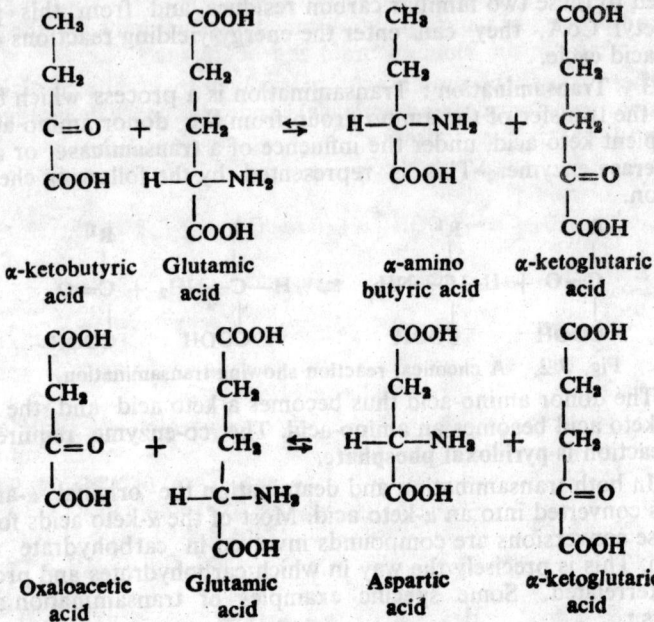

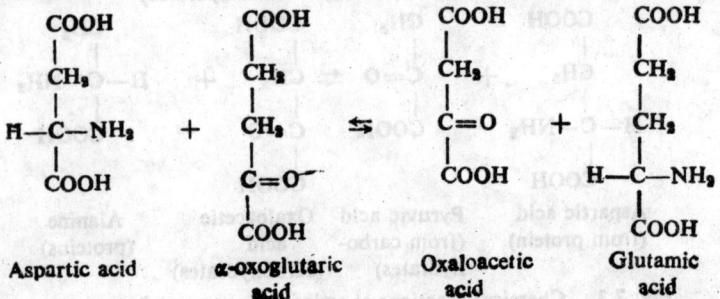

α-keto acids other than pyruvic acid may react with glutamic acid. Two such compounds are α-keto butyric acid and oxaloacetic acid.

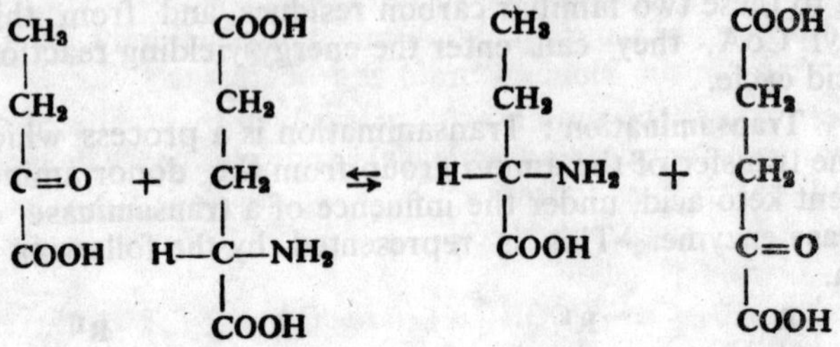

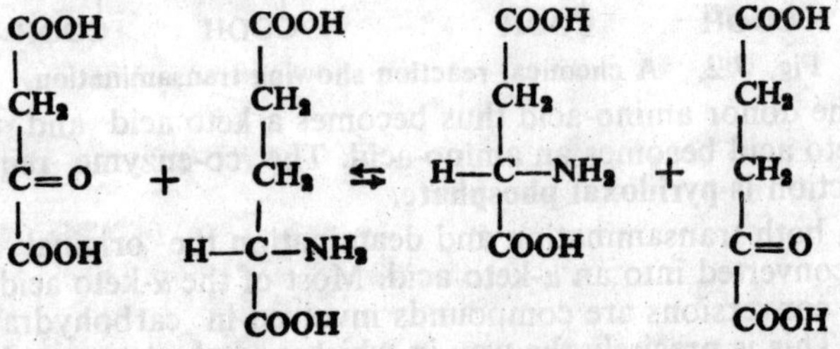

$$
\begin{array}{ccccc}
\underset{|}{CH_3} & \underset{|}{COOH} & & \underset{|}{CH_3} & \underset{|}{COOH} \\
\underset{|}{CH_2} & \underset{|}{CH_2} & & \underset{|}{CH_2} & \underset{|}{CH_2} \\
\underset{|}{C=O} \; + & \underset{|}{CH_2} & \rightleftharpoons \quad H-\underset{|}{C}-NH_2 \; + & \underset{|}{CH_2} \\
COOH & H-\underset{|}{C}-NH_2 & COOH & \underset{|}{C=O} \\
& COOH & & COOH
\end{array}
$$

α-ketobutyric   Glutamic   α-amino   α-ketoglutaric
   acid   acid   butyric acid   acid

$$
\begin{array}{ccccc}
\underset{|}{COOH} & \underset{|}{COOH} & & \underset{|}{COOH} & \underset{|}{COOH} \\
\underset{|}{CH_2} & \underset{|}{CH_2} & & \underset{|}{CH_2} & \underset{|}{CH_2} \\
\underset{|}{C=O} \; + & \underset{|}{CH_2} & \rightleftharpoons \quad H-\underset{|}{C}-NH_2 \; + & \underset{|}{CH_2} \\
COOH & H-\underset{|}{C}-NH_2 & COOH & \underset{|}{C=O} \\
& COOH & & COOH
\end{array}
$$

Oxaloacetic   Glutamic   Aspartic   α-ketoglutaric
   acid   acid   acid   acid

There are, however, certain limitations to the reaction while most amino-acids may act as donor or recipient and must be either α-oxoglutaric acid or oxaloacetic acid or pyruvic acid. It is important to note that all of these keto acids are components of the tricarboxylic acid and are, therefore, common metabolites in the cell. The amino-acids formed from them are glutamic acid, aspartic acid and alanine respectively. There are, thus, three main types of transaminase, the most important reaction is that involving glutamic acid its corresponding keto acid, α-oxoglutaric acid. For example, the reaction between aspartic acid and α-oxoglutaric acid is catalyzed by the aspartic amino transferase, sometimes referred to as glutamate oxaloacetate transaminase.

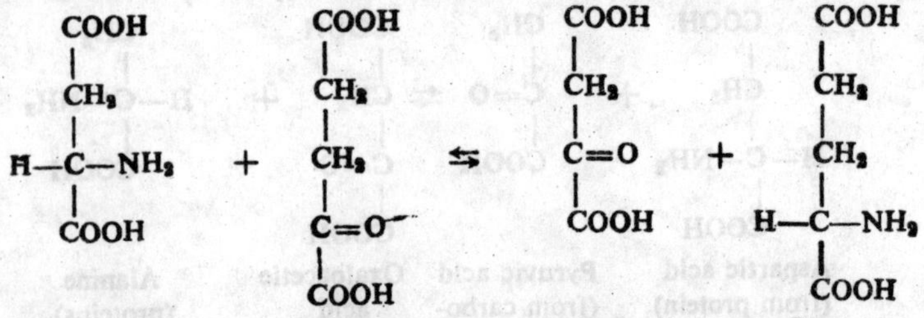

$$
\begin{array}{ccccc}
\underset{|}{COOH} & \underset{|}{COOH} & & \underset{|}{COOH} & \underset{|}{COOH} \\
\underset{|}{CH_2} & \underset{|}{CH_2} & & \underset{|}{CH_2} & \underset{|}{CH_2} \\
H-\underset{|}{C}-NH_2 \; + & \underset{|}{CH_2} & \rightleftharpoons \quad \underset{|}{C=O} \; + & \underset{|}{CH_2} \\
COOH & \underset{|}{C=O} & COOH & H-\underset{|}{C}-NH_2 \\
& COOH & & COOH
\end{array}
$$

Aspartic acid   α-oxoglutaric   Oxaloacetic   Glutamic
                 acid   acid   acid

These reactions reveal that transamination represents the major mechanism leading toward the eventual disposition of nitrogen as waste products and also results in the production of carbon compounds which may be metabolized for energy purpose.

The process of transamination takes place chiefly in the liver but also occurs in the kidneys, brain and heart, etc.

The process of amino-acid catabolism by combined action of an **amino transferase (transaminase)** and **glutamate dehydrogenase** may be summarized as follows :

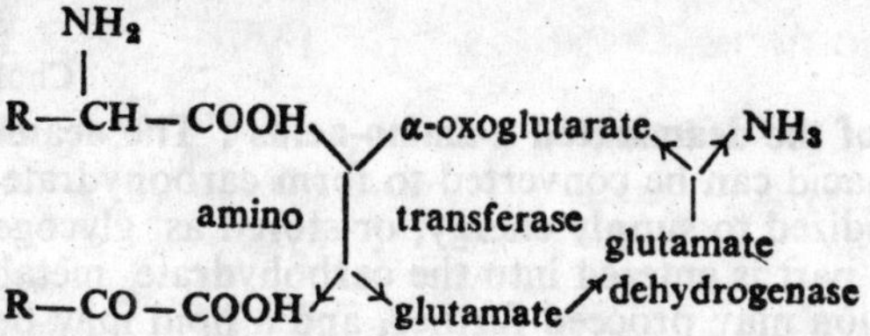

Fig. 7·4. A chemical reaction showing the mode of catabolism of amino-acids.

4. Decarboxylation : It is a process by which —COOH group leaves a primary amine as CO_2 and is probably not an important pathway for amino-acid metabolism in man. This process is carried out with the help of decarboxylases which are found in a variety of animal tissues. Pyridoxal phosphate is required as a co-factor.

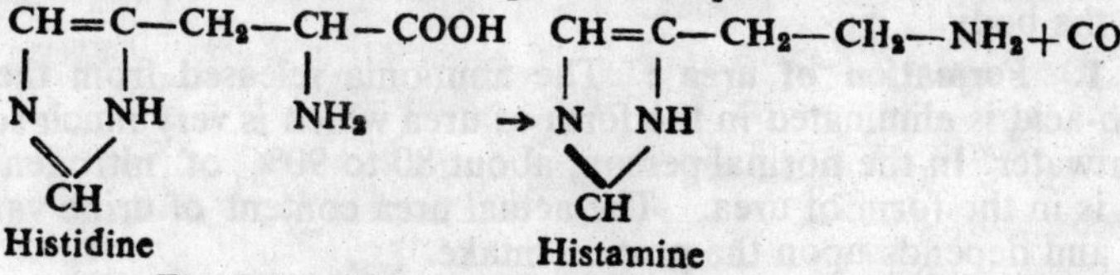

Fig. 7·5. A chemical reaction showing the process of decarboxylation.

5. Transmethylation : Transmethylation is the process whereby methyl groups are transferred from one compound to another. In this way, the body can synthesize some essential compounds. Methionine is an efficient supplier of methyl groups.

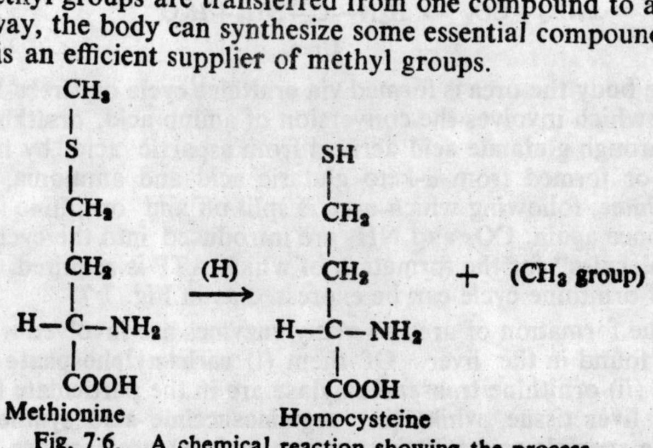

Fig. 7·6. A chemical reaction showing the process of transmethylation.

These methyl groups may be used in the formation of choline, creatine and other important compounds needed by the body, for example :

$$HO.CH_2.CH_2-NH_2 + (CH_3\ groups) \xrightarrow{\text{may be converted into}} \left[HO.CH_2.CH_2-\overset{\displaystyle CH_3}{\underset{\displaystyle CH_3}{\overset{|}{\underset{|}{N}}}}-CH_3\right]^{+}\overset{-}{OH}$$

Choline

Fate of the deaminated α-amino-acids : The deaminated part of the amino-acid can be converted to form carbohydrate such as glucose, or oxidized to supply energy, or stored as glycogen. Once the deaminated part is entered into the carbohydrate metabolic pathway the conversion may proceed further, and a lipid may be formed and stored as depot fat. The deaminated part may be reconverted into another α-amino-acid via carbohydrate pathway and resynthesized into a particular protein. The various deaminated portions may be decomposed further and excreted in several forms.

Fate of the nitrogen : In the deamination process the nitrogen in the form of amino group is released which undergoes various chemical reactions to form nitrogen wastes so that they may be removed from the body.

1. **Formation of urea :** The ammonia released from the α-amino-acid is eliminated in the form of urea which is very much soluble in water. In the normal person, about 80 to 90% of nitrogen in urine is in the form of urea. The actual urea content of urine varies daily and depends upon the protein intake.

A simple reaction representing the urea formation is as follows :

$$2NH_3 + CO_2 \rightarrow H_2N-\overset{\displaystyle O}{\overset{\|}{C}}-NH_2 + H_2O$$

Urea

In the body the urea is formed via **ornithine cycle** or **Krebs-Henseleit cycle** which involves the conversion of amino-acid, **ornithine** to **citrulline** through **glutamic acid** derived from **aspartic** acid by transamination or formed from α-keto glutaric acid and ammonia, and then to **arginine**, following which **urea** is split off and ornithine is regenerated once again. CO_2 and NH_3 are introduced into the cycle by **"carrier molecules"** for the formation of which ATP is required. The reactions of ornithine cycle can be expressed as in Fig. 7·7.

For the formation of urea so many enzymes are involved which are usually found in the liver. Of them (i) **carbamylphosphate synthetase** and (ii) ornithine **transcarbamylase** are in the particulate fraction of the liver tissue, while (iii) **argininosuccinic acid synthetase** (iv) **arginino succinic acid splitting enzyme** and (v) **arginase** are present in the soluble fraction. Energy is needed for the formation of urea which is provided by 3 molecules of ATP for each molecule of urea formation.

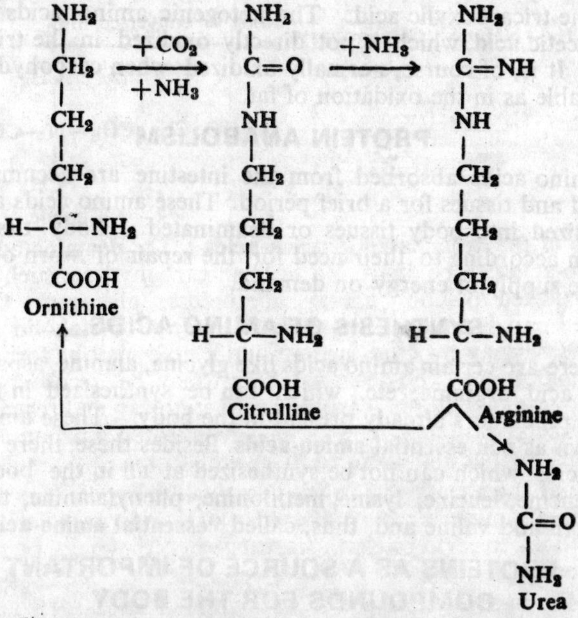

Fig. 7·7. Ornithine cycle (formation of urea).

2. Formation of ammonium salts : About 4–5% of the nitrogen excreted in urine is in the form of ammonium salts. The ammonium ion comes from the ammonia produced during deamination of α-amino acids. Apparently, the body converts most of this into urea but some is simply acidic to form NH_4^+.

$$NH_3 + H^+ \rightarrow NH_4^+$$

Fig. 7·8. Formation of ammonium ion.

Creatine and creatinine are the essential nitrogen compounds which are generally present in urine.

Fate of the sulphur : There are some amino-acids such as cystine, methionine and cysteine which contain sulphur in their constitution. After deamination of these amino acids the sulphur is removed from the deaminated part and is usually oxidized to sulphates. These sulphates may be eliminated as inorganic sulphates such as potassium sulphates or as conjugated sulphates in the detoxication mechanism.

Fate of the nucleic acids : When the nucleic acids are completely hydrolyzed phosphoric acid, simple monosaccharides and nitrogen bases such as purines, pyrimidine and pyridines are resulted. The monosacchrides may be involved in the carbohydrate metabolism and phosphoric acid may be utilized in the synthesis of phospholipids or inorganic phosphates. The purines such as adenine and guanine are oxidized to uric acid and eliminated in the urine but the metabolic fate of the pyrimidines is not yet completely known.

Fate of α-keto acids : Each α-keto acid follows a special metabolic pathway to a compound which can be completely oxidized by way of the tricarboxylic acid. The ketogenic amino acids give rise to acetoacetic acid which is not directly oxidized in the tricarboxylic acid. It is, of course, normally oxidized when carbohydrates are not available as in the oxidation of fat.

PROTEIN ANABOLISM

Amino acids absorbed from the intestine are accumulated in the blood and tissues for a brief period. These amino acids are either resynthesized into body tissues or deaminated. Body tissues may take them according to their need for the repair of worn out tissues or for the supply of energy on demand.

SYNTHESIS OF AMINO ACIDS

There are certain amino acids like glycine, alanine, aspartic acid, glutamic acid, arginine, etc., which can be synthesized in the body from the precursors already present in the body. These amino acids are known as **non-essential amino-acids**. Besides these, there are other amino acids which can not be synthesized at all in the body, these are isoleucine, leucine, lysine, methionine, phenylalanine, threonine, tryptophan and valine and, thus, called **"essential amino-acids."**.

PROTEINS AS A SOURCE OF IMPORTANT COMPOUNDS FOR THE BODY

Most of the hormones such as insulin, adrenocortico trophic hormone are protein in nature. These proteins are synthesized by the endocrine glands. These glands pick up the proper amino acids from the metabolic pool of amino acids, in correct amounts and prepare these special protein structures. If the endocrine glands are not furnished with a proper supply of starting α-amino acids or if they are unable to synthesize a proper supply of the protein hormone, a serious endocrine deficiency may result. Likewise, if the glands are overactive and synthesize too much protein hormones, the body may exhibit serious disorders

ROLE OF LIVER AND KIDNEY IN PROTEIN METABOLISM

Liver and kidneys play important role during the protein metabolism because most of the metabolic reactions take place in these organs. When the blood flows through liver amino-acids are partly retained and a **"reserve protein"**, which can be easily utilized by the body during limited protein consumption, is synthesized from them in the liver. Like liver the kidneys also play important role in protein metabolism. In the kidney ammonia is liberated from the amino-acids and is used for neutralizing the acids. The latter are voided along with urine in the form of ammonium salts. These are kidneys which eliminate the nitrogenous end products of protein metabolism. If kidneys fail to perform function properly, then these products can not be removed from the body and, thus, are accumulated in tissues and blood.

The increased accumulation of these nitrogenous products may lead to the death of the individual.

REGULATION OF PROTEIN METABOLISM

There are certain factors which regulate the protein metabolism. The secretion of thyroid gland what is known as thyroxine greatly influences the intensity of protein metabolism.

The nature of the consumed food also exerts considerable influence on the protein metabolism. Meat food gives rise to greater amounts of uric acid, creatinine and ammonia. The amount of ammonia formed during the metabolism of amino-acids in kidneys depends upon the acid base balance in the body ; more of it is formed in acidosis, less in alkalosis.

Adrenocorticotrophic hormone also regulates the protein metabolism.

CARBOHYDRATE METABOLISM

Carbohydrates are known as chief energy giving compounds of the body as they supply the major portion of the daily energy requirement of the normal individual. On an ordinary diet more than half of the total daily calories usually come from this source. In addition to being oxidized as a source of energy, they may be transformed to glycogen or to supply the carbon chain for certain amino-acids or to be converted into fats. Man's daily carbohydrate requirement is 450–500 gm.

The chief sources of carbohydrates are the food such as flour, all cereals, rice, potatoes, bread, milk, fruits including cane sugar, grape sugar, malt sugar and milk sugar and vegetables etc. These normally contain complex carbohydrates which are known as saccharides.

In the digestive tract the complex carbohydrates or saccharides contained in the food are subjected to the action of the enzymes present in the saliva, pancreatic and intestinal juices. These are hydrolyzed to monosaccharides which are soluble in water and are absorbed through the intestinal mucosa into the blood stream. In our everyday life, the principal carbohydrates in the diet are starch, glucose, sucrose and lactose. Complete hydrolysis of these carbohydrates produces the monosaccharides, glucose, galactose and fructose which are carried directly to the liver by the portal circulatlon. In the liver, the three monosacchrides may be converted into one simple sugar, glucose, the primary monosaccharide of the blood stream and body. In the liver and the muscles the excess of glucose is converted into glycogen. The glycogen in the liver is a reserve material and when necessary it is reconverted by the action of enzyme into glucose which passes into the blood and is transported throughout the body.

The glucose content in the blood remains constant (70–100 mg per 100 ml of blood). If the glucose content increases from the normal glucose content of the blood the liver takes up the excess glucose from the blood and converts it to glycogen. All the changes involved in the conversion of glucose into glycogen constitute **carbohydrate anabolism.** If the glucose content decreases from the normal content of the blood, a part of the liver glycogen is brokendown and liberated into the blood as glucose. In this way fluctuations in the blood glucose level are kept within fairly narrow limits by the liver cells.

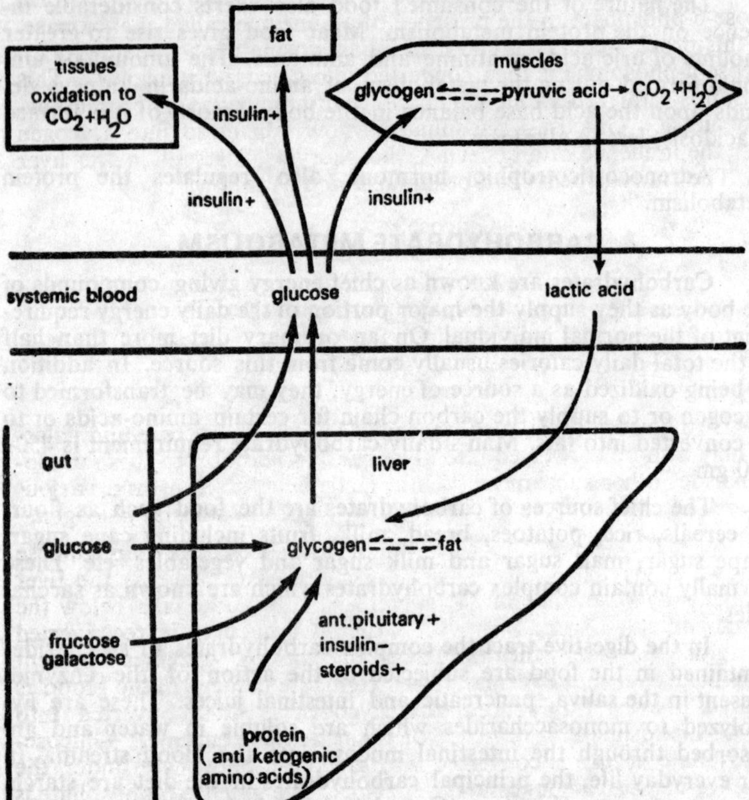

Fig. 7·9. Diagrammatic representation of carbohydrate metabolism.

The glycogen on its breakdown liberates energy. A particularly large amount of glycogen is brokendown when the muscles are working, the energy liberated is used for mechanical work and as a source of heat. During the process of breakdown of glycogen or glucose (monosaccharides) carbon dioxide and water are formed as by products. They are duly eliminated by lungs and the skin. The destruction changes involved in the breakdown of glucose (monosaccharides) and resulting in the production of energy constitute **carbohydrate catabolism.**

The carbohydrate metabolism can be studied under the following heads :

1. Glycogenesis, 2. Glycogenolysis, 3. Glycolysis.

1. **Glycogenesis :** The process by which glycogen is synthesized is known as **glycogenesis**. This process usually takes place in liver and is carried out with the help of certain enzymes present in the liver cells. During this process glucose is first phosphorylated to form glucose-6-phosphate in the presence of an enzyme called **hexokinase.** For this process energy is supplied by adenosine triphosphate (ATP). Glucose-6-phosphate is then converted into glucose-1-phosphate. This reaction is catalyzed under the influence of an enzyme called **phosphoglucomutase**. Glucose-1-phosphate is now converted into glycogen under the influence of an enzyme, **phosphorylase** present in the liver cells. The entire process can be summarized as follows :

Glucose
↓ hexokinase
Glucose-6-phosphate
↓ phosphoglucomutase
Glucose-1-phosphate
↓ phosphorylase
Glycogen

Like glucose other monosaccharides such as fructose and galactose may also be converted into glycogen. In addition to these monosaccharides other substances such as lactic acid, glycerol and various amino-acids may also be utilized by the liver to synthesize glycogen.

2. **Glycogenolysis :** The breakdown of glycogen into glucose is known as **glycogenolysis**. This process also takes place in the liver cells but only when the glucose content of the blood falls below the normal content. During this process the liver glycogen is reconverted into glucose under the influence of various enzymes present in the liver and, thus, normalcy of the glucose content in the blood is restored. During this process the glycogen is first converted into glucose-1-phosphate in presence of **phosphorylase** enzyme. Glucose-1-phosphate in turn is converted into glucose-6-phosphate under the influence of **phosphoglucomutase** enzyme. The glucose-6-phosphate is now converted into glucose in the presence of liver **phosphatase**. The entire process can be summarized as follows :

Glycogen
↓ phosphorylase
Glucose-1-phosphate
↓ phosphoglucomutase
Glucose-6-phosphate
↓ liver phosphatase
Glucose+inorganic phosphate

The conversion of glucose into glycogen and glycogen back into glucose is entirely under the control of hormones secreted by the endocrine glands. For example, the islets of Langerhans present in pancreas produce a hormone called insulin, it controls the maintenance of glucose level in the blood. When the glucose content falls below normal in the blood, the liver is influeneed by the insulin hormone which in turn converts glycogen into glucose and, similarly, if the glucose content goes up, the liver, in turn, converts glucose into glycogen.

3. **Glycolysis :** The term **glycolysis** is used to refer to the anaerobic phase of the intermediary metobolism of carbohydrates whereby glucose or glycogen is converted to lactic acid. During the conversion of glucose or glycogen into lactic acid certain amount of chemical energy is released which is used by the cells to carry on other activities. Lactic acid is also broken down at the later stage into CO_2 and water. During this breakdown energy is again released. The sequence of reactions is frequently referred to as the **"Embden Meyerhof pathway"**.

In the process of glycolysis six-carbon glucose units are separated from the larger molecule by the introduction of a–PO_4 group. Specific enzymes accomplish the transfer of a —PO_4 radical from the cellular store of phosphates to a glucose unit on the glycogen molecule. As a result, a molecule of glucose-1-phosphate is formed (the 1 designates the position in the chain of the carbon to which the —PO_4 is attached). The remaining portion of the glycogen molecule can further split by a series of such phosphorylations, each of which produces a molecules of glucose. Glucose-1-phosphate is then rearranged, so that the phosphate is attached to other end of the carbon chain. The molecule is then called glucose-6-phosphate. Further reactions of the glycolysis can be grouped under the following four stages :

1. **Stage (activation) :** It is commonly known as preparatory stage during which symmetric glucose molecule is converted to the almost symmetrical fructose 1, 6 diphosphate by donation of two phosphate groups from ATP, adenosine triphosphate. The conversion is brought about under three stages :

(i) Glucose is phosphorylated to glucose-6-phosphate in the presence of **glucokinase** or hexokinase enzyme.

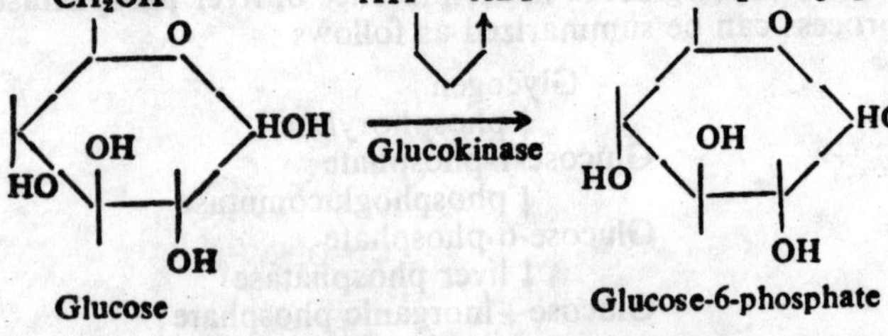

(ii) Glucose-6-phosphate undergoes an internal rearrangement to give fructose-6-phosphate under the influence of an enzyme called **glucose phosphate-isomerase (phosphoglucoisomerase)**.

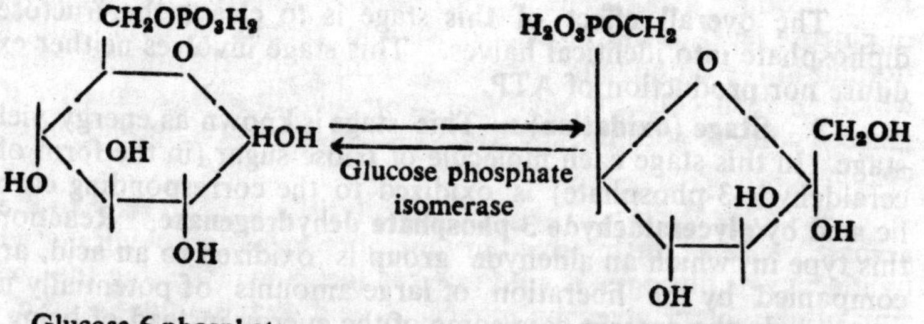

Glucose-6-phosphate Fructose-6-phosphate

(iii) Fructose-6-phosphate undergoes further phosphorylation in the presence of **phosphofructokinase** and gives fructose 1, 6-diphosphate.

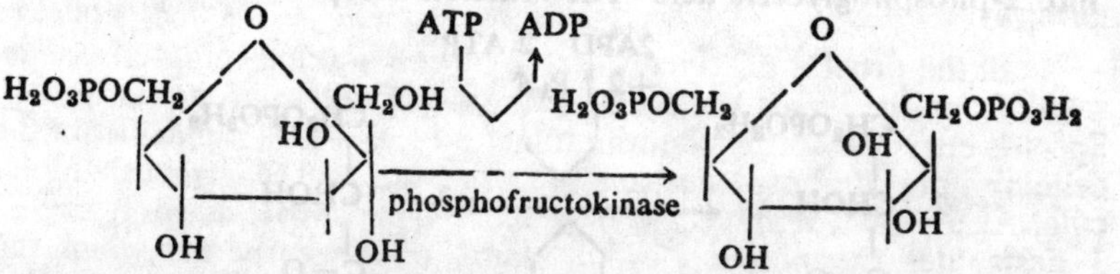

Fructose-6-phosphate Fructose 1-6-diphosphate

The overall effect of these reactions is to convert one molecule of monosaccharide into one molecule of fructose diphosphate. During these reactions one or two molecules of ATP are converted into ADP depending upon the starting substrate. ATP is an essential catalysts in these reactions.

2. Stage (cleavage) : During this stage fructose 1-6-diphosphate is cleaved to form two molecules of triose, a three carbon sugar, each of which contains a phosphate group. These are glyceraldehyde-3-phosphate and dihydroxy acetone phosphate. This reaction is catalyzed in the presence of enzyme, **fructo-aldolase**. These two are interconvertible in the presence of the enzyme **triosephosphate isomerase.** The entire reaction is represented as follows :

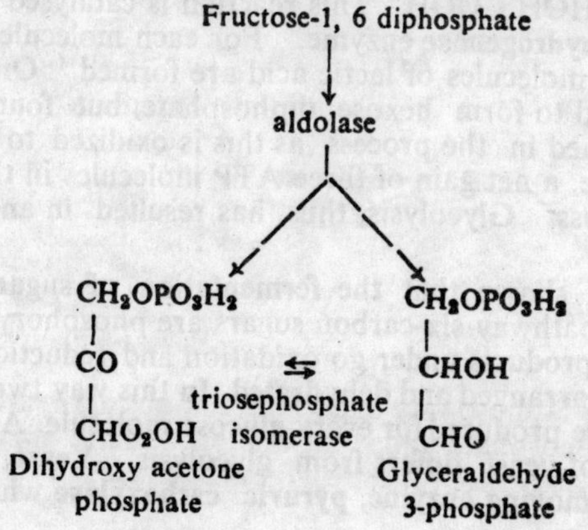

Dihydroxy acetone phosphate Glyceraldehyde 3-phosphate

The overall effect of this stage is to cleave the fructose 1-6 diphosphate into identical halves. This stage involves neither expenditure nor production of ATP.

3. **Stage (oxidation) :** This stage is known as energy yielding stage. In this stage each molecule of triose sugar (in the form of glyceraldehyde-3-phosphate) is oxidized to the corresponding carboxylic acid by **glyceraldehyde 3-phosphate dehydrogenase**. Reactions of this type in which an aldehyde group is oxidized to an acid, are accompanied by the liberation of large amounts of potentially useful energy. In the present case some of the energy instead of being wasted as heat is used to form ATP from ADP+P_i (P_i indicates one molecule of inorganic phosphate).

After the oxidation, the glyceraldehyde-3-phosphate is converted into 2-phosphoglyceric acid The reaction is represented as follows :

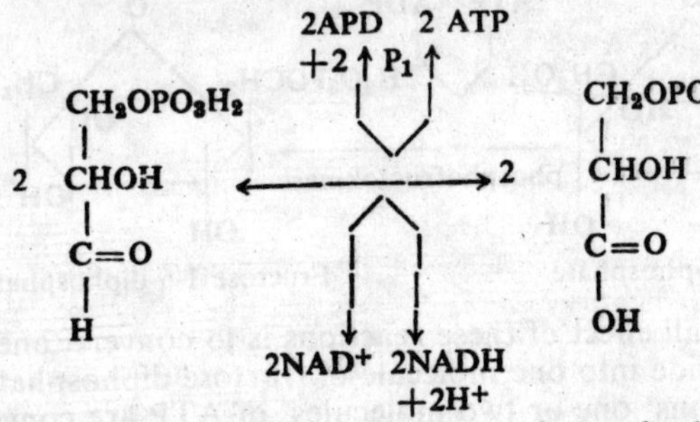

Glyceraldehyde-3-phosphate 3-phosphoglyceric acid

4. **Stage :** During this stage the oxidized three carbon compound (3-phosphoglyceric acid) is converted by means of a number of transformations into pyruvic acid ($CH_3COCOOH$). In one phase of this transformation water is removed from the molecule, and as this happens, energy is again freed from the carbohydrate molecule and can be used to form a molecule of ATP.

Glycolysis in muscle is completed by the reduction of three carbon pyruvic acid by NADH (reduced NAD) with the formation of lactic acid, $CH_2CHOH.COOH$. This reaction is catalysed in the presence of **lactic dehydrogenase** enzyme. For each molecule of glucose 6-phosphate, two molecules of lactic acid are formed. One molecule of ATP is required to form hexose diphosphate, but four molecules of ATP are formed in the process as this is oxidized to lactic acid. There is, therefore, a net gain of three ATP molecules in this portion of glycolytic process. Glycolysis, thus, has resulted in an energy release.

It has been shown that the fermentation of sugar by yeasts follows a similar pathway six-carbon sugars are phosphorylated; and rearranged, their products under go oxidation and reduction reaction and are further rearranged and dehydrated. In this way two molecules of pyruvic acid are produced for every glucose molecule. At this point the fermentation of yeast differs from glycolysis. Yeast possesses a carbon dioxide removing enzyme, **pyruvic carboxylase** which acts on pyruvic acid to remove carbon dioxide and produce a two carbon compound, acetaldehyde (CH_3CHO).

It is interesting that the fermentation reaction and the glycolysis reaction are identical to the pyruvic acid step, at which point they diverge. The glycolysis ends up with lactic acid and fermentation ends up with ethanol and carbon dioxide.

Pyruvic acid is an important compound because it can form acetyl CoA in a very complex series of reactions. In a simplified form, this reaction is an oxidative decarboxylation (*e.g.*, carbon dioxide is lost, and essentially oxygen is added).

$$\underset{\text{Pyruvic acid}}{CH_3-\overset{\overset{\large O}{\|}}{C}-COOH}+CoA\xrightarrow{\text{other factors}}\underset{\text{Acetyl CoA}}{CH_3-\overset{\overset{\large O}{\|}}{C}-S-CoA}+CO_2$$

Acetyl CoA is one of the most important chemical in the body because it is a link between carbohydrates and lipids and is also involved in many reactions in living system. The inter-relationship of carbohydrates, lipids and proteins is as follows :

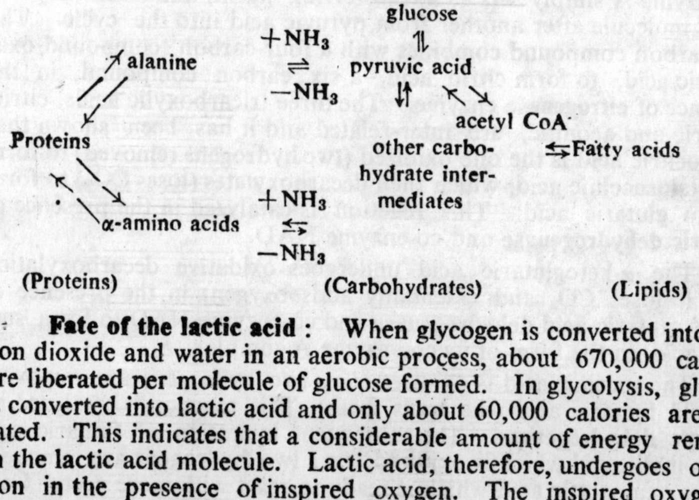

Fate of the lactic acid : When glycogen is converted into carbon dioxide and water in an aerobic process, about 670,000 calories are liberated per molecule of glucose formed. In glycolysis, glucose is converted into lactic acid and only about 60,000 calories are liberated. This indicates that a considerable amount of energy remains in the lactic acid molecule. Lactic acid, therefore, undergoes oxidation in the presence of inspired oxygen. The inspired oxygen is carried by the blood (as oxyhaemoglobin) to the tissues needing it. About 20—30% of the lactic acid formed during glycolysis is oxidized to carbon dioxide and water. This liberetes a rather large amount of energy, which is used to convert ADP molecules into the very energy rich ATP's.

The lactic acid which is not oxidized is converted back into glycogen, with the help of the energy furnished by the oxidation of the other lactic acid. There is some conversion of lactic acid to glycogen in the muscles, but most of lactic acid molecules are carried by the blood into the liver, where the conversion takes place (glycogenesis).

KREBS CYCLE OR TRICARBOXYLIC ACID CYCLE OR CITRIC ACID CYCLE

Glucose (carbohydrate) metabolism does not stop at the point when lactic acid is formed but continues until the end products, carbon dioxide and water are formed. The process whereby a portion of the lactic acid is oxidized, has been proposed by KREBS and elaborated by GREEN and others. This is a cyclic process which is influenced by several enzymes, since citric acid is very evident in the process it is referred to as the "**citric acid cycle**". It is also known as "KREBS **cycle**" since KREBS invented this entire system of chemical changes. In this process pyruvic acid keeps going into the cycle and three carbon dioxide molecules keep coming out as energy is liberated.

During this cycle the pyruvic acid is oxidized into CO_2 and water and maximum energy is liberated. The entire cycle can be summarized as follows :

Each molecule of pyruvic acid undergoes oxidative decarboxylation to form acetyl CoA. This reaction is catalyzed in the presence of pyruvic dehydrogenase enzyme, and co-enzyme A and NAD. As acetyl CoA enters the cycle, it separates into its component parts. Co-enzyme A simply acts as a transferring agent, and can carry one acetyl molecule after another from pyruvic acid into the cycle. The two carbon compound combines with a four-carbon compound oxaloacetic acid, to form citric acid, a six carbon compound, in the presence of **citrogenase** enzyme. The three tricarboxylic acids, citric, isocitric and aconitic, are inter-related and it has been shown that the isocitric acid is the one oxidized (two hydrogens removed) to form the oxalosuccinic acid, which then decarboxylates (loses CO_2) to form α keto glutaric acid. This reaction is catalyzed in the presence of **isocitric dehydrogenase** and co-enzyme NAD.

The α-ketoglutaric acid undergoes oxidative decarboxylation (*i.e.*, it loses CO_2 and essentially adds oxygen) in the presence of **α-keto glutaric acid dehydrogenase** and co-enzyme NAD to form succinic acid, in the form of its co-enzyme A complex.

The succinic acid is oxidized (*i.e.*, two hydrogens are removed) to yield fumaric acid (the **trans** form). This reaction is catalyzed by **succinic dehydrogenase**. The subsequent hydration of fumaric acid yields malic acid which is oxidized (*i.e.*, two hydrogens are removed) to give oxaloacetic acid which is ready to react with more acetyl CoA to repeat the cycle. This reaction is catalyzed by **malic dehydrogenase** enzyme and co-enzyme NAD.

This shows that each molecule of glucose and two molecules of pyruvic acid are formed. Each of these then produces three molecules of CO_2, making a total of six for the glucose molecule. At the same time twelve hydrogen atoms of the molecule are being removed and combined with oxygen to form water. In the oxidation of pyruvic acid through KREBS cycle 30 ATP molecules are formed.

ALTERNATIVE PATHWAY OF GLUCOSE CATABOLISM

For a number of years, the classical EMBDEN-MEYERHOF scheme

of glycolysis was considered to be the sole pathway for the conversion of glucose to pyruvic and lactic acid but the discovery of glucose-6-phosphate dehydrogenase in yeast by WARBURG in 1931 and of 6-phosphogluconate dehydrogenase a few years later has established that

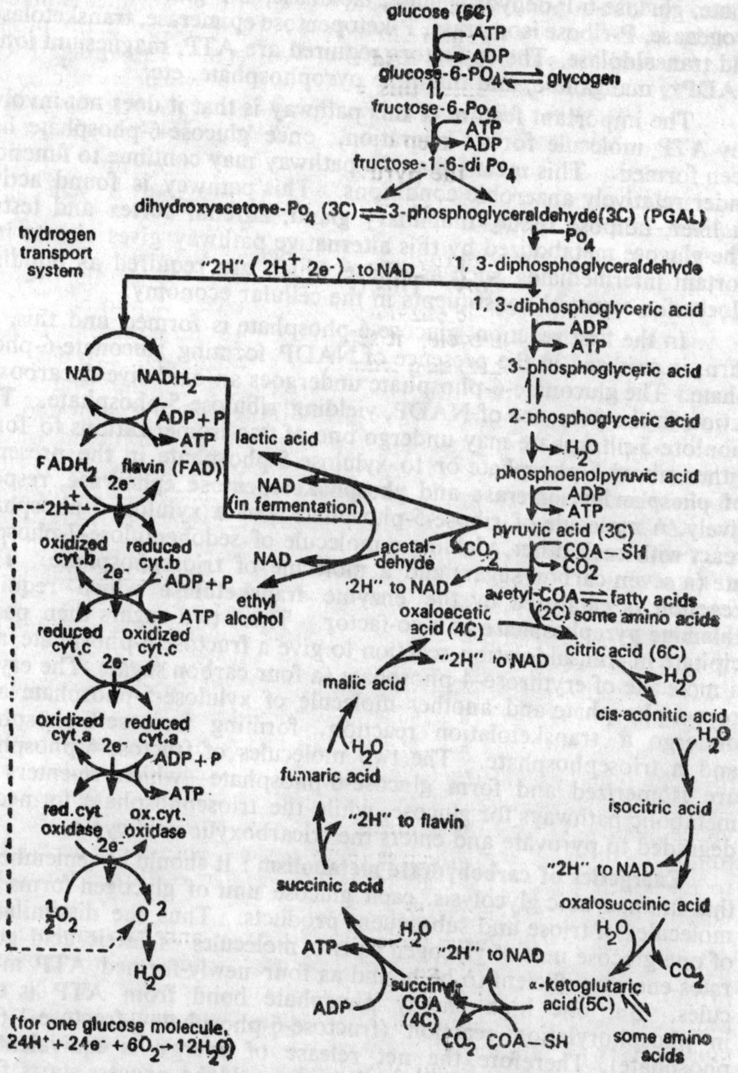

Fig. 7·10. Glycolysis, Krebs cycle, and hydrogen transport system.

a number of alternative pathways of glucose catabolism exists, one of the most important of these sometimes called the WARBURG-DICKENS **pathway**, the **phosphogluconate shunt**, or the **hexose monophosphate shunt.** This pathway generally serves two important roles : as a source of reducing equivalents for the reduction of $NADP^+$ to NADPH, and as the mechanism for synthesis and disposal of pentoses. This pathway does not lead to the direct fission of glucose molecule into triose molecules. It involves enzymes such as glucokinase, glucose-6-P-phosphate, glucose-6-P-dehydrogenase, lactonase, 6-P-gluconic acid dehydrogenase, P-ribose isomerase, P-ketopentose epimerase, transketolase, and transaldolase. The co-factors required are ATP, magnesium ions, $NADP^+$, manganese, and thiamine pyrophosphate, etc.

The important feature of this pathway is that it does not involve any ATP molecule for its operation, once glucose-6-phosphate has been formed. This means that the pathway may continue to function under relatively anaerobic conditions. This pathway is found active in liver, adipose tissue, mammary gland, adrenal cortex and testis. The glucose metabolized by this alternative pathway gives rise to important intermediates, such as ribose which are required as building blocks for essential constituents in the cellular economy.

In the first reaction, glucose-6-phosphate is formed and this, in turn, is oxidized in the presence of NADP forming gluconate-6-phosphate. The gluconate-6-phosphate undergoes an oxidative decarboxylation in the presence of NADP, yielding ribulose-5-phosphate. The ribulose-5-phosphate may undergo one of two isomerizations to form either ribose-5-phosphate or to xylulose-5-phosphate in the presence of **phosphoriboisomerase** and **phosphoketopentose epimerase**, respectively. A molecule of ribose-5-phosphate and a xylulose-5-phosphate react with each other, yielding a molecule of sedoheptulose-7-phosphate (a seven carbon sugar) and a molecule of triosephosphate. This reaction is catalyzed by the enzyme **transketolase** which requires **thiamine pyrophosphate** as a co-factor. These two sugars then participate in transaldolation reaction to give a fructose-6-phosphate and a molecule of erythrose-4-phosphate (a four carbon sugar). The erythrose-4-phosphate and another molecule of xylulose-5-phosphate also undergo a transketolation reaction, forming fructose-6-phosphate and a triosephosphate. The two molecules of fructose-6-phosphate are isomerized and form glucose-6-phosphate which re-enters the metabolic pathways for glucose, while the triosephosphate formed is degraded to pyruvate and enters the tricarboxylic acid cycle.

Energetics of carbohydrate metabolism : It should be remembered that in anaerobic glycolysis, each glucose unit of glycogen forms two molecules of triose and subsequent products. Thus the dissimilation of one glucose unit of glycogen to two molecules of lactic acid generates energy sufficient to be bound as four newly-formed ATP molecules. But one high energy phosphate bond from ATP is used in phosphorylation reaction (fructose-6-phosphate→fructose-1-6-diphosphate). Therefore, the net release of energy is equivalent to 3ATP=approx. 36 K cal/mole of glucose. If the process starts from free blood glucose, the net release is only 2ATP molecules, since one high energy phosphate bond from ATP is used in the hexokinase reaction.

When pyruvate is transformed into acetyl CoA and CO_2, the oxidation of the NADH provides 6 moles of ATP per glucose equivalent. Since complete oxidation of each mole of acetyl CoA yields 12 moles of ATP, 24 moles of ATP are generated per glucose equivalent.

The complete aerobic dissimilation of two molecules of pyruvic acid (*i.e.*, one glucose unit) thus results in the generation of 38ATP molecules which are equivalent to 266 kilo calories of utilizable energy. The following table is showing the details of production of ATP molecules.

Table : 7·1. ATP production from glucose.

Glucose→fructose 1, 6-diphosphate	− 2
2 Triosphosphate→2, 3-phosphoglyceric acid	+ 2
2 NAD→2 NADH→2NAD$^+$	+ 6
2 phosphoenolpyruvic acid→2 pyruvic acid	+ 2
2 Pyruvic acid→2 acetyl CoA+2CO_2	
2 NAD$^+$→2 NADH→2 NAD$^+$	+ 6
2 Acetyl CoA→4 CO_2	+ 24
Net : $C_6H_{12}O_6$+6 N→6 CO_2+6H_2O	+ 38

ATP molecules are important to ensure a proper functioning of the body, the body must maintain a satisfactory supply of ATP or means of ATP, because it is responsible for the chemical reactions.

REGULATION OF CARBOHYDRATE METABOLISM

The metabolism of carbohydrate is controlled by hormones secreted by the endocrine glands, of which the pancreas, the anterior pituitary lobe and adrenal cortex appear to be of major importance.

1. **Pancreas :** Islets of LANGERHANS of pancreas secrete a hormone what is known as **insulin**. Insulin was first of all obtained by BANTING BEST and MACLEOD in the year of 1921. Opinions differ about the mechanism of insulin control of carbohydrate metabolism. It is believed that it aids in the conversion of sugar to glycogen and in the utilization of sugar with the formation of the intermediate products of its metabolism. Some say that it simply accelerates carbohydrate oxidation, while others regard its function as primarily concerned with the synthesis of carbohydrate from non-carbohydrate precursors or possibly with the conversion of carbohydrate to fat.

2. **Anterior pituitary lobe :** According to HOUSSAY the growth hormone and to some extent the adrenocorticotrophic and thyroid stimulating hormone control the carbohydrate metabolism as they are diabetogenic. Growth hormone mobilises free fatty acids from the adipose tissue and thus favours ketogenesis. It also causes secretion

of blood glucose elevating factor from the pancreas other than glucagon.

3. **Adrenal cortex :** The hormones produced by the cortical part of the adrenal gland also influence carbohydrate metabolism (according to LONG and LUKENS).

Recent findings state that besides these glands there are other glands which also regulate the carbohydrate metabolism. These are thyroid glands which secrete thyroxine which not only increases the rate of absorption of carbohydrates from the intestine but also brings about increase in the rate of metabolism of carbohydrates.

The functions of these glands are under the control of the nervous system which, in the final analysis, regulates the entire carbohydrate metabolism.

FAT METABOLISM

Fats constitute an essential component of all protoplasm. Like carbohydrates and proteins they are also energy giving substances. Carbohydrates and proteins on being oxidized yield about four calories per gram where as fats about nine calories, it is evident that fats are more ideal for the storage of energy. Fats have more energy it is because, that they are more reduced than carbohydrates and proteins, *e.g.*, the ratio of hydrogen to carbon is greater in fats than in carbohydrates or proteins.

Fats in the body are stored under the skin in tissues called **adipose tissue** or **depot fat**. During extreme starvation considerable amounts of fat can be extracted from these adipose tissues where they are stored as energy fuel or energy source. When too much fat is deposited, obesity results. In addition to being an energy storehouse, the fat tissue functions as an insulator and provides some physical support to internal organs. There is a evidence that they also serve some important non-caloric metabolic functions. In this connection BURR and BERNES concluded that *"there are ample reasons for recommending that the fats intake be not reduced much below the normal established by habit"*. This shows that the fat content of the diet may influence such diverse processes as the digestibility and absorbility of other food stuffs in the gastro-intestinal tract, and the rate of calcification of the bones. Fats of the diet serve as a vehicle for the fat soluble vitamins, their presence in the protein free diet exerts a favourable effect on nitrogen retention, as well as on the course of pregnancy and lactation.

During the process of digestion the fats contained in the food are first simplified into substances such as glycerol, phosphoric acid and the fatty acids, under the influence of the enzymes present in pancreatic and intestinal juices (with the participation of bile). The glycerol and phosphoric acid (usually in the form of a salt) are soluble in water and easily absorbed. The fatty acids containing about 10 carbon atoms or more, are not very much soluble, even as sodium salts. The bile salts combine with the fatty acids (saponification) and thus render them soluble and suitable for absorption.

Various theories have been formulated about the nature of fat absorption. It is believed that the fatty acids are absorbed through the villi, and as they pass through the intestinal mucosa they are converted into phospholipids before resynthesized into simple glycerides. Combination with phosphoric acid seems an essential step before many reactions can occur. This phosphoric acid combination furnishes the necessary energy to make the reaction go.

The bulk of the fat of the food in the lymphatics appears in the form of colloidal particles with variable diameter called as **chylomicrons.** A small portion of the absorbed fat, consisting mostly the chain fatty acids, is sent directly to the liver via the blood stream. However, most of the absorbed fat passes into the lymphatics in the form of chylomicrons. The fat of lymphatics finally enters the blood circulation via the thoracic duct and may then make its way to the liver.

Fat stores : After absorption fat is either oxidized to release energy or stored in the body until required. The extent to which fat is stored or oxidized is in some measure under the control of endocrine and nervous systems. Considerable amounts of fat are stored in subcutaneous depots all over the body, these are known as **adipose tissue** or **fat depots.** Fat storage is not only confined to the subcutaneous depots but it also occurs around the intestine, heart, kidneys and in between the muscle fibres. When the body feels the requirement of fat, the same is withdrawn from these depots.

All fat in the body does not come from dietary food but it is readily synthesized from carbohydrate and any excess of carbohydrate in the diet which is not immediately oxidized or stored as glycogen is converted to fat.

It has been shown that the fatty acids of the stored fat are in a constant state of flux. The body can some what modify dietary fat by converting the fatty acids into other fatty acids, but the animal fat tends to resemble; to a great extent, the dietary fat.

According to SCHOENHIMER and RETTENBERG animals consume not only the food fat but also depot fat in equal amount, therefore, depot fat, is not a inert storage fat but it is constantly involved in metabolic processes. The saying mean is that depot fat very well resembles with dietary fat.

It has been shown by the experiments that even depot is undergoing a relatively rapid chemical interchange. The changes usually involve a transformation of one fatty acid into another, even though the synthesized fatty acid may be abundantly supplied in the diet. One of the most common of these transformations is the desaturation reaction which leads to the conversion of a saturated fatty acid into an unsaturated fatty acid.

$$CH_3\,(CH_2)_7-CH_2-CH_2-(CH_2)_7COOH \quad \text{Stearic acid}$$
$$+2H \uparrow \qquad \downarrow -2H$$
$$CH_3-(CH_2)_7-CH=CH-(CH_2)_7COOH \quad \text{Oleic acid}$$

Fig. 7·11. An example of desaturation

These reactions usually take place in the presence of oxygen, which aids in the removal of the two hydrogens. If oxygen is not available then conversion takes place through other way. It has been shown that in animal tissues, the unsaturated fatty acids such as oleic, linoleic and linolenic are not interconvertible. This means that oleic acid can not form linoleic acid and linoleic acid can not form linolenic acid. However, each particular unsaturated fatty acid can become more unsaturated within its own family of related acids. In this latter sequence of reactions a polyunsaturated fatty acid of two additional carbons is formed. For example,

$C_{17}H_{33}COOH$

does go / adds two carbons — does not go

$CH_3(CH_2)_7(CH{=}CHCH_2)_3(CH_2)_2COOH$ $\quad$ $CH_3(CH_2)_4CH{=}CHCH_2CH{=}CH(CH_2)_7COOH$

5, 8, 11 eicosatrienoic acid ($C_{19}H_{33}COOH$) Linoleic acid ($C_{17}H_{31}COOH$)

This example explains that any given unsaturated fatty acid can

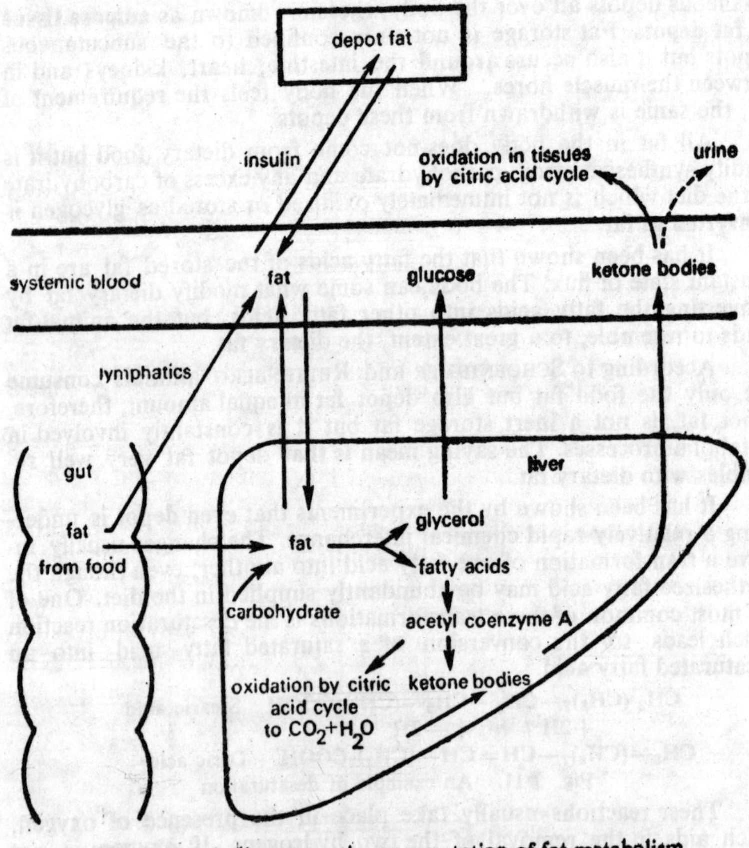

Fig. 7·12. A diagrammatic representation of fat metabolism.

form a specific polysaturated fatty acid with two more carbons and with two or more unconjugated double bonds, the additional double bonds appear toward the carboxyl group end of the molecule. However, a conversion of the one saturated fatty acid into another is quite common.

FAT IN THE LIVER

The liver plays an important role in the metabolism of fat as most fat metabolism takes place in this organ. The normal liver contains about 4% of lipids, consisting of mostly neutral fats and phospholipids. Free fatty acids, cholesterol and other steroles or steroides are also present in small amounts. Under fasting or untreated diabetes conditions where lipid must be metabolized in greater than normal amounts because of the unavailability or inability of carbohydrate to supply its normal quota of the caloric requirement, the lipid content of the liver may increase to 30 or 40%, giving rise to the condition termed **"fatty liver"**, The oxidation and synthesis of fatty acids take place readily and rapidly in the liver. Cholesterol, cholesterol ester and bile salt synthesis also take place in the liver. In the liver the carbon chain of fatty acid can be lengthened or shortened, saturated or desaturated. The fat in liver remains very active metabolically.

BREAKDOWN (OXIDATION) OF FATS

1. **Oxidation of glycerol** : Glycerol liberated after hydrolysis of neutrat fats can be metabolized via conversion to glycerol phosphate and dihydroxy acetone-phosphate, which enters the glycolytic pathway and may either be converted to glycogen or oxidized to CO_2 and H_2O.

Oxidation of fatty acids : Muscles of the body oxidize normally the glucose (glycogen) as their energy fuel but fatty acids may also be oxidized when glucose (glycogen) concentration falls below normal. Fatty acids on their oxidation form **acetyl CoA** or **"active acetate"** which later enters the citric acid cycle.

There are several explanations of how fatty acids are brokendown by oxidation. Out of them the most important one is the classical theory of β-oxidation elaborated by Knoop. According to this theory fatty acid chains are oxidized by the removal of two carbon atoms at a time. The carbon atom in the β-position to the carboxyl group is assumed to be attacked with the formation of the corresponding **β-keto acid**, then the two terminal carbon atoms are split off as acetic acid. A new carboxyl group (—COOH) is formed at the site of the keto (=CO) grouping so that a fatty acid remains with two carbon atoms fewer than the original. Again the new β-carbon atom is attacked and two more carbon atoms are split off. Thus, the fatty acid is brokendown by the removal of two carbon atoms at a time, until finally the stage of aceto-acetic acid (β-keto butyric acid) is reached.

According to the modern views of Green, Lynen and Kennedy the oxidation of fatty acid involves the following five chemical reactions which take place in the cell mitochondria.

1. Activation : During this reaction the fatty acid is converted into active fatty acid in the form of acetyl CoA derivative. Atleast three activating enzymes (**thiokinases**) take part in this reaction. One acts on acetic and propionic acids, the second on acids having four to twelve carbon atoms and the third on acids having more than twelve carbon atoms. The thiokinases require a supply of ATP and Co-enzyme A, while the theophorases require NAD, as in the reaction by which α-ketoglutarate is converted to succinyl CoA. They do not require ATP. Succinyl CoA can also be formed from succinate and GTP by a thiokinase.

2. Desaturation : Once the fatty acid has been activated, it can be dehydrogenated in the α, β-position by acyl dehydrogenases. These enzymes contain a flavoprotein, flavin adenine nucleotide, and they show specificity in relation to the chain length of the fatty acid. The reaction is not reversible by the same enzyme but may be reversible in the presence of reductases which require NADPH.

$$R-CH_2-CH_2-\overset{O}{\overset{\|}{C}}\sim SCoA \;\underset{FAD \;\rightarrow\; FADH_2}{\overset{NADP^+ \;\leftarrow\; NADPH+H^+}{\rightleftharpoons}}\; R-CH-CH-\overset{O}{\overset{\|}{C}}\sim SCoA$$

3. Hydration : In the next reaction the compound takes up one molecule of water under the influence of an **enoyl CoA hydratase** to form β-hydroxy acyl CoA derivative.

$$RCH=CH-\overset{O}{\overset{\|}{C}}\sim SCoA+H_2O \rightleftharpoons R-\overset{OH}{\overset{|}{C}H}-CH_2-\overset{O}{\overset{\|}{C}}\sim SCoA$$

4. Oxidation : During this reaction the β-hydroxy acyl CoA derivative is oxidized to a keto group in the presence of **β-hydoxyl-acyl dehyrogenases** and NAD.

$$R-\overset{OH}{\overset{|}{C}H}-CH_2-\overset{O}{\overset{\|}{C}}\sim SCoA+NAD \rightleftharpoons R-\overset{O}{\overset{\|}{C}}-CH_2-\overset{O}{\overset{\|}{C}}\sim SCoA+NADH+H^+$$

The NADH and FADH formed in the dehydrogenation steps are oxidized through the hydrogen transport system and provide part of the energy, in the form of ATP, which is obtained from fatty acid oxidation.

5. Thiolytic cleavage : The final step in the process of β-oxidation in cleavage of the β-keto derivative by a molecule of Co-enzyme A. The reaction is similar to hydrolysis, involving the sulphydryl group of CoA instead of water and is called **thiolysis**. The enzymes involved in this reaction are thiolases and appear to be non-specific, converting β-keto acyl CoA esters from C_4 to C_{18}.

After the oxidation of fatty acid a molecule of acetyl CoA and a molecule of activated fatty acid which is two carbon shorter than the fatty acid at the start are formed. The activated fatty acid may again be degraded by repetition of the process, starting at the second reaction since activation is not necessary. By successive repetitions of the proceess the entire fatty acid chain can be converted to acetyl CoA.

Fate of the acetyl CoA : The acetyl CoA produced after the oxidation of fatty acid is very much identical with acetyl CoA formed from carbohydrate by of way pyruvate and amino acid. Most of it combines with oxaloacetate to form citrate and is oxidized via the citric acid cycle (tricarboxylic acid cycle). The final oxidation pathway for fat and carbohydrates is, therefore, same and the end products are CO_2 and H_2O with the production of ATP. The co-enzyme A released during citric acid cycle enters the cycle again with another molecule of fatty acid, while the reduced FAD and NAD are reoxidized by the usual hydrogn transport systems.

Some acetyl CoA is used in the formation of cholesterol of fatty acids and of aceto-acetate. Small amounts of acetyl CoA are used in various acetylation processes. Further aspects of acetyl CoA metabolism are discussed below.

Ketosis : Under normal metabolic conditions when acetyl coenzyme A, produced after the oxidation of fatty acids, pyruvate and from other sources, is not required for the synthesis of cholesterol or fatty acid or acetyl derivatives combines with oxaloacetic acid and is oxidized to CO_2 and H_2O through tricarboxylic acid cycle. While some of the oxaloacetic acid is of course regeneratd in the operation of tricarboxylic acid cycle, this amount may be supplemented by additional supplies formed in the liver by the carboxylation of pyruvic acid derived from carbohydrate metabolism.

However, in circumstances when the metabolism of carbohydrates is impaired or operating at loss level such as in diabetes mellitus, starvation or prolonged subsistence on low carbohydrate diet, the fate of the acetyl CoA is altered by two reasons, (i) the oxaloacetic acid available to combine with acetyl CoA is in limited supply and (ii) a much greater proportion of the body's energy needs is being supplied by the oxidation of fatty acids, leading to the production of acetyl CoA in greater than normal amounts. Because of this combination of circumstances (large amounts of acetyl CoA and small amounts of oxaloacetic acid) acetyl CoA metabolism proceeds to a greater extent than normally via a different route. This consists of the condensation of two molecule of acetyl CoA to form **aceto-acetyl CoA** which, in turn, is hydrolyzed by **deacylase** in the liver to yield acetoacetic acid which may be reduced to β-hydroxybutyric acid in the presence of **β-hydroxybutyric dehydrogenase** and reduced NAD or decarboxylated to form acetone.

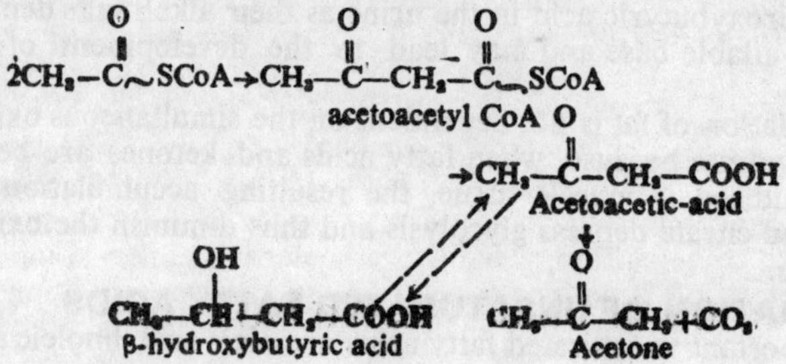

These three compounds are collectively known as **"ketonebodies"** or **"acetone bodies"** and the process is called **"ketogenesis"**. They are disposed of by oxidation in the process of **"ketolysis"** which takes place mainly in the extra-hepatic/tissues especially muscles. In this process the acetoacetic acid is activated by succinyl Co-enzyme A produced by operation of the citric acid cycle to form acetoacetyl CoA which can act as immediate precursor of acetyl CoA which in turn, is oxidized in usual way to CO_2 and H_2O. The ketone bodies are normal end products of fatty acid oxidation in the liver, but the amount formed is relatively small. The acetoacetic acid produced in the liver is not further utilized by the organ, except during fasting. Other tissues, however, readily metabolize acetoacetic acid to CO_2 and H_2O and there appears to be no impairment in this respect in diabetes (SOSKIN). If adequate amount of carbohydrate is available, the liver apparently prefers carbohydrate oxidation as a source of energy, and ketone body production is small. Carbohydrate is, therefore, an **"antiketogenic"** substance. In the diabetic where there is impaired glucose metabolism, the operation of the citric acid cycle is impaired by a decrease in oxaloacetic acid. The degradation of fatty acids, however, continues uninterruptedly, and the concentration of acetyl CoA which requires oxaloacetic acid to enter the citric acid cycle, increases. As a results acetyl CoA is shunted into the formation of acetoacetyl CoA and subsequently into ketone bodies.

The term **"ketosis"** is applied to the condition in which ketone bodies accumulate in the blood (**ketonaemia**) and appear in urine (**ketonuria**). This condition appears most commonly in starvation and clinical or experimental diabetes mellitus when the carbohydrate metabolism is very low. The ketosis of diabetes mellitus is accompanied by profound metabolic disturbances which lead to gradually depending coma and finally to death.

It is now clear that the accumulation of ketone bodies may be due to an abnormally high rate of formation of acetyl CoA together with a diminished capacity of its disposal. The capacity for dispersal is limited by the availability of oxaloacetic acid and is, therefore, dependent upon carbohydrate metabolism.

It was felt at one time that ketosis was harmful and that it could be controlled by the proper ratio in the diet of ketogenic material (fats) to antiketogenic material (carbohydrates). This view is no longer held. The major effect of ketosis on the animal body appears to be in relation to acid-base balance, excretion of large amounts of acetoacetic acid and β-hydroxybutyric acid in the urine as their alkali salts deplete the body of available base and may lead to the development of an **acidosis**.

The oxidation of fat is not dependent on the simultaneous oxidation of carbohydrate because, when fatty acids and ketones are being extensively oxidized in muscle tissue, the resulting accumulation of acetyl CoA and citrate depress glycolysis and thus diminish the oxidation of glucose.

OXIDATION OF UNSATURATED FATTY ACIDS

The important unsaturated fatty acids are oleic acid, linoleic acid and arachidonic acid, etc. These fatty acids are metabolized much more slowly than the saturated fatty acids. The oleic acid before undergoing its oxidation is probably activated by conversion to Co-enzyme A thioester and, than, is oxidized in the β-position as usual. After the removal of three acetyl groups during oxidation the compound is left as $CH_3(CH_2)_7CH{=}CH.CH_2CoSCoA$, which is β, γ-unsaturated acetyl CoA not an α, β-compound.

Linoleic, linolenic, arachidonic and other vital polyunsaturated acids are found in high concentrations in some structural lipids particularly in the phospholipids and cholesterol esters.

It has been noticed that animals which feed exclusively on fat free diet fail to grow and develop skin and kidney lesions and are infertile as well. Administration of some of these vital polyunsaturated fatty acids not only restores the growth but also cures the deficiency manifestations. So they are called **"essential fatty acids"**.

METABOLISM OF CHOLESTEROL

Cholesterol is of major significance because of its relationship to many physiologically active steroids, sex hormones, adrenal cortex hormones, bile salts, etc., which are present in our body. It is insoluble substance and along with other substances, tends to precipitate in and along the lining of the blood vessels, thereby restricting the flow.

Ingested cholesterol is absorbed along with other lipids. It is normally present in blood to the extent of 150 to 250 mg per 100 ml, being equally distributed between the cells and the plasma. In the cells cholesterol occurs in free form, while in the plasma about 75% is found in the form of cholesterol esters.

Cholesterol is synthesized in the body from two-carbon units in the form of acetyl CoA formed either from fatty acids or from the metabolism of the carbohydrate through pyruvate. Two molecules of acetyl CoA condense to form acetoacetyl CoA which reacts with third molecule of acetyl CoA to form β-hydroxy β methyl glutaryl CoA which in turn, gives rise to the important intermediate compound called **mevalonic acid** which is activated by two molecules of ATP to yield **5-diphospho-mevalonic acid** (**mevalonic acid-5-pyrophosphate**). The 5-diphosphomevalonic acid in the presence of ATP loses CO_2 and water to form isopentenyl pyrophosphate which can also exist in an isomeric form 3, 3-dimethylellyl pyrophosphate. These compounds are said to be the forerunners of many important biological compounds including carotenoid pigments and cholesterol. One molecule of 3, 3 dimethylellyl pyrophosphate now reacts with one of isopentenyl pyrophosphate to yield geranyl pyrophosphate which with another molecule of isopentenyl pyrophosphate forms farnesyl pyrophosphate, with the removal of inorganic pyrophosphate at each stage. The two molecules of farnesyl pyrophosphate finally condense to form the hydrocarbon squalene which by ring closure and loss of methyl groups is readily converted into cholesterol by enzymes present in the liver.

It has been estimated that 1·5 to 2·0 gm cholesterol is synthesized daily in the body of human. The excess of cholesterol is eliminated from the body chiefly in the faeces. In the liver, cholesterol is reduced to **dihydrocholesterol** and coprostenol which pass into the intestine via the bile duct and are not reabsorbed.

FATTY ACID SYNTHESIS

The natural fatty acids found in the food have an even number of carbon atoms and range generally from 4 to 24 carbon atoms in length. The fatty acids in the body may arise from the diet or may be synthesised from carbohydrates as well. Acetyl CoA is the building block for the fatty acid chain. Therefore, any substance which can yield acetyl CoA may serve as a source of fatty acids. The acetate position of acetyl CoA supplies the carbon atoms of the fatty acid chain, and the chain is, in an overall sense, built up essentially by a process of successive additions of this two-carbon fragment to an existing chain. This accounts for the fact that naturally occurring fatty acids have an even number of carbon atoms.

Several processes have been recognized for the synthesis of fatty acids. One process, which takes place in liver mitochondria is essentially the reverse of the β-oxidative pathway. The principal exception is that saturation of the α-β double bond requires NADPH whereas desaturation requires FAD.

Most fatty acid synthesis takes place in the cell cytoplasm by a process which involves CO_2 fixation. The process proceeds as follows :

The acetyl CoA is converted to malonyl CoA by fixation of CO_2, *e.g.*, carboxylation. The malonyl CoA then reacts with the Co-enzyme A derivative of a fatty acid to form the Co-enzyme A derivative of a new fatty acid containing two additional carbon atoms, *e.g.*, the malonyl CoA reacts with acetyl CoA to form **acetoacetyl Co-enzyme A**. The entire reaction can be represented as follows :

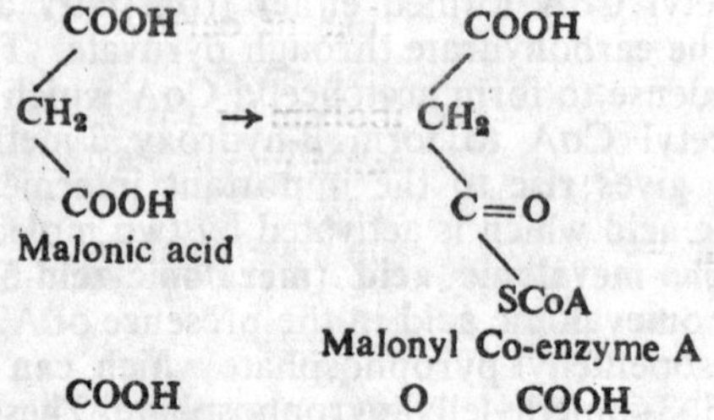

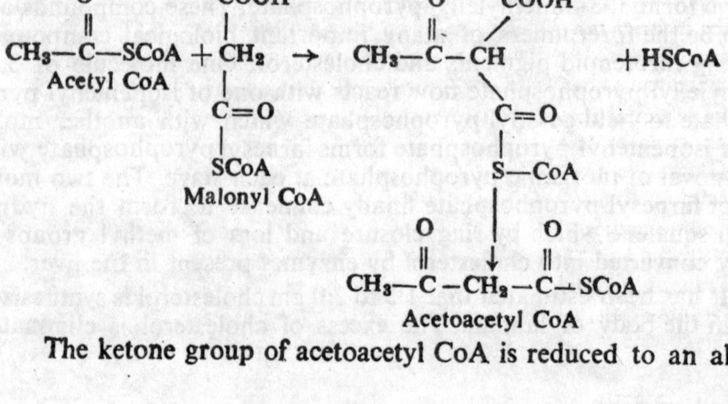

The ketone group of acetoacetyl CoA is reduced to an alcohol, the alcohol is dehydrated and the double bond is reduced to give butyryl CoA.

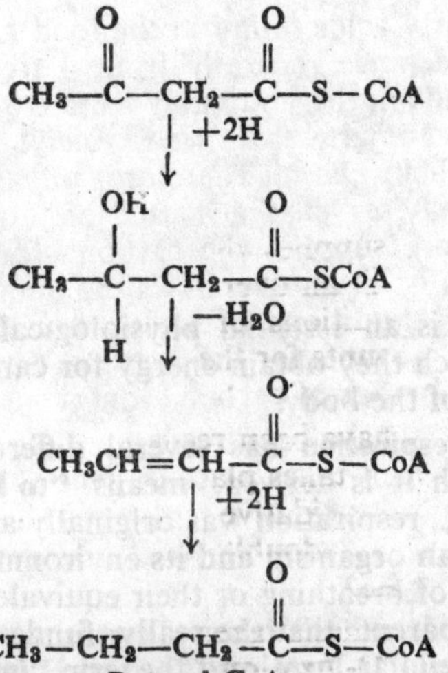

$$CH_3-\overset{\overset{O}{\|}}{C}-CH_2-\overset{\overset{O}{\|}}{C}-S-CoA$$

$$\downarrow +2H$$

$$CH_3-\underset{\underset{H}{|}}{\overset{\overset{OH}{|}}{C}}-CH_2-\overset{\overset{O}{\|}}{C}-SCoA$$

$$\downarrow -H_2O$$

$$CH_3CH=CH-\overset{\overset{O}{\|}}{C}-S-CoA$$

$$\downarrow +2H$$

$$CH_3-CH_2-CH_2-\overset{\overset{O}{\|}}{C}-S-CoA$$

Butyryl CoA

Butyryl CoA, thus formed, then reacts with another malonyl CoA and repeats the entire sequence of the reactions given above. In this way, the fatty acids are synthesized, two carbons being added with each sequence of reactions, which are repeated, etc.

REGULATION OF FAT METABOLISM

The metabolism of fats like that of carbohydrate is influenced by the nervous system and hormones secreted by endocrine glands. Of the endocrine glands, the metabolism of fats is regulated by, the hypophysis (pituitary gland), the sexual glands (testes and ovaries) the thyroid, the pancreas and the adrenals. The anterior lobe, pituitary secretes a hormone which fosters the accumulation of ketone bodies in the body.

Revision Questions

1. What is metabolism ? Discuss various types of metabolism.
2. Describe in brief the metabolism of carbohydrates.
3. Describe different theories of oxidation of fatty acids in animals body.
4. Write what you know about the protein metabolism.
5. Describe in detail the Krebs tricarboxylic acid cycle.
6. Write short notes on :
   (i) Alternative pathway,
   (ii) Adipose tissue,
   (iii) Ketone bodies,
   (iv) Nitrogen equilibrium,
   (v) Transamination,
   (vi) Deamination,
   (vii) Glycolysis,
   (viii) Glycogenolysis,
   (ix) Glycogenesis.

CHAPTER 8

Respiration

Respiration is an essential physiological activity of all living organisms by which they obtain energy for carrying out all other metabolic activities of the body.

The term **respiration** has several different usages. The Latin word, from which it is derived, means **"to breathe"** or **"exhale"** and, in this sense, respiration was originally applied to the exchange of gases between an organism and its environment. It referred to the obvious activities of breathing or their equivalent. As the years went by, it became apparent that the really fundamental exchanges were occurring at the cellular level, and the term **"internal respiration"** was often applied to this phase of gaseous exchange. At present, the adjective **"internal"** has been dropped, and respiration is frequently applied to cellular processes of energy production. So in its true physiological sense, **respiration is a chemical activity taking place within the protoplasm of the cell and results in the liberation of energy**.

The salient features of respiration are the intake of oxygen and output of the carbon-dioxide. The oxygen is used in the oxidation of digested food in the cell to liberate energy. Carbon dioxide is produced as a result of the oxidation of food materials. Its presence in the body is harmful, therefore, it is removed from the body during this activity.

In small animal like protozoans and sponges, the oxygen is taken directly from the air or from the watery medium surrounding them into all parts of their structures and carbon dioxide is given out from all their parts directly into the surrounding medium. It is because the cells of all parts are in direct contact with the environment and an exchange of gases between cells and their surrounding environment occurs directly.

In insects air is delivered directly to the tissues through the tracheae but in larger forms the cells are deprived of direct contact with the external environment, it is because of their complex structure. They, therefore, require the aid of respiratory and circulatory systems to permit satisfactory gaseous exchange and distribution of oxygen to all parts of the body. In these animals the process of respiration involves the following steps, according to G.S. CARTER.

1. **External respiration :** External respiration is usually defined as **"breathing"**. It refers to those mechanisms by which oxygen is brought into the body from the environment and carbon dioxide is expelled from the body into the environment. The exchange of

gases takes place at the respiratory surface which may be integument, gills, tracheae or lungs.

2. Transport of respiratory gases : This phase of respiration involves the transportation of oxygen from the respiratory surface to the body tissues and carbon dioxide from the tissues to the respiratory surface. In higher animals the transportation of respiratory gases is effected through blood, the components of which are very much sensitive to the respiratory gases.

3. Internal or tissue respiration : This phase of respiration includes all forms of oxygen consumption by the cells and production of carbon dioxide in the cells as a results of oxidative processes which lead to liberate energy for biological work. In other words it refers to the sum of enzymatic reactions both oxidative and non-oxidative by which energy is made available to maintain the other vital activities.

ROBERT HOOK was the first who gave the real understanding of respiration in the living organisms. No doubt PRIESTLEY discovered oxygen but it was LAVOISIER who established the idea that during respiration, oxygen is used and carbon dioxide is released by some life processes which now collectively called **"respiration"**

KINDS OF RESPIRATION

The essential feature of respiration is intake of oxygen which is used in the oxidation of digested food, but in many instances living processes can go on without oxygen being involved. As such the respiration is of following two types depending upon the availability of oxygen :

1. Aerobic respiration : When respiration involves the uptake of oxygen, then this type of respiration is called **aerobic respiration.**

2. Anaerobic respiration : When respiration does not involve the uptake of oxygen then this type of respiration is called **anaerobic respiration** and life in the absence of oxygen is called **anaerobiosis.** Such type of respiration is generally found in the parasitic worms like *Ascaris.*

SOURCES OF OXYGEN

Oxygen for respiration can be derived from two distinct sources : (1) free available oxygen from the air and (2) oxygen dissolved in the water. These are the only sources of oxygen of animals.

RESPIRATORY ORGANS AND MODE OF EXTERNAL RESPIRATION FOUND AMONG ANIMALS

The organs which are concerned with the gaseous exchange, *e.g.*, intake of O_2 and output of CO_2 and have greater rate of gas exchange per unit area than the general body surface are referred to as **respiratory organs.** Thus, respiratory organ may be a part of special region of the body or may be an organ specifically meant for this purpose such as lung.

In animals various types of respiratory organs are found. These may differ in their structure but all have a large surface of contact with the surrounding environment and are richly supplied with blood vessels and capillaries which ensure rapid gaseous exchange between the external environment (either air or water) and the blood.

The respiratory organs may be classified by their physiology or by their morphology. If they work in water they are gills, if in air, lungs. They may project outwards or they may be invaginations. The most important respiratory organs of the animals are as follows :

1. **Integument :** There are many animals such as protozoans, sponges, coelenterates, helminthes and amphibians in which integument or skin functions as a respiratory organ. The integument of these animals is richly vascularized and remains moist all the time so that oxygen from the surrounding environment can pass into the blood through simple diffusion.

In aquatic animals a circulatory mechanism is present to ensure free flow of water over the respiratory surface. This is effected by cilia present (in most of the cases) over the respiratory surface or integument.

In *Chaetopterus* and *Nereis* specialized parapodia are found which move in a fan-like manner, thereby maintaining a constant stream of water over the integument.

2. **Gills :** Gills are the respiratory surfaces of a number of aquatic animals including the chordates which all at one time or another had gills or gill-slits during their development. Gills are found in lamellibranchs, molluscs among others, also in many crustacean and fishes.

Typically gills are filamentous structures richly supplied with blood capillaries. They are located inside the body when derived from the anterior part of the alimentary canal as in fishes, or outside the body when they are outgrowths of the body surface as in amphibian larvae, *Polypterus* and other invertebrate larvae.

The gills found inside the body are referred to as **internal gills** where as the gills found outside the body are referred as to **external gills**. Both are meant for respiration.

In crustaceans such as the crayfish, lobster and crabs, etc., the gills are filamentous outgrowths from certain of the segmental appendages and are enclosed in a chamber of chitin carapace. These gills are usually ventilated by the paddle-like movements of special appendages such as the scaphognathites. In Malacostraca specialized branches of appendages (epipodites) are found which function for respiration. Numerous aquatic insects have gills but these are usually abdominal or caudal. Tracheal gills occur in nymphs of Odonata, Trichoptera and in larvae of some beetles but these gills are supplied with fine tracheae instead of blood capillaries.

Molluscs possess a variety of gills. The lamellibranchs possess two pairs of ctenidia which are variously modified. In *Chiton* there are 6 to 80 gills in each pallial groove. In gastropods the gills or ctenidia are relatively simple and are located in the mantle cavity but in cephalopods the gills are large and are richly vascularized.

In echinoderms finger-like evaginations of the coelomic cavity, the so called dermal papillae or dermal branchiae, serve as respiratory organs.

Protochordates also possess gills as their respiratory organs. In *Amphioxus* as many as 90 pairs of gills are present in series along the pharynx.

In fishes true gills are found which are usually covered. They are ventilated by breathing movements of mouth and operculum. In certain fishes accessory respiratory organs are also found which are respiratory in function like gills. In *Clarias* these are the branch extensions of the gill arches bearing numerous papillae. In *Anabas* these are labyrinthiform organs, in *Ophiocephalus* supra-branchial chamber while in *Saccobranchus* pharyngeal lung.

As soon as water comes in contact with gills, simultaneously gaseous exchange takes place at the gill surface, *e.g.*, oxygen is absorbed and carbon dioxide is released into the environment. Thus external respiration is effected.

3. **Lungs :** Lungs are referred as the chief respiratory organs of all the land living vertebrates such as reptiles, birds and mammals. Virtually all lungs are parts of or outgrowths from the alimentary canal and are richly vascularized. Most lungs are aerial, a few are water filled. Ventilation and diffusion lungs are distinguished according to the presence or absence of mechanisms for air renewal.

AIR LUNGS

1. **Diffusion lungs :** Such type of lungs are found in most invertebrates including scorpions, pedipalps, spiders, chilopods and tropical snails. These lungs as they are filled with air by simple diffusion are termed as "**diffusion lungs**". These lungs are more primitive when compared with ventilation lungs, because of the absence of ventilating mechanism.

2. **Ventilation lungs** : The lungs which are filled in with air by some sort of ventilating mechanisms, are referred to as "**ventilation lungs**". These are found in most vertebrates and are highly elastic. Two different types of ventilation are found among vertebrates; a positive pressure types in which air is forced into the lungs by swallowing or buccal movements as in the frog and a negative pressure system in which air is drawn in by increasing the space around the lungs as in mammals. Birds have extensive air sacs.

3. **Gas bladders and lungs of fishes :** In some teleost fishes a bag-like structure of variable shape is found above the oesophagus and below the vertebral column. This bag-like structure is called "**air bladder**" or "**gas bladder**". It is primarily a hydrostatic organ but in some fishes it serves as respiratory organ as it contains much oxygen which is usually used in hypoxic conditions. It is believed that it is a forerunner of the vertebrate lung.

In fishes the gas bladders are of two types, **open** found in physostome teleosts and **close** found in the physoclistous fishes. Both types of gas bladders can secrete gas into the bladder.

4. **Lungs of pulmonate gastropods :** Pulmonate gastropods also possess lungs but these are in the form of cavities in the mantle whose walls are well supplied with blood vessels. These animals are devoid of gills and depend solely upon pulmonary chamber to meet their respiratory requirements.

5. **Alimentary mucosa :** A number of tropical fishes respire by means of the intestinal or gastric mucosa, because in these fishes the intestinal or gastric mucosa becomes richly vascularized. The air is swallowed in and during its transit through the intestine exchange of gases takes place and the residual air is sent out through the anus.

WATER LUNGS

Such type of lungs are found in most invertebrates. These are alternately filled with or emptied of water. The water is drawn into and expelled from the body rhythmically and in doing so gaseous exchange takes place, oxygen is absorbed and carbon dioxide is given out.

In holothurians the respiratory organs are the respiratory trees which are filled with sea water by cloacal pumping and contractions of the trees themselves.

4. **Tracheae** : Tracheal respiration is characteristic of insects and also of onychophorans, some spiders, isopods, diplopods and chilopods, during which air is carried directly to the metabolizing cells without blood. Tracheal system consists of a large number of inter-concected small tubes, the **tracheae** which usually open outside through minute pores called spiracles which are located on either side of the body. The air is pumped into and outside the body through these spiracles by the movements of the body and the gaseous exchange takes place directly in the individual cell. Single spiracle may serve for both inspiration and expiration, but more usually there are numerous spiracles, some of them (commonly the anterior ones) for inflow and others for outflow of the air.

The tracheae divide and redivide to form minute branches called **tracheoles**. These ramify through the different tissues of the body. In some cases such as crickets, *Dytiscus* and *Apis*, the tracheal tubes are dilated to form **air-sacs**.

HOW GASEOUS EXCHANGE TAKES PLACE AT THE RESPIRATORY SURFACE ?

Gaseous exchange, *e.g.*, intake of oxygen and output of carbon dioxide, always takes place at the respiratory surface because the surface is richly vascularized. The respiratory surface may be a integument, a lung or a gill, the description of which have been given above.

As soon as the source of oxygen (atmospheric air or water) comes in contact with the respiratory surface by any sort of ventila ting mechanism,gaseous exchange, takes place, *e.g.*, oxygen is absorbed into the body and carbon dioxide is released into the environment whatever it may be.

It is believed that gaseous exchange takes place by **"simple diffusion"** which is caused due to the partial pressure of the respiratory gases. Gases move from their high partial pressure to low partial pressure. In the environment (air or water) the partial pressure of the oxygen is comparatively high than the body, therefore, the oxygen is diffused from the environment into the body through the respiratory surface. Similarly the partial pressure of the carbon dioxide in the blood is high than the immediate environment, therefore, it is diffused outside the body through the respiratory surface.

The partial pressure of any gas in gaseous mixture is proportional to its concentration in the mixture. For example, oxygen comprising approximately 21% of the air at a barometric pressure of 760 mm Hg exerts a partial pressure of about 160 mm Hg ($0{\cdot}21 \times 760$). This is often called the **"oxygen tension"** of air.

TRANSPORT OF RESPIRATORY GASES

In single celled animals like protozoans the respiratory gases are directly diffused into or outside the body but in complex animals like mammals the respiratory gases (O_2 and CO_2) are transported with the help of circulatory system in which the blood is circulatory fluid (in most of the cases particularly in higher animals). Blood itself is not the carrier of the respiratory gases but contains a respiratory pigment which actually acts as the carrier of the respiratory gases because this pigment has a special affinity for respiratory gases. The respiratory pigment in its nature varies in different animals. The different respiratory pigments found in the animals are as follows :

1. Haemoglobin, 2. Haemerythrin, 3. Haemocyanin, and 4. Chlorocruorin.

Of these the first two pigments are found either in solution or in corpuscles in the plasma of the blood and are widely distributed where as the rest two are in solution in the plasma of the blood. In addition to the above mentioned pigments there are certain other miscellaneous pigments which are sporidically met in the animals. These are **pinnaglobin, echinochrome, vanadium** and **molpadium,** their functions are not clear so far.

1. Haemoglobin : It is the most important and commonest respiratory pigment which is widely distributed in the animals. It is found in the dissolved condition in the plasma of the blood of molluscs, annelids and arthropods but in the corpuscles (R.B.C.) in the chordates. It is even found in the body fluid of some protozoans. It is also found in the muscles of birds and mammals and is known as **myoglobin**.

Haemoglobin is a red coloured pigment formed of two distinct components, the **haeme** and **globin**. The haeme is iron containing component or the prosthetic group of iron while the globin is the conjugated protein. Each molecule of haemoglobin contains four haeme groups, each of molecular weight 872. An oxygen molecule may unite reversibly by combining with one of the four iron atoms that is attached by valency bonding to four pyrrole groups that make up the haeme molecule. A remaining sixth valency bond of iron apparently attaches with the globin. The percentage of haemoglobin in the blood varies from animal to animal. In the human blood every 20 cc of blood approximately contains 10 gms of haemoglobin.

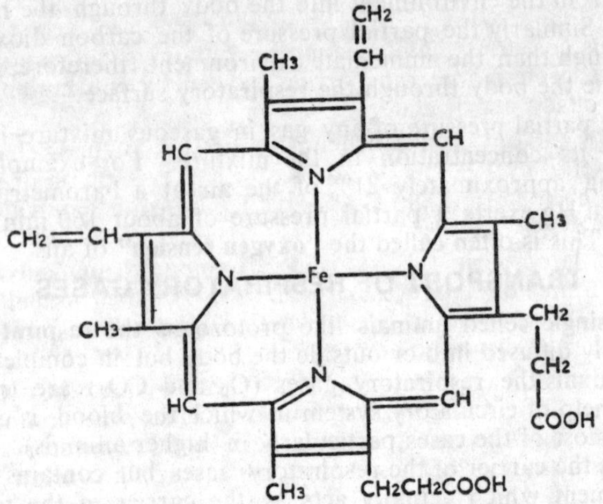

Fig. 8·1. Basic haeme structure of the haemoglobin molecule.

The most important property of haemoglobin is that it has a special affinity towards respiratory gases, O_2 and CO_2. It combines at normal temperature with oxygen and carbon dioxide when it comes in contact with them and also readily dissociates itself from them. The haemoglobin when combined with oxygen is called **oxy-haemoglobin.** The reaction is represented as follows :

$$\text{Hb} + O_2 \leftrightarrows \text{HbO}_2$$

Haemoglobin + Oxygen   Oxyhaemoglobin

Oxyhaemoglobin on reaching at cellular level, where partial pressure of the oxygen is very low, dissociates into free oxygen and reduced haemoglobin. This is really a reduction process. Thus the reaction is reversible as shown above.

The amount of oxygen to which haemoglobin combines depends upon the partial pressure of oxygen available. At maximum oxygenation human haemoglobin will combine with upto four oxygen molecules which it does at a partial pressure of oxygen of approximately 70 mm Hg. The saying mean of this is that the haemoglobin fully combines with oxygen at that partial pressure at which it is available in the environment. Similarly it must dissociates at the partial pressure of oxygen existing in the animal tissues.

2. **Haemerythrin :** This is also a iron containing pigment of

violet colour, found commonly in the polychaetes, worm *Magelona*, in the sipunculid worms *Sipunculus*, *Dendrostomum* and *Glofingia*. This is found in corpuscles. It is less efficient in its oxygen carrying capacity when compared with haemoglobin.

3. **Haemocyanin :** This pigment is next in importance to haemoglobin. It is found in many crustaceans and some mollscus such as cephalopods (squids). It is not confined to corpuscles like haemoglobin but exists freely in the blood plasma.

This is colourless or blue copper containing pigment which is capable of accepting and transporting oxygen by conversions of part of the cuprous to cupric state.

The physiology and biochemistry of this pigment have been studied in detail by REDFIELD. It has the same shape of dissociation curve as of haemoglobin but is not able to transport equivalent volumes of oxygen.

4. **Chlorocruorin :** This is green coloured pigment found in polychaete annelids particularly of the families Sabellidae and Serpulidae. It is green in dilute solutions but red in concentrated ones. It occurs in dissolved condition in plasma.

This pigment was discovered by MILNE-EDWARDS in polychaetes in the year of 1838. Later DUJARDIN and QUADRIFAGES confirmed the occurrence of this pigment in other annelids. Its respiratory properties have been studied by LANKESTER and according to him the pigment exists in two forms, one oxidized and the other reduced.

Functions of respiratory pigments : The most important functions of respiratory pigments are as follows :

1. They are the carrier of oxygen. In the absence of respiratory pigments the blood could carry the oxygen only in solution and the amount carried will be very low.

2. All the pigments have great affinity towards the respiratory gases, to whom they can combine at high partial pressure and dissociate at low partial pressure.

The quantity of either of the respiratory gases combining with any of the respiratory pigment depends upon the partial pressure of that gas. The increased pressure results in the increased fixation of that gas and it begins to decline when it reaches close to the saturation point. The saturation point is reached in the alveoli due to maximum partial pressure.

3. These pigments exhibits reversibility in their action, the reversible nature of their action can be expressed by the following equation.

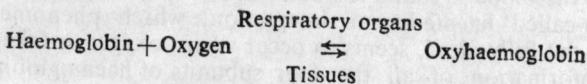

$$\text{Haemoglobin} + \text{Oxygen} \underset{\text{Tissues}}{\overset{\text{Respiratory organs}}{\rightleftarrows}} \text{Oxyhaemoglobin}$$

TRANSPORT OF OXYGEN

1. **Removal of oxygen from respiratory surface :** Oxygen from the external environment is diffused into blood through the respiratory

surface which is permeable to respiratory gases. A small quantity of diffused oxygen is carried by the respiratory pigment present in the blood. It has been estimated that oxygen capacity of haemoglobin in cubic centimeters per 100 ml of blood is 9·0 in fishes, 12·0 in amphibians, 9·0 in reptiles, 18·5 in birds and 25·0 in mammals. In annelids and molluscs it is 6·5 and 1·5 respectively. The quantity of oxygen carried by haemoglobin is affected by the partial pressure of the oxygen in the blood plasma.

As the oxygen diffused into the blood, it combines with the respiratory pigment of the blood to form a temporary compound. If the respiratory pigment is haemoglobin than it combines with oxygen to form **oxyhaemoglobin**. Thus oxygen is carried by the blood in the form of oxyhaemoglobin.

2. Diffusion of oxygen from the blood into the tissue cells : As it reaches the different body tissues which are in need of oxygen, the oxyhaemoglobin present in the blood dissociates into free oxygen and reduced haemoglobin which is ready to combine with molecules of oxygen. The oxygen, so liberated, is used in the oxidation of digested food to liberate energy in the tissue cells. The decomposition of oxyhaemoglobin in the tissue cell is caused due to low partial pressure of the oxygen and high partial pressure of carbon dioxide in the tissue cells.

Oxygen dissociation curves : Each haemoglobin molecule, as being composed of four globin-haeme units, combines with four molecules of oxygen, enabling whole blood to carry about 60 times as much oxygen as could be transported by an equal volume of water or plasma. The amount of oxygen carried by the haemoglobin molecules in the blood is related to the partial pressure of oxygen (Po_2) in the blood. By exposing blood at a constant pH to different partial pressures of oxygen and, following equilibration, determining the oxygen content of the red cells, a curve representing the combining capacity at varying partial pressures of oxygen is obtained. Oxygen content may be expressed as per cent saturation. If, after exposure to pure oxygen, the oxygen content of haemoglobin was found to be 19 ml/100 ml of blood and this blood is said to be 100% saturated, a similar sample exposed to a lower Po_2 and containing 9·5 ml O_2/100% ml would be 50% saturated. Such plots are called **oxygen dissociation curves**. These curves serve to express the importance of oxygen tension on both loading and unloading (Fig. 8·2). In fig. 8·2 there is a series of curves which are sigmoid in shape rather than linear expressing that at high Po_2 (100 mm Hg) the blood is 100 saturated. As the Po_2 drops, the oxygen in the haemoglobin molecules is given up as a result of which the blood is found less saturated. This is believed to be due to the so-called **haeme-haeme** interaction, which phenomenon, although not yet fully clear, seems to occur when oxygen is taken up and the conformation of all the four subunits of haemoglobin, or more specially the β-chains, is altered. There is then some separation of the subunits which, in turn, increase the rate of uptake of oxygen by the four haeme groups.

The curves in fig.8·2 also show the influence of different carbon dioxide pressures in the dissociation of oxyhaemoglobin of human blood. Thus, if curve *O* is followed from right to left, when there is no carbon dioxide the blood is fully saturated with oxygen at 100 mm oxygen pressure; at 40 mm oxygen pressure it is about 96% saturated; at 20 mm ; it is 83% saturated ; and at 0 mm it does not contain oxygen. This shows that when the Po_2 increases, the above reaction proceeds to the right and oxyhaemoglobin is formed in great amount, as in the lungs. When Po_2 decreases, as in the tissues, the reaction proceeds toward the left and more oxygen is liberated. As the carbon-dioxide partial pressure increases the dissociation curves are shifted to the right. This states that if more carbon dioxide is present the haemoglobin can hold less oxygen, a phenomenon known as the **Bohr effect**, after its discoverer. The decrease in oxygen saturation with an increase in carbondioxide partial pressure may be due to some change in the conformation of the α- and β-chains of the globin moiety of haemoglobin by addition of hydrogen ion concentration associated with the increase in the carbon dioxide partial pressure.

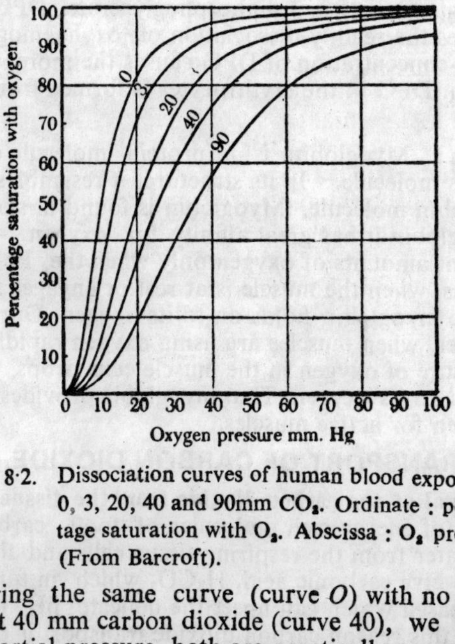

Fig. 8·2. Dissociation curves of human blood exposed to 0, 3, 20, 40 and 90mm CO_2. Ordinate : percentage saturation with O_2. Abscissa : O_2 pressure (From Barcroft).

Comparing the same curve (curve *O*) with no carbon dioxide and the one at 40 mm carbon dioxide (curve 40), we see that, at 100 mm oxygen partial pressure, both are practically completely saturated with oxygen; *i.e.*, the haemoglobin is almost all present as oxyhaemoglobin. At 90 mm oxygen partial pressure, which is the pressure in the arteries, they are still nearly the same, curve *O* being about 99% and curve 40 about 95% saturated. At 40 mm oxygen partial pressure, which is the pressure of veins, the *O*-curve still shows about 95% saturation while curve 40 is down to 72% saturation ; *i.e.*, the presence of 40 mm carbon dioxide has caused the oxyhaemoglobin to dissociate 23% of its oxygen. This shown that the effect of carbon dioxide pressure is just opposite that of oxygen pressure, and both have desirable physiologic effect. In the tissues with low oxygen and high carbon dioxide partial pressures, oxyhaemoglobin dissociates more readily and oxygen becomes available for tissue needs. In the lung the Po_2 is high and oxyhaemoglobin is formed readily despite the high carbon dioxide pressure.

pH, temperature, and the presence of electrolytes are also some other factors which influence the transport of oxygen by haemoglobin to a great extent. Slight decrease in pH (more acidic) increases the dissociation of oxyhaemoglobin. Thus the slightly more acid pH in tissues due to carbon dioxide favours the release of oxygen to the tissues. The slight increase in temperature and the presence of electrolytes also have the similar effect on the dissociation of haemoglobin and transport of oxygen by blood.

Very recently BENESCHS, 1968, found that there are certain **organic phosphate compounds**, mainly **diphosphoglyceric acid,** which have a marked effect on the oxygen-binding power of haemoglobin. The high concentration of 2, 3 diphosphoglycerate (DPG) in the erythrocytes initiates the readily dissociation of oxyhaemoglobin. Conversely, the low-concentration of DPG causes the more production of oxyhaemoglobin. DPG in the erythrocytes is formed from glucose and phosphate.

Myoglobin : Myoglobin is a protein molecule containing an iron group in its molecule. In its structure, it resembles a single unit of the haemoglobin molecule. Myoglobin is found in skeletal muscles and like haemoglobin it has great affinity for oxygen. It begins to release significant amounts of oxygen only when the Po_2 falls below 20 mm Hg. Thus, when the muscle is at rest or engaged in only moderate activity, the myoglobin holds on to its oxygen. During strenuous exercise, however, when muscles are using oxygen rapidly and when the partial pressure of oxygen in the muscle cells drops toward zero, myoglobin gives up its oxygen. Thus myoglobin provides an additional reserve of oxygen for active muscles.

TRANSPORT OF CARBON DIOXIDE

1. **Removal of the carbon dioxide from the tissue cells :** Due to the oxidation of energy rich molecules of food, carbon dioxide is released with water from the respiring tissue cells and the two combine together to give carbonic acid, H_2CO_3 which in turn ionises and H^+ ions are released which can upset the delicate pH balance of the organism. For this reason carbon dioxide transport is often involved with some form of buffering. In marine invertebrates blood proteins assist as buffers for H^+ ions but in the vertebrates the role of haemoglobin in buffering and CO_2 transport is vital.

The production of carbon dioxide in the tissue cells results in high tissue carbon dioxide partial pressure which causes its diffusion from the cells into the blood through the capillary walls.

In vertebrates the carbon dioxide is transported in blood in two ways ; through the plasma and through the blood corpuscles. Nearly

80% of the carbon dioxide in the plasma is transported in the form of the bicarbonates of sodium and potassium, the remaining nearly 20% is transported in the form of carbamino compound of haemoglobin in the red blood cells.

On arrival at the lungs the bicarbonates re-enter the red blood corpuscles and once again combine with H^+ ions from the dissociation of the H.Hb forming carbonic acid. This carbonic acid breaks in the presence of **carbonic anhydrase** into carbon dioxide and water which are liberated into the alveoli and then to the atmosphere.

2. Removal of carbon dioxide from the respiratory surface to the external environment : As the carbon dioxide, in the form of bicarbonates, carbamino-haemoglobin and carbonic acid, reaches the respiratory surface it starts diffusion from the blood into the external environment. The movement of carbon dioxide from the blood into the environment is due to low partial pressure of the carbon dioxide at the respiratory surface.

The oxygen and carbon dioxide transports are closely associated in vertebrates because the presencc of carbon dioxide in the blood causes a decrease in the amount of oxygen that can be carried at any given partial pressure. The effect of this is to stimulate the release of O_2 in the tissue cells where the CO_2 pressure is high and the loading of O_2 at the respiratory surface where it is low.

The tissue cells, thus, continuously consume oxygen and produce carbon dioxide. Owing to this the tension of oxygen in the cells falls while the tension of carbon dioxide rises. As a result oxygen diffuses from the blood into the tissue cells and carbon dioxide in the opposite direction.

CELLULAR RESPIRATION

The oxygen liberated from the dissociation of oxyhaemoglobin in the cell, is used in the oxidation of energy rich molecules as a result of which carbon dioxide and water along with certain amount of energy are formed. So all the reactions taking place in the cell to liberate energy constitute **"cellular respiration"** or **"tissue respirations"**. Much of the cellular respiration takes place in the mitochondria of the cells. This is expressed by the following equation :

$$C_6H_{12}O_6+6O_2 \rightarrow 6CO_2+6H_2O+\text{Energy}$$

The various chemical reactions involved in energy release or cellular respiration can be grouped under two heads :

1. Chemical reactions proceeding in the absence of oxygen, these reactions are collectively referred to as **glycolysis** (anaerobic phase of cellular respiration).

2. Chemical reactions taking place in the presence of oxygen, these reactions are collectively referred to as KREBS **cycle** or **tricarboxylic acid cycle** or **citric acid cycle** (aerobic phase of cellular respiration).

GLYCOLYSIS

Glycolysis is the anaerobic phase of cellular respiration. The entire process of glycolysis is completed under the following stages :

1. Stage (activation) : During this stage the basic substrate like glucose is activated, after undergoing phosphorylysis in the presence of ATP (adenosine triphosphate) and enzyme **phosphorylase** (**glucokinase** or **hexokinase**), to form glucose-6-phosphate. Glucose-6-phosphate undergoes an internal rearrangement in the presence of an enzyme **phosphogluco isomerase** to form fructose-6-phosphate. Fructose-6-phosphate undergoes further phosphorylation in the presence of **phospho fructokinase** enzyme to form fructose 1, 6-diphosphate.

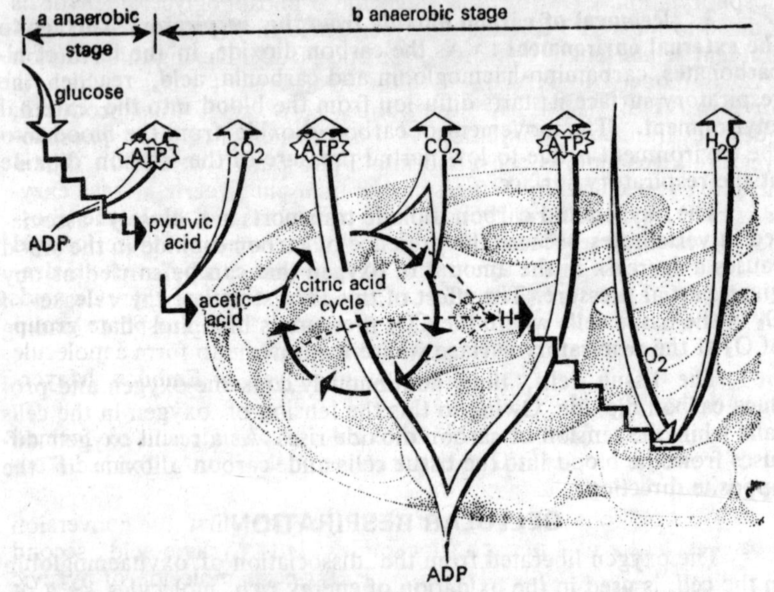

Fig. 8·3. A diagrammatic representation of the chemical changes that occur in cellular respiration.

The effect of these reactions is to convert one molecule of monosaccharide (glucose) into one molecule of fructose diphosphate. In these reactions no oxidation occurs but one or two molecules of ATP are converted to ADP (adenosine diphosphate) depending upon the starting substrate. Adenosine triphosphate, *i.e.*, ATP is an essential catalyst in this entire process. The various reactions of this stage can be summarized as follows :

Glucose
↓ hexokinase (glucokinase)
Glucose-6-phosphate
↓ phosphoglucoisomerase
Fructose-6-phosphate
↓ phosphofructokinase
Fructose 1, 6, diphosphate

2. Stage (cleavage) : In this stage the fructose 1, 6 diphosphate produced in previous stage, splits into two separate compounds,

3 phosphoglyceraldehyde and dihydroxy acetonephosphate, in the influence of enzyme **fructo aldolase.** Dihydroxyacetone phosphate is also converted into 3 phosphoglyceraldehyde in the presence of **triose-isomerase** enzyme because the 3 phosphoglyceraldehyde can only take part in the subsequent reactions of the glycolysis. The result of this stage is the conversion of one molecule of hexose-diphosphate to two molecules of triose phosphate.

3. **Stage (oxidation) :** Third stage of glycolysis is an oxidative phosphorylation. During this stage the 3-phosphoglyceraldehyde is oxidized to form 1, 3 disphosphoglyceric acid in the presence of **glyceraldehyde phosphate dehydrogenase** and NAD. During this reaction NAD is reduced to NADH with the liberation of energy which is stored in ATP. The 1, 3, diphosphoglyceric acid, thus produced, undergoes transphosphorylation in the presence of enzyme **phosphoglycerate kinase** and is converted into 3-phosphoglyceric acid. The 3 phosphoglyceric acid in the presence of **phosphoglyceric mutase** enzyme is rearranged to form 2-phosphoglyceric acid and, thus, the phosphate residue moves from C–3 to C–2. The 2-phosphoglyceric acid under the influence of the enzyme **enolase (phosphopyruvate hydratase)** loses the elements of water to form phospho (enol) pyruvic acid. The phospho (enol) pyruvic acid then loses its phosphate group to ADP in the presence of **pyruvate kinase** enzyme to form a molecule of pyruvic acid, it is an end product of **glycolysis** or **Embden Meyerhof pathway**.

In the end of glycolysis two molecules of pyruvic acid are formed because two molecules of 3-phosphoglyceraldehyde are formed from the fructose 1, 3. diphosphate molecule.

The results of the glycolysis are three fold ; first the conversion of each molecule of 3 phosphoglyceraldehyde to pyruvic acid ; second the synthesis of two molecules of ATP for each molecule of pyruvic acid formed ; and third, the reduction of one molecule of NAD into NADH for each molecule of pyruvic acid formed.

KREBS CYCLE OR TRICARBOXYLIC ACID CYCLE OR CITRIC ACID CYCLE

The end products of glycolysis still contain much of the energy of the glucoce molecule, therefore, these products enter the KREBS cycle or tricarboxylic acid cycle so that they may be oxidized through the oxidative reactions of the KREBS cycle to liberate energy. In this cycle NADH formed in the oxidation of the organic acid is oxidized through a series of steps involving the cytochrome system as the final link with oxygen. Each of these steps represents free energy decrease.

The energy which is liberated during KREBS cycle is conserved in the same way as it is conserved in glycolysis, *e. g.*, by conversion to high energy phosphate compounds.

The pyruvic acid formed during the process of glycolysis, undergoes oxidative decarboxylation in the presence of **pyruvic dehydrogenase** and co-enzyme thiamine pyrophosphate (co-carboxylase) to form acetyl CoA. The acetyl CoA in the presence of **citrate synthetase** reacts with oxaloacetic acid to form citric acid. In this reaction no oxidation or decarboxylation is involved but a molecule of water is required to hydrolyse the linkage between the acetyl group and co-enzyme A.

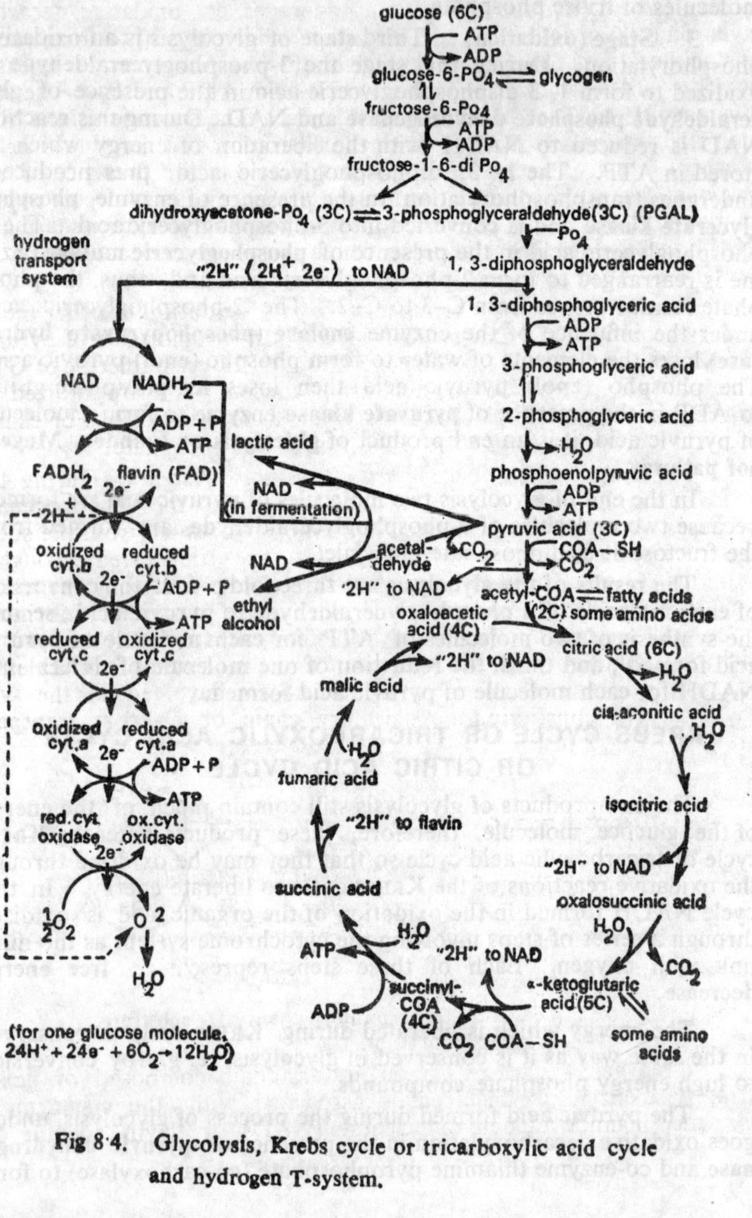

Fig 8·4. Glycolysis, Krebs cycle or tricarboxylic acid cycle and hydrogen T-system.

In this way the Co-enzyme A, of which only a small amount is present in the tissue cells, is liberated and can react with more pyruvic acid. In the mean time, the citric acid undergoes an internal rearrangement in the presence of enzyme **aconitate hydratase,** in which the hydroxyl group migrates to an adjacent carbon atom to give isocitric acid. The isocitric acid is then oxidized (two hydrogens removed) in the presence of **isocitrate dehydrogenase** to form the oxalosuccinic acid, which then decarboxylates (loses CO_2) to form α-ketoglutaric acid. The α-ketoglutaric acid then undergoes an oxidative decarboxylation (*i. e.*, it loses CO_2 and essentially adds oxygen) to form succinic acid and free Co-enzyme A. During this reaction energy is released which is utilized indirectly to form ATP from ADP and P_1. The succinic acid is oxidised to yield fumaric acid. The succinic acid is first dehydrogenated in the presence of flavoprotein and **succinate dehydrogenase** to give fumaric acid. Fumaric acid in the presence of the enzyme fumarate hydratase (fumarase) takes up water molecule to give malic acid which is dehydrogenated by **malate dehydrogenase** and NAD to give oxaloacetic acid which is ready to react with more acetyl CoA to repeat the cycle.

The citric acid cycle is repeated for the oxidation of other molecule of pyruvic acid. During KREBS cycle for one molecule of pyruvic acid, two molecules of NADH, one of NADPH, one molecule of GTP and one molecule of succinic acid are produced. All are utilized as energy carriers. Ultimately the energy is located in ATP. For example, GTP converts ADP to ATP by transfer of phosphate.

Respiratory chain or **electron transport system :** During the oxidation and reduction processes of KREBS cycle, the dehydrogenase enzymes remove hydrogen atoms or pairs of electrons from the substrate. The electrons or hydrogen atoms are transferred through a series of enzymes untill they are oxidized to water. During the transfer process of hydrogen atoms from one enzyme to another, enormous amount of energy is released which is sufficient to form the energy rich phosphate bond in ADP molecule and thus several molecules of ATP are synthesized. The enzymes which are involved in the synthesis of ATP constitute the **respiratory chain** or **electron transport system** (Fig. 8·5.).

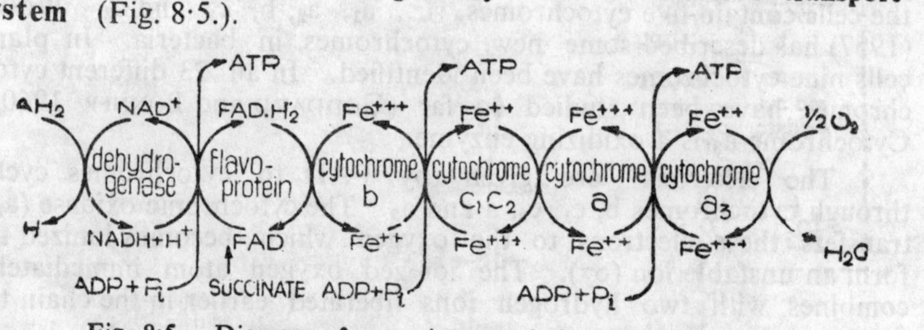

Fig. 8·5. Diagram of a respiratory chain AH_2 substrate, A oxidized substrate.

Transport system : The respiratory chain composed of dehydrogenase enzymes which remain associated with the coenzyme I, *viz.*, nicotinamide adenine dinucleotide (NAD) or flavin adenine dinucleotide (FAD), cell pigments known as cytochromes, non-haeme iron, copper and coenzyme Q. All these enzymes of respiratory chain are localized in the stalked particles (F_1 particles) of the mitochondria. They are probably arranged in a definite sequence like a mass production factory. The start of the respiratory chain is localized at the base of the stalks where it can collect reduced NAD (NADH) molecules diffusing across the outer space where the KREBS cycle occurs. Reduced molecules enter the chain while those leaving it are fully oxidized to CO_2 and H_2O. The various stages of respiratory chain are as follows :

1. The dehydrogenase enzymes remove the hydrogen atoms from the substrate at various stages of KREBS cycle. The hydrogen atoms become ionized or broken up into protons (+) and electrons (e^-).

$$2H \rightarrow \underset{\text{Protons}}{2H^+} + \underset{\text{Electrons}}{2e^-}$$

The protons or hydrogen ions reduce the coenzyme (NAD) part of the dehydrogenase enzyme.

$$NAD + 2H^+ \rightarrow NADH + H^+$$

The NADH molecules pass through the space between the inner and outer membranes of the mitochondria. The NADH molecules act as a link between the Krebs cycle enzymes which are on the outer side and the respiratory chain enzymes which occur on the inner side. The NADH molecules of the Embden Meyerhof path way also come in the mitochondria and enter in the respiratory chain system.

2. The NADH is oxidised into NAD^+ by transferring its hydrogen to the flavoprotein enzymes (flavin adenine dinucleotide, FAD) which acts as a hydrogen carrier. From the flavoprotein each hydrogen atom is discharged into the cell fluid as a hydrogen ion and the electrons are passed on to cytochromes.

3. The cytochromes are the iron containing cell pigments which occur universally in aerobic cells and act as enzymes. They receive electrons from the flavoproteins or other carrier enzymes and pass them to cytochrome oxidase which activates oxygen. Usually the cells contain five cytochromes, *viz.*, a_1, a_2, b, c_1 and c_2. WOOD (1957) has described some new cytochromes in bacteria. In plant cells nine cytochromes have been identified. In all 23 different cytochromes have been studied so far (GODDARD and BONNER 1960). Cytochrome a_3 is a oxidizing enzyme.

The electrons are given by FAD to cytochromes cycle through cytochromes b, c_1, c_3, a and a_3. The cytochrome oxidase (a_3) transfers these electrons to the oxygen which becomes ionized to form an unstable ion (o^-). The ionized oxygen atom immediately combines with two hydrogen ions liberated earlier in the chain to form water molecules.

According to certain authors, *viz.*, AMBROSE and EASTY (1970) and ROBERTIS, NOWINSKI and SAEZ (1970) the electrons before passing through cytochromes pass through the coenzyme Q (uniquinone). But this stage is still little understood.

Oxidative phosphorylation : The cytochromes of respiratory chain act as carriers of electrons. When a cytochrome transfers electrons to other cytochrome then enormous amount of energy is released. This energy is traped by ADP and inorganic phosphate molecule to form one molecule of ATP.

$$ADP + Pi + energy \rightarrow ATP \text{ (i-inorganic)}$$

The process of ATP formation is known as **oxidative phosphorylation** because phosphate is added to ADP using energy from the oxidation. In the process of oxidative phosphorylation 36 molecules of ADP are used and 36 molecules of ATP are synthesized.

In the end it can be calculated that in the oxidation of one molecule of glucose or glycogen, two molecules of ATP are used during phosphorylation in glycolysis or Embden Meyerhof pathway and 40 molecules of ATP are released (synthesized) during the Embden Meyerhof pathway and Krebs cycle. Therefore, when one molecule of glucose is oxidized by the aerobic cells it releases 38 molecules of ATP.

$$C_6H_{12}O_6 + 6O_2 + 38\ ADP + 38\ Pi \xrightarrow[\text{Enzymes}]{\text{Respiratory}} 6CO_2 + 12H_2O + 38\ ATP \text{ or } (4,56000 \text{ calories}).$$

ENERGY CARRIERS AND STORES IN THE CELL

Energy liberated during cellular respiration is stored in the adenosine triphosphate, *i. e.*, ATP. The energy is carried in the third phosphate bond which on hydrolysis yields adenosine-diphosphate, ADP, phosphate and 33,000 joules.

$$ATP \leftrightarrows ADP + P\ 33,000 \text{ joules per molecule}$$

The high energy phosphate bond may also be transferred to and stored in other instances, thus in vertebrates muscle creatine is converted to creatine phosphate and in some invertebrates arginine is converted to arginine phosphate. This releases the ADP to collect further energy-rich phosphates.

ATP molecules are involved in the synthesis of complex molecules of the cells. They are important in the provision of energy for active intake and secretion across membranes against diffusion gradi ents. Muscle contraction and nerve conduction depend upon ATP.

FACTORS AFFECTING THE RATE AND PRODUCTS OF RESPIRATION

There are several factors which influence the rate and products of respiration. The most important factors are as follows :

1. **Temperature :** It is one of the factors which influences the rate of respiration because enzymes are temperature dependent and respiration involves a series of enzyme catalyzed reactions. An increase in the temperature results an increase in the rate of respiration until the point of denaturisation of the enzyme occurs.

For warm-blooded animals the optimum temperature of metabolic activity, including respiration is approximately 37°C. Other poikilothermic animals and plants may have different optima.

2. **The respiratory substrate :** The type of substrate being oxidized by a particular animal at any time has an effect on the proportion of gases consumed and produced.

3. **The availability of oxygen :** The rate and products of respiration are also influenced by the availability of oxygen because the concentration of oxygen determines which chemical pathway the pyruvic acid, produced at the end of anaerobic glycolysis, will follows. As a general rule above concentrations of two per cent it passes via acetyl CoA into the KREBS cycle while below this concentration it is converted to alcohol and carbon dioxide.

RESPIRATORY QUOTIENT

The ratio of carbon dioxide liberated to that of oxygen taken up in unit time is termed the "**respiratory quotient**" or in other words the relative proportion between the volume of carbon dioxide released and the volume of oxygen absorbed in unit time during the respiration of the animal is termed **"respiratory quotient"** or **R. Q.** This is expressed by the following equation :

$$R.\,Q. = \frac{\text{Volume of } CO_2 \text{ given out in unit time } t}{\text{Volume of } O_2 \text{ absorbed in unit time } t}$$

The volume of a gas, under given conditions of temperature and pressure, is proportional to the number of molecules that it contains so that the respiratory quotient is also the ratio of the number of molecules of CO_2 and O_2 taking part in the over all reaction by which the carbohydrates, fats and proteins are oxidized.

Respiratory quotients have been worked for different substrates. For carbohydrates the R.Q. is $6\ CO_2/6O_2 = 1$, while for a fat the R.Q. is $57\ CO_2/80\ O_2 = 0{\cdot}71$ and for protein is 0·8. Thus for animal using all three substrates R. Qs. would vary between 0·7 to 1·0. For man the usual figure is about 0·85.

The quantitative determination of R. Q. can be very useful for us in suggesting what the organism respiring at any time but great care must be taken in its interpretation because tissues with R. Qs. of 1·0 and 0·7 obviously have different chemistry, but two with values of 0·8 are not necessarily similar, for while one may be using a mixture of carbohydrates and fats.

The respiratory quotient seems to be influenced by changes in temperature in certain animals as HALL has reported about the increase in the respiratory quotient of turtles from 0·52 at 1°C to 0·75 at 29°C.

ROLE OF MITOCHONDRIA IN RESPIRATION

Mitochondria are the structures found in thousands in the cell cytoplasm. They play vital role in the process of respiration because all the enzymes involved in respiration are found in them. The enzymes concerned with KREBS **cycle** and with the oxidation of fatty acids are found in the matrix of these structures, while the hydrogen acceptors, oxidases and phosphorylases are found in the membrane and cristae system of mitochondria.

There is some evidence that the sequence of stage in cellular respiration is paralleled by the spatial arrangements of the enzymes within the mitochondria. There may be as many as 100,000 sets of oxidizing enzymes in a single mitochondria.

Mitochondria are commonly known as **"the power-houses of the cell"** because they take the respiratory substrate and generate ATPs which are released to the cell on demand of energy because these compounds contain high energy phosphate bonds.

CONTROL OF RESPIRATION

Normally the rate of respiration is determined automatically. However, several factors appear to be involved, the most important one is certainly the proportion of carbon dioxide in the blood (according to LAVOISIER and PFLUGER). It has been established that an increase in the rate of breathing movements occurs when the carbon dioxide content of the inspired air is markedly increased, although the oxygen content is hardly changed.

GALEN reported that brain plays a cardinal role in respiration as he found that high sectioning of the spinal cord led to a cessation of respiration. Later Russian scientist N. MISLAVSKY in 1885 reported the presence of **"respiratory centre"** in the medulla of the brain which maintains and controls the rate of respiration and the chief factor influencing the centre is the carbon dioxide content of the blood.

According to MARSHALL and HALL vagus nerve also controls the respiration upto some extent. KUSSMAUL and TENNER postulated that the **"venosity"** of the blood of the cerebral circulation also influences the respiration.

RESPIRATION IN MAMMAL

In mammal for respiration there is a special system of respiratory organs. The respiratory organs, according to their functions, may be divided into two groups. The first group includes the **air passages** through which air travels in reaching the blood stream : the nostrils, nasal passages, larynx, trachea, bronchi, branchial tubes and lungs. The second group is concerned with the mechanics of breathing, that is, with changing the size of the thoracic cavity. This group includes the ribs, the rib muscles (internal intercostal muscles and external intercostal muscles) the diaphragm, and the abdominal muscles.

One of the most important structural characteristics of respiratory organs of mammals is that the walls of most of them have a bony or cartilaginous skeleton and, therefore do not collapse; that is why they always contain air. All air passages are lined with mucous-membrane and have ciliated epithelium.

1. The nostrils and nasal passages : The nostrils are the openings of the nasal cavities and lies above the mouth. The two nostrils are separated by the septum.

The nasal cavity is the first part of the respiratory system. Air enters the nasal cavity through the nostrils. The nostrils and the nasal cavity are lined internally with mucous membrane which has ciliated epithelium. In the mucous membrane there are mucous glands which secrete mucus onto its surface and the dust and microbes brought in with air adhere to the mucus. The cilia of ciliated epithelium continuously vibrate in the direction opposite to the inhaled air and thus, help to keep the air passages clean from dust and microbes. The length of nasal passages helps in warming and moistening the air before it enters the trachea.

2. **The larynx and the trachea :** From the nasal cavity the air passes through the pharynx and enters the **trachea,** the so called wind pipe. It is a hollow tube about 12 cm long extending throughout the length of neck and a part of thoracic cavity. Its wall is protected externally by skeleton which is formed by half rings of cartilage articulated by ligaments. Its posterior wall is soft and consists of a connective tissue membrane and is quite closely connected with the oesophagus. It is internally lined with mucous membrane in which smooth muscles and glands which secrete **mucus** are found. Externally it is converted with **a connective tissue membrane**.

The upper end of trachea is communicated with buccal cavity through an opening called **glottis**. Glottis is guarded by a cartilage flap called the **epiglottis**. During swallowing the end of the trachea is closed by the epiglottis and, thus, keeps food out of the respiratory tract. At other times the trachea remains open to permit breathing.

The **larynx** or **Adam's apple,** is the enlarged upper end of the trachea situated in the neck on the level of the fourth to the sixth cervical vertebrae. Anteriorly it is covered by the muscles of the neck situated below the hyoid bone; laterally it adjoins the lobes of the thyroid gland and large vessel of the neck, and behind it is the pharynx.

The larynx is the sound producing organ which contains **vocal cords** in its interior. It is lined with mucous membrane and covered with ciliated epithelium. The skeleton of the larynx is also formed of cartilage.

3. **The bronchi :** The trachea after entering the thoracic cavity on the level of the fourth and fifth thoracic vertebrae divides into branches what are known as **bronchi**. One bronchus extend to each lung and subdivides into countless small **bronchial tubes** which, in turn, divide into many fine tubes called **bronchioles**. These bronchioles end in **air sacs** which are made up of protrusions called alveoli and compose most of the lung tissue. This gives the lungs a spongy texture. Accompanying the main bronchial tubes are branches of the pulmonary arteries and veins which also branch repeatedly, their ends being jointed by capillary network situated on the outside of the air-sacs.

The walls of the bronchi have the same structure as those of the trachea. The right bronchus is comparatively wider but short than the left one and is continuous with the trachea.

4. **Lungs :** The lungs are a pair of conical hollow organs

housed on either side of the heart in the pleural cavities of the thoracic cavity and are connected to the pharynx by the **trachea.** They are spongy and consist of the bronchioles and some seven hundred million minute **air-sacs** or **alveoli** and an extensive network of the blood vessels and the capillaries, held together by connective tissue.

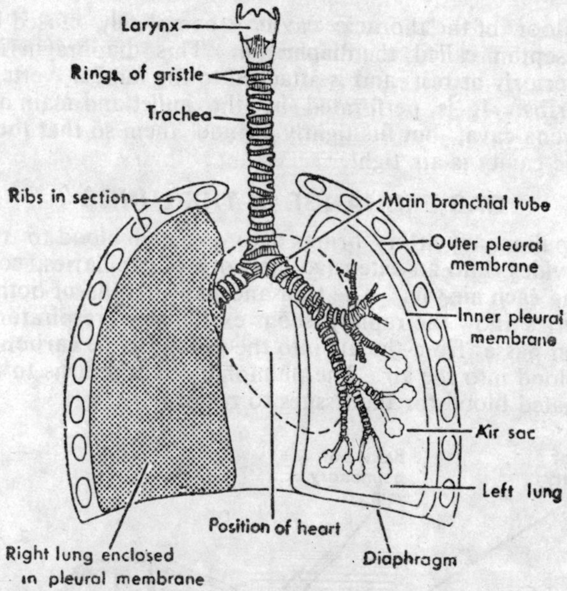

Fig. 8·6. The organs concerned with breathing and external respiration of mammals (man).

Each air sac or alveolus is approximately 0·1 mm in diameter and has a thin wall 0·5μ in thickness. The walls of the air-sacs are elastic and are supplied with capillaries from the pulmonary artery and are kept moist by secretion. Through these thin walls gases are exchanged between the capillaries and the air-sacs. Thus the lungs provide enough surface to supply oxygen by way of blood for the needs of the cells having no direct access to air. The total area of the air sacs in the lungs is about two thousand square feet or more than one hundred times the body's surface area.

Each lung is covered externally by a **pleural membrane** which is turned back on itself to line the inside of the thoracic cavity. The two layers are separated only by a thin film of liquid (**the pleural fluid**). This secretion is secreted by the glands located in the pleural membrane and acts as a lubricant, permitting the lungs to move freely in the thoracic cavity during breathing.

5. The thoracic cavity and the diaphragm : The thoracic cavity is hollow cavity and distinguished into two **pleural cavities,** each enclosing a lung. It is barrel-shaped in man and roughly triangular or conical in the rabbit. It is bounded at the back by back-bone (vertebral column) and at the front by the breast bone (sternum), while the sides are formed by the ribs. The ribs are provided with **internal intercostal muscles** and **external intercostal muscles.** By the contraction of these muscles the volume of the thoracic cavity is reduced or increased which results in inhalation or exhalation of the air.

The floor of the thoracic cavity is completely closed by a thin muscular septum called the **diaphragm**. This diaphragm is dome-shaped anteriorly at rest and is attached to the lumber vertebrae and posterior ribs. It is perforated by the gullet and main aorta and posterior vena-cava, but fits tightly around them so that the floor of the thoracic cavity is air tight.

GAS EXCHANGE IN THE LUNGS

The pulmonary artery brings deoxygenated blood to the lungs. There it divides into an extensive network of capillaries, completely surrounding each air-sac. The thin and moist walls of both air-sacs and capillaries allow the rapid gaseous exchange of respiratory gases, *e.g.*, oxygen passes from the air into the blood, while carbon dioxide from the blood into the air. The pulmonary vein returns to the heart the oxygenated blood for the tissues to respire.

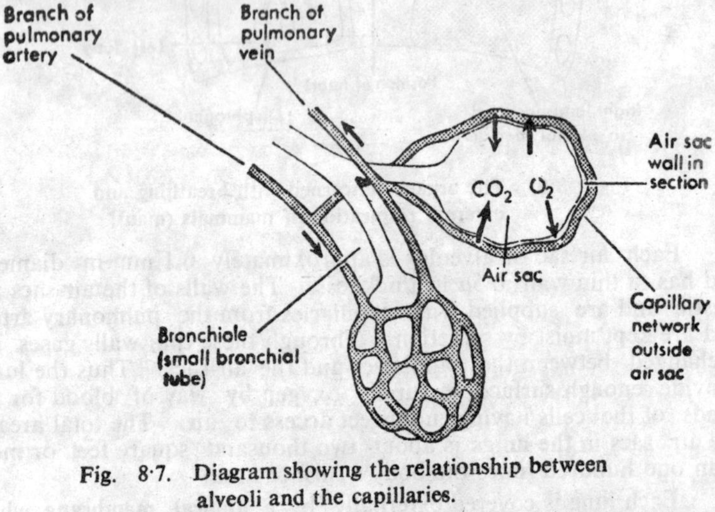

Fig. 8·7. Diagram showing the relationship between alveoli and the capillaries.

The gaseous exchange in the lungs depends upon the difference between the oxygen and carbon dioxide pressures in the pulmonary alveoli and the venous blood flowing to the lungs. The pressure of oxygen in the air-sacs of the lungs is higher than that in the lung capillaries, while the pressure of carbon dioxide in the lung capillaries is higher than that in the air sacs. That is why oxygen is passed from the air into the blood and carbon dioxide from the blood into the air which is, later, exhaled.

The absorbed oxygen combines with the haemoglobin of the red·blood corpuscles to from a temporary compound, the **oxyhaemoglobin**. In the tissues where the concentration or pressure of oxygen is low, oxyhaemoglobin releases its oxygen and is converted into reduced haemoglobin which is again ready to combine with oxygen. The oxygen, thus, liberated is used by the tissue cells to oxidize the digested food.

TRANSPORTATION OF RESPIRATORY GASES BY THE BLOOD

The blood with the help of respiratory pigment, the haemoglobin, transports oxygen from the lungs to the tissues and CO_2 from the tissues to the lungs. The greater part of the oxygen is in the form of unstable chemical compound called **oxyhaemoglobin**. The oxygen dissolved in the plasma of the blood flowing through the lungs combines with the haemoglobin of red blood corpuscle and forms oxyhaemoglobin. Oxygen will keep on dissolving in the blood until all the haemoglobin has changed to oxyhaemoglobin. When respiratory air is respired at normal conditions 96% of the haemoglobin is changed to oxyhaemoglobin with the result that the red blood corpuscles contain 60 times as much oxygen as the plasma contains.

GASEOUS EXCHANGE AT TISSUE LEVEL

Gaseous exchange in the tissues is governed by the same principle as in the lungs. Oxygen passes from the region of high partial pressure to the region of low partial pressure. As oxygen leaves the plasma the oxyhaemoglobin changes back to haemoglobin thereby ensuring an adequate concentration of oxygen in the plasma.

The carbon dioxide forming during oxidative processes of cellular respiration passes into the tissue-fluid and establishes a high partial pressure there. The partial pressure of carbon dioxide is much lower in the blood flowing through the blood capillaries of the organs and so carbon dioxide is diffused from the tissue-fluid into the blood.

The carbon dioxide is not only dissolved in the plasma but also enters into chemical combination with the haemoglobin of red blood corpuscles and the plasma salts. This enables all the CO_2 formed in the tissues to be carried away. Blood that has released its oxygen and is saturated with CO_2 is called **venous blood**. **Venous blood** flows to the lungs where carbon dioxide is expelled into exhaled air.

MECHANISM OF BREATHING (BREATHING MOVEMENTS)

Many people suppose that the lungs draw in air, expelled and enlarge the chest, but this is not true. The lungs are spongy, airfilled sacs which do not have muscle tissue and cannot expand or contract of their own accord. Breathing is accomplished through changes in the volume and air pressure of the thoracic cavity.

1. Inspiration or inhalation : Intake of air or inspiration takes place when the volume of the thoracic cavity is increased and the pressure is decreased. To increase the volume of thoracic cavity the **diaphragm** and **external intercostal** muscles of the ribs take vital part under the infiuence of nerve-impulses coming from the respiratory centre of the brain. Enlargement of the thoracic cavity involves the following movements which occur simultaneously.

(i) The external intercostal muscles of the ribs contract and pull ribs upward and outward and thereby increasing the anterio-posterior and transverse size of the thoracic cavity.

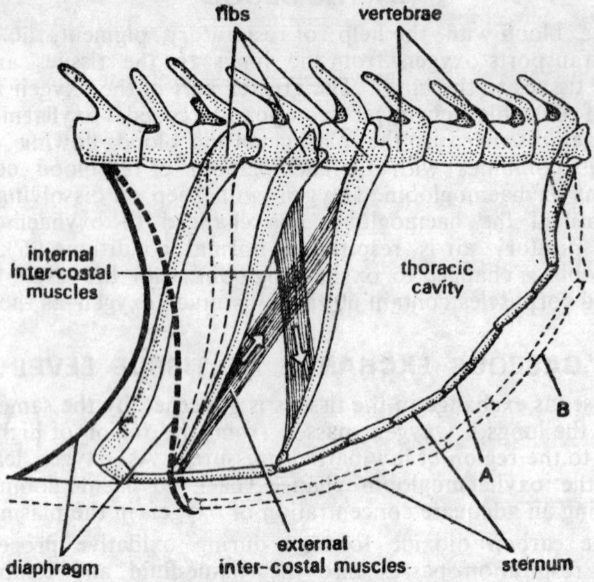

Fig. 8·8. Diagram showing the location of external intercostal muscles and internal intercostal muscles of the ribs.

(ii) The muscles of the resting dome-shaped diaphragm contract. This action straightens and lowers the diaphragm and increases the size of the thoracic cavity from below.

(iii) The abdominal muscles relax and allow compression of the abdominal organs by the diaphragm.

As a result of these muscular movements the size or the volume of the thoracic cavity is increased. Since the pleural cavity contains no air and the pressure in it is negative, the lungs expand simultaneously with the increase of the size of the thoracic cavity. As the lungs expand or inflat the air pressure in them drops (it falls below the atmospheric pressure) and atmospheric air rushes into them through the air passages just to equalize the outer and inner pressures.

Hence an inhalation involves a contraction of the muscles, an increase in the size or volume of thoracic cavity, an expansion of the lungs with a drop of pressure inside them and entrance of atmospheric air into the lungs through the air passages. Inhalation is followed by exhalation.

2. **Expiration or exhalation :** The expelling of air or exhalation from the lungs, takes place when the size and the pressure of the thoracic cavity are reduced. This action involves the following movements :

(i) The ribs take their original position as a result of contraction of the internal intercostal muscles and because of their own weight.

(ii) The diaphragm relaxes and rises to resume its original position.

(iii) The compressed abdominal organs push up against the diaphragm.

As a result of these movements the size of the thoracic cavity is reduced, the lungs become compressed, the pressure in them rises (become higher than the atmospheric) and air rushes out through the air passages.

INTERNAL RESPIRATION OR CELLULAR RESPIRATION

The oxygen liberated in the tissue cells is utilized in the oxidation of food molecules to liberate energy. The various chemical reactions involved in the cellular respiration have already been discussed earlier.

Revision Questions

1. What is respiration ? Describe the process of respiration in mammal.
2. Describe various types of respiratory organs found among the animals.
3. Describe the structure and functions of respiratory pigments found in animals.
4. Describe with suitable diagram the process of internal or tissue respiration.
5. Write short notes on :
   (i) Glycolysis,
   (ii) Transport-system,
   (iii) Respiratory quoteint.

CHAPTER 9

Excretion

The metabolic activities of the body are accompanied by liberation of energy and formation of a variety of by-products, most of them are useful to the body, while others are harmful and, thus, are eliminated from the body of the animal. A single substance may sometimes be an excretion product and at other times, an indispensible metabolite. Water is a by-product of metabolism and must often be excreted in large amounts to avoid a serious condition of **edema.** On the other hand in some animals the metabolic water is the only available source of water, therefore, in them it must be rigorously conserved. The carbon dioxide is a metabolic by-product of cellular respiration but it is also an important component in the synthetic and regulatory machinary of animals and plants. The same is true about urea, a prominent constituent of the urine in many animals, sometimes discharges useful physiological functions. If the blood urea in man rises above about 0·05% (normal values 0·01 to 0·03) a pathological condition of **uremia** develops, but the elasmobranch fishes actively retain urea for the purpose of osmotic regulation and have normal blood urea values of 2·0 to 2·5% (SMITH 1953). The above facts show that **no concise definition of excretion can be based solely on the chemical nature of the materiai removed and it is better to define excretion in very general way as the process of separation and elimination of water soluble waste products of cellular metabolism**. In other words excretion can be defined as a process by which the by-products of cellular metabolism are so treated that they take no further part in the metabolism. These by-products or waste products are removed from the body in aqueous solutions, therefore, water constitutes the bulk of the excreta by weight.

The excretory processes play a most important role in maintaining the relative constancy of the body's internal environment without which life is impossible. If the excretory processes fail to eliminate these metabolic wastes from the body, the same may be accumulated in the bodv. This accumulation disturbs certain delicate acid-base balances in the body and also upsets the osmotic-relationships between blood and lymph and the tissues. This may lead to even the death of the individual after a short period of their accumulation.

Disorders of the excretory processes kill higher animals much faster than do food deficiencies.

In single-celled animals like protozoans and animals like sponges and jelly fishes excretion occurs directly, *i.e.*, their metabolic wastes are discharged directly into the surrounding medium through their body surface. To some extent, excretion occurs in a more indirect way by the secretion of waste products into a vacuole which is later extruded from the cell. In higher animals these products are not discharged directly into surrounding medium because every cell of the body is not in a direct contact with the surrounding environment, therefore, in them each cell discharges its metabolic wastes into the tissue fluid, which in turn reaches the blood stream. The blood transports these metabolic wastes in the excretory organs which eliminate them outside the body.

SUBSTANCES EXCRETED IN ANIMALS

The metabolic wastes which are excreted in the animals are of several kinds and vary from animal to animal and in the same animal from time to time. The difference in the nature of waste products in animals is correlated with the metabolic processes taking place in the body of the animals.

For convenience the metabolic wastes which are excreted by the animals may be grouped under the following heads :

1. **Respiratory waste products :** Carbon dioxide and water are the by-products of catabolism of all classes of food stuffs. In small animals carbon dioxide is eliminated directly into the environment through the body surface but in higher animals it is eliminated almost exclusively with the expired air through the lungs. Excess of water is eliminated in the form of urine and sweat.

2. **Nitrogen containing waste products :** The nitrogen containing waste products are derived partly from the deamination of the excess amino-acids taken in with the diet (exogenous source) and partly from the break-down of the animals own proteins and nucleic acids and also miscellaneous compounds (endogenous source).

The amount of waste nitrogen is determined by the utilization of protein for energy and by the rate of break down and turnover of body cell constituents.

Some important specific nitrogen containing waste products of the animals are as follows :

(i) **Ammonia :** It is one of the important nitrogenous compounds which is very toxic to the body. It is very soluble and can only be excreted in very dilute solution. It is the major nitrogen excretory substance of many aquatic animals such as crustaceans. Ammonia is formed as a result of oxidative or hydrolytic deamination of amino acids.

(ii) **Urea** : It is a common excretory product and is less toxic in nature in comparison to other nitrogenous compounds. It is generally found in aquatic animals such as fishes and those terrestrial forms such as mammals which are not well adapted to water conservation.

How urea is produced ? The formation of urea always takes place in the liver and it is produced from an amino acid **arginine** in the presence of **arginase** enzyme. CLEMENTI was the first who pointed out the presence of **arginase** enzyme in the liver of ureotelic animals. The entire process of urea formation is carried out in a cyclic chain of chemical reactions during which arginine is formed again and again by using ammonia. This cyclic chain of chemical reaction of urea formation is known as "KREBS-HENSELEIT **cycle**" as KREBS and HENSELEIT were the first who observed this cyclic chain of chemical reactions of urea formation. KREBS-HENSELEIT **cycle** is also known as **"ornithine cycle"** because ornithine aminc acid takes a vital role in the urea formation.

During this cycle ornithine first combines with ammonia and carbon dioxide so as to form **citrulline** and eliminates water. Citrulline thus produced combines with ammonia and water to form arginine which is catalyzed in the presence of **arginase** enzyme to produce ornithine and urea. Ornithine once again combines with ammonia and carbon dioxide and, thus, repeats the cycle. CO_2 and ammonia are introduced into the cycle by **"career molecules"** for the formation of which ATP is required.

Recent findings state that; no doubt, arginase enzyme plays a vital role in the urea formation but other enzymes, like **carbamoyl-phosphase, synthetase, ornithine transcarbomylase, arginosuccinic acid synthetase** and **arginosuccinic acid splitting enzyme,** also take part to great extent in the urea formation.

(iii) **Uric acid** : Uric acid and its salts, the urates, are relatively insoluble and much less toxic than ammonia and urea, therefore, they can be stored in the body. It is the major nitrogenous product of those animals which conserve water as one of the ways of survival on land, such as, birds, terrestrial reptiles, some snails and insects, etc.

It is the only nitrogenous excretory product which can be removed from the body in solid form, it, thus, permits nitrogen excretion without loss of water.

(iv) **Amino acid :** In certain animals like molluscs *(Limnaea, Paludina)* and echinoderms (*Pentacentrotus*) the excess of amino acids is removed as such without undergoing any further change. Amino acids are formed as a result of hydrolysis of proteins in the alimentary canal.

(v) **Other nitrogenous compounds :** Other nitrogenous excretory products include allantoin and allantoic acid. These are insoluble and are used during embryonic development by amniotes with shelled eggs.

There are also endogenously derived nitrogenous waste products such as guanine and adenine from nucleic acid break down and creatine from the creatine of muscles, which are excreted with other nitrogenous products in the urine.

3. **Mineral ions as waste products** : The excess mineral ions taken in with the diet are also excreted by one means or another. In the vertebrates ionic composition of urine is controlled by hormones secreted by the adrenal glands.

Sodium, potassium, calcium, magnesium, chloride and ammonia are the essential mineral ions of the animals. The ionic conrentration

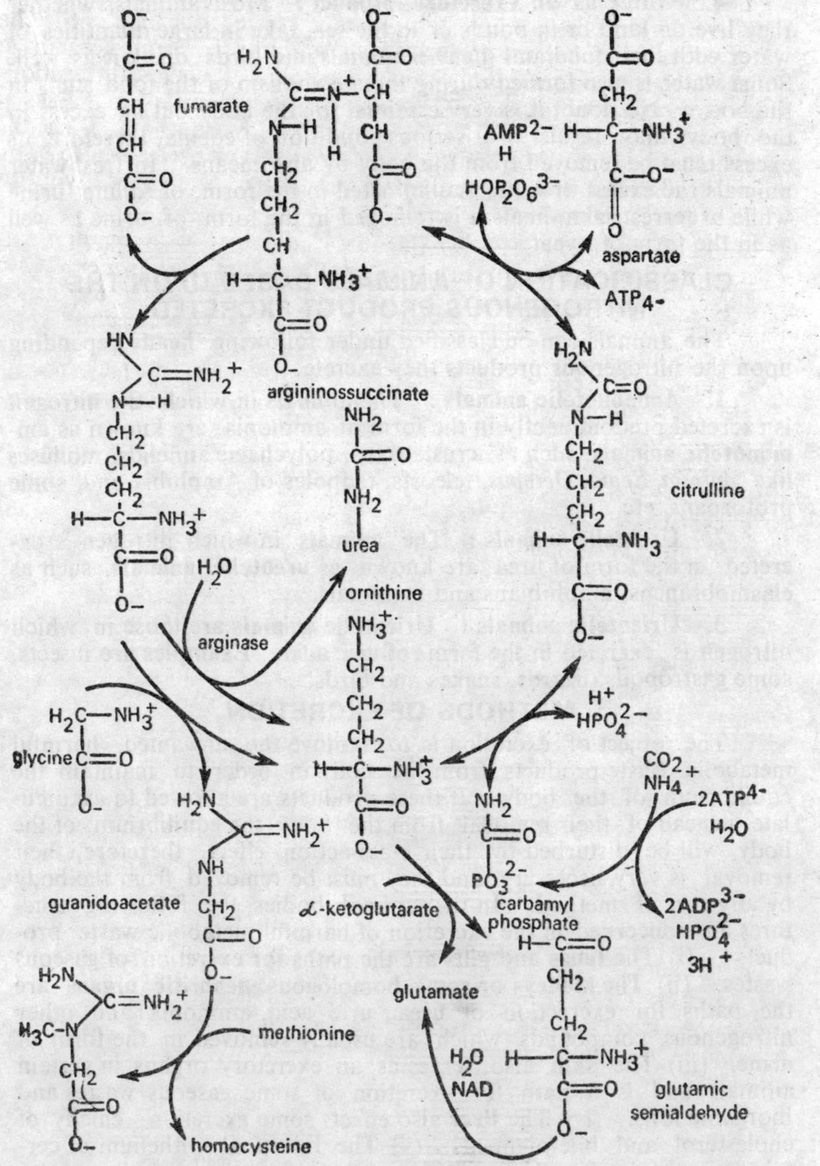

Fig. 9·1. The Krebs ornithine cycle in relation to urea formation. This system of reactions is a normal part of amino acid catabolism and anabolism in all animals, but in those which form urea as a nitrogenous end product, the system has become especially active.

of them may differ in different animals corresponding to the different environments in which they are found.

4. **Water as an excretory product :** Most animals, whether they live on land or in ponds or in the sea, take in large quantities of water with their food and many mammals and birds drink it as well. Some water is also formed during the metabolism of the food stuffs in the body. No doubt it is very essential for the body but its excess in the body may result in a serious condition of edema, therefore, its excess must be removed from the body by any means. In freshwater animals the excess of water is eliminated in the form of dilute urine while in terrestrial animals it is removed in the form of urine as well as in the form of sweat.

CLASSIFICATION OF ANIMALS BASED UPON THE NITROGENOUS PRODUCT EXCRETED

The animals can be classified under following heads depending upon the nitrogenous products they excrete.

1. **Ammonotelic animals :** The animals in which the nitrogen is excreted predominently in the form of ammonia, are known as **ammonotelic animals**, such as, crustaceans, polychaete annelids, molluscs like *Aplysia*, *Sepia*, *Octopus*, teleosts, tadpoles of Amphibia and some protozoans, etc.

2. **Ureotelic animals :** The animals in which nitrogen is excreted in the form of urea are known as **ureotelic animals**, such as elasmobranchs, amphibians and mammals.

3. **Uricotelic animals : Uricotelic animals** are those in which nitrogen is excreted in the form of uric acid. Examples are insects, some gastropods, lizards, snakes and birds.

METHODS OF EXCRETION

The object of excretion is to remove the unwanted, harmful metabolic waste products from the body in order to maintain the equilibrium of the body. If these products are allowed to accumulate instead of their removal from the body, the equilibrium of the body will be disturbed by their mass-action effect, therefore, their removal is very necessary and they must be removed from the body by any sort of method. In the animal bodies, the following structures are concerned in the excretion of harmful metabolic waste products : (i) The **lungs** and **gills** are the paths for excretion of gaseous wastes. (ii) The **kidneys** or some homologous **nephritic organs** are the paths for excrection of urea, uric acid, ammonia and other nitrogenous compounds which are usually removed in the form of urine. (iii) The **skin** also serves as an excretory organs in certain animals and is a path for excretion of some gaseous wastes and inorganic ions. (iv) The **liver** also effects some excretion, chiefly of cholesterol and bile-pigments. (v) The intestinal epithelium of certain animals also functions as an excretory organ because it excretes certain inorganic constituents which are present in excess in the body. (vi) The salivary, mammary and tear glands upto some extent also serve as vehicles of excretion of traces of wastes and may excrete a significant amount of foregin substances.

No doubt all the mechanisms are of equal importance because all operate, of course, so as to tend to preserve the steady state of the blood and the body as a whole but the kidneys deserve a special discussion as they can eliminate any excess of non-volatile substances that tend to make the blood either too acidic or too alkaline. They are also chief organs for regulation of the osmotic pressures of the blood and other body fluids and for the maintenance of the comparative constancy of the internal environment of the living cells.

PHYSIOLOGICAL PROCESSES INVOLVED IN EXCRETION

In the highly organised animals both invertebrates and vertebrates, excretion involves three well-defined physiological processes, **filtration, reabsorption** and **secretion**. In **filtration** non-colloidal solutions are filtered out by hydrostatic difference through a semi-permeable membrane from the body fluid into a space connected with the exterior. The filtrate then passes through a tube or other space lined with cells capable of actively transporting molecules from it back into the body fluids (**reabsorption**), or of secreting additional substances from the body fluid into the filtered fluid (**secretion**). Obviously, at the cellular level, there are but two processes, **diffusion** and **active transport** but it seems likely that the processes are similar at the cellular level.

EXCRETION IN DIFFERENT ANIMALS

In small animals like protozoans no special excretory organs are found, therefore, in them excretion occurs directly, *e.g.*, metabolic wastes are directly removed into the surrounding environment as they are in direct contact with environment. In higher animals, on the other hand, each cell of the body is not in direct contact with the environment, therefore, in them these metabolic wastes are removed from the body with the help of specific excretory organs which are exclusively meant for the purpose of excretion of metabolic wastes. How excretion is affected in different animals that has been given below :

Excretion in Protozoa : There is no uniformity in the production of nitrogenous wastes in protozoans. *Amoeba* forms uric acid, *Glaucoma* and *Didinium* ammonia and *Paramecium* and *Spirostomum* urea according to some workers and ammonia according to other workers.

In Protozoa no special organs of excretion are found. Excretion in most of the protozoans takes place by simple diffusion through the body surface. Some protozoans have contractile vacuoles which are essentially the organs of water balance but these organs may also serve as excretory organs according to SHAW, 1960. The contractile vacuoles undoubtedly eliminate the excess of water, which probably contain waste products in solution, in this way to a certain extent they assist in excretion, but they can not be responsible for the whole or even a major part of it. Analysis of the fluid removed from the contractile vacuole of *Spirostomum* shows that it eliminates near about 1% of the total urea produced.

Excretion in Porifera : Like protozoans, porifers excrete their nitrogenous wastes in the form of ammonia or urea. In these animals the excretion is effected by direct diffusion from the tissues in the surrounding medium. These animals also possess contractile vacuoles which are supposed to serve a part of excretion.

Excretion in Coelenterata : In coelenterates different types of nitrogenous wastes are met such as uric acid, it is found in *Anemonia sulcata*, urea, it is found in sea anemones and ammonia, it is chief product in actinians.

In coelenterates also, there are no specialized organs of excretion and it is assumed that waste products escape by simple diffusion through the cell membrane as they do in the Protozoa. In actinians it has been shown that injected substances, such as, carmine, accumulate in certain regions and this may mean that these places are especially active in excretion.

Excretion in Platyhelminthes : In Platyhelminthes the nitrogenous wastes, no doubt, are excreted by simple diffusion, yet a well-organised excretory system is found which consists of a series of tubes which branch and end into **flame cells.** The lumen of each flame cell is continued with the excretory tube and contain cilia. The movement of these cilia sweeps water into the excretory tubes.

The chief nitrogenous waste product of planarians, *Fasciola hepatica*, *Taenia pisiformia* is ammonia, while *Trichinella spiralis* removes one third of its nitrogen as ammonia and rest as amino acids and peptides.

Excretion in Nematoda : In Nematoda ammonia is the chief nitrogenous excretory product, but fair quantity of urea is also produced. Some species produce amino acids and amines.

In Nematoda two types of tubular system are found for excretion. Cilia and flame cells are completely lacking. In many free living species there are one or two ventral gland cell each with a terminal ampulla, which are said to be excretory in function. In parasitic forms there is a canal formed from a single cell, usually shaped like a capital **'H'** and opening ventrally. In some, one lateral link is lost, while in others such as *Ascaris*, the anterior portion is missing, giving a system like a capital of *Rhabditis* have H-shaped system and two glands as well.

Excretion in Annelida : Polychaetes like *Aphorodite* and medicinal leech excrete most of their nitrogen in the form of ammonia, while various species of *Lumbricus* and *Pheretima* produce both ammonia and urea. Other substances like amino acids creatine and small quantities of purines are also produced.

The **chlorogogen cells** of the gut wall were for long said to be the excretory organs of the annelids. No doubt, within the chlorogogen cells amino acids assimilated from the gut are deaminated and various nitrogenous excretory substances such as ammonia, urea and uric acid are formed but how these substances find their way to the exterior, is not clear so far.

Now it has been established that nephridia are the chief excretory organs of the annelids. These are tubular structures and are found in pairs nearly in all the segments of the body. Their walls are single-layered epithelia which often show evidence of being capable of active transport. They may be assumed to secrete materials into the tubular fluids or extract materials from them. The nephridia, which are closed at the inner end, are called protonephridia while those which open into the coelom by a ciliated funnel, **nephridiostome** or **nephrostome**, and termed **metanephridia**. Both types of nephridia open to the exterior through **nephridiopores** and frequently enlarge into a storage bladder just before they discharge.

The nephridia in most of the annelids, like *Pheretima*, are of three types; (i) **septal nephridia** (ii) **pharyngeal nephridia** and (ii) **integumentary nephridia**. Septal nephridia the so called typical metanephridia are found along the septa of the body. Pharyngeal nephridia are found in the pharyngeal region of the alimentary canal. Integumentary nephridia are found attached to the integument. Their number varies from 200 to 250 in all the segments of the body. These are **exonephric** as they are directly communicated with the exterior. These nephridia pick up the nitrogenous wastes from the fluid of coelomic cavity through the nephrostomes and carry them outside the body.

Excretion in Arthropoda : In some arthropods such as crustaceans, the excretory products are ammonia, while in others like insects the excretory products are uric acid. Clothes moths excrete most of their nitrogen in the form of ammonia but rest in the form of urea and amino-compounds.

In arthropods various types of excretory organs are found to remove the nitrogenous wastes from the body. The **antennary** gland and **maxillary** gland are said to be the chief excretory organs of Malacostraca and Entomostraca respectively. The MALPIGHIAN tubules are the chief excretory organs of most insects and also of Myriapoda and Arachnida. Their morphological arrangement and histological structure vary in different animal. At one end they are communicated with the gut and at the other end they lie in the blood sinuses from where they extract the wastes. They excrete both wastes and carbon dioxide which they receive from the blood in solution. In insects there are subsidiary organs of which the chief is the fat body, the parietal layer of which stores a sufficient amount of uric acid.

Excretion in Mollusca : The lamellibranchs excrete most of their nitrogen in the form of ammonia and amino-compounds. Urea and purines are also excreted in fair quantities. The cephalopods excrete ammonia, trimethylamine-oxide, amino compounds and smaller quantities of urea and purines. *Helix pomtia*, eliminates ammonia, amino compound, urea, uric acid and other purines where as *Limax agrestes* excretes chiefly urea and *Aplysia limacina* excretes ammonia, amino-compounds, purines and urea.

The excretory organs of molluscs are coelomoducts opening

from the pericardium to the exterior. The cells of the inner portions generally have a brush border, while of the distal parts are ciliated. In lamellibranchs, snails and cephalopods all the three physiological processes, *e.g.*, filtration, reabsorption and secretion take place simultaneously during the course of excretion of waste products but the sites of these processes vary. In cephalopods filtration takes place in branchial heart appendage, where as in snails probably in kidney sac. The cilia and body contractions help to eliminate the waste fluid.

Excretion in Echinodermata : The nitrogenous wastes of echinoderms are ammonia, amino-compounds, urea and uric acid. In echinoderms no specialized organs of excretion are found. Excretion takes place through wandering **phagocytic cells**, the so called **amoeboid cells** or **amoebocytes**.

Excretion in Vertebrates : The nitrogenous wastes that are excreted in vertebrates vary from animal to animal. Certain teleosts, tadpoles of amphibia and crocodiles excrete most of their nitrogen in the form of ammonia. Elasmobranchs, amphibians and mammals produce urea as their chief excretory product where as in lizards, snakes and birds uric acid is eliminated as their chief nitrogenous compound. Other products like creatine, creatinine, trimethylamine oxide, guanine and inorganic ions are also excreted in vertebrates.

The chief excretory organs of the vertebrates are the kidneys which are of three types depending upon their position and stage of development in the life cycle of the particutar animal. These kidneys are known as **pronephros, mesonephros** and the **metanephros** according to the English embryologist FRANCIS BOLFAUR. **Pronephros** is embryonic kidney in all vertebrates except in *Bdellostoma* and *Myxine*, where it is the functional excretory organ in the adult. The **mesonephros** is the functional kidney of the *Petromyzon*, fishes and the amphibians. It is functional excretory organ during the embryonic stage of the reptiles, birds and the mammals, while the **metanephros is the functional excretory organ of the adult reptiles, birds and the mammals.**

The kidney of adult vertebrate is made up of a large number of structural and functional units what are known as **uriniferous tubules or renal tubules** or **nephrons**, each consisting of a MALPIGHIAN **body** and a tubule. The BOWMAN'S **capsule** enclosing the glomeruli forms a malpighian body and there are about a million of such bodies in each kidney. The **uriniferous tubule** or **nephron** resembles to a great extent in structure and physiology with the nephridium of the invertebrates, therefore, nephridia are supposed to be the forerunners of the vertebrate kidneys.

Other structures like integument and gills may also be involved in excretion. In frog a negligible quantity of nitrogenous waste, carbon dioxide and water goes through integument. In the teleosts most of the ammonia and urea that are formed in the metabolism of foodstuffs are lost through the gills but in elasmobranchs the gills are not permeable to urea.

Most of the knowledge about excretion has come from the

mammals, therefore, we will discuss the excretion in mammals in detail so that one may able to get the real idea of the excretion.

EXCRETION IN MAMMAL (MAN)

Nitrogenous wastes of the mammals : The most important nitrogenous waste of the mammals is the urea but uric acid, creatine, inorganic salts like chlorides, sulphates and phosphates of sodium, potassium and ammonium are also eliminated. These waste materials are removed in solution in the form of urine. The kidneys are the chief excretory organs of the mammals.

KIDNEYS—THE PRINCIPAL EXCRETORY ORGANS OF MAMMALS (MAN)

The **kidneys** are reddish brown paired structures which lie along the posterior abdominal wall on either side of the vertebral column in the lumber region of the boody. Each kidney is bean-shaped and is

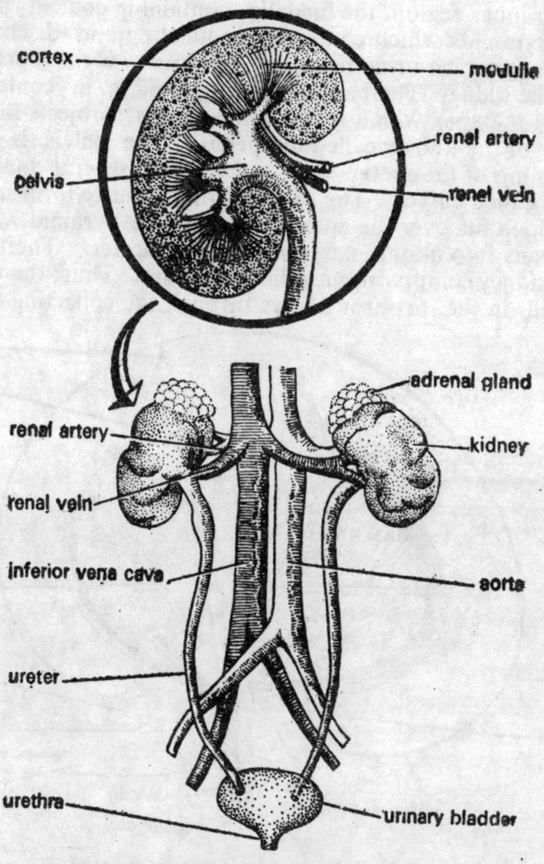

Fig. 9·2. Excretory organs of man.

about 4″ long 2·5″ wide and 1·5″ thick. The two kidneys are not situated at the same level. The right kidney is slightly on a lower level than the left one. It is due to the fact that the right side of the abdominal cavity is occupied by the liver. They are said to be **retroperitoneal** because they lie behind the peritoneal lining of the abdominal cavity. They are held in place by the padding of fat which surrounds them and by pressure from the other organs.

Each kidney is enclosed in a thin, tough, fibrous, whitish capsule. The outer susface of each kidney is convex, while the inner one is concave. In the concave depression there is an opening called the **hilus renalis** through which the blood vessels, nerves and lymphatics enter or leave the kidney. Inside the kidney the hilus expands into a central cavity called **the renal sinus** which contains the major and minor **renal calyces,** the **renal pelvis, nerves** and **blood vessels.**

If we study a longitudinally cut section of the kidney, we find two distinct regions of the kidney, an outer firm region containing the **uriniferous tubules** or **nephrons** which manufacture the urine, the **cortex** and an inner region, the **medulla**, containing conical projections, the **renal pyramids** which contain the collecting and discharging tubules which carry the urine from the nephrons in the cortex to the pelvis of the kidney. The base of each pyramid is in contact with the cortex and the apex which is called **the papilla**, projects into the sac-like cavity of the kidney called the **pelvis.** The pelvis is the expanded beginning of the **ureter** and consists of several major cap-like structures called **calyces**. The calyces branch out to form smaller ones, each of which fits over the apex or papilla of a pyramid. The pelvis, in turn, leads into a long, narrow tube, the **ureter**. The ureters (one for each kidney) empty into a **urinary bladder**. Thus, the urine which is secreted in the nephrons flows through the collecting tubules and

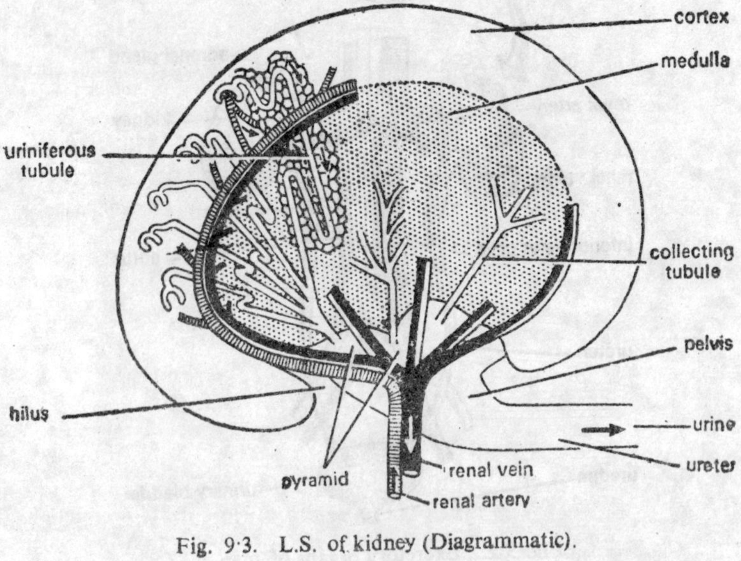

Fig. 9·3. L.S. of kidney (Diagrammatic).

openings of the papilla into the calyx of the pelvis and thence through the ureter to the urinary bladder which on contraction removes the urine outside the body time to time.

Histology of the kidney : Histologically the cortical part of the kidney is made up of a large number of tiny filters called **nephrons** or **uriniferous tubules** which are referred to as the structural and functional units of the kidney. Each human kidney is said to contain about one million such units. MARCELLO MALPIGHI was the man who first of all established the presence of renal corpuscles, the so called nephrons, in the kidneys. BOWMAN in 1842 studied the anatomy of the nephrons and pointed out a cup-like structure in each nephron which is now called as BOWMAN'S **capsule.**

NEPHRON AND ITS STRUCTURE AND BEHAVIOUR

Each nephron consists of a small, cup-shaped structure called BOWMAN'S **capsule**. From the capsule a tiny tubule comes out and takes a tortuous course called the **proximal convoluted tubule**, then dips down to form a wide HENLE'S **loop**, whose ascending limb forms a second tubular coil called the **distal convoluted tubule** which enters a large straight tube called the **collecting tubule**. The collecting tubule receives the tubules of many nephrons. The collecting tubule carries the urine towards the **pelvis**.

In the cavity of BOWMAN'S **capsule** is a network of capillaries

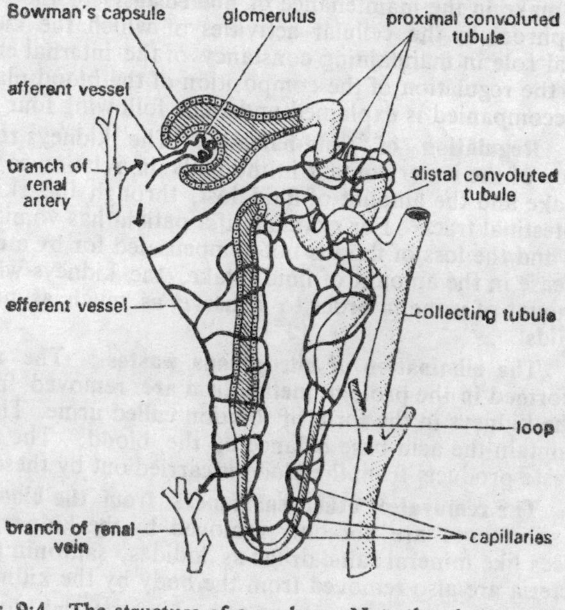

Fig. 9·4. The structure of a nephron. Note the close relationship of the nephron and the blood vessels by which materials are reabsorbed.

called the **glomerulus.** An **afferent** arteriole from a branch of renal artery forms the glomerulus. The capillaries forming the glomerulus at the exit of BOWMAN's capsule unite to form an **efferent arteriole** which goes towards the proximal convoluted tubule and forms another network of capillaries **peritubular capillaries** around the proximal convoluted tubule, the loop of HENLE and the distal convoluted tubule. These capillaries then form **venules** which join together to form the branch of the renal vein which removes the venous blood of the kidney.

The afferent arterioles, glomeruli and efferent arterioles contain arterial blood. It should be noted that efferent arterioles have smaller diameters than afferent arterioles. This increases the pressure in the glomerular capillaries, which is important for the process of urine formation. The arterial blood changes to venous blood, while blood flows through the capillaries found around the proximal convoluted tubule, the loop of HENLE and the distal convoluted tubule.

It is to be noted that formation of urine always takes place in the uriniferous tubules or nephrons where as the collecting tubules do not take any part in the urine formation but they simply convey the urine to the pelvis via pyramids for removal.

FUNCTIONS OF NEPHRON

The nephrons carry out all the important contributions the kidneys make in the maintenance of **homeostasis** of the body. It is these nephrons by the cellular activities of which the kidneys play their vital role in maintaining constancy of the internal environment through the regulation of the composition of the blood plasma. How this is accompanied is explained under the following four functions :

1. **Regulation of fluid-balance :** The kidneys regulate the constancy of the water content in the body in relation to amount of fluid intake and the amount of fluid lost, through the skin and the gastrointestinal tract. For example, if a patient has vomiting and diarrhoea and the loss of fluid is not compensated for by a corresponding increase in the amount of fluid intake, the kidneys will excrete a little amount of urine in order to conseve as much as possible, the body fluids.

2. **The elimination of nitrogenous wastes :** The nitrogenous wastes formed in the protein metabolism are removed from the body by the kidneys in the form of solution called urine. Thus, the kidneys maintain the acid-base balance of the blood. The filtration of these waste products from the blood is carried out by these nephrons.

3. **The removal of other substances from the blood** : Nitrogenous substances are not only removed by the kidneys but other substances like mineral salts; drugs as iodides, santonin and arsenic and bacteria are also removed from the body by the kidneys.

The kidneys participate in regulation of arterial blood pressure by secreting a hormone **venin** which is used by the body to maintain haemostasis.

4. **The maintenance of acid-base balance :** The kidneys assist to a great extent in the maintenance of acid-base balance of the body after removing the various inorganic salts and nitrogenous wastes of the blood.

NEPHRON PHYSIOLOGY AND MECHANISM OF URINE FORMATION

The processes by which the nephrons remove the nitrogenous wastes as urine from the blood without losing at the same time valuable small molecules, water and ions, are as follows :

1. **Glomerular filtration :** When the blood in the afferent arteriole enters the glomerulus a part of the water and some dissolved constituents of the blood of low molecular weight like nitrogenous wastes, glucose and mineral salts filter out through the capillary walls into the surrounding BOWMAN'S **capsule,** by a process called **glomerular filtration.** This phenomenon was established by RICHARDS and his associates in the year of 1924. According to LUDWIG glomerular filtration is caused due to changes in blood pressure and, thus, the dynamics of the renal corpuscle (nephron) were established in physical terms of blood pressure, osmotic pressure and pressures of the capsular fluid. This shows that the process of glomerular filtration in the renal corpuscles or nephrons is made possible by the high blood pressure in the glomerular capillaries and the osmotic pressure exerted by the contents on either side of the membrane. A sharp drop in blood pressure leads to decrease urinary excretion.

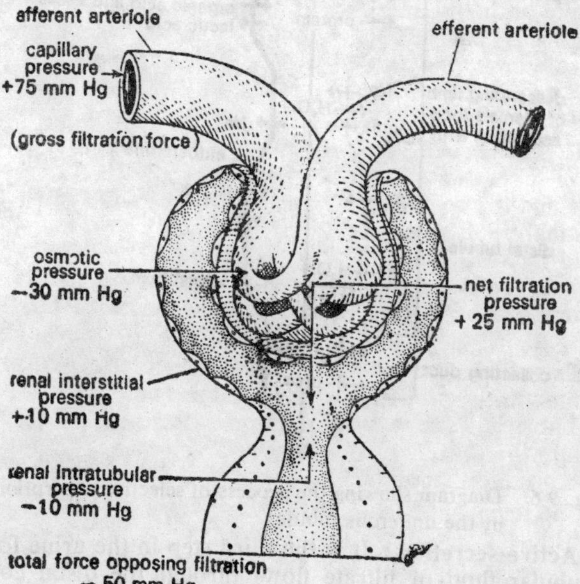

Fig. 9·5. The nephron showing the pressures which operate to produce a net filtrations pressure of about 25 mm Hg.

The glomerular filtrate resembles the blood plasma in its chemical composition, except for the absence of large molecules. About 180 liters (approximately 45 gallons) of fluid are filtered from the plasma into BOWMAN'S capsule from the glomerular vessels every 24 hours. (In other words, the entire plasma volume is filtered about 60 times a day). However, only about 1 to 1½ liters of urine are produced every day.

2. Tubular reabsorption : This is the second step in the urine formation. The glomerular filtrate in BOWMAN'S **capsule** flows on through the convoluted tubule into the collecting tubule and thence into the pelvis of the kidney and down the ureter to the bladder. As the glomerular filtrate flows through the proxmial convoluted tubule some water, physiologically important solutes like glucose, amino acids and inorganic salts like sodium chloride and sodium bicarbonate are reabsorbed into the blood in the capillaries around this portion of the tubule, according to FORSTER, 1961. The remaining constituents in the filtrate are waste products to be excreted.

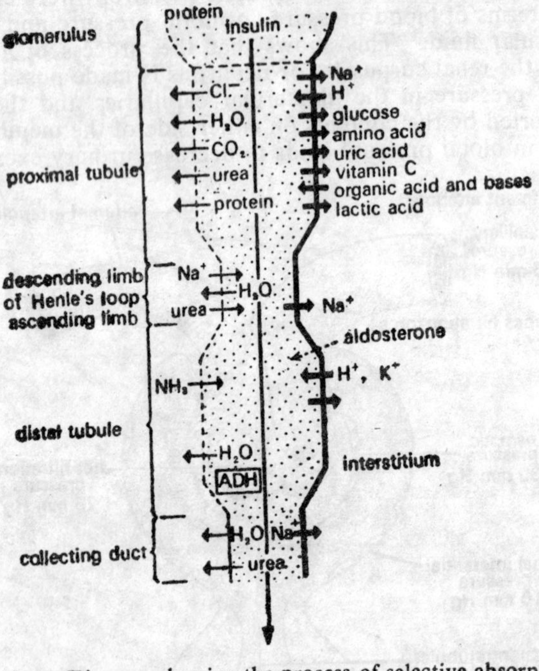

Fig. 9·6. Diagram showing the process of selective absorption in the uniferous tubule.

3. Active secretion : It is the third step in the urine formation. As the tubular fluid or filtrate flows through the distal convoluted tubule the unwanted substances which could not filtered out in the glomerulus, are actively secreted by the tubular wall into the filtrate from the blood.

As a result of this entire process, homeostasis of the blood is maintained and all the waste products remained in the tubular fluid constitute urine which is ready for excretion from the body.

Mammals produce a hypertonic urine. Urine remains isotonic in the proximal convoluted tubule but becomes progressively more hypertonic as it slowly descends the loop of HENLE. Within the ascending limb of this loop it becomes gradually less hypertonic. Within the distal tubule it is either hypotonic or once more isotonic to the surrounding tissue fluids. As the urine passes through the

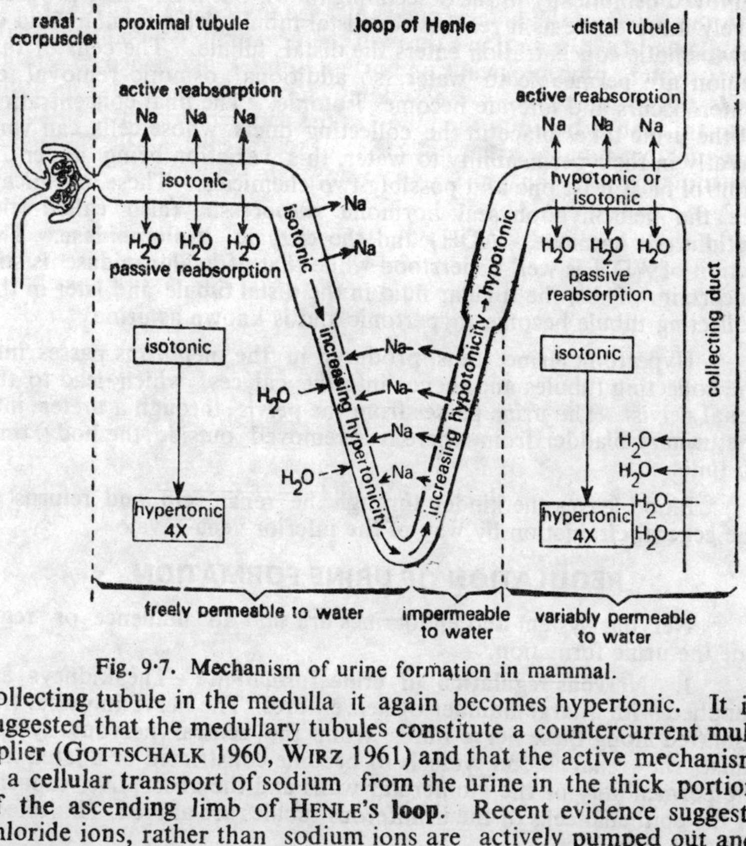

Fig. 9·7. Mechanism of urine formation in mammal.

collecting tubule in the medulla it again becomes hypertonic. It is suggested that the medullary tubules constitute a countercurrent multiplier (GOTTSCHALK 1960, WIRZ 1961) and that the active mechanism is a cellular transport of sodium from the urine in the thick portion of the ascending limb of HENLE's **loop**. Recent evidence suggests chloride ions, rather than sodium ions are actively pumped out and that sodium ions follow passively. The basic features of the mechanism by which hypertonic urine is formed are as follows : The glomerular filtrate that enters BOWMAN's capsule and the contents of the proximal tubule are isosmotic with the plasma of the blood. About 75% of the original filtrate volume is reabsorbed by the time the tubular fluid enters the HENLE's loop. This reabsorption of water occurs osmotically, as a result of dilution of the filtrate due to active uptake of sodium by the cells lining the walls of tubule.

The descending tubules of HENLE's loops are permeable to

diffusion of water, sodium and chloride ions. As the tubular fluid passes through HENLE's loop, it enters regions of increasing osmotic, sodium and chloride ion concentrations. Water then is osmotically withdrawn and sodium and chloride ions are actively pumped into the tubule, keeping tubular fluid concentrations close to those existing outside the tubules. The cells of the ascending limb of HENLE's loop are impermeable to diffusion of water and chloride ions. However, these actively transport chloride ions outward. As a result, the tubular fluid remains reduced in volume (about 50% of its volume was removed osmotically in the descending limb), but it becomes progressively more dilute as it reaches the distal tubule. This tubular fluid of low osmotic concentration enters the distal tubule. The cells of this region are permeable to water, so additional osmotic removal of water occurs and filterate becomes isotonic. The final concentration of the urine takes place in the collecting ducts whose cells can vary greatly in their permeability to water, this variation being under the control of at least one and possibly two chemicals. These chemicals are the neurohypophyseal hormone **vasopressin** (also called the **antidiuretic hormone**, ADH) and the enzyme **hyaluronidase**. The action of ADH is well understood while that of hyaluronidase is still uncertain. Thus, the tubular fluid in the distal tubule and later in the collecting tubule becomes hypertonic and is known as urine.

Hypertonic urine, thus, produced in the nephrons passes into the collecting tubules and thence into the calyces which lead to the renal pelvis. The urine passes from the pelvis through a ureter into the urinary bladder from where it is removed outside the body time to time.

Blood leaves the kindey through the renal vein and returns to the general circulation by way of the inferior vena-cava.

REGULATION OF URINE FORMATION

Nervous system and endocrines are said to influence or regulate the urine formation.

1. **Nervous regulation of urine formation :** The kidneys are supplied with a large number of nerve fibres. The nerve impulses transmitted along these nerve fibres from the central nervous system cause the renal blood vessels to become constricted or dilated and the permeability of the glomerular walls and the absorptive capacity of the epithelial cells of the uriniferous tubules are altered. This affects in urine formation.

2. **Endocrine regulation of urine formation :** There are few hormones secreted by the endocrine glands which control or regulate the urine formation. The **antidiuretic hormone** (ADH) secreted by the pituitary gland regulates the amount of water excreted in urine. ADH acts on the membranes of the collecting ducts of the nephrons and apparently increases their permeability to water. If ADH is absent, the collecting ducts are relatively impermeable to water and so less water passes back into the blood from the urine by osmosis before it enters the ureters. In short, ADH allows the kidney to conserve

water. In amphibians, it acts on the skin to increase the uptake of water from the surrounding medium.

Thyroxine, the thyroid hormone, is also said to control the urine formation. It reduces the reabsorption of water in the uriniferous tubules which leads to **diuresis.**

Aldosterone, a hormone of adrenal glands, also influences the urine formation as it regulates the amount of sodium and potassium retained in the blood and those excreted.

URINE

Urine of man is a dark yellow or dark amber colour fluid, the colour of urine depends upon its concentration whether it is dilute or concentrated. Man excretes an average of about 1·5 litres of urine in 24 hours. The specific gravity of urine is slightly higher than that of water and is 1·015 to 1·020.

Fresh urine has a characteristic not pleasant odour. State urine has an odour of ammonia due to the breakdown of the urea to release ammonia. Its pH value may range from 6 to 8 depending upon the diet.

Chemical composition : Chemically urine is composed of water and organic and inorganic substances. Water constitutes about 95% of the urine where as other substances 5% of the urine. The organic substances include chiefly the urea, uric acid and creatine. The inorganic substances include the salts, chiefly the chlorides, sulphates and phosphates of sodium, potassium and ammonium.

Besides these above mentioned substances the urine of a healthy person may also contain gases (CO_2) single leucocytes and desquammated epithelial cells of the urinary tract.

Revision Questions

1. What is excretion ? Describe the physiology of excretion in a mammal.
2. Trace the path of urea through the kidney of human.
3. Describe various physiological processes involved the process of excretion.
4. Write short notes on :
   (i) Urea,
   (ii) Excretory products,
   (iii) Ammonotelic animals,
   (iv) Urotelic animals,
   (v) Uricotelic animals,
   (vi) Urine.

CHAPTER 10

Osmoregulation

Water is an essential constituent of all living things present on earth. It is the universal biological solvent and the medium in which most of the cellular reactions of metabolism occur. Animals as well as plants contain a large amount of water both in their cells and in their extra cellular fluids. In an animal's body water may enter as the result of drinking, of food intake, of metabolic reactions that yield water as their by products, and of osmosis caused by concentrations gradients across the exposed semipermeable membranes of the animal. Water may leave the organism in the form of excreta, through evaporation from exposed body surfaces, through sweat or other glands, by osmosis through the skin, and by exócytosis. It is vitally important to an organism that its water content, both intracellular and extracellular, be nearly constant, or at least that it should vary only within a moderately small range, so that its tissues or its body fluids may not be diluted or concentrated beyond the limits of its tolerance This more or less steady state of water contents in the body can only be maintained when there is an equality between the amount of water entering the body and the amount of water leaving the body, that is, water in=water out. The process by which the movement of water and its volumes in the body are regulated is known as **osmoregulation**. The term osmoregulation was coined by HOBER in 1902.

Like water, the concentration of many dissolved substances in the body fluids and in the blood must also remain within narrow limits of tolerance because they regulate the movement of water as a result of osmosis. These substances, that are needed for proper functioning of tissues and cells of the body, are controlled by a variety of mechanisms whose activities fall under the study of **ionic regulation**. Since water and ionic regulation are inseparable activities in organisms, by extension the term osmoregulation is sometimes used to include both activities.

Different organisms have different limits of tolerance both for water and salt gain and for water and salt loss because of different environmental conditions. Animals whose body fluids and blood have the same concentration as that of surrounding medium (**isotonic**) never face the problem of osmoregulation as long as they live in such a medium. Most marine organism, for examples have the same concentration of the body fluids as present in the sea water. Animals which live in a medium of a lower salt concentration (**hypotonic**) have to face the difficulty of dilution of blood and body fluids due to endosmosis and hence they have evolved special mechanisms to get rid of the excess water that has entered their body and diluted their body fluids. Animals which live in a medium of high salt concentration (**hypertonic**) have to face the difficulty of excessive exosmosis which may cause the shrinkage of the body fluids and blood. To avoid this they have to evolve regulatory mechanisms. In terrestrial animals water is lost by evaporation from the general body surfaces and hence mechanisms to conserve water are developed in them.

Aquatic animals that have only a limited tolerance to changes in the osmotic concentrations of the external environment or are restricted to a narrow range of salinity, usually to full sea water, are called **stenohaline**. Most animals fall into this category. Animals which can tolerate a wider range of osmotic concentrations or are tolerant of a variety of salinities, are called **euryhaline**. When aquatic animals are transferred from their natural environments to different waters or salinities, they are usually put under conditions of osmotic stress. Some cannot survive such a change and, therefore, in nature are prevented from migrating to areas where the salinity of the water is greater or less than that of their natural habitats. Others have proved more able to endure and survive considerable degrees of osmotic stress, either as **osmoconformers** or **osmorgulators**. **Osmoconformers** are those animals which are osmotically labile (dependent) and whose body-fluid concentrations change with the medium and that, having a high tissue tolerance, can survive such changes as long as their basic metabolic functions can proceed effectively at the dilutions or concentrations to which they are thus subjected. **Osmoregulators** are those animals which are osmotically stable (independent) and are able to maintain their internal osmotic concentration at constant level (or nearly so) despite of changes in that of their external environments. Gradations between these two extremes of lability and stability in respect to internal concentration are frequent, and an organism may conform to external conditions in one situation and regulate in another. But, in general, osmoconformers can tolerate greater variations in their internal environments than can osmoregulators and osmoaegulators can tolerate greater variations in their external environments than can osmoconformers.

Some animals change in volume (due to alterations of water content) as the external osmotic concentration changes. These are known as **volume conformers**. Others which maintain a constant volume despite of external osmotic changes, are known as **volume regulators**.

Animals tend to maintain an optimum osmotic concentration for a given environment. Many animals, upon return to normal environment after a period of dehydration, take up water, and after a period of hydration, lose water, until the osmoconcentration reaches the "optimum" for the particular animals.

MECHANISM OF OSMOREGULATION

In nature the animals live in diverse habitats, *e. g.*, sea, fresh water and land. To maintain their internal osmotic concentration to an "optimum" or at a constant level (or nearly so) they have developed various types of adaptations. One of these is the reduction in the area of permeability by investing as large a part of the body as possible with same kind of impermeable covering, such as skeleton or a cuticle. Another is the reduction in the premeability of the exposed areas, a physiological protection against loss or gain of water and lose of valuable solution that might be affected by a reduction in the number of carrier molecules or their inactivation under certain conditions. Secretion of water or salts, either inward or outward, against a concentration gradient is also a means of their acquisition or disposal. Loss of salts and water may be compensated by their absorption through some part of the excretory system, through the gut or from the external medium through some specialized surface area of the body. Food and drink are the primary, but not the only, source of salt and water replacement for terrestrial animals which can only survive if their elimination of water is reduced to a minimum. This is affected by two ways : (i) by resorption of water from the tubules of the excretory system, if this is present ; and (ii) by biochemical adaptations leading to the production of end products of nitrogenous metabolism of low solubility, whose removal, therefore, does not require any considerable volume of water.

OSMOREGULATION IN FRESH-WATER ANIMALS

All fresh-water animals, whether they are invertebrates or vertebrates, are hypertonic to the surrounding water because the body fluids are more concentrated than the surrounding water. Therefore, there is a tendency for the osmotic inflow of water inside their body and the outward diffusion of salts (=inorganic ions) from their body surface which is not completely impermeable. If this process goes on without some restoration mechanisms, the useful salts of the body will be lost and it will result into haemodilution. This problem has been solved in different animals differently. The problem of excess of water has been solved in two ways : by storing the excess water that enters the body through exposed areas, or by pumping it out. The latter is the more effective way. The problem of salts loss from the body has been solved by the development of active transport systems designed to bring salts into the body through specialized regions of the gills, skin, rectum, or other regions of the gastro-intestinal tract.

Procerodes (*Gunda*) *ulvae*, a triclad flatworm living in the tidal zone of estuaries of small streams and fresh-water *Hydra* are the best examples in which water storage has been demonstrated, They have storage vacuoles in their body by which they are able to retain the excess of water for many hours and after that they disintegrate entirely.

The fresh water protozoans such as *Amoeba, Euglena* and *Paramecium* have solved the problem of osmoregulation by their contractile vacuoles which expell the excess of water. These vacuoles are formed by coalition of smaller vacuoles near the nucleus and are disintegrated to remove excess of water at the outer surface of the body.

All other fresh water organisms except the coelenterates and sponges, to remove excess water from the body, have developed kidneys, or other excretory organs capable of producing a copious urine, hypotonic to body fluids. The kidneys of amphibian, most bony fishes and elasmobranchs have very small nephrons and large glomeruli which are adapted for filtering out and eliminating the excess water in the form of dilute (hypotonic) urine. It has been suggested that the renal systems of fresh-water organisms usually contain cells that can remove sugar, salts and other useful materials from the urine and restore them to body fluids. To maintain the high osmotic concentration of the body fluids than that of the surrounding water, work must be done in order to absorb salts against the concentration gradient, that is, active transport must be involved. Fresh-water fishes and some crustaceans have special type of cells in their gills which can reabsorb salts from the environment and thus carrying out the function.

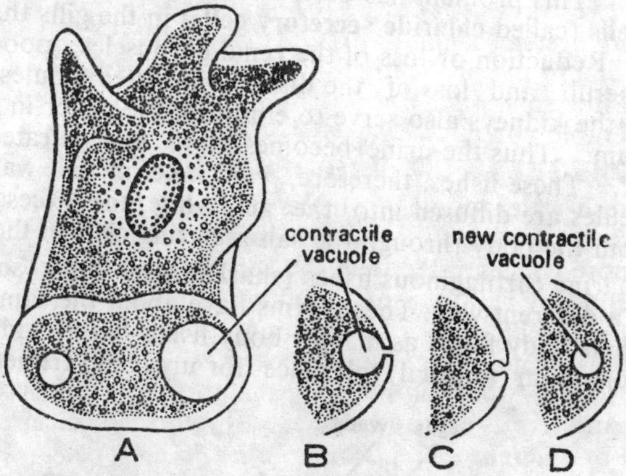

Fig. 10·1. Section through the body of *Amoeba* showing osmoregulation.

In general, the fresh water organisms show the following adaptations for osmoregulation :

(i) They possess semipermeable areas for respiratory and digestive purposes.

(ii) The rest of the body remains covered by a protective and impermeable covering.

(iii) They possess excretory organs like kidneys and nephridia that are capable to secrete very dilute (hypotonic) urine.

(iv) They possess special type of cells either in the gills, or in gut, or in some parts of the kidney that can remove salts from the environment and restore them to body fluids to maintain high osmotic concentration of them.

OSMOREGULATION IN MARINE ANIMALS

Marine animals such as hag fishes, bony fishes, elasmobranch fishes, sea birds (gulls and petrels), marine reptiles (turtle, crocodiles, snakes. lizards) and marine mammals (whales, porpoises and seals) are hypoosmotic to their surrounding environment because their body fluids have an osmotic concentration about one half of that of seawater and so they tend to lose water by osmosis and conserve salts by diffusion. Their need is to conserve water and remove salts and thereby keep their body fluids from becoming too concentrated. This problem has been solved in different ways by above mentioned animals.

In hag fishes, for example, body fluids are about as salty as the salt waters of the surrounding ocean, and so are isotonic with them. Hence they do not tend to lose water by osmosis. In bony fishes the body fluids are hypotonic in relation to the surrounding environment having an osmotic pressure only about one-third that of sea water. Thus they have the problem of losing so much weter to their environment that the solutes in the body fluids become too concentrated and the cells die. This problem has been solved by the evolution of **special gland cells** (called **chloride secretory cells**) in the gills that excrete excess salt. Reduction or loss of the renal corpuscles, poorly developed glomeruli and loss of the distal convoluted tubules from the nephrons in the kidneys also serve to eliminate the water in the urine to a minimum. Thus the urine becomes more concentrated and is more scanty. These fishes, therefore, can take in salt water. The water molecules are diffused into the gut, while the excess salts are removed from the body through the salt secreting cells in the gills.

The marine cartilaginous fishes (sharks, rays) have solved this problem in a different way. These forms have about the same amount of salts in their body fluids as marine bony fishes but in addition they have developed an unusual tolerance for urea and trimethylamine

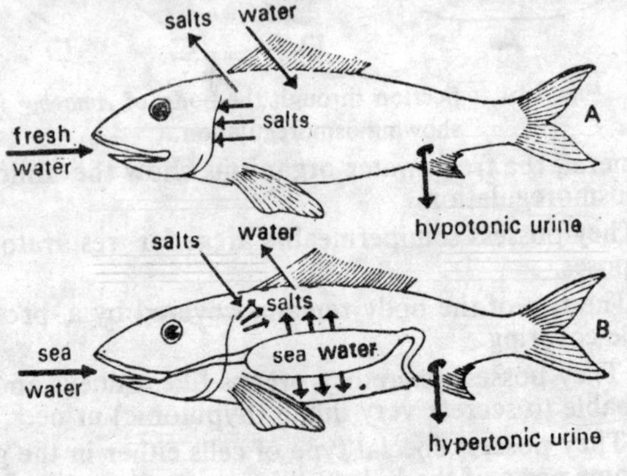

Fig. 10·2. Osmoregulation A—Fresh-water fish ; B—Marine fish.

oxide, so instead of constantly pumping them out, as do all other fishes, they retain a high concentration of them in the blood and body fluids. This high concentration of urea and trimethylamine oxide makes their body fluids almost isotonic in relation to sea water, hence they do not tend to lose water by osmosis. These animals have, however, solved the problem of extra salt by eliminating the same by rectal glands which are capable to secrete salts (BURGER, 1962). These animals retain numerous large glomeruli in their kidneys like the fresh water forms, and they excrete large amounts of dilute or isotonic urine.

Sea birds such as gulls and petrels live at sea generally feed on marine animals which have high salt concentration and occasionally drink sea water. They, therefore, have an identical problem of marine fishes, *i. e.*, osmotic loss of water from the body and influx of salts into the body. Similarly there are marine reptiles such as turtles, crocodiles, snakes and lizards which are hypoosmotic to their environment, *i. e.*, sea water. They also have the same problem of loss of water from the body and infusion of high salts into the body. The kidneys of both marine birds and reptiles are not so efficient that they could excrete high concentrated urine but they have solved the problem of excess salt elimination by means of specialized salt glands (K. SCHMIDT-NIELSEN and FANGE, 1958; K. SCHMIDT-NIELSEN, 1960). In marine birds these salt glands are located in the head while in sea turtle and other reptiles close to the eyes. These glands are capable of desalting. The salt glands, containing a rich supply of capillaries, consist of thousands of branching tubules which remove salt from the blood and drain the resulting concentrated salt solution into the nasal cavities or directly to the exterior.

Marine mammals, such as whales, porpoises and seals also have the same problem of water loss. They lose water through excretion and through moisture in the exhaled air, but this cannot be replaced by drinking sea water. Instead the water is obtained through oxidation of food, since water is a by-product of such oxidation. Also, water is conserved through a great concentration of the urine. The females have an additional water loss in their milk when they are nursing young, but this loss is counterbalanced by a concentration of milk. The milk of these mammals is about ten times as concentrated in fat as is cow's milk.

So in general the following physiological adaptations are found among marine animals to meet the osmotic problems :

1. To prevent the loss of water due to exosmosis the entire animal body remains covered by a thick cuticle (invertebrates) or scales or thick body wall (vertebrates).
2. Reduction or lose of the renal corpuscles, poorly developed glomeruli and loss of the distal convoluted tubules from the renal system serve to conserve large amount of water in marine bony fishes.
3. Marine bony fishes also have special gland cells in their gills which remove excess salts from the body.
4. The marine elasmobranch fishes are capable to retain a large concentration of urea and other dissolved salts in their body fluids which keep the body fluids either isotonic or hypertonic to the sea water and thus, do not tend to lose water by osmosis. If there is any excess of water in the body that is removed by the action of well developed glomeruli of the kidneys.

OSMOREGULATION IN TERRESTRIAL ANIMALS

The terrestrial habitat lacks both water and salts in the surrounding medium (air), therefore, the terrestrial animals often face the problem of both water and salt losses. Water loss may be due to evaporation of water from the general body surface or through sweat or through urine. Salt loss may be due to sweat and urine. This problem can only be solved by remaining in water and salt balance, that is by maintaining a balance between water and salt loss and water and salt gain. Therefore, the terrestrial animals usually drink large amount of water and develop various devices for the conservation of tissue water and feed on such food materials which are rich in salts so that the deficiency of salts may be compensated. In terrestrial animals such as reptiles, birds, mammals and some crustaceans and annelids, the physiological adaptations to meet the osmotic problem are the following :

(i) For water conservation the body is covered by means of a water proof covering which prevents the evaporation of water from the body. The water proof body covering may be of horny scales (as in certain reptiles), feathers (as in birds) and hair or fur (as in mammals).

(ii) The loss of water through urine and faeces is checked by various mechanisms. In reptiles and birds the water from the faeces is absorbed by rectum and cloaca. Further, in birds and mammal the water from the urine is absorbed by HENLE's loop in the uriniferous tubules of the kidneys. The terrestrials animals such as birds, snakes and lizards excrete a semisolid urine containing uric acid crystals, thus minimizing water loss. Uric acid is quite insoluble in water, and it can be excreted without the use of much water. The amount needed is the necessary to flush the uric acid into cloaca. Within the cloaca most of the water is reabsorbed and the waste is given off in paste form.

(iii) The animals of extreme terrestrial habitat, the dry hot desert, have developed special means of water conservation. They have adapted to a lack of drinking water as they obtain water entirely from the metabolic reactions of which the water is by product, and the water contained in foods. For example, the **kangaroo rat** is found in some of the hottest and driest regions of the south western United States. It eats only dry seeds and never drinks water. It has no sweat glands, so loses no water in sweat, but it must remain in cool burrows during the day to minimize the loos of moisture through the lungs. It produces very dry faeces and excretes a very concentrated urine. With such adaptation, oxidative metabolism supplies all the water needed. Similarly camels can go for long periods without water in hot, dry deserts. Their primary adaptation is their ability to withstand extreme body dehydration. A camel can lose as much as 40 per cent of the water in its body fluids and still survive, yet a man will die if he loses as much as 20 per cent. A dehydrated camel appears very thin as if from starvation, but when he drinks, the body fluids are restored to their normal volume and his body size will return to normal in a short time.

(iv) In other animals the osmoregulation is performed by sweat glands, mouth, tongue, lungs and kidneys.

OSMOREGULATION IN PARASITES

The endoparasites have no problem of osmoregulation because they have the same concentration of salts in their body fluids as in host, *i.e.*, they live in isotonic media. The parasitic protozoans do not have any contractile vacuoles in their body. The parasitic flat worms, round worms, etc., possess less developed excretory system.

CONTROL OF OSMOREGULATION

LEVER and his associates (1961 a, b) after conducting the experiments on freshwater pulmonate snail *Limnaea stagnalis*, observed that hormones are responsible for water and salt balance in these invertebrates which regulate in accordance with environmental demands. According to KAMEMOTO (1964) these hormones are secreted in the brain particularly in earthworms. Pleural glands of adult snails are also said to control the water and salt balance.

In vertebrates the hormones which control the osmoregulation are the **arginine vasotocin** and **vasopression** from the posterior pituitary (according to GORBMAN and BERN, 1962) and **aldosterone** from adrenal cortex. Their regulation operates at the level of the surface membranes (gills, integument, urinary bladder), at the level of the kidney and also on the special glands of extra renal salt excretion (gills, orbital glands, rectal glands etc.).

The hormonal regulation of water and salt balance is an integrated process dependent on the co-operative action of many hormones. For example, amphibian skin is said to be responsive to hormones of posterior pituitary (arginine vasotocin and vasopressin in the presence of aldosterone of adrenal cortex (CHESTER JONES, 1957).

Besides these above mentioned hormones, there are also other factors which have indirect effect on the water and salt balance. These are thyroid hormone (**thyroxine**) and parathyroid hormones (**parathormone** and **paracalcitonin**) (HICKMAN, 1959 ; MATTY and GREEN, 1963).

In some fishes caudal neurosecretory system is also involved in the regulation of salt and water balance as it secretes active neurosecretions (ENAMI, 1959 ; TAKASUGI and BERN, 1962 ; MAETZ *et al.*, 1964). The pineal apparatus is also believed to control water and salt balance (GORBMAN and BERN, 1962).

Revision Questions

1. What is osmoregulation ? Describe in brief the physiological adaptations involved in osmoregulation in different animals.
2. Describe various factors which control the osmoregluation.
3. Write short notes on :
   (i) Euryhaline animals,
   (ii) Stenohaline animals,
   (iii) Osmoconformers,
   (iv) Osmoregulators,
   (v) Volume regulators,
   (vi) Volume conformers.

CHAPTER 11

Circulation

The circulatory system is the most important system of the body as it ensures the exchage of substances between all the tissues of the body and the external environment and the transport of various substances from one bodily organ to another. It contains a circulatory medium and a network of continuous tubes called vessels. In unicellular organisms such as *Amoeba* and *Paramecium*, the circulatory system is completely absent as in these organisms oxygen and other useful substances including food can easily diffuse to all parts of the cytoplasm from the environment and waste products such as carbon dioxide and ammonia can easily be eliminated from the body surface. In *Hydra* and other coelenterates as well as in flatworms the transport of oxygen and other substances to the cells and elimination of waste products is achieved by making the gastrovascular cavity a highly branched structure. The food taken into such a cavity is circulated to all parts of the branching system by the movements of the body and the cells of the body can pick up required substances and remove unwanted substances. These animals, thus, do not require any special sort of circulatory system. In higher animals, on the other hand, because of their complex body organisation there is an imperative need of some sort of circulatory or transport system through which various substances may be transporated to different parts of the body which are not in direct contact with the external environment.

WILLIAM HARVEY (1628) was the first who discovered the function of heart and the circulation of the blood. He stated that the heart was a pumping organ provided with valves to maintain the flow of blood in one direction only, that blood was distributed to the oragns by means of deep lying vessels which he called **arteries** and the blood was returned to the heart by more surperficial vessels which he called **veins**.

He showed this by a simple experiment employing a bandage applied to the upper arm of one of his patients. The bandage was tightened until the pulse at the wrist could not be felt. The patient's arm after a lapse of few period became cold. On loosening the bandage a little to release the pressure on the deeper arteries, the patient felt a sensation of warmth returning to his arm while the veins below the bandage swelled. Therefore, Harvey concluded that the blood was flowing down the arteries from the heart but could not return in the veins since the bandage was still compressing them.

HARVEY did not know of the existence of capillaries as there was no microscope at that time. The capillaries were discovered by MALPIGHI in 1661 after HARVEY'S death. Later in 1732 HALES succeeded in to determine the blood pressure.

FUNCTIONS OF CIRCULATORY SYSTEM

Although each part of the ciculatory system has its individual functions, the system as a whole performs the following functions :

1. It provides essential chemical substances to the cell of the body which are needed for their metabolism.

2. It transports chemical waste products to those organs from where they can be excreted from the body.

3. It plays a most significant role in the prevention of and defense against invasion of infectious micro-organisms.

4. It assists in maintenance of normal body temperature and homeostasis of the tissue fluids that is the fluid, electrolyte, and acid-base balance.

CIRCULATORY MEDIA FOUND IN ANIMALS

Different types of circulatory media have been observed in different animals which are as follows :

1. **Hydrolymph :** This type of circulatory medium is found in lower category animals such as nematodes, ectoprocts and rotifers. It is a watery fluid having no respiratory pigment. It carries nutritive substances to organs and tissues and is also connected with the removal of waste products.

2. **Haemolymph :** It is a common circulatory medium which is generally found in arthropods and many molluscs. It is comparatively less watery than the hydrolymph. It is rich in proteins and serves as blood in higher animals. It not only transports nutritive substances to the tissues and remove waste products but it also serves respiratory function as it has haeme, a respiratory pigment, in its compsition.

3. **Lymph :** It is a colourless liquid contained in a system of branching tubes called the lymphatics or lymph vessels. It is similar to plasma in the composition as it, like blood plasma, contains proteins, lymphocytes and some granulocytes, but the concentration of these substance is much lower than the blood. It contains no red cells. One of the most important functions of lymph is to transport materials to and fro in the tissue. It is generally found in almost all the animals including invertebrates and vertebrates.

4. **Blood :** It is the chief circulating medium of the animals including the invertebrates and the vertebrates, with the exception of *Amphioxus* and *Leptocephalus*. It transports different substances from one organ to another within the body. It ensures humoral intercommunication of all organs and serves also to maintain uniformity of temperature in the body. In all its capacities it functions in the regulation and maintenance of the constancy of the normal environment of the body. Although, various chemical substances continuosly

enter or leave the blood during its circulation the over all composition of the blood remains remarkably constant.

STRUCTURE AND COMPOSITION OF BLOOD

The blood of lower animals is a thin watery solution, while of higher animals including mammals is a viscous complex tissues fluid of red colour. It is salty in taste, the specific gravity of which ranges from 1·050 to 1·060. It is made up of two principal components—the fluid **plasma** and the **blood cells** or **corpuscles** which are found suspended but unattached to one another in the plasma.

Plasma : The plasma of blood represents the intercellular substance and is a complex sticky aqueous solution of straw-colour Its forms 55% of the total volume of the blood particularly in the case of human being. It contains a liquid called serum and a coagulative substances called fibrinogen. It also has an anti-coagulant, or anti-prothrombin or heparin. Its specific gravity is about 1.027, while that of blood is 1.055.

The chemical composition of plasma shows great variation in different vertebrates due to different amount of water and essential solid constituents. It is composed of water, inorganic and organic constituents in solution, the concentration of these remains at an almost constant level. However, the concentration of other substances that blood transports fluctuates within limits, according to the activities and demands of different parts of the body. These substances include oxygen and carbon dioxide, the products of metabolic activity, the products of partial decomposition of food, and the secretions of endocrine glands.

Inorganic components of the plasma : The inorganic components of the blood plasma found dissolved in water are the chlorides, bicarbonates, phosphates and sulphates of sodium, potassium, calcium and magnesium. Minute amounts of iron, phosphorus and iodine are also found. They are absolutely essential to the blood and to the normal functioning of the body tissues. Without calcium compounds blood cannot clot in a wound.

Organic components of the plasma : The most important organic components of the blood plasma are a variety of complex plasma proteins such as the fibrinogens, the albumins and the globulins. The globulins exist in α_1^-, α_2^-, β^-, and γ^-globulins. The average normal concentrations of the main plasma proteins in g per 100 ml are :

albumin, 4·8; globulins, 2·3; fibrinogen, 0·3.

Thus, the albumins are present in the greatest concentration and they contribute most to the osmotic pressure. The fibrinogens are essential for the clotting of the blood and the globulins generally called serum globulins, are protein antibodies that provide immunity to various disease. These plasma proteins also act as efficient buffers and so help to maintain the pH of the blood at an almost constant level very close to neutrality.

A number of proteins with specific physiological functions have been isolated from the globulin fraction by electrophoresis and these are prothombin, plasma thromboplastin, isohaemagglutinins angiotensinogen, immune globulins and anterior pituitary hormones.

In addition to those plasma proteins certain, other organic subtances are also found in the plasma which may be classified under following heads :

1. **Nutrients** and **catalysts :** These include glucose, fats, amino acid, vitamins and hormones.

2. **Waste products :** These include urea and other nitrogenous products produced after the breakdown of ingested proteins or cellular proteins from cells wornout or destroyed by infection.

Blood cells or **corpuscles :** The cellular elements of vertebrate blood are the **erythrocytes** or **red cells** containing haemoglobin; the **leucocytes** or **white cells** having no haemoglobin; and, in all but the mammals, the **thrombocytes.** These are essential for proper functioning of the body.

The cellular elements of blood are short-lived and their destruction and replacement go on constantly during the life of an animal. The sites of their formation potentially possible in regions of the body were there is diffuse mesenchyme, have become more and more localized during the course of the vertebrate evolution. In some of the lower chordates, connective tissue within the walls of the stomach or intestine is the principal region of the formation of cellular elements of the blood. In higher chordates, the bone marrow has become the principal site in the adult. The formation and disintegration of blood cells is a balanced process and the result is a relatively constant number of cells circulating at all times.

Erythrocytes : Erythrocytes are nucleated in all vertebrates execpt mammals and contain haemoglobin in the cytoplasm. In lower vertebrates (fish, Amphibia, reptiles and birds) these are oval or lentil in shape but in mammals these are biconcave discs with an average diameter of 6–9 microns. These cells are produced in the red bone marrow of bones such as the ribs, vertebrate and skull.

The red cells or erythrocytes during their development, are large colourless and have larges nuclei. Normally by the time, they are to be released into the blood stream they have lost the nuclei and have accumulated haemoglobin that is why they are technically corpuscles and not cells.

The average span of these cells is 20–120 days after this time they are removed by the liver and the spleen where globin part of these cells is converted to the protein end product, urea. The liver and spleen extract the iron and conserve most of it for making more red blood cells or corpuscles. The dye is converted to red pigment called biliverdin and is excreted from the body in the bile which the liver manufactures and sends down to the intestine via the bile ducts. At the same time certain valuable substances are released into the blood stream which are used in the manufacturing of the new blood red cells.

The production of red cells or corpuscles requires an intake of food which provides sufficient amounts of amino-acids, iron and traces of copper. Two vitamins of the B-complex participate in the formation of RBC—folic acid and vitamin B_{12}. Vitamin C is necessary to convert folic acid to its active form.

The presence of antigens A, B and the Rh factor in the red blood cells accounts for blood types. The individual blood cells are pale yellowish in colour but when they are aggregated they appear to be reddish in colour. The red colour of the blood is entirely due to the presence of **haemoglobin** in the cells. Haemoglobin is a complex protein made up of 95% globin and 5% haematin, it is commonly known as respiratory pigment. Venous blood is purplish red and arterial blood is bright scarlet. The variation in colour of blood is due to the difference in amounts of oxygen absorbed by the haemoglobin in the two cases. Each red cell is enveloped externally by a thin membrane which is composed of lecithin and cholesterol. Inside the cell there is a elastic substance called stroma, in the meshes of which iron pigment haemoglobin is present. The blood of woman contains an average of about 5 million red cells per cubic millimeter, while in man about 5·5 million per cubic millimeter.

Functions of erythrocytes : Erythrocytes contain haemoglobin in their cytoplasm and haemoglobin has great affinity for respiratory gases, therefore, these perform the following important functions :

1. Oxygen transport from lungs to tissues.
2. Carbon dioxide transport from tissues to lungs.
3. Maintenance of blood pH as the haemoglobin acts as a buffer system.
4. These maintain the viscocity of blood.

Energy requirement : Mature erythrocytes entering the blood stream from their site of formation actually do not have any nuclei, mitochondria and ribosomes in their composition and nor use oxygen because tricarboxylic acid cycle does not function in them. However, the erythrocytes require a certain amount of energy in order to maintain their individuality and structural integrity for that the energy is supplied to them by glycolytic splitting of sugar into lactic acid.

White blood corpuscles or leucocytes : White blood corpuscles (W.B.C.) do not have any pigment in their structure, therefore, they are colourless and called as **white blood corpuscles** or **leucocytes.** They are found less in number in the blood as compared to the red blood corpuscles, the ratio being about one white blood corpuscle to every 6 hundred red blood corpuscles. These are formed in the red bone marrow and in the lymph glands. Normally there are between 600 to 800 white blood corpuscles in one cubic millimeter of blood as against four and a half to five million red blood corpuscles.

These corpuscles contain nuclei and somewhat larger in size in comparison to the erythrocytes. These vary in size. They are actively motile and capable of amoeboid movement. Their chief function is to check the invasion of the small organisms like bacteria and other foreign substances into the body. This they do in two ways :

1. Some white blood corpuscles engulf the invading organisms

in the manner of *Amoeba* ingesting a particle of food, the manner is commonly known as **phagocytosis**. This phenomenon was first discovered by METSCHNIKOFF and named **phagocytosis** (Gr. *phagcin*-to eat).

The white blood corpuscles as they take part in phagocytosis, are referred to as **phagocytes**.

2. Other white blood corpuscles secrete into the plasma **antibodies** which are chemical substances capable of neutralizing poisonous (toxic) chemicals in the blood such as the poisonous waste materials passed out by bacteria.

The mature white blood corpuscles can be grouped into two main categories : **granular leucocytes** or **granulocytes** and **agranular leucocytes** or **agranulocytes** depending upon whether their cytoplasm contains visible granules or not.

Granulocytes : In these corpuscles the cyloplasm is granular and the nuclei are large and irregular in shape with two or more lobes. These are produced in bone marrow and are of three types according to their staining properties :

(i) Eosinophils, (ii) Basophils, and (iii) Neutrophils.

(i) **Eosinophils :** These corpuscles have granular cytoplasm and are stained with acid dyes such as **eosin**. These cells have lobed nuclei which stain deep purple. These constitute 2% to 3% of the total leucocytes count and are 10 to 15μ in diameter. The important function of eosinophils is to bring about destruction and detoxication of toxins of protein origin.

(ii) **Basophils :** In these corpuscles the cytoplasm contains coarse granules which stain blue. These cells also have lobed nuclei which stain dark purple. These are about zero to four per cent of the total leucocytes count and are 10–15μ in diameter.

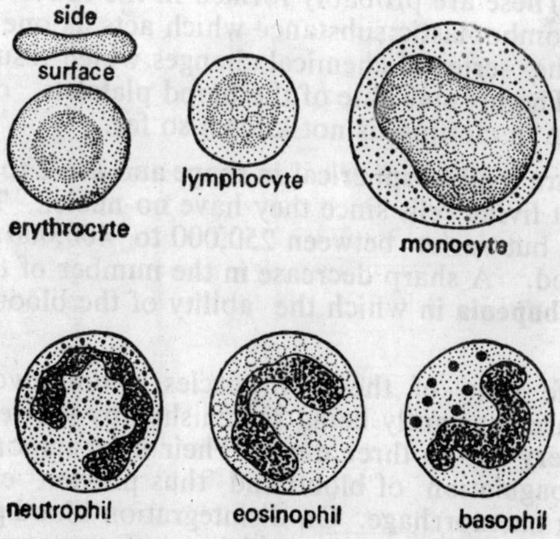

Fig. 11·1. Types of blood cells in man.

(iii) Neutrophils : The neutrophils, so called because they show up with a particular neutral dye. They are commonly called **polymorphonuclear leucocytes** as they have many shaped nuclei. The cytoplasm of these corpuscles contains fine granules and take very faint basic stain. These corpuscles form 65% to 70% of the leucocytes count and are about 12 to 15 μ in diameter. These show amoeboid movement and can move upto 40μ per minute. When they come across the bacteria or any other foreign bodies, a contact is made with them and are then engulfed. Once engulfed these bacteria or foreign bodies are digested and destroyed.

2. Agranulocytes : These corpuscles do not have any granular cytoplasm. The nuclei are simple and without lobes. The corpuscles can be grouped under two heads :

(i) Lymphocytes, and (ii) Monocytes.

(i) Lymphocytes : These are small cells and constitute 20% to 30% of the total leucocytes count. They are of variable size, the size may range from 8μ to 16μ. They are less motile in comparison to other leucocytes. They have large nuclei either spherical or indented in shope. The amount of cytoplasm in a lymphocyte is much less as compared to the size of its nucleus. The important function of these corpuscles is to produce antibodies. **They are of two types : large lymphocytes and small lymphocytes depending upon the size of the corpuscles.**

(ii) Monocytes : These are comparatively larger cells, 12–20μ in diameter and constitute 4% to 8% of the total leucocyte count. They have either bean-shaped or oval nuclei. These are motile in nature and can engulf the bacteria. The amount of cytoplasm in these cells is much more as compared to nucleus.

Blood platelets or **thrombocytes :** These are small, colourless, flat granular corpuscles which are much smaller than the red blood corpuscles. These are probably formed in the red bone marrow and contain a thromboplastic substance which acts as one of the enzymes involved in the series of chemical changes which result in clotting of the blood. The normal fate of the blood platelets, other than their distintegration in clotting, is not known so far.

These are oval to spherical in shape and are 2 to 3μ in diameter. These are not living cells since they have no nuclei. Their number is not constant but varies between 250,000 to 400,000 per cubic millimeter of blood. A sharp decrease in the number of thrombocytes is called **thrombopenia** in which the ability of the blood to clot is impaired.

The life span of these corpuscles is only two to three days, hence, they are constantly being replenished by the red bone marrow cells after every two to three days. Their most important function is to help in coagulation of blood and thus prevent excessive loss of blood during haemorrhage. On disintegration blood platelets liberate serotonin of 5-hydrotryptamine which causes vasoconstriction and thus acts as a haemostatic agent.

GENERAL FUNCTIONS OF BLOOD

Blood has many functions, the most important cnes are summarized below :

1. **Transport of oxygen and carbon dioxide :** Blood transports oxygen from the respiratory surfaces (lungs) to the tissues and carbon dioxide from the tissues to the respiratory surfaces and thus helps in respiration. Oxygen enters the blood through the lungs and carbon dioxide is eliminated from the blood mainly through the lungs.

2. **Transport of food :** Blood carries soluble food from the intestines, first to the liver and then to the whole of the tissues where it is required for cellular activities. The nutritive substances transported by the blood are glucose, amino acids, polypeptides, fats, vitamins, minerals and water.

3. **Transport of waste products :** Waste products are being constantly produced by all the cells of the body. They are harmful to the body, therefore, they require immediate elimination. Blood transports these wastes to the kidneys, lungs, skin and intestine so that they may be eliminated.

4. **Chemical co-ordination :** Endocrine glands of the body produce hormones which are distributed by the blood to the vital tissues, and in various ways harmonise or co-ordinate the working of the body.

5. **Maintenance of pH :** The plasma proteins of blood act as buffer system and thus prevent any shift in pH of the blood. This is because of amphoteric property of proteins.

6. **Water balance :** Blood maintains water balance to a constant level by bringing about constant exchange of water between circulating blood and tissue field.

7. **Transport of heat :** The blood allows the transfer of heat from the deeper tissues to the surface of the body where it can be lost.

8. **Defence against infection :** Blood contains corpuscles which possess properties of phagocytosis and special products called antibodies which combat the bacteria and thus plays protective role after neutralizing their toxins.

9. **Temperature regulation :** Blood maintains the body temperature to a constant level after distributing heat within the body.

10. **Support :** Being under pressure in the arteries the blood helps to support the tissues.

11. **Blood loss :** Blood prevents the excessive loss of blood in injury as it has the power of co-agulation.

CLOTTING OF BLOOD

The blood possess the ability to clot. Normally when the blood flows through the blood vessels does not clot but usually clots after escaping from the blood vessels or on coming into contact with the external environment, it changes into a jelly-like mass called **blood**

clot or **haemostatic plug** and blood is than said to be clotted or coagulated. The haemostatic plug or blood clot is formed initially by aggregation of blood platelets; later fibrin deposition occurs and the fully formed blood clot contains also red cells and leucocytes within the fibrin meshwork. The haemostatic plug is relatively stable except in haemophilia in which fibrin formation fails to occur. The rapidity with which the process takes place and the degree of consistancy of jelly produced vary greatly in different species of animals and sometimes in the animals of the same species but at different times.

The exact mechanism of blood clotting is not known. Many theories have been put forward to explain this process. All these theories differ only in the exact details but agree on the few main steps. The first theory of blood clotting, which later became classic, was put forward by ALEXANDER SCHMIDT. Originally (in 1861) it was formulated as a chemical theory but subsequently (in 1872) as an enzyme one. In contrast to the classic theory of blood coagulation as a biphasic process (phase I—thrombin formation ; phase II—fibrin formation) the **current theory** recognises three main phases in the process (phase I—thromboplastin formation ; phase II—thrombin formation ; and phase III—fibrin formation).

BLOOD CLOTTING FACTORS

Biochemical techniques of isolation and purification, immunological methods of identification have shown that blood clotting has the character of a **"chain reaction"** in which apart from the four factors constituting the basis of the classic theory (factor I—fibrinogen ; factor II—prothrombin ; factor III—thromboplastin ; and factor IV—calcium), a number of other factors are also involved. We will first consider these factors and then we will discuss the mechanism of blood clotting.

Factor I (fibrinogen) : VIRCHOW (1845) first of all pointed out that fibrinogen is a plasma protein of high molecular weight (400,000—500,000) which actively takes part in the clotting of blood. Fibrin is the end product of the blood clotting reaction. Latest discoveries show that fibrinogen in its native state consists of particles of mycella which unite at the time of clotting to form fibrin threads of about 0·4μ long. Most fibrinogen is synthesized in the liver.

The biological feature of fibrinogen is that it clots and the clotting occurs in the presence of a specific enzyme, **thrombin**.

Factor II (prothrombin): This factor was discovered by SCHMIDT in 1863. Prothrombin is a glycoprotein synthesized in the liver. Vitamin K is essential for its formation. Inactive prothrombin is converted into active thrombin in the presence of thromboplastin and accelerators and calcium ions. The amount of thrombin formed is proportional to the initial level of prothrombin. However, its insufficient amount is always present in the circulating flood in connection with continuous latent microcoagulation. Prothrombin has the molecular weight about 69,000, while that of thrombin isapproximately half that of prothrombin, *i.e.*, about 33,000.

Factor III (thromboplastin) : It is a lipoprotein found in blood platelets and tissue cells. Its role was discovered and studied by SCHMIDT. It is secreted in its inactive form called prothromboplastin (KUDRYASHOV, 1948) in the tissues. Under the action of a powerful activator, **proconvertin** contained in the plasma tissue, prothromboplastin is transformed into active thromboplastin.

Modern investigators regard the action of thromboplastin as enzymatic, catalyzing the process of conversion of prothrombin into thrombin in the presence of factors V, VII and X, Ca^{++} and phospholipids.

Factor IV (calcium ions) : Although the role of calcium ions has not been definitely established, its importance in the process of haemostasis is generally recognized. It is not directly involved in the reaction of thrombin formation but it is required for the formation of prothrombin activator, and for the formation of insoluble fibrin clot.

Factor V (labile factor) : This factor was first of all described by OWREN in 1947 and is essential for conversion of prothrombin to thrombin by tissue extract and plasma factors. Factor V is absent from serum being consumed during blood clotting.

Factor VII (stable factor, autoprothrombin I) : This factor is required for the formation of prothrombin activator by tissue extract and is present in serum as well as plasma as being not consumed during blood clotting. Its deficiency is rarely occured but is frequently induced by oral anticoagulated drugs of the coumarin type.

Factor VIII (antihaemophilic globulin, antihaemophilic factor) : This factor is required for the formation of prothrombin activator from blood constituents and is found absent from blood serum being consumed during blood clotting. The deficiency of this factor may cause haemophilia.

Factor IX (Christmas factor, plasma thromboplastin component, autoprothrombin II) : This factor is also required for the formation of prothrombin activator from blood constituents. It is found in plasma and it is activated during clotting so that the activity in serum is much greater than in plasma. The deficiency of this factor is associated with a congenital haemorrhagic state resembling haemophilia (**Christmas disease**).

Factor X (Stuart-power factor) : This factor is found in plasma and serum and is responsible for haemorrhagic state is deficiency.

Factor XI (plasma thromboplastin antecedent) : This factor is found in plasma and serum and is required for formation of prothrombin activator from blood constituents. Its deficiency causes a haemorrhagic state.

Factor XII (Hageman factor) : This factor is also found in plasma and serum and is required for the formation prothrombin activator from blood constituents. In the deficiency of this factor blood clots very slowly.

Factor XIII (fibrin stabilizing factor) : This is a plasma protein which causes polymerization of soluble fibrin to **produce** insoluble fibrin. Its deficiency causes haemorrhagic state.

Blood platelets are not true cells but cell fragments formed in the bone marrow, lungs, liver and spleen. These tend to clump readily, cling to rough surfaces and liberate **phospholipids** which are essential for clotting in the absence of tissue extract.

MECHANISM OF BLOOD CLOTTING

The basic reaction involved in the clotting of blood is the conversion of the soluble protein fibrinogen into the insoluble protein fibrin by means of thrombin enzyme. Fibrinogen protein exists in the circulating blood as such, thrombin does not, but is formed from an inactive circulating precursor, prothrombin, when the blood is shed. The conversion of prothrombin into thrombin takes place in the liver and is dependent on the presence of Ca^{++} and of factors which are derived from damaged tissues, disintegrating platelets and from the plasma itself. The formation of prothrombin depends on the absorption of adequate amount of vitamin K.

According to recent findings all the thirteen above mentioned factors are involved in the blood clotting, apart from the blood platelets. Most of the factors undergo transformations to become active and thus constitute **enzymatic cascade** (BIGGS and MACFARLANE, 1966). In this enzymatic cascade, the activated form of a factor catalyzes the activation of the next factor. Very small amounts of the initial factors are needed because of the catalytic nature of the activation process. The numerons steps yield a large amplification, assuring a rapid response to trauma (injury or wound).

Clotting involves the interplay of **intrinsic** and **extrinsic** pathways. The intrinsic pathway involves the reactions of factor XII (hageman factor), factor XI (plasma thromboplastin antecedent), factor IX (Christmas factor), factor VIII (antihaemophilic factor) and factor X (Stuart factor) of plasma. This intrinsic pathway ends with the formation of active Factor X (active Stuart factor) which in the presence of factor V (Proaccelarin v), calcium ions and phosphoglyceride converts the inactive factor II (prothrombin) into active factor II, *i.e.* thrombin. Active thrombin converts the soluble protein fibrinogen into insoluble protein.

Extrinsic pathway involves the reactions of factor VII (tissue thromboplastin), factor IV (calcium ions) and phospholipids. These factors come from the trauma or disintegrating cells. These factors along with active factor X of intrinsic pathway take part in conversion of inactive prothrombin into active thrombin. Factor XIII on activation by thrombin converts fibrin into the form of hard clot.

The various processes taking place in the clotting of blood can be summarized under the following heads :

Conversion of prothrombin to thrombin : The conversion of prothrombin into active thrombin is carried out by the proteolytic action of active factor X. This conversion is accelerated by factor V, which is not itself an enzyme. Factor V can be regarded as a

modifier protein. In addition, calcium ion and phospholipid surface promote the conversion of prothrombin into thrombin. Factor X is activated by the proteolytic enzymes of both intrinsic and extrinsic pathways.

2. Proteolysis of fibrinogen : Fibrinogen, a highly soluble molecule in the plasma, is converted into insoluble fibrin monomers by the proteolytic action of thrombin. Thrombin hydrolyzes four **arginine-glycine peptide bonds** of fibrinogen, one in each of the two α chains, and one in each of the two β chains, resulting is liberation of two peptides, termed **fibrinopeptides** A and B. Peptide A contains 18 amino acid residues; B has 20, including tyrosine-O-sulphate. A fibrinogen molecule devoid of these fibrinopeptides is called **fibrin monomere** which contains four new amino-terminal glycine residues; each of the liberated fibrinopeptides contains carboxyl-terminal arginine.

3. Fibrin monomer aggregation (formation of soft clot) : The fibrin monomers formed from fibrinogen protein on action of thrombin, spontaneously associate to form fibrin, possibly as a consequence of electrostatic attraction or of hydrogen bonding between groups unmasked by the removal of the peptides. The formation of fibrin occurs in stages and depends on factors such as pH and ionic strength but is independent of the presence of thrombin. In the beginning the fibrin formation occurs end-to-end with formation of primitive fibrils, followed by a side-to-side polymerization of these fibrils to form coarse fibrin strands.

The fibrinogen molecules do not undergo polymerization because of the presence of an unusual negatively charged derivative of tyrosine namely **tyrosine-O-sulphate** in the fibrinopeptide B. The presence of these and other negatively charged groups in the fibrinopeptides probably keeps fibrinogen molecules apart. **Their release by thrombin gives fibrin monomers a different surface-charge pattern, leading to their specific aggregation.**

4. Formation of hard clot : The formation of highly insoluble fibrin clot occurs in the presence of calcium ions and an enzyme, **fibrinase**. Fibrinase is derived from factor XIII which is known as **fibrin stabilizing factor**. This reaction takes place in the presence of thrombin. Factor XIII is found in blood platelets as well as in plasma.

$$\text{Fibrin—}CH_2\text{—}CH_2\text{—}\overset{\displaystyle O}{\overset{\|}{C}}\text{—}NH_2 + {}^{+}H_3N\text{—}CH_2\text{—}CH_2\text{—}CH_2\text{—}CH_2\text{—Fibrin}$$

$$\downarrow \textbf{Transamidase}$$

$$\text{Fibrin—}CH_2\text{—}CH_2\text{—}\overset{\displaystyle O}{\overset{\|}{C}}\text{—}\underset{\displaystyle H}{\underset{|}{N}}\text{—}CH_2\text{—}CH_2\text{—}CH_2\text{—}CH_2\text{—Fibrin} + NH_4^{+}$$

Cross-linked fibrin

The clot produced by the spontaneous aggregation of fibrin monomer is quite fragile. It is subsequently stabilized and strengthened by the formation of **covalent cross-links** between the side chains of different molecules in the fibre. Thus formation of cross-links between α chains of different fibrin molecules converts the fibrin into its final insoluble polymeric form or **hard clot**. This cross-linking reaction is catalyzed by a transamidase enzyme and ammonia is liberated during this reaction. In this cross-linking reaction, in fact peptide bonds are formed between specific glutamine and lysine side chains in a transamidation reaction.

The steps for the conversion of fibrinogen to fibrin may be summarized as follows :

1. Proteolysis : Fibrinogen $\xrightarrow{\text{Thrombin}}$ Fibrin monomer + peptides
2. Polymerization : Fibrin monomer $\longrightarrow$ Fibrin polymer
3. Clotting : Fibrin polymer $\xrightarrow{\text{Factor XIII}}$ Insoluble fibrin clot.

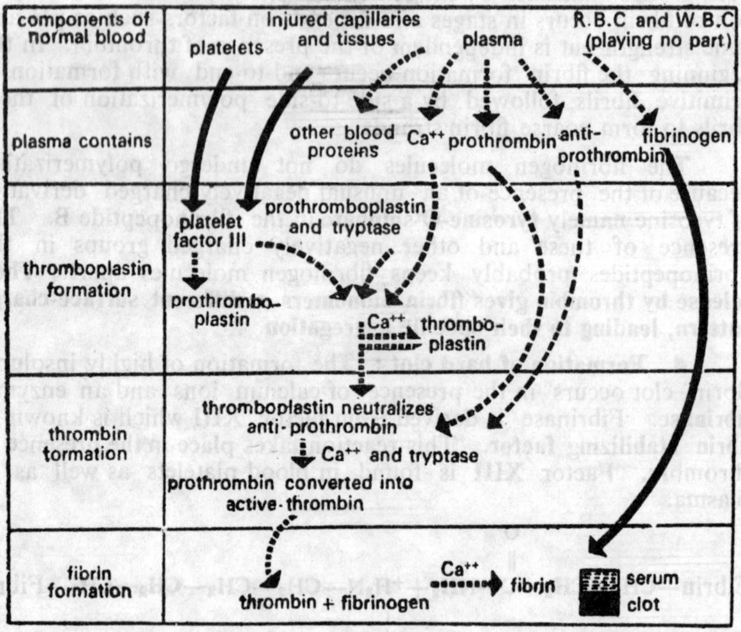

Fig. 11·2. Diagram showing the mechanism of blood clotting.

Human blood clots within 3 to 4 minutes. Heat accelerates blood clotting while cold sharply slows it down.

Anticoagulants or **Anticlotting compounds :** These are compounds which by their combined action maintain the blood in a fluid state after neutralizing the effect of coagulation factors. **Pavlov** discovered these physiological anticoagulants in the year of 1887.

The physiological anticoagulants include inhibitors of all the three main coagulation phases, *i.e.*, antithromboplastins, antithrombins and fibrinolysin. It has recently been demonstrated that there are several inhibitors for each blood coagulation factor. **Heparin** and **fibrinolysin (plasmin)** are the most thoroughly studied anticoagulants and have most practical significance.

Heparin is a very powerful anticoagulant which retards the conversion of prothrombin into thrombin. In addition, it influences phases of the formation of thromboplastin and fibrin. It is synthesizd in the liver. **Hirudin** is also a powerful anticoagulant. These two heparin and hirudin are the anticoagulants of animal origin. There are some other synthetic anticoagulants such as dicumarol, peleutan, etc., which block the synthesis of prothrombin in liver and thus prevent blood clotting.

TRANSFUSION AND BLOOD GROUPS

When the blood of a person is introduced into the body of other is called **transfusion.** Transfusion is generally given when the person either lacks the blood in his body or lost most of the blood of his body in any sort of injury. The practice of transfusion was tried as early as 18th century in France and later in England. Some cases of transfusion were successful but in others fatal transfusion reactions resulted. This led the people to think that why did some people die after a transfusion, while others did not show any such reaction, although the same sterile technique of administering the blood was followed in all cases. It could not have been infection, therefore, there must be something in the blood which made the two blood incompatible. In what way they were incompatible; what actually happened ?

The reason for adverse result of transfusion was not understood until about 1900. Thereafter, it was found that in the transfusion reaction the red blood cells clumped together or agglutinated and then underwent dissolution or haemolysis. This so affected the kidneys that they could no longer perform their function of removing the waste materials of the body and the patient dies of toximia.

The agglutination of red blood cells is due to protein substances in the red blood corpuscles coming in contact with incompatible substances in the blood plasma. It is an antigen-antibody reaction. The substances which found in red blood cells are called **antigens** or **agglutinogens** where as which are found in blood plasma called **antibodies** or **agglutinins.** Antigens are proteins capable enough to stimulate the production of speeific antibodies. Antibodies on the other hand, are the substances produced in the body in response to counteract with foreign antigens and react specifically to particular antigen.

In the human blood two types of antigens namely A and B are found in the red blood cells and two types of antibodies namely a (anti A) and b (anti B) are found in the blood plasma. The blood antigens are glycoproteins while the antibodies are specific proteins in the plasma and are called natural antibodies. Agglutination of red blood cells occurs when antigen **A** comes in contact with agglutinin or antibody a and when antigen **B** comes in contact with agglutinin or antibody b.

Based on the mutual compatibility of these antigens and antibodies KARL LANDASTEINER dicovered four blood groups in the human blood in the year of 1900. The are :

Type A has the antigen A and antibody b
Type B has the antigen B and antibody a
Type AB has both A and B antigens but no antibody.
Type O has both a and b antibodies but no antigen.

These above blood groups show that only these antibodies or agglutinins can remain in the blood whose corresponding antigen or agglutinogen is not present.

It will be evident by looking at the above blood groups that blood of group A can be given only to those person who do not possess antibody a, it means only in A and in AB. Similarly the blood of group B can be given only to B and AB. AB blood can be given to only AB recipient, because in all other types there is always some agglutinin. O type can be given to O and to any other type because it has no antigen at all. A person with O blood is called a **universal donor,** while AB is called the **universal recipient**. The possibility of transfusion among these four blood groups is as follows :

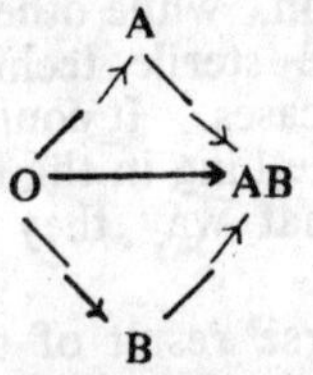

OTHER BLOOD GROUPS

Besides classical blood groups A, B, and O there are many other new blood groups which have been identified very recently. These are as follows : (1) Rh (2) MNS. (3) P, (4) Lutheran, (5) Kell, (6) Lewis, (7) Duffy and (8) Kidd. Among these blood groups Rh group is very important from clinical point of view as it is related with the haemolytic transfusion reaction and heamolytic diseases of new born children.

Recent findings show that blood groups are not as simple as those discussed above. For example the A antigen is composed of the antigen A_1 and A_2 and there are more than 12 Rh groups. New blood groups are coming into light day by day.

Blood grouping is a valuable aid in blood transfusion forensic (legal) medicine and to determine the possible paternity of a child.

Thus the study of blood groups has been a source of interest to pathologists, biologists, geneticists, anthropologists and statistician.

Rh FACTOR

In the 1940 LANDSTEINER and WEINER discovered the presence of another factor in the red blood cells of *Rhesus* monkey and some humans which they called it as **Rh factor or antigen.** This symbol for Rh antigen stands for '*Rhesus*'—a species of monkey which serves as the experimental animals for the discovery of this factor. This Rh factor is found in 85% white population.

If a person's red blood cells contain the Rh antigen he is said to be **Rh positive.** If he has no Rh antigen then he is **Rh negative.** Normally the blood of a Rh negative person does not carry any antibody that can react with this antigen or factor but such bodies may be produced if Rh negative blood comes across Rh positive blood. If Rh positive blood is transfused into the body of Rh negative person the Rh positive red blood cells serve as foreign bodies to the Rh negative person and so his body proceeds to produce antibodies called **anti-Rh agglutinin** to destroy the foreign bodies, *i.e.*, the Rh positive red blood cells. One transfusion is not apt to be serious but repeated transfusions may result in severe reaction and ultimately the death of Rh negative person may be occured. The safest procedure, for the Rh negative person, is to receive only Rh negative blood. 93% Indian population is Rh positive, while the rest 7% is Rh negative.

Rh factor is very important factor as it is inheritable and consequently presents a serious problem if a Rh negative woman marries a Rh positive man. When these two mate and woman conceives a Rh positive foetus in her uterus, Rh positive factor will pass into the circulatory system of the mother through the placenta and evoke Rh antibodies formation in the mother. As the amount of Rh antibodies produced during first conception is small, the first Rh positive child will be normal but if this woman again becomes pregnant with Rh positive foetus the results will be disastrous because more Rh antibodies will be produced and some of these will pass through the placenta into the body of foetus. Rh antibodies came from mother will react with Rh antigens present in the foetus red blood cells, thereby, causing agglutination of red cells which may result in the death of foetus or new born child. The death of newly born child may be avoided by a continuous transfusion, *i.e.*, all the blood of child is replaced by some donor's blood.

TYPES OF CIRCULATORY SYSTEMS

The circulatory systems of animals, the main function of which is to transport substances from one bodily organ to another within the body, can be classified in the following way :

1 Intracellular transport : This type of transport mechanism is the chraracteristic feature of the lowly-organised organisms such as protozoans; the protoplasm of these animals constantly shows streaming movement in a definite course which supplements not only the diffusion of the substance into and outside the body but also proper dispersal of them throughout the cell cytoplasm. This can be examplified by definite course of the movements of food vacuoles in the ciliates. Protoplasmic streaming also occurs in most, if not all, cells of the metazoans. It is most evident in amoeboid cells.

2. Transport through external medium : In sponges and coelenterates water serves as circulatory medium. In these animals water along with dissolved nutrients and oxygen passes through definite channels and cilia, flagella and muscular activity of the body assists in maintaining the regular circulation of water.

3. Transport through fluid filled body spaces : This type of transport mechanism is found in those animals which are having primary body cavities such as **pseudocoeloms** of nematodes, ectoprocts and rotifers. The fluid contained in these primary cavities is made to propell by the muscular movements of the body. As the fluid moves, the nutritive substances and oxygen contained in it also move from one place to another and the cells of different organs pick up required substances and oxygen from the fluid. Some transport occurs in the mesodermally lived coelom of echinoderms, annelids, ectoprocts and some chordates. In arthropods and molluscs the coelom is very much reduced to gonads and kidneys, the circulatory medium in these animals is haemocoelomic fluid which runs along with haemocoel.

4. Open circulation of haemolymph : In many invertebrates such as most arthropods, molluscs and ascidians, the presence of circulatory system though has come into existance but is of **open type**. In these cases the circulatory medium is **haemolymph** which is often called blood. It flows through a system of vessels in no more than a part of its circular path, in a part of the path the blood flows out into extensive tissue spaces. Such a system is called an open system. There may be two types of open systems. In one the tissue spaces, called sinuses, are lined by a cellular membrane which keeps the blood separated from the tissue cells. In the other type the blood spaces called lacunae are not lined by a cellular membrane and blood bathes the tissues directly. Both sinuses and lacunae may exist in the same system. The haemolymph circulates mainly by virtue of muscular movements of the appendages. Rudimentary heart is, of course, found in these animals but it does not play any role in the conduction of haemolymph.

Aquatic isopods have well-developed vessels and indistinct lacunae; in terrestial isopods the vessels are reduced and lacunae are well differentiated.

5. Transport through lymph channels : In vertebrates tube-like structures are found called lymphatics. These structures connect the intracellular spaces with the blood vascular system and coverage on veins and form a lined network which may be as extensive as the capillary bed. In some amphibians and a few teleosts lymph hearts are also found. Fats and other substances are transported by these vessels.

6. Blood vascular system : The blood vascular system is a system of delicate branched tubes connected with each other. These tubes are known as blood vessels. This system is found in oligochaetes many polychaetes, leeches, cephalopod molluscs, holothurian echinoderms and vertebrates. It is of closed type and appeared first in annelids. The circulatory medium is blood which in this systems is pumped to the tissues through the arteries where it comes into intimate association with the tissues by capillaries. The blood returns to the heart by some sort of closed return vessels and thus blood is circulated again and again.

COMPONENTS OF CIRCULATORY SYSTEM

In higher animals the circulatory system includes the following components :

1. Heart, and 2. Blood vessels.

Heart : The heart is a great organ of the body whose usual function is to contract periodically in order to pump either haemolymph or blood to various parts of the body. It pumps the blood at one end and receives it at the other end by the returning vessels or the veins. To keep the blood circulating in a uniform direction it is provided with valves which prevent the backward flow of the blood.

The rhythmic contraction of heart may be triggered entirely through nervous impulses (**neurogenically**), or independently of the nervous system through modified muscles (**myogenically**) or through a myogenic mechanism which subject to regulation by nerves.

Morphologically the following four distinct types of heart have been recognised in the animals :

1. Pulsating hearts, 2. Tubular hearts,
3. Ampullar or booster hearts, and 4. Chambered hearts.

1. Pulsating hearts : In annelids and *Amphioxus* many blood vessels are contractile in nature and show rhythmic peristaltic waves, such vessels referred to as pulsating hearts. The peristaltic activity of these vessels actually imparts motion to the blood. This pushes the blood forward which is circulated throughout the body. In general the pulsations are not as regular as the beats of the chambered heart.

2. Tubular hearts : Such type of hearts are found in arthropods. They are in the form of contractile muscular tubes. In some cases there is an specialized area in the dorsal blood vessel which functions as heart. These are usually anchored at several corners and receive blood through paired ostia which are provided with valves.

In *Branchipus*, *Artemia* and insects the heart is tubular which extends for a considerable length of the body. In crustaceans the heart is pulsating muscular sac (MAYNARD, 1960).

Tubular hearts are invariably surrounded by pericardium. In many insects the heart is suspended by variously arranged **alary muscles** which maintain tension on the heart and by their contraction lateral openings (**ostia**) on the heart are closed. Contraction of the alary muscles usually coincides with contraction of the heart. These contractions together with the closing of ostia result in the expulsion of blood to tissues in the anterior region of the animal. Another result of the contractile actions is the creation of negative pressure within the pericardial chamber. This vaccum, along with muscular movements throughout the body provides the means by which a new supply of blood may enter the heart from the haemocoel.

From the heart one or two arteries generally take their origin towards the anterior end but sometimes posterior and lateral arteries also take their origin from the heart.

3. **Ampullar** or **booster hearts :** Such type of hearts are found in some crustaceans, insects, and other animals like cephalopods. These are in the form of dilated portions of the blood vessels and are commonly known as "**blood pumps**" or "**booster hearts**". In crustacea these are compressed by contractions of somatic muscles which have their origin and insertions outside the heart and run through the heart or its wall or lie in close proximity to it. These somatic muscles are secondarily adapted to the problems of circulation ; in some cases they contract rhythmically (MAYNARD, 1960). In insects they are connected with the circulation of wings (Diptera and Odonata), the antennae (Orthoptera) and the legs (Hemiptera).

Lymph hearts of fishes, amphibians and reptiles are other examples of such types of hearts. These hearts help in collecting the tissue and discharging the same into veins at many points. These are composed of striated anastomosing fibres and may have valves which prevent the back flow of lymph.

In cephalopods branchial hearts is found in the form of circular structure beneath the gills which simply pumps the blood into the gills.

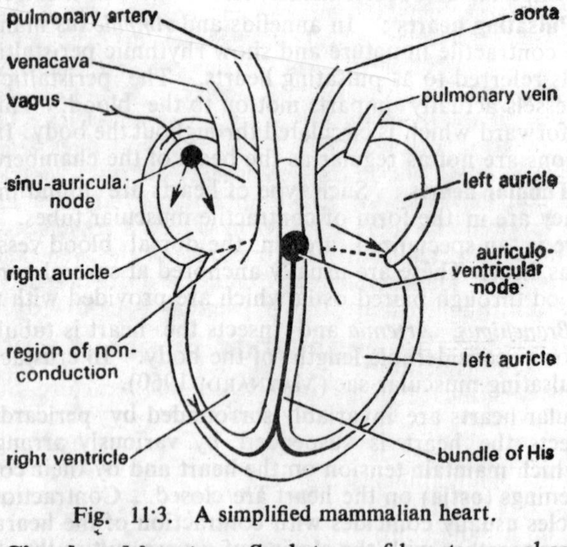

Fig. 11·3. A simplified mammalian heart.

4. **Chambered hearts :** Such type of hearts are characteristic feature of most vertebrates and molluscs. In vertebrates these hearts are made up of either two chambers (as in fishes) or three chambers

(as in amphibians) or four chambers (as in birds and mammals). The heart in each case is provided with valves which regulate the direction of the flow of blood and also prevent backward flow of blood.

In fishes the heart is an 'S' shaped structure having only two chambers; one auricle and one ventricle. These two chambers lie in the same plane. In amphibians the heart is three chambered consisting two auricles and one ventricle. The two auricles are separated into left and right with the help of interauricular septum, while the ventricle is an undivided single chamber. The oxygenated blood from the lungs returns to the left auricle while the deoxygenated blood coming from different parts of the body enters the right auricle. As there is only one ventricle the oxygenated and deoxygenated blood get mixed in the ventricle. In most reptiles the heart is also three chambered having two completely divided auricles and an incompletely divided ventricle. Higher reptiles such as crocodiles show four chambered heart consisting of two auricles and two ventricles. In birds and mammals the heart is distinctly four chambered consisting of two auricles and two ventricles. Each ventricle has a valve at its inlet, and a valve at its outlet. The inlet valves are termed the **atrioventricular valves**. On the left side it is also known as the **mitral valve** while on the right side it is known as the **tricuspid valve**. The mitral valve has two cusps while tricuspid three cusps. The outlet valves have three cusps and are known as the semilunar valves. The valve on the left side of the heart is also known as the **aortic valve** while that of the right side **pulmonary valve**. These are structures which allow the blood to flow in one direction only.

Blood returning from the body tissues enters the right auricle through two large veins, the superior and inferior venae cavae. Blood returning from the lungs enters the left auricle through the pulmonary veins. The auricles expand as they receive the blood. Both auricles then contract side by side, pushing the blood through the open valves (**atrioventricular valves**) into the ventricles. Then ventricles contract simultaneously; the atrioventricular valves become closed by pressure of the blood in the ventricles. The right ventricle propels the blood into the lungs through the pulmonary arteries and the left ventricle propels it into the aorta, from which it travels to the other body tissues. The valves in the openings of the ventricles into the pulmonary artery (pulmonary valve) and the aorta (aortic valve) close after the ventricles contract, thus preventing backflow of blood.

PHYSIOLOGICAL PROPERTIES OF HEART MUSCLES

The heart of vertebrates is a chambered structure, the walls of which are composed of **cardiac muscles** or **myocardium,** held together by strands of connective tissue. The inner and outer surfaces of the heart are covered by layers of epithelial cells called **endocardium** and **epicardium** respectively. The enire heart is surrounded by a transparent covering called pericardium. The muscles of molluscan heart and of cephalopods and gastropods are of striated type. The ampullary hearts are composed of spongy tissue of feebly striated endothelial cells.

The heart muscles of all animals exhibit some important physiological activities which are summarized af follows :

1. Excitability and contractility : The heart muscles contract and expand rhythmically throughout life. These are effected by heat, chemicals, mechanical and chemical stimuli but while contracting there is no effect of any of these stimuli. This non-responding period is called "**absolute refractory period**. When the contraction is over there is a brief period in which only a strong stimulus is effective, while the weaker ones have no effect at all.

2. Rhythmicity : These muscles keep on working rhythmically due to their own property without the help of any external factor.

3. Conductivity : The heart muscles are syncytial structure on account of which the stimulation received at any part of the wall of the heart spreads rapidly to all parts without stopping. This conduction in mammalian heart is very swift and the basis for the swiftness are the **bundles of His** and **Purkinje fibres**.

4. All and none law : The heart muscles do not contract when the strength of the stimuli is inadequate but contract maximally if the strength of stimuli is adequate. This property of heart muscles is known as **all and none law**.

5. Action of vagus on heart : The vagus nerve supplying the heart inhibits the heart movements and, thus, it balances the effect of sympathetic nerve which accelerates the heart beats.

HEART ACTION

The work of heart consists of rhythmic contractions and relaxations of the auricles and ventricles. A contraction phase of a chamber of the heart is called a **systole** and a relaxation phase **diastole.** The contractions and relaxations of the different parts of the heart take place in a definite order. The auricles and ventricles never contract simultaneously, the auricles contract first followed by ventricular systole. Ventricular systole is always followed by ventricular diastole. and so on. The sequence of one systole followed by one diastole is termed a **cardiac cycle.** It lasts eight-tenths of a second. Thus the action of heart can be divided in the following three phases :

Phase I : Simultaneous contraction of both auricles with the blood passing from the auricles into the ventricles which are relaxaing.

Phase II : Simultaneous contraction of ventricles ; the blood is forced into the aorta and the pulmonary trunk, while the auricles are relaxing.

Phase III : The ventricles relax and the auricles are also relaxed. This phase of cardiac activity is called the **general pause.** During the general pause blood enters the auricles from the venous vessels.

All these three phases constitute a single cycle of the heart action. The heart action, thus, follows a cyclic pattern.

HEART BEAT

The contraction of the heart (systole) and the relaxation of the heart (diastole) constitute heart beat. Heart beat can be defined as a propagated wave of muscular contraction. The heart is capable to contract rhythmically for some time even after it has been removed from the body. Hence the contractions of the heart muscles may be conditioned by processes occurring in the heart itself.

Each time the heart beats, each ventricle pumps out about 70 ml blood. This volume is termed the **stroke volume**. The heart beats about 70 times a minute (in man) and this is termed the **heart rate**. The stroke volume and heart rate on multiplication give the volume of blood pumped out by each ventricle per minute. This termed the **cardiac out put**. This may be represented as follows :

$$C.O. = H.R. \times S.V.$$

where C.O. stands for cardiac out put, H.R. stands for heart rate and S.V. stands for stroke volume.

The initiation and co-ordination of the heart beat was a subject of controversy for many years and led to extensive comparative investigations. In many and probably all vertebrates we are now convinced that the beat arises in the heart muscle or myocardium itself and is in no way dependent on nervous tissue. The most convincing evidence for this view originally was that the embryonic heart begins to pulsate long before any nervous tissue invades the heart. The studies of DAVIS, FRANCIS (1946) and PRAKASH (1957) also suggest that rhythmicity or heart beat is myogenic in origin and they pointed out the presence of specialized **impulse conducting cells** in the heart. These cells are quite different from the general cardiac muscle fibres. These cells constitute **nodule tissue.**

In lower vertebrates such as fishes and amphibians the nodule tissue is located in the **sinus-venosus** and, therefore, the impulse of beat originates in it. In higher animals such as birds and mammals, the nodule tissue is located in the **sinu-auricular node** which has a rich capillary blood supply. The sinu-auricular node is found at the point where superior vena-cava empties into the right auricle. This region is also known as **pace-maker**, because it is the region which initiates the heart beat.

At regular intervals a wave of contraction originates at the sinu-auricular node and spreads all along the auricles. It is picked up by a similar mass of tissue, the **auriculo-ventricular node** situated in the right auricle near the ventral part of the interauricular system. Branched fibres extend from the auriculo-ventricular node which for some distance run together as a single bundle (**Bundle of His**) and than separated into the right and left branches, each branch extending through the wall of the ventricle on the corresponding side and then it reaches the apex of the ventricle whence it gives out branches (**Purkinje fibres**) forming a network which spreads in the entire walls of the ventricle.

The excitation wave from the auriculo-ventricular node spreads along the bundle of His and Purkinje fibres and thus exciting the muscles of the ventricle with the result the two ventricles with their all parts contract simultaneously.

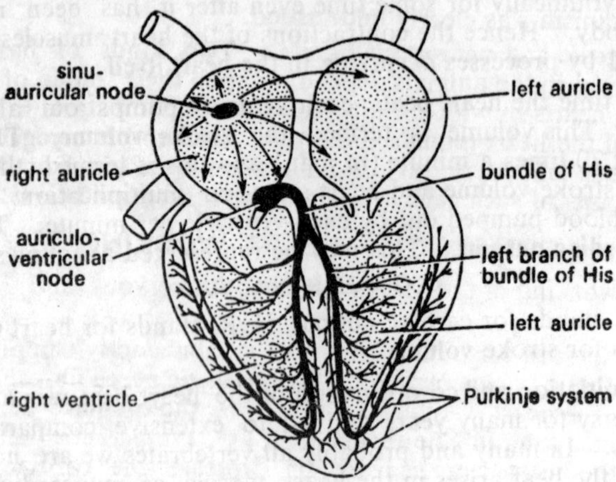

Fig. 11·4. Diagram showing the location of Pace-makers in the mammalian heart.

If the **bundle of His (atrio-ventricular bundle)** does not function properly due to any reason, then the heart beat, which originates at the **sinu-auricular node,** will spread to the auricles and to the **auriculo-ventricular node**, but will not reach the ventricles. This condition is termed **heart block**. The ventricle stop beating, and the circulation of the blood ceases. If the heart block is restricted to one half of the bundle of His, the condition is termed **bundle branch block**. The contraction wave can still spread from the **auriculo-ventricular** node to one ventricle and this ventricle will contract first. The contraction wave will then spread by the process of conduction to other ventricle which will contract a short time later.

The rate of conduction of impulse of contraction through bunble of His is very fast about 5 mm/sec.

Recently Mott (1957) observed that in elasmobranchs the **sinus-venosus**, the **auriculo-ventricular junction** and **truncus arteriosus** all show pace making activity. No distinct areas of pace making cells have been recorded in the myogenic hearts of invertebrates where rhythmic activity is assumed to be an inherent property of the cardiac muscles (Krijgsman and Divaris, 1955).

Recently A. J. Carlson has established that rhythmicity of heart is not only myogenic in origin but in some cases it is neurogenic in origin. A. J. Carlson conducted so many experiments on the heart of the xiphosuran arthropod, *Limulus polyphemus* and observed that the heart of this animal has on its dorsal surface a ganglionic mass of nerve cells. On removing this mass the rhythmicity of the heart ceases. This shows that the heart beat in this animal is neurogenic since it originates in the nervous tissue.

MAYNARD, 1961 tried to study such pace making ganglion in some other neurogenic hearts but KRIJGSMAN, 1952 in his classical review concluded that *Limulus* was the only arthropod in which the heart mechanism was clearly understood.

Myogenic and **neurogenic hearts :** Two kinds of hearts have been observed in the animals depending upon their mode of contraction. The hearts in which the wave of contraction starts in the muscle fibres of the heart (**nodule tissue**) are said to be **myogenic heart,** while those in which contraction wave takes its origin from thc nerve cells or groups of such cells are said to be **neurogenic hearts.**

CONTROL OF HEART WORKING

The working of the heart is under both nervous and hormonal control.

1- Nervous control : The heart is abundantly supplied with parasympathetic (vagus nerve) and sympathetic nerve fibres. In most vertebrates the heart is myogenic, therefore, the impulse of contraction originates itself in the heart and the sympathetic nerve fibres supplying the heart can increase or decrease the cardiac activity, while parasympathetic nerve fibres on the other hand inhibit the cardiac activity.

2. Hormonal control : The activity of the heart is also controlled by certain hormones. The most important hormones which control the heart activity are : **nor adrenaline** and **adrenaline.** These hormones increase and decrease the heart activity like sympathetic nerve fibres. Recently it has been established that the influence of these hormones on the work of the heart is closely connected with nervous regulation as these hormones are produced at the tips of the vagus and sympathetic nerve fibres only when they are excited.

According to RINGER inorganic ions like calcium and potassium ions also control the activity of the heart. High potassium ions follow a diastole arrest, while high calcium ions a systole rrest particularly in the frog heart while reverse is true in many invertebrates and vertebrates according to PROSSER and BROWN, 1961. According to HEILBRUNN, 1952, pH is also a important factor which influences the heart activity.

HEART SOUNDS

The various valves present at the outlet and inlet of ventricles determine the direction of flow of blood in the heart. Their action creates the **heart sounds.** The heart beat begins as the auricles contract, forcing the blood into the ventricles. As the auricles relax, the **tricuspid** and **bicuspid valves (mitral valve)** snap shut. This creates the **first heart sound** or **"lubb".** Now the ventricles contract and the blood is forced into the pulmonary artery and the aorta. As the ventricles relax, the two **semi-lunar valves** close, creating the **second heart sound**; or **"dup".** Since the interval between the closure of the atrioventricular valves (first heart sound) and the closure of the semi-lunar valves (second heart sound) is shorter than the interval between the closure of the semilunar valves and the next closure of the atrio-ventricular valves, the beating of the heart has a characteristic rhythm **lubb-dup-pause-lubb-dup-pause** (1–2-pause—1-2-pause). This cycle of beats occurs, on the average, about 72 times a minute. If any of the four valves is damaged, as from rheumatic fever, blood may leak back through one of the valves, producing the noise characterized as a "**heart murmur**" (a "ph—f—f—t" sound).

2. **Blood vessels :** There are three types of blood vessels—**arteries, veins** and **capillaries**. Functionally arteries carry the blood away from the heart to the tissues, while veins bring it from the tissues towards the heart. Capillaries connect the arteries and the veins.

Arteries : The blood vessels which carry the blood away from the heart to the tissues are called arteries. The blood in these blood vessels flow at high pressure which is created by the peristaltic movements of the muscles of the arteries. The arteries divided into thinner **arterioles** which branch into extremely thin and small **metarterioles,** these divide to form capillaries.

The wall of an artery consists of three layers, an outermost **tunica externa** or **adventitia** of connective tissue having longitudinal elastic and collagen fibres ; a thick middle **tunica media** made of circular smooth muscles and elastic fibres and an inner most **tunica interna** or **intima** made of endothelial cells and an elastic membrane. In between the tunica intima and tunica media there is a sheath of elastic tissue called **internal elastic lamina**, it completely surrounds the artery. There is also a similar membrane in between tunica media and tunica externa and is called **external elastic lamina**. The elasticity of the artery is mainly due to the presence of elastic fibres in its wall.

Veins : Veins are the blood vessels which carry blood towards the heart. These collect the blood from the capillaries in the tissues and empty into the heart. These are thin-walled structures with little musculature. The veins like arteries are made up of the same three layers but the tunica media poorly developed and hence the veins are thin-walled. Tunica externa or adventitia is relatively better developed in veins than the arteries. The external elastic lamina is less developed. In larger veins the tunica externa, some connective tissue and elastic fibres form valves which are crescentic pockets arranged in groups of three so that they may prevent a back flow of blood. The larger arteries and veins have their own small blood vessels in their coats called **vasa-vasorum** which supply nourishment and oxygen to these blood vessels.

Capillaries : Functionally capillaries are the most important part of the circulatory system. These are thin-walled and smallest of vessels which anastomose and form a very fine network of minute tubes. These are supplied with fine nerve ending and are so minute that they can be seen only under the microscope. Their walls are made up of a single layer of endothelial cells. The lumen of capillaries varies from $7{\cdot}5\mu$ to 0·0075 mm in diameter.

The capillaries are formed as a result of the branching of the arterioles. The capillaries reunite to form venules and venules form veins.

BLOOD PRESSURE

Blood pressure is the force with which blood pushes against the walls of the blood vessels. It is generally measured in terms of how high it can push a column of mercury. It varies at different points in the circulatory system. When the ventricles contract the pressure of the blood inside the blood vessel is highest and this pressure is termed as **systolic pressure**. In a young man it is usually 120 millimeters of mercury (120 mm Hg). When the ventricles relax, no blood is ejected in the arteries and it will tend to come to zero, but it is prevented to fall to such a low level due to elastic recoil of vessels. The process of blood when the ventricles are contracting is termed as **diastolic pressure** and it is about 80 mm Hg in a young man.

Constriction of the arteries by loss of elasticity or by the formation of fatty deposits within their walls increases the blood pressure. Gravity also affects the blood pressure.

GENERAL PLAN OF CIRCULATORY SYSTEM

In the vertebrates the circulatory syetem is of closed type as blood flows through the closed blood vessels, while in most invertebrates it is of open type. It is very simple in fishes but it is of complex type in birds and mammals. In fishes the heart is two chambered but sinus venosus and conus arteriosus are also found attached to the heart. Blood from the tissues enters a sinus venosus through the veins. It passes through a sinu-auricular aperture to the auricle, a relatively thin-walled but large chamber, thence it is propelled through auriculo-ventricular valve to the ventricle. Vigrous ventricular contraction forces the blood into the bulbous (or conus) arteriosus through a third set of valves. In elasmobranchi and dipnoi fishes there are several sets of valves in the conus. Blood from the conus arteriosus goes into the gills for oxygenation through blood vessels and then into dorsal aorta for distribution. Blood from the tissues is collected by the veins and comes back into the heart. Blood passes through the heart only once in its complete circuit so this is known as **simple circuit of circulation**.

Dipnoi or lung fishes show some advancement in the circulatory system. In them the heart has become three chambered as auricle has become partially divided into right and left auricles. Deoxygenated blood coming to the sinus venosus enters the right auricle, while the oxygenated blood coming from the air bladder enters the left auricle. In Amphibia and reptiles the heart is three chambered but to some extent double circulation is maintained. Birds and mammals have a complete double circulation. This situation is achieved by the development of the complete ventricular septum, with the pulmonary artery originating from the right ventricle and a single arota originating from the left.

In birds and mammals the heart is four chambered consisting of two auricles and two ventricles. From the left ventricle aorta takes its origin which is divided into different arteries which supply different

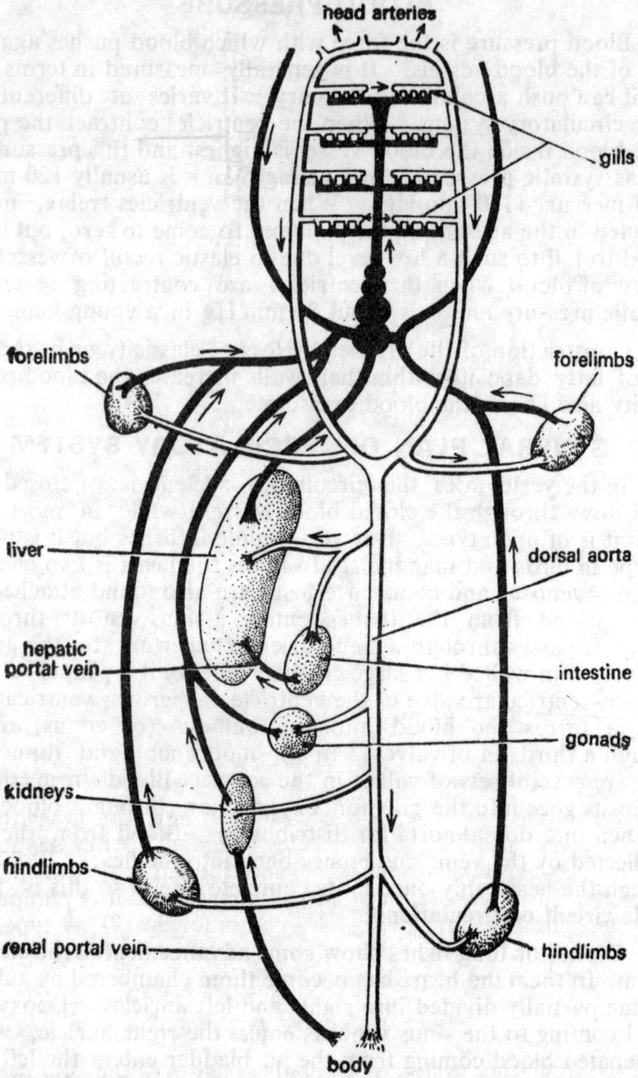

Fig. 11·5. Diagram showing the course of blood flow in fish.

parts of the body. Blood from different parts of the body is collected by tiny veins which in turn join to form larger veins. The veins of the posterior part of the body empty into the inferior vena-cava, while of anterior region empty into the superior vena-cava. Both these larger veins empty into the right auricle. The blood from the right auricle goes into the right ventricle through auriculo-ventricular

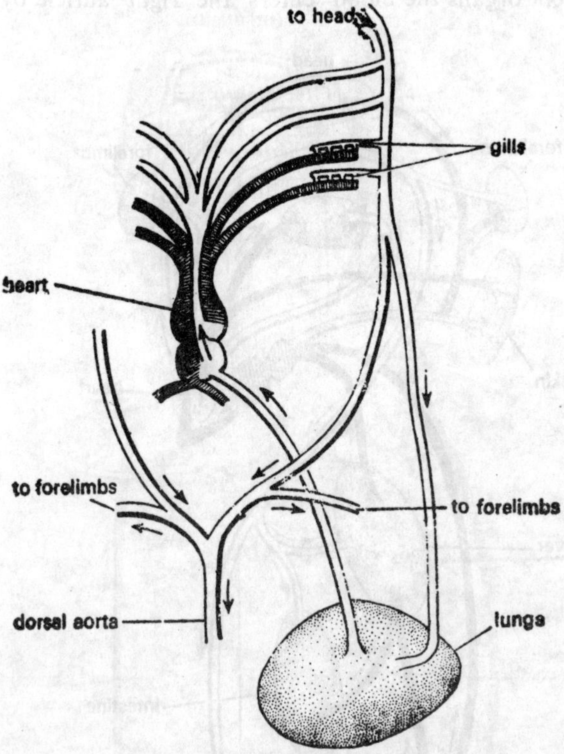

Fig. 11·6. Diagram showing the course of blood flow in lung fish.

valve. From the right ventricle it is pumped into the lungs through the pulmonary artery where it is oxygenated. After oxygenation blood through pulmonary vein comes into the left auricle from where it is transferred into the left ventricle from where it is pumped into the aorta. The blood thus circulates continuously. This type of circulation in which the blood passes twice through the heart during one complete circuit is called a **double circulation**.

CIRCULATORY ROUTE

The circulatory system of animals consists of the following circulatory routes :

1. The systemic route, 2. The pulmonary route, and 3. The portal route.

1. The systemic route: The systemic route begins with the aorta which takes its origin from the left ventricle. This route carries blood rich in oxygen plus all the other substances the cells need. Blood from the left ventricle is pumped into aorta and then through its branches to every organ of the body except the lungs. After passing in different organs the blood enters the right auricle by way of

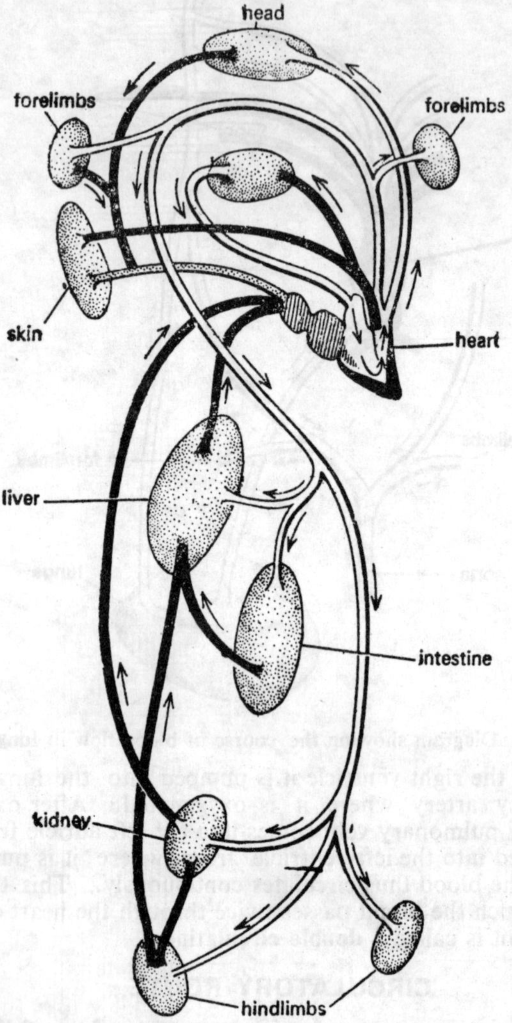

Fig. 11·7. Diagram showing the course of blood flow in the amphibian.

superior and inferior venae-cavae. As the blood circulates through the tissues it gives up its oxygen to the cells and take carbon dioxide.

Thus the systemic route includes the vessels through which the blood travels from the left ventricle to the organs and from the organs to the right auricle or atrium.

2. The pulmonary route : The pulmonary route begins with the pulmonary trunk which arises from the right ventricle and conveys

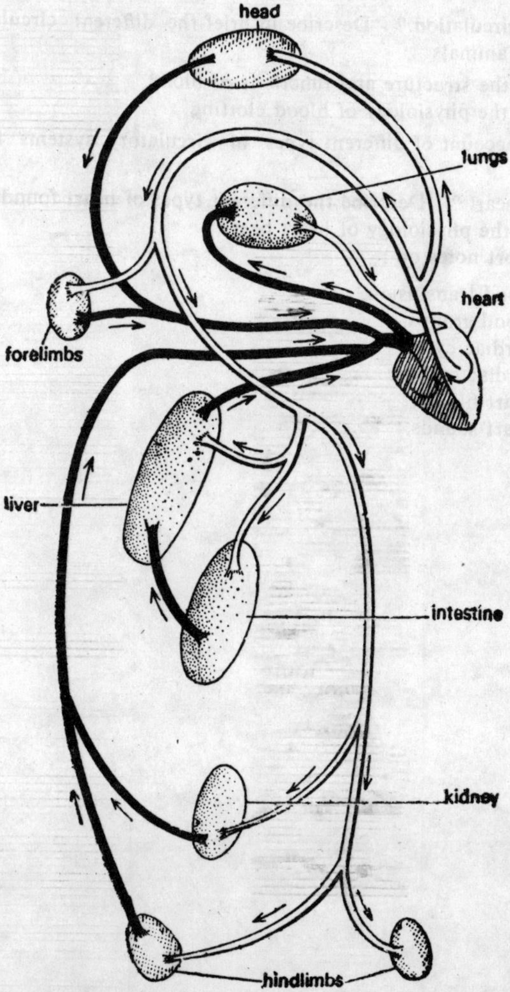

Fig. 11·8. Diagram showing the course of blood flow in mammal.

venous or deoxygenated blood to the lungs. The arterial blood flows from the lungs through the pulmonary veins into the left auricle. In other words the pulmonary route includes vessels through which blood moves from the right ventricle to the lungs and from the lungs to the left auricle.

3. **The portal route :** The portal route detours the blood from the digestive tract organs, the spleen and pancrease through the liver on its way back to the heart.

Revision Questions

1. What is circulation ? Describe in brief the different circulatory media found in animals
2. Describe the structure and functions of blood.
3. Describe the physiology of blood clotting.
4. Give an account of different types of circulatory systems found in animals.
5. What is heart ? Describe the different types of heart found in animals
6. Describe the physiology of heart beat.
7. Write short notes on :
   - (i) Blood transfusion,
   - (ii) Blood groups,
   - (iii) Cardiac cycle,
   - (iv) Cardiac output,
   - (v) Heart block,
   - (vi) Heart sounds.

CHAPTER 12

Muscles and Muscle Contraction

Muscles are the active part of the motor apparatus of higher animals. These contribute most of the total weight of the body of organisms. These are essentially machines for converting chemical energy into mechanical work. These also adjust the body against the environmental changes. The movement of the body as a whole from one point in space to another or the movement of a limited part of the body in respect to the body itself is also brought about by the muscles. These also maintain the particular posture of the animal against the effects of gravity. In addition vital processes such as contraction of heart, constriction of blood-vessels, breathing, peristalsis of the digestive tract are also accomplished by muscles. All these functions result from the contraction and relaxation of muscles.

Muscles consist of thousands of elongated fibres or cells organised in a variety of ways and bound together by connective tissue. These have the power to contract on getting the stimuli of any sort transmitted down the nerve that innervates them. The particular kind of contraction exhibited by an innervated muscle depends largely upon the pattern of its innervation. Although some muscles are capable of contracting independently of nervous stimulation, most muscles contract after a neuromuscular transmitter diffuses across a synaptic cleft from a nerve fibre to a muscle fibre. The excitory or inhibitory effects of a transmitter are coupled to contraction and relaxation respectively.

The contractility of muscle is due uniquely to contractile proteins which are present in all muscle fibres but the organisation of these proteins differs in different muscles. Different kinds of contractile units are found accordingly. These produce heat during muscular activity.

PRINCIPAL TYPES OF MUSCLES

Three principal types of muscles have been observed so far in the animals which may differ in structure (histologically), in location (anatomically), in function (physiologically) and in their manner of innervation (neurologically). These are striated, smooth and cardiac muscles.

1. **Striated muscles :** These muscles are multinucleate. These clearly display longitudinal and cross striations of alternating light

isotropic (weakly fibringent) and dark **anisotropic** (strongly fibringent) bands and are, thus, named **striated muscles**. These muscles nearly in all higher animals (vertebrates) are mostly found attached to the skeleton of the body and are involved with skeleton movements, therefore, for this reason they are called **skeletal muscles.** These muscles are also called as **voluntary muscles** because their contractile activities are under the control of central nervous system rather than the autonomic nervous system. These muscles are found not only in the higher category animals but also in lower animals such as annelids, arthropods, molluscs and non-segmented worms.

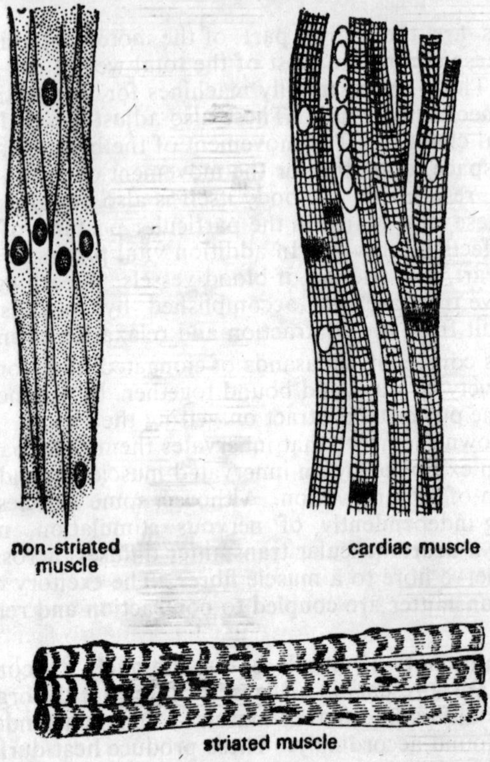

Fig. 12·1. Principal types of muscles.

Ultrastructure of striated muscle : Striated muscles are of complex structure having a large number of muscle fibres of different length (upto 12 cm). The muscle fibres are usually found **parallel** to each other and are grouped in bundles or **fasciculi.** Each muscle is composed of many such bundles. The fasciculi are arranged in a particular pattern to accomodate the movements of a limb or other body region. The muscle fibres, fasciculi and the muscle as a whole are invested by connective tissue that forms a continuous frame work. The outer part of this frame work, the **epimysium,** covers the whole muscle. Thin collagenous partitions extending inward to surround the individual fasciculi comprise the **perimysium** and the delicate network of connective tissue that invests the individual muscle fibres is the **endomysium**. The connective tissue serves to bind these individual contractile fibres together and to integrate their action.

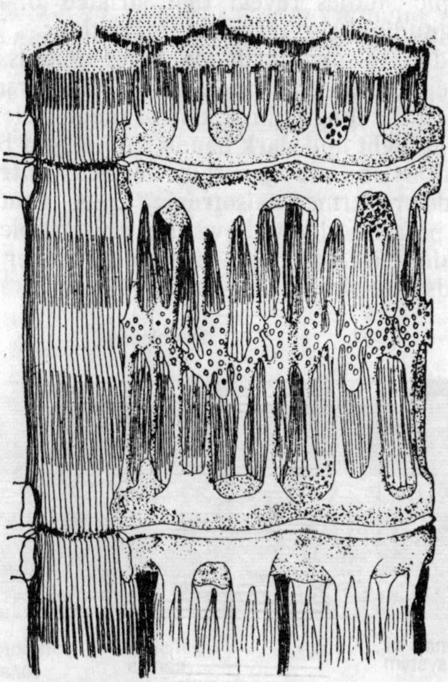

Fig. 12·2. A detailed structure of the striated muscle.

The striated or skeletal muscles are attached to bones by means of tendons which are found at the ends of muscles. A tendon is composed of dense, white, fibrous (inelastic) connective tissue. An expanded tendon consisting of a fibrous or membranous sheet is called an **aponeurosis.**

Each striated or skeletal muscle has blood vessels and afferent (sensory) and efferent (motor) nerves. Blood passes along the blood vessels, delivers nutrients to the muscles and carries away their waste products. The nerves link the muscles and the central nervous system.

The striated or skeletal muscle fibres are elongated cylindrical cells of variable size. The length of these fibres or cells may vary from a few millimeters to few centimeters (upto 10 cm), while their thickness may vary from 10μ to 100μ in diameter and depending on location the length of a muscle fibre may extend from few microns to centimeters. Each musele fibre is bounded externally by a tough, exceedingly thin elastic membrane called the **sarcolemma.** Inside the sarcolemma there is a semifluid substance called the **sarcoplasm.** In the sarcoplasm a large number of delicate bundles of longitudinally running myofibrils of 2μ in diameter are embedded. They are a sort of intracellular organoids of the muscle fibres. Each muscle fibre is multinucleate and the nuclei are usually found just beneath the sarcolemma in the sarcoplasm.

Microscopic studies reveal that striated or skeletal muscles show, both longitudinal and transverse striations. The longitudinal striations are due to myofibrils or myofilaments composed of the contractile proteins, actin and myosin. The transverse striations occur periodically along the longitudinal axis of the muscle fibre, these are due to alternating **light** and **dark bands**. The light bands are widest transverse striations and are generally termed **isotropic** or **I bands,** while dark bands are termed **anisotropic** or **A bands**. Microscopic examination in polarized light shows that anisotropic or A bands are birefringent or double refractive, while on the other hand the light bands or I bands give simple refraction.

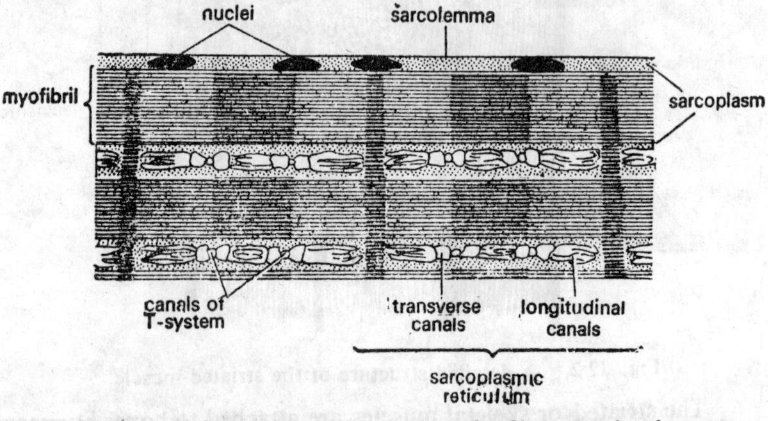

Fig. 12·3. Longitudinal section of a portion of a fibre of striated muscle.

Electron microscopic studies reveal that the alternating light and dark bands are produced by the arrangement of two types of protein filaments. The thick filaments (about 110Å in diameter and 1·60μ in length in fixed vertebrate muscle) lie parallel to one another about 450Å apart and form the anisotrophic band, they are probably made of myosin protein. The thin filaments (about 50Å in diameter and ·05μ in length in vertebrate muscle) are disposed in an orderly array between the thick filaments, they are probably made of actin protein. Both of these types of filaments appear within the A band on each side of a centrally bisecting **H band** which contains only thick myosin filaments, the centres of which are slightly thickened, giving the image of another band termed the **M band**. Finally in the centre of each I band there is a band of dense amorphous material termed the **Z-disc**. Dense Z-discs cross the centre of each I band, thereby, dividing the myofibril into smaller units as **sarcomeres**. A sarcomere, thus, is a repeating unit of the myofibril and represents the distance from one

Z-disc to the next. It is regarded as the unit of contractile activity in the striated muscle fibres of vertebrates and arthropods. The length of a sarcomere is related to contractility. In general the length of a sarcomere is inversely proportional to the speed at which a fibre contracts and relaxes.

The two different kinds of filaments are jointed together by **cross-bridges** which are structurally a part of the thick filaments and which project outward. These cross-bridges are thought to be significant in the sliding or ratchet-like action of the filaments which according to a popular theory of contraction, is the significant event in the shortening of the muscle (H. E. HUXLEY, 1958, 1960, A. F. HUXLEY and J. HANSON, 1960).

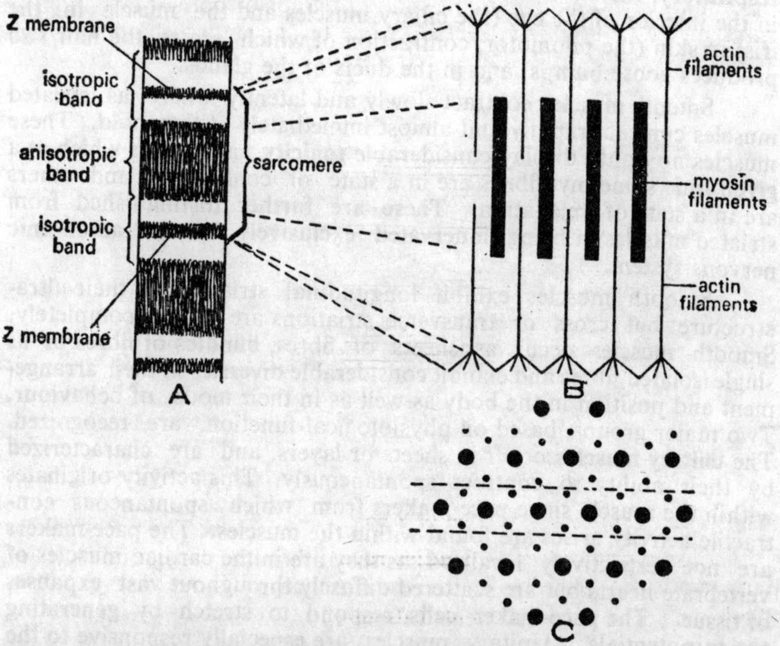

Fig. 12·4. A—A portion of a single myofibril as it appears under an ordinary microscope ; B—Showing the arrangement of actin and myosin filaments ; C—Transverse section through the filaments.

T-system : Recently it has been observed that two separate systems of channels exist within striated fibres. The channels of first system lie parallel to the long axis of the fibre and are delimited from transversely oriented channels of the second system (T-system, T for transverse). The channels of the T-system are invaginations of the sarcolemma which in most vertebrates run transversely but in many invertebrates these, no doubt, run transversely but exhibit considerable amount of branching. This T-system probably serves to convey electrical excitation rapidly into the muscle fibre or cell.

The longitudinal set of tubules in the striated muscle fibre is distinctly separated from the transverse set. The longitudinal set is homologous to **endoplasmic reticulum** of other types of cells and for this reason is called the **sarcoplasmic reticulum**. In addition mitochondria and cytochromes organelles are also found in the striated muscle fibres.

2. Smooth muscles : In vertebrates these muscles are also known as **involuntary** and **visceral**; involuntary because the activities of these muscles are not under the control of will and, thus, are controlled by the autonomic nervous system, and visceral because these muscles are found predominantly in visceral tissues such as in the walls of the four great tracts of hollow organs—the circulatory, the respiratory, the alimentary and the urinogenital. They are also found in the interior of the eye (the ciliary muscles and the muscle in the iris) in skin (the pilomotor, contraction of which erects the hair and produces goose bumps) and in the ducts of the glands.

Smooth muscles contract slowly and latently where as striated muscles contract rapidly and almost immediately at threshold. These muscles no doubt display considerable **tonicity,** a state in which at a given time some myofibrils are in a state of contraction and others are in a state of relaxation. These are further distinguished from striated muscles in being innervated exclusively by the **autonomic** nervous system.

Smooth muscles exhibit longitudinal striations in their ultrastructure but cross or transverse striations are lacking completely. Smooth muscles occur as sheets of fibres, bundles of fibres or as single isolated fibres and exhibit considerable diversity in their arrangement and position in the body as well as in their mode of behaviour. Two major groups, based on physiological function, are recognized. The **unitary muscles** occur in sheets or layers and are characterized by their ability to contract spontaneously. This activity originates withih the muscle since pace-makers from which spontaneous contractile activity arises are found within the muscles. The pace-makers are not restrictively localized as they are in the cardiac muscles of vertebrate hearts but are scattered diffusely throughout vast expanses of tissue. The pace-maker cells respond to stretch by generating action potentials. Unitary muscles are especially responsive to the mechanical stimulus of stretch but their activity is modulated and coordinated by neural influences. The muscles of gastro-intestinal tract and the ureters are the good examples of such types of muscles. The **multiunit muscles** are those smooth muscles which do not contract spontaneously, which are unresponsive to stretch and which have a multiple innervation by motor nerves so that each cell may be innervated.

Some smooth muscles such as muscles of urinary bladder resemble both unitary and multiunit muscles in that they may react spontaneously as well as reacting to motor nerve impulses.

The smooth muscle fibres are innervated by the vegetative nervous system, *viz.*, sympathetic and parasympathetic nervous system.

The primary function of these muscles is to contract rhythmically and the precise cause of the same is not known. These muscles show rhythmic contraction even when they are denervated. Sympathetic nervous system stimulates the contraction of smooth muscles in some organs, while it depresses their activity in others. Similar role is played by parasympathetic nervous system but sympathetic and parasympathetic nervous systems are opposite in function.

By rhythmic contraction and relaxation these muscles in the wall of alimentary canal and in the urinary and reproductive organs propel the contents of these organs forward (peristalsis). In the walls of the blood vessels they govern the amount of blood passing through a blood vessel and thus its distribution. These muscles in the form of strong circular band or ring (sphincter) control the opening and closing of a tube orifice.

Ultrastructure of smooth muscles : The smooth or unstriated muscle fibres are smaller than the fibres of striated muscles fibres. These muscle fibres not only exhibit sharp contrast with striated or skeletal muscles but differ from each other in their structure, location in the body, mode of action, innervation and general function in the animal economy. The fibres of smoooth or unstriated muscles are elongated, they are 50–100μ long and 2–5μ in diameter. Unlike the striated muscles these are spindle-shaped with two pointed ends and a centrally thickened part, the **belly**. Each muscle fibre contains a single elongate and narrow nucleus in the centre of the belly. Each fibre is bounded from the inside to the outside by a **sarcolemma,** a **basement membrane** and a **coat of connective tissue.** When these muscles occur in sheets and bundles the basement membrane and connective tissue are discontinuous in some areas and at such points the sarcolemmas of two separate cells form a **tight junction**. These **tight junctions** are probably areas of low electrical resistance between the cells which serve to transmit the excitations from one cell to another. The **sarcoplasm** is granular in nature and is free from strips or striations. A poorly developed **sarcoplasmic reticulum** has been detected in electromicrographs of these muscle fibres. No **sarcomeres** are evident in smooth muscle cell, however, faint longitudinal striations appear which run along the long axis of the cell, these are actually myofilaments of about 50 to 80A° in diameter. These myofilaments or myofibrils are made of actin and myosin. The contractile protein content and energy rich compound concentration are present much lower then they are present in the striated muscles.

3. **Cardiac muscles :** These muscles are found in the heart of animals (vertebrates). The fibres of these muscles are involuntary in nature as their activities are not under the control of will. The fibres of these muscles are both longitudinally and transversely striated as are striated or skeletal muscles but its fibres exhibit branching and contain myofibrils and filaments of actin and myosin are arrayed similarly to those in skeletal muscles. Likewise, the mechanisms of contraction are essentially the same as those in skeletal muscle but these muscles differ histologically from striated muscle of vertebrates in some ways (i) these muscles are uninucleated whereas striated muscles are multi-

nucleated, (ii) these are automatically innervated, while striated muscles are innervated by peripheral fibres whose cell bodies lie within the central nervous system and they interdigitate to form what are known as **intercalary discs**, these are actually cell membranes that separate individual cardiac muscle cells from each other. Thus cardiac muscle is not a syncytium. Despite the lack of syncytium cardiac muscle fibres transmit impulses from one fibre to another throughout the muscle without the need of booster stimuli from nerves. This suggests that cardiac muscle fibres have pseudosyncytial arrangement through which cardiac muscle is allowed to contract sequentially from fibre to fibre and, thus, the cardiac muscle is said to be functional syncytium.

In the end we can say that the cardiac muscles have one character of voluntary (striated) muscles being striated and one character of involuntary muscles, being not under the control of will.

GENERAL PROPERTIES OF MUSCLES

All the three types of muscles, *viz.*, striated, smooth and cardiac like other tissue (nervous tissue) possess some common porperties which are as follows :

1. **Excitability** : The basic and the most important functional property of all the three types of muscles is **excitability**, *i.e.*, the ability to respond to stimulation and become active. Striated muscles are, however, more excitable as compared to other muscles. These respond to different stimuli such as mechanical, thermal, chemical or electrical and in the living body nerves are responsible for bringing about the stimulation.

2. **Conductibility** : Once a part of the muscle fibre is stimulated by a stimulus of adequate strength, it is conducted within no time to all its other parts. This property is called as **conductibility**. Conduction is much faster in striated muscles as compared to other types of muscles. In cold blooded animals like the frog the rate of conduction is 3–4 meters per second, while in warm blooded animals it is 6—12 meters per second.

3. **Contractibility** : All the three types of muscles possess the property of contraction and relaxation and these two actions together constitute the **twitch**'. Muscles can be stimulated by mechanical, thermal, chemical or electrical stimuli. Nerves are responsible for bringing about the stimuli in the body. When a muscle is contracted it shortens and becomes thicker but its total volume barely changes. Contraction is stronger as well as faster in striated muscles as compared to other types of muscles.

4. **Tonicity** : All the muscles in the body at a given time are never found in a perfectly relaxed state. Although showing no outward signs of activity, they are in a state of mild contraction which causes them to resist being stretched. This activity of muscle is known as **muscle tonus.**

5. **Tensility** and **elasticity** : All the muscles possess the property of **tensility**, *i.e.*, the ability to **stretch** (to a limited extent). When

the cause producing the stretching of the muscle disappears, the muscle resumes its former state (relax), this property is called **elasticity**.

6. **Threshold** or **liminal, sub-liminal** and **supra-liminal stimuli** : All the muscles contract only when they receive the stimulation of a certain strengh. The lowest limit of stimulus capable to bring contraction in muscles is called the **threshold** or **liminal stimulus**. A stimulus weaker than the threshold (which fails to bring contraction of the muscles) is called **sub-liminal stimulus** and the stimulus which is stronger than the threshold is called **supra-liminal stimulus.**

7. **Refractory period :** After stimulation there is a brief period during which the muscle does not remain in a excitable state, this period is called **refractory period** or the period of relaxation of muscles.

8. **The period of latent excitation and contraction of muscles :** The period between the application of the stimulus and beginning of the muscular contraction is called the **period of latent excitation** or the **latent peiod**. The period during which muscle remains in a contraction state is called the **period of contraction** (**shortening**) of the muscle.

Muscle innervation : All the three types of muscles differ markedly in their innervation. The smooth or involuntary muscles and the cardiac muscles are automatically innervated, while striated muscles receive the impulses directly, without any relaying, from the central nervous system. In most cases the nerves sub-divide to form minute branching ramifications which innervate the most of the muscle fibres. The small terminal branches or ramifications come into close contact with the muscle cells or fibres. The naked cell membranes of the nerve endings are in direct contact with the sarcolemma of the muscle fibre, but the nerve endings do not actually penetrate the muscle cells except in the highly specialized dipteran fibrillar muscles (TIEGS, 1955). Such specialized structures are known as the **motor end plates**. Each muscle cell may have one or more motor end plates. These motor end plates have long been known in vertebrate skeletal muscles but are not recognised in vertebrates smooth or cardiac muscles nor in the slow-fibre junctions of the lower vertebrates. Among the invertebrates motor end plates have been recognized in only a few insect skeletal muscles (HAYLE, 1955, and TIEGS, 1955) and in the segmental muscles of nereid polychaetes (DORSELT, 1964).

According to CONTEAUX, 1960 and PROSSER and BROWN, 1961 the vertebrate motor end plates vary in detail but appear similar in that the myelin sheath of the axon ends near the muscle, while the nerve ending covered only by its plasma membrane, expands into an irregular area that sits in synaptic gutters or troughs formed as depressions in the muscle fibre. The sarcolemma lining the gutters or troughs is thrown into a series of folds which in vertical sections, look like tiny perpendicular rods, the **palisades**. The plasma membranes of muscle and nerve seem to be in direct contact. Both the exoplasm and the sarcoplasm of the end plate contain numerous mitochondria, indicating the region of high metabolic activity.

Each muscle cell may have one or more motor end plates. **An** individual nerve fibre with all the muscle fibres it innervates constitutes the functional neuro-muscular unit called the **motor unit**.

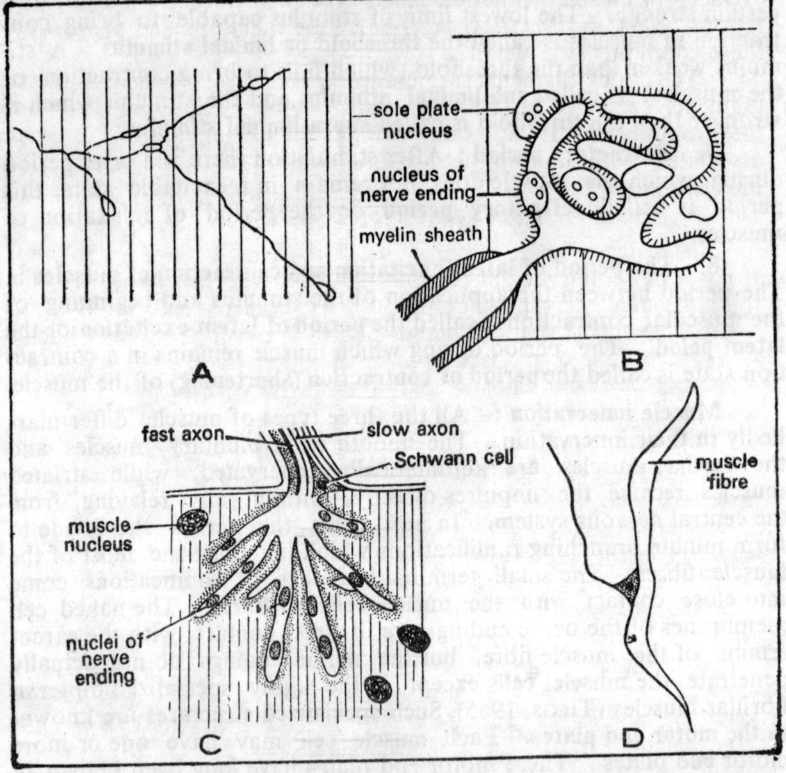

Fig. 12·5. Diagram showing the morphology of neuro-muscular junctions. A—From deep muscle of the tench; B—From a mammal C—From locust muscle; D—Single nucleated ending on muscle fibre from the pharynx of *Dendrocoelum*.

Molecular components of muscles : The molecular components of muscles associated with the contractile machinery of muscles include (1) contractile proteins of which myosin, actin and tropomyosin predominate, (2) enzymes and other proteins such as troponin that are associated with contraction, relaxation and energetics,(3) fats or carbohydrates as the ultimate source of energy, (4) nucleotides particularly ATP, ADP, ITP (inosine triphosphate) and phosphogens as more immediate sources of energy for contraction and (5) ions, the most important of which are sodium, potassium, magnesium, calcium and chloride.

Myosin, actin, tropomyosin and troponin are proteinaceous components of the myofilaments which are generally said to be the building blocks of the myofibrils. Each of these molecular components can be separated from others by virute of their solubility properties and their chromatographic behaviour.

These molecular components of muscles have been intensively studied by many biochemists. KHUNE was the first who prepared extracts of muscle which he called myosin. Recently BAILEY 1956, HODGE, 1958 and MAMMAERTS studied the functions and properties of these components in detail.

The thick filaments are composed of myosin whereas the thin filaments are made up of polymers of actin. The exact location of the tropomyosin is not clear but it seems clear that this protein is associated with the thin actin filaments probably at regularly spaced intervals along them.

CONTRACTILE PROTEINS

Myosin : This is the protein of thick filaments of the myofibril. It has a molecular weight of about 500,000 which is about 10 times the molecular weight of actin. Its molecule is much more complicated than the actin and one of its most important properties is that it is an ATPase, *i.e.*, it will enzymatically hydrolyse ATP to form ADP and inorganic phosphate. The reaction is activated by calcium ions but inhibited by magnesium ions.

Myosin in its structural constitution probably consists of two major subunits and two-three minor ones. The larger subunit has a molecular weight of about 200,000, whereas the smaller one a molecular weight of about 20,000. Even if the molecule has three of the smaller subunits, the total molecular weight does not exceed the value of 500,000 obtained for whole molecule. Whatever the resolution of this particular discrepancy, it is obviously true that the molecule has a highly elongated form. The long straight tail section of the molecule has the two α-helical subunits intertwined to make a larger helix, and has a width of only 15 to 20Å. At one end each of the main subunits appears to coil up to form a globular head region, which includes one of the small subunits as well. This gives the molecule two head pieces approximately 200Å long and 50Å in diameter. It is this part of the molecule which serves as an ATPase and which has an affinity for actin. This information on the molecular structure has come from the fact that a brief digestion of myosin by trypsin splits it into two different types of particles. These derivatives are called light and heavy meromyosin. The light meromyosin aggregates to form filaments and has no enzyme activity. The heavy meromyosin, on the other hand, is water soluble and will split ATP. Thus, the light meromyosin, seems to be part of the double-stranded tail region of the myosin, whereas the heavy meromyosin represents the rest of the molecule (including both head piece, a portion of double-stranded tail, and the small subunits).

The thick filaments in the myofibrils appear to consist almost entirely of myosin molecule which have their tail regions organised in a parallel array. The cross bridges which stick out from the thick filaments at regular intervals are now thought to contain the globular head pieces of the myosin molecules. This part of the molecule also contains a region with an affinity for actin, and this is probably concentrated at the tips of the cross bridges where they approach the thin filaments.

Actin : This is the protein of thin filaments of myofibril. It has molecular weight about 4,000 and N-acetylaspartic acid as its N-terminal residue. It contains one molecule of 3-methyl-histidin per mole of protein and is a single polypeptide chain. It exists in a globular form, **G-actin** in the absence of salts, which has a high affinity for calcium and can form dimers. In the presence of salts there is a transformation of G-actin to a fibrous state F-actin which is highly polymerized. Each form of actin has a high affinity for ATP.

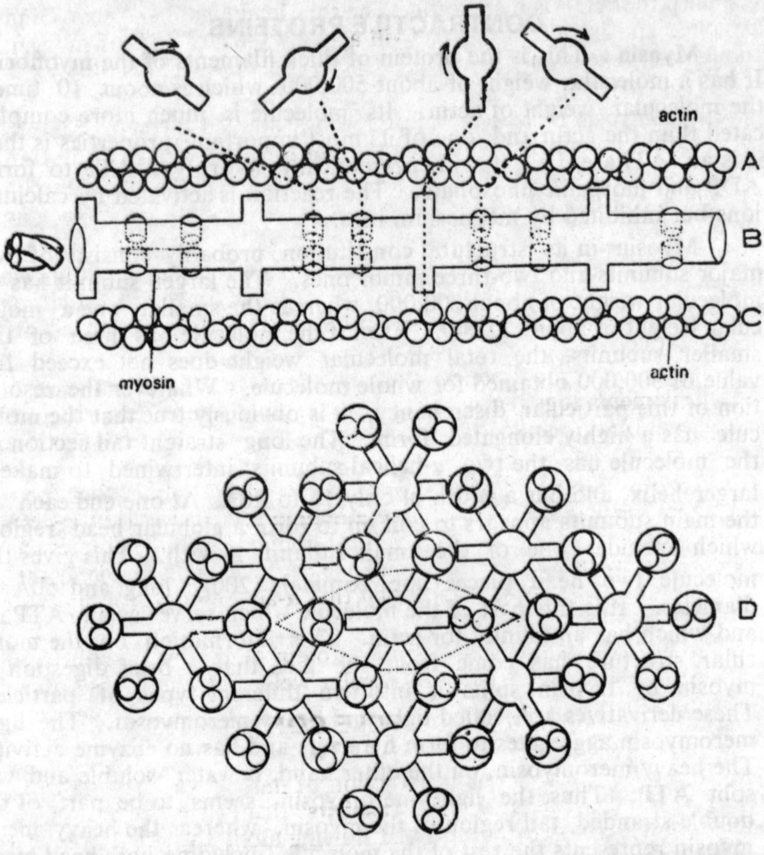

Fig. 12·6. Diagrammatic representation of two actin filaments (AC) and myosin filament; B—in a sartorius myofibril of a frog. The actin filament is the double helix, each strand of which is composed of a globular G-actin molecule arranged linearly ; D—an end view of the two types of filaments to illustrate their probable spatial inter-relationship.

According to SELBY and BEAR, 1956 the thin filaments of all muscles consist principally of the fibrous form of actin and they also pointed out that the actin monomeres are arranged either in a net-like structure or in helics. Recently HANSON, 1963 and LOWY, 1963 has reached the conclusion that F-actin consists of two chains of monomeres connected together in a double helical form.

The most interesting feature of this protein is that it undergoes polymerisation in solution in the presence of ATP, Mg^{2+}, and KCl.

$$n\ \text{G-actin} + n\ \text{ATP} \rightarrow (\text{F-actin} + \text{ADP})_n + n\ \text{phosphate}$$

The ADP produced in this reaction remains firmly bound to each unit of the actin polymer, and no further hydrolysis of ATP takes place until the ADP is displaced. The thin filaments can also bind Ca^{2+} ions, and in this form they will activate the ATPase of myosin, although, as mentioned earlier, it now appears that this binding is due to tropomyosin.

Tropomyosin : It is the most recently studied protein component of the myofibrils which is present together with actin in thin filaments and forms a specific complex with F-actin in vitro. Its molecule is rod-like, 400Å long and 20Å diameter with a molecular weight 70,000.

It has two subunits in the form of α-helices in an extended coiled—coil conformation. When added to actomyosin solution tropomyosin inhibits its calcium activated ATPase activity but not its magnesium activated ATPase activity.

Troponin : This protein has globular molecules and is found in the thin filaments together with actin and tropomyosin. It promotes the aggregation of purified tropomyosin. Its biological function is not clear.

In addition to these above proteins there are some other proteins found in the muscles, these are **α-actinin** and **paramyosin**. α-actinin protein has a molecular weight of about 160,000 and interacts strongly with actin, causing cross-linkage of the F-actin filaments and gel formation. It has been established that α-actinin and tropomyosin are contained in the Z-lines.

Paramyosin is the contractile protein of muscle myofibrils of annelids and molluscs. It has a molecular weight of about 151,000.

MECHANISM OF MUSCLE CONTRACTION AND RELAXATION

In the ancient time the contraction of muscle was thought to be caused by the movement of "**animal spirits**" distilled in the brain from the materials derived from food and flowing along the tubular nerves to inflat the muscles, much as a ballon can be inflated with air. The early biophysicists of the 18th century were able to disprove this theory by demonstrating that when a muscle contracts its volume does not change to any large extent ; shortening is accompanied by thickening.

In the 19th century several hypotheses have been put forward to explain the mechanism of contraction (shortening) and relaxation (elongation) of muscles but out of them only two hypotheses are currently under investigation. The older of these and perhaps the simpler to visualize, postulates a folding (during contraction) and unfolding (during relaxation) of the fibrous protein thread of which the muscles are composed of. This theory has been proposed in different ways but all variants assume that the contractile elements remain extended, while there is a mutual repulsion of like charges along the molecule. The changes associated with contraction (hydrolysis of ATP and activity of Mg^{++}) produce equal number of positive charges among the negative ones causing the contractile elements to fold or coil, thereby shortening the muscle. Many facts have been made to support this folding and unfolding theory, but these facts now seem less popular than the more recently suggested theory 'the sliding filament theory'.

The sliding filament theory : The sliding filament theory was independently formulated by H. E. HUXLEY and J. HANSON and by A. F. HUXLEY and R. NIEDERGERKE. This theory was most strongly supported by the arrangements of myofilaments. This theory states that the contractile units of muscle appear to be composed of thick and thin filaments which normally overlap somewhat in the relaxed muscle. When the muscle contracts on stimulation, these filaments do not change in length but merely slide over one another, *i.e.*, the thin (actin) filaments slide in the spaces between the myosin filaments with the result that I or light bands shorten, while there is no change in the A or dark band. However, disappearance of H zone in dark band may be observed. This is because of closing in of actin filaments. All these changes are associated with closing in of two Z lines of a sarcomere which is in proportion to muscular contraction. It is thought that the cross bridges on the thick filaments might pull the thin filaments in a kind of ratchet action, while muscle is contracted but during relaxation these cross bridges disappear. This indicates the presence of active sites on the actin filaments into which the cross bridges temporarily hook to pull the filaments a short distance and then release them. It means the contraction and relaxation of muscle are brought about by the repetition formation and breakage of cross bridges respectively between thick filaments of A band and thin filaments of I band.

Histological changes during muscle contraction : As you know that muscle fibre is composed of numerous myofibrils and each myofibril is composed of a number of sarcomeres. A sarcomere represents the contractile unit of the muscle. As soon as muscle is stimulated by any sort of stimulus it brings about electrical changes and the membranes are depolarized followed by repolarization. These electrical changes tend to liberate calcium ions into the sarcoplasm from T-sarcotubular system to L—sarcotubular system. The calcium ions, thus liberated bring about the sliding of the actin filaments in the spaces between the myosin filaments. The sliding of actin filaments in spaces between the myosin filaments requires a certain amount of energy and the energy for the same comes from the break down of the ATP into ADP. As the muscle is contracted two Z-lines come closer, light band and H zone disappear but no change takes place in dark A band.

Molecular basis of muscle contraction : The most important molecular basis of muscle contraction is the repetitive formation and breakage of cross bridges which are the only structures responsible for the contraction of muscle. As these cross bridges are formed in between the thick filaments of A band and thin filaments of I band, the two sets of filaments start movement past one another resulting in the shortening of the sarcomere. It has been established that cross bridges are formed only in the absence of ATP, while they disappear in the presence of ATP.

How sliding in the filaments takes place ? The cross bridges which take their origin from the myosin (thick) filaments are supposed to consist of heavy or H-meromyosin which shows ATPase activity and its active enzymatic sites are presumed to be nearer the base of the cross bridges. Because of the presence of Mg^{++} ions the base and the distal end of the cross bridge are supposed to be negatively charged owing to the presence of ionised ATP formed through a complex with Mg^{++} ions. The mutual repulsion of these two sets of negatively charge would keep the cross bridges in an extended state, and the ATP away from the active sites on the H-meromyosin. As the calcium ions are released from the T-sarcotubular system and L-sarcotubular system during activity, form chemical links between the distal end of the cross bridge and the adjacent thin filaments. On release of calcium ions the negative charge on the distal end is abolished with the result mutual repulsion between the two ends of the cross bridge is ended. As the mutual repulsion is over cross bridges are formed and shortening of the muscle is resulted. Because of the shortening the ATP at distal end is brought close to the active sites at the base of bridge. The ATP splits by the meromyosin ATPase which may lead the breaking of calcium-actin complex link, but it is rephosphorylated by creatine phosphate (CP). The ionized ATP is thus reformed and the electrostatic repulsion between the two ends of the bridge is restored. This process goes on again and again until the calcium ions are exhausted.

Role of calcium ions in muscle contraction : Calcium ions play significant role in the muscle contraction. These activate the interaction of myosin and actin, but apparently only through the intervention of tropomyosin and troponin. We noted earlier that the tropomyosin is bound uniformly to actin and that troponin complexes with tropomyosin at specific, repeating sites about 400Å apart on the actin filament. It is probable that troponin in the non-active muscle serves as a regulatory protein to prevent the interaction of myosin and actin. However, troponin is also a calcium receptive protein. Under the influence of an increase in free calcium ion concentration, troponin is thought to undergo a structural change. This abolishes the previous repression of the myosin-actin interaction and contraction ensues. Relaxation is brought about when the free calcium ion concentration is reduced.

Chemical changes during muscle contraction : When muscle contracts certain chemical changes of distinct nature take place simultaneously. The knowledge about the study of chemical changes involved in the muscle contraction begins with von HELMHOLTZ (1845) idea that in the liberation of energy in the form of mechanical work as it occurs in muscle the energy must come from some pre-existing energy. In the year of 1907 FLETCHER and HOPKINS declared that the lactic acid is formed during anaerobic contraction of vertebrate muscle. A few years latter PARNAS showed that the lactic acid originates from the glycogen. This discovery led to an enormous source of activity, both biophysical and biochemical. On the biophysical side A.N. HILL began his studies on the energetics of muscular contraction, while on the biochemical side OTTO MEYERHOF began the studies of glycolysis which ultimately led to our present rather complete understanding of this complex and important process.

The immediate energy for the contraction of the muscle is derived, whether directly or indirectly is uncertain, from the breakdown of the ATP to ADP, while the ultimate source is the combustion of the carbohydrates (glucose, glycogen and others) present in the muscle. However, there is a strong evidence that lipids are also utilized at times. These lipids exist in the form of free fatty acids which can diffuse rapidly from the blood stream into the muscle or derive from the muscle's fat stores. All the enzymes required for the oxidative metabolism of the free fatty acids are found in the muscle. The availability of oxygen determines the degree to which free fatty acids constitute a source of energy.

When ATP is not available in sufficient amount during the muscle contraction, **creatine phosphate (phosphocreatine)** serves as a supplemental source of energy which through the action of a transferase enzyme restores the depleted ATP. The sequence of chemical changes that take place during muscle contraction are as follows :

1. **Conversion of adenosine triphosphate into adenosine diphosphate :** The first and the important chemical change, that takes place during muscle contraction, is the conversion of the adenosine triphosphate into the adenosine diphosphate. This conversion is brought about by the enzyme, **adenosine triphosphatase** (ATPase) present in the muscle. During this conversion one molecule of phosphoric acid is removed from the adenosine triphosphate which supplies the immediate energy to the muscle for contraction. This reaction can occur anaerobically.

2. **Break down of creatine phosphate (phosphocreatine) :** The next step is the break down of creatine phosphate present in the muscle to produce creatine and phosphoric acid. The phosphoric acid molecule combines with adenosine disphosphate (ADP) and forms ATP.

$$\text{Creatine phosphate} \rightarrow \text{Creatine} + \text{Phosphoric acid}$$

$$\text{Phosphoric acid} + \text{ADP} \rightarrow \text{ATP}.$$

3. **Break down of muscle glycogen :** The glycogen present in the muscle after reacting with phosphoric acid liberated during the break down of ATP into ADP, is converted into glucose phosphate.

Glycogen+Phosphoric acid→Glucose phosphate.

4. **Formation of fructose diphosphate :** The glucose phosphate, after undergoing various chemical reactions, is converted into fructose diphosphate.

$$\text{Glucose phosphate} \xrightarrow[\text{reaction}]{\text{Enzymatic}} \text{Fructose diphosphate.}$$

5. **Formation of lactic acid :** Fructose diphosphate after undergoing various chemical changes is converted into lactic acid. During the formation of lactic acid three molecules of ATP are formed.

Fructose diphosphate→Lactic acid+3 ATP.

6. **Resynthesis of creatine phosphate :** During periods of inactivity or less intense activity, the creatine is rephosphorylated by ATP (enzymatic reactions are reversible) produced in intermediary

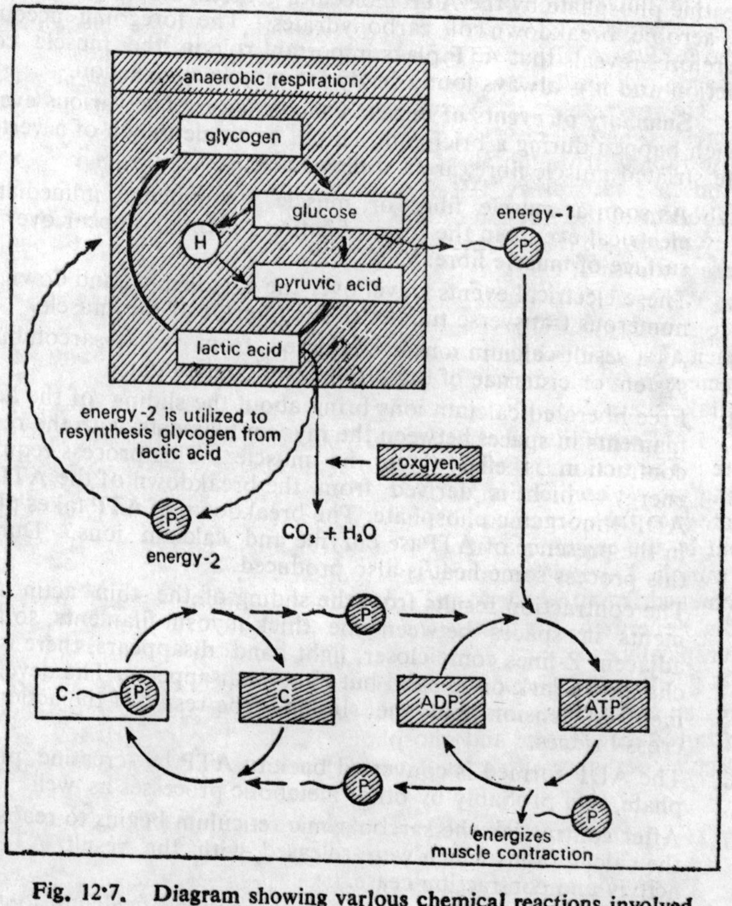

Fig. 12·7. Diagram showing varlous chemical reactions involved in the liberation of energy for muscle contraction.

metabolism. The energy for the resynthesis of creatine phosphate is derived from the breaking down of glycogen to lactic acid. 1/5 part of the lactic acid is oxidized in the presence of oxygen with the result energy along with carbon dioxide and water is produced, while remaining 4/5 part of the lactic acid is resynthesized to form glycogen in the liver, muscles and other tissues of the body. The energy for the body and for the resynthesis of glycogen is derived from the break-down of the 1/5 part of lactic acid produced during the glycolytic process of the glycogen.

Thus, as the muscle contracts due to sliding of the actin filaments in the spaces between the myosin filaments, simultaneously breakdown of ATP into ADP takes place with the result heat is liberated. The ADP thus formed is regenerated to ATP by an energy rich phosphate bond denoted by creatine phosphate. Creatine is converted back to creatine phosphate by the ATP molecules generated during anaerobic or aerobic breakdown of carbohydrates. The foregoing account, therefore, reveals that ATP plays important role in the muscle contraction and it is always found ready for muscle contraction.

Summary of events of muscle contraction : The various events which happen during a brief contraction, or single twitch of a vertebrate striated muscle fibre, are as follows :

1. As soon as muscle fibre or muscle is stimulated, immediately electrical events in the form of action potentials appear over the surface of muscle fibre.
2. These electrical events travel over the sarcolemma and down the numerous transverse tubules to the interior of the muscle.
3. As a result calcium ions are released from the T-sarcotubular system or cisternae of the sarcoplasmic reticulum.
4. Free liberated calcium ions bring about the sliding of the actin filaments in spaces between the myosin filaments, with the result contraction is effected in the muscle. This process requires energy which is derived from the breakdown of the ATP to ADP+inorganic phosphate. The breakdown of ATP takes place in the presence of ATPase enzyme and calcium ions. During this process some heat is also produced.
5. The contraction results from the sliding of the thin actin filaments in spaces between the thick myosin filaments, so that adjacent Z-lines come closer, light band disappears, there is no change in dark or A band but H zone disappears. The development of tension and the sliding is the result of the action of cross-bridges.
6. The ADP formed is converted back to ATP by creatine phosphate, and probably by other metabolic processes as well.
7. After contraction the sarcoplasmic reticulum begins to reabsorb the calcium ions which were released, with the result ATPase activiy and contraction cease.
8. When contraction ceases due to reabsorption of calcium ions by the sarcoplasmic reticulum, the myofibrils are stretched out

again through the action of antagonistic muscles. ATP must be present.

9. Creatine formed during the contraction is rephosphorylated in the presence of ATP, some heat is produced.
10. ADP formed during the rephosphorylation of creatine into creatine phosphate, is converted back into ATP by oxidative phosphorylation in the mitochondria. During this conversion more heat is produced.
11. Any lactic acid generated during contraction is reoxidized to pyruvic acid, and the reduced nucleotide formed, is ultimately oxidized by the mitochondria to produce more ATP.

Thermal changes or **heat production :** As and when muscles contract a considerable amount of heat is produced. Some heat is produced even at rest. It is for this reason that body remains hot during muscular exercise. According to HILL heat production occurs in three successive phases. The **initial heat** which is also known as **maintenance heat** is generated when the ATP breaksdown into ADP. This initial or maintenance heat is essential for keeping the muscle in a state of readiness for contraction. The initial heat is followed by **shortening heat** which is generated when shortening occurs in the muscle. The shortening heat is directly proportional to the amount of shortening. The shortening heat is followed by **delayed heat** or **recovery heat** which is generated when muscle tends to recover or relax The delayed heat can be further subdivided into an anaerobic phase and an aerobic phase. Heat in the anaerobic phase arises from the breakdown of glycogen to lactic acid but is of small magnitude compared to the aerobic phase, perhaps as little as one twentieth. Aerobic delayed heat is produced for an extended period, upto thirty minutes, and represents the oxidation of the fuels to CO_2 and H_2O. Heat, of course, is energy wasted through the inefficiences in chemical processes. Nevertheless, this heat may be of great value in the maintenance of body temperature.

Other changes : Besides heat production there are several other physical changes that take place during muscle contraction. These are changes in volume, electrical resistance, optical properties, the production of sound and the production of energy which is very negligible.

MECHANICAL PHENOMENA

Isotonic and isometric contractions : The muscles of the body some times contract purely isotonically and sometimes isometrically. When the muscle shortens and performs mechanical work, without undergoing any change in its tone, such a contraction of muscle is known as **isotonic contraction** (*iso*, same ; *tonus*, tension). In other words isotonic is that contraction when the resistance offered by the load is less than the tension developed in the muscle. The muscles of legs which propel the body by alternate contraction and relaxation, as in walking, the contractions are isotonic. On the other hand when the length of the muscle does not change but its tone is considerably increased during activity, such a contraction is known as **isometric** contraction (*iso*, same ; *metric*, length). During isometric contraction, no doubt, tension is developed in the muscle but the muscle fails to perform mechanical work. The muscles of the arm and hands in holding an object and the muscles of the trunk and legs in supporting the body in its erect position against the force of gravity are the examples of muscles which contract isometrically.

Summation : When a muscle is stimulated by a single subliminal stimulus, no contraction occurs. Yet, if two or more of these inadequate stimuli that are just below threshold intensity are given in rapid succession, a muscle contraction is evoked. This is what is known as summation of subliminal stimuli. Similarly the phenomenon where in one contraction is added to a previous one to produce a great shortening of the muscle is called summation.

Tetanus : Tetanus is a type of contraction caused by repeated brief stimuli at a frequency such that each successive stimulus comes after the refractory period of the preceding one or in other words it is a sustained contraction of muscle due to the fusion of many twitches following each other in rapid succession ; the external cause lies in the large number of stimuli presented to the muscle in a unit of time.

Staircase phenomenon or treppe : When a muscle is stimulated with single shock of constant strength at a frequency of about 1/sec, a series of contraction is obtained in which first few twitches of the series increase successively in amplitude. This is known as the **staircase phenomenon** or **treppe.**

Fatigue : If a muscle is stimulated repeatedly at intervals not close enough to produce tetanic contraction, it does not contract. The muscle which does not respond to stimuli at all, is said to be in a state of **fatigue**. This is due to accumulation of lactic acid in the muscle.

Revision Questions

1. Give an account of different types of muscles found in animals.
2. Describe in brief the functions and properties of muscles.
3. Explain in brief the physiology of muscle contraction and relaxation.
4. Write short notes on :
   (i) T-system,
   (ii) Motor end plates,
   (iii) Isotonic and isometric contractions,
   (iv) Summation,
   (v) Staircase phenomenon or treppe.

CHAPTER 13

Nervous System

In multicellular organisms the various activities of the body are controlled or regulated by means of two different systems. One is chemical system which includes specific substances called hormones secreted by endocrine cells. Hormones are usually carried throughout the organism's body in its circulatory system and cause responses in those cells possessing appropriate receptor sites. Hormonal control is usually used to regulate slower processes such as metabolism, activity of smooth muscles and the transport of materials across membranes. In single-celled organisms, chemical transmission must play an important regulatory role, and in micro-organisms rapid responses could be obtained with this type of mechanism or system.

The other system of controlling the various body activities is the nervous system in which electrical signals are generated in the form of nerve impulses which are usuallly associated with the fast responses to environmental stimuli.

Although in the past nervous and endocrine systems were treated as separate entities, it is now apparent that the two are closely integrated in the whole organism. Not only the chemical system or endocrine system is itself under the control of the nervous system, but also many nerve cells (by which the nervous system is composed of) are specialized to secrete or store neurohormones (or neurohumors) that activate certain effector cells. Further, the transmission of signals from nerve cell to nerve cell, or from nerve cell to effector cell, is often accomplished by chemical, not electrical means.

Specialization of cells for the functions of information reception, transmission, coordination and integration evolved early in the development of animal phyla (KAPPERS, 1929 ; KAPPERS, *et al.*, 1936 ; BASS, 1959).

The nervous system has become increasingly more complicated in its structure and functions in the animal organisms due to continuous changes in the conditions of existence of animal organisms. In a complex organism it plays the leading role in regulating all physiological processes and in effecting the connections of the organism with the external environment.

ROLE OF NERVOUS SYSTEM

The important roles played by the nervous system in the animal organisms in general are the following :

1. It regulates and controls the activities of the different organs and

of the entire organism. Muscular contraction, glandular secretion, heart action, metabolism and the many other processes continuously operating in the body of the animal organisms are controlled by the nervous system.

2. The nervous system links the various organs and systems, co-ordinates all their activities and ensures the integrity of the organism.

3. It helps in maintaining the unity of the organism and its external environment. All outside stimuli are perceived by the nervous system through the sense organs.

The stimuli in the form of nerve messages (nerve impulses) travel along nerve from the receptors to the central nervous system (brain and spinal cord). Here the impulses are passed on to other nerves which lead from the central nervous system to organs which can make a suitable response, such as muscles which contract, or glands, which secrete. The nervous system is thus responsible for co-ordinating the various parts of the body so that all may work together for the benefit of the whole animal. It also enables the body to react admirably even to the little changes in the environment.

GENERAL INFORMATION

On the structure of the nervous system : The nervous system of higher category animals includes the brain, spinal cord, and nerves. The brain and spinal cord constitute the central part of the nervous system or as it is customarily referred to the **central nervous system.** The brain gives rise to 10–12 pairs of cranial nerves and the spinal cord to 31 pairs of spinal nerves. These nerves give off branches to the different organs and tissues. The nerves and their branches constitute the **peripheral nervous system**. Although the nervous system is divided into central and peripheral parts, these parts form a single system.

The neuron theory : The fundamental unit of nervous system is the nerve cell or **neuron** (the term "nerve" is reserved for bundles of fibrous processes arising from many nerve cells). Some early histologists including CAMILLO GOLGI (1873) thought that the nervous system was a syncytial reticulum of branching fibrous processes with nerve cell bodies lying at nodal points—**reticular theory** theory (for example, GOLGI, 1898). As late as 1929, HELD argued for a syncytial network of neurons and felt that nerve axons invaded the cytoplasm of the cells they contacted.

The neuron doctrine : According to this concept the nervous system is composed of individual cells (now called neurons). These are discrete and separate entities and are functionnl units of the nervous system. From this concept it has become clear that there is no direct continuity between the protoplasm of one cell and the next and that definite points of contact between nerve fibres can be recognized. This concept is applicable to the nervous system of all animals. A syncytial organisation is recognized only in some of the giant fibres of invertebrates, the enteric plexus of the leech, and perhaps at certain points in nerve nets (BULLOCK, 1959b : MACKIE, 1960).

This concept was staunchly supported by the findings of HARRISON (1907) but it was championed by the work and arguments of RAYMON Y CAJAL (1909–11, 1934). HARRISON, by his simple hanging drop method, observed growing neurons and noted that the processes extending from them had enlarged ends, from which fine filaments often extended. These processes grew out of the nerve cell body into the medium, and it appeared that all cells were separate. It might be noted that the mechanisms by which neurons grow and extend fibrous processes during embryological development are still not known. Nor are the mechanisms known by which neurons make the proper connections to other neurons or to effector cells (LEVI-MONTALCINI, 1965). Recently LE GROS (1963) has tried to observe the connections in the nervous system.

The neuron doctrine was finally substantiated by the electron microscope, which showed that neurons, like most other cells, were bounded by a unit membrane whose presence prevents cytoplasmic continuity between adjoining cells of the nervous system. The gap separating neurons is small ($\approx$75 to 400Å wide), and in some cases there is a fusion of the outer layers of two unit membranes at neuronal junctions.

The generalized neuron : The characteristic cells of which the nervous tissue is composed are called nerve cells or neurons which differ greatly from each other in size and outward apperance. Neurons consist of a cell body—the perikaryon or soma and associated processes that arise from it.

1. The nerve cell body—the perikaryon or soma : It is a irregular shaped structure in the centre of which there lies a spherical **nucleus** with prominent **nucleolus** and fine **chromatin granules** but there is no **centriole** that is why the mature nerve cells or neurons can not divide. Besides these other typical cell organelles are also contained in the cytoplasm of the cell body or soma. These are **Nissl bodies** composed of ribonuclear proteins and are thought to be responsible for the synthesis of proteins—possibly the enzymes concerned with acetylcholine synthesis. The electron microscopic studies reveal that the cytoplasm of the nerve cell body or soma contains parallel arrays of the **endoplasmic reticulum** with **cisternae** and **ribosomes** (BLOOM and FAWCETT), **mitochondria and Golgi-complex,** therefore the soma appears to be the side of normal cellular metabolism. There is also present a reticular network of neurofibrils of unknown function.

The nerve cell body or soma is covered with a limiting membrane, it is a part of the bioelectrical potential generating mechanism in some neurons. This membrane may be the site of connections with other neurons, while the cytoplasm of the soma is not involved at all directly in the bioelectrical phenomenon.

2. The nerve processes : Since the neuron is especially differentiated to transmit impulses through the nervous system or through the organism's body or to make connections with other neurons or with effector cells, it bears a number of processes or projections or branches, these include the dendrites (singular dendron), axon and colleterals (these are side branches of axon).

Dendrites or dendron : The branch processes or projections coming out from the cell body of the neuron all except one are usually shorter and known as **dendrites** or **dendron**. These processes

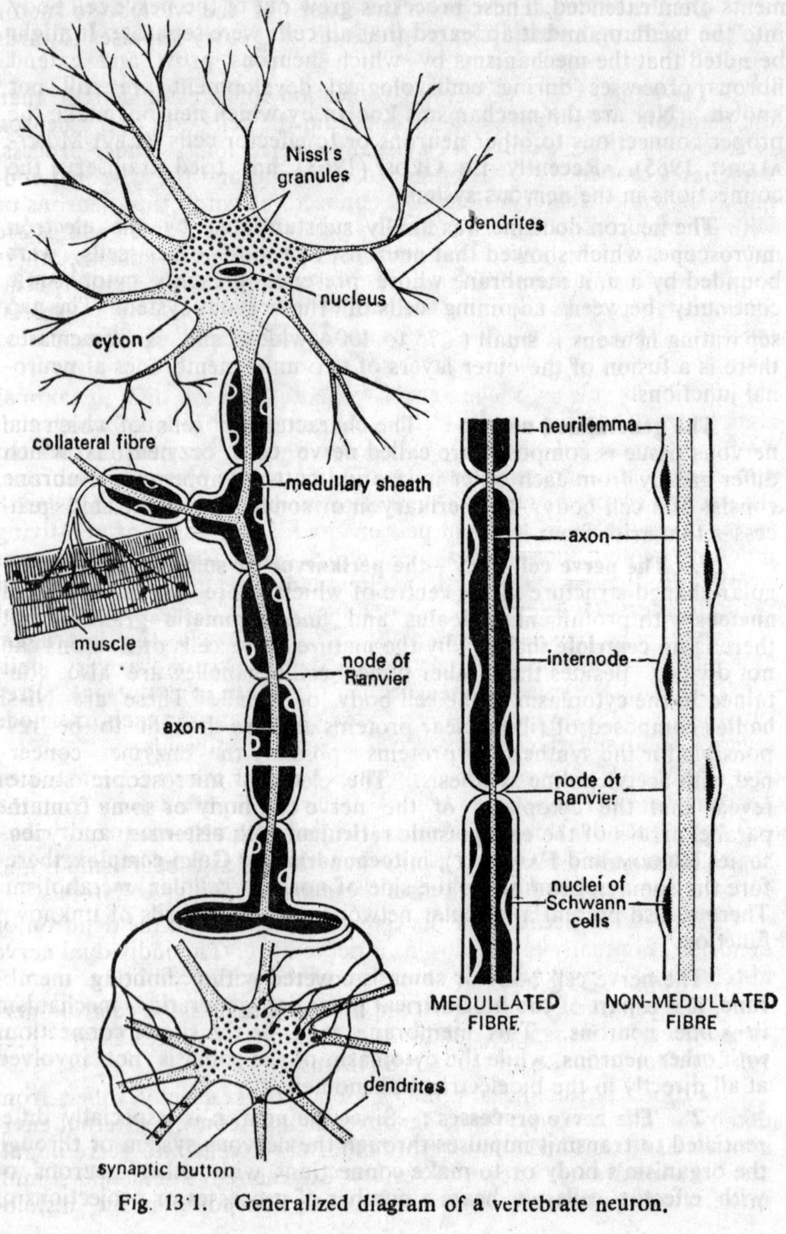

Fig. 13·1. Generalized diagram of a vertebrate neuron.

further have numerous finer processes or branched ends by which they contact the processes or projections of other cells. Many dendrites possess minute spines for such contact or connections. The functional junction between the dendritic region of one neuron and the axon ending of another neuron is the **synapse**.

Dendrites or dendron on the basis of the vertebrate motor neuron, were defined as fibrous processes that carry impulses towards the cell body.

The axon : The process or projection of greater length than the dendron or dendrites coming out from the cell body of the neuron is known as the **axon**. It is usually single, long, slender process, sometimes branched. It is branched at its end and these axonal endings, now called **telodendric**, make connections with other neurons or with effector cells. They exist in a variety of types. Some endings are knob-like structures called **synaptic knobs** or **boutons terminaux**, others are filamentous, ribbon-like endings ; still others have a single specialized ending such as the end plate found at junctions with skeletal muscles. Many axons have simple, morphologically undifferentiated endings.

The protoplasm of the axon is called **axoplasm** and at axonal endings, may contain concentrations of mitochondria, numerous small membrane-bounded vesicles, and a collection of enzymes associated with the chemical transmitters. In addition to the plasma membrane, the axon is covered over by a myelin or medullary sheath. This sheath varies in thickness in different neurons and is composed of non-living lipid and protein formed by the SCHWANN **cells**. This sheath does not form a continuous cylindrical envelope but instead it is broken up at regular intervals by means of definite constrictions, the **nodes of** RANVIER. The region between two nodes is known as the **internode**. Outside the myelin or medullary sheath there is one more very delicate covering known as the **neurilemma** or **sheath of** SCHWANN. Presumably one SCHWANN cell provides the sheath for each internode region. This sheath contains nuclei at regular intervals.

Axons of the neurons, on the basis of the vertebrate motor neuron, were defined as processes that carry impulses away from the cell body but now this definition is true only for a few neuron types.

Collaterals : In some neurons the axons also bear small and delicate side branches or processes these are termed as **collaterals.**

The axon because of its greater length than the dendron or dendrites, is usually referred to as a nerve fibre. The individual nerve fibre is, thus, a protoplasmic extension of the nerve cell body.

The cell bodies of neurons are always found in the central nervous system or in ganglia, while the collections of processes or fibres run together to make up the nerves themselves.

From a physiological point of view, axons usually differ from dendrites in two ways. First, dendrites usually convert external environmental stimuli and stimuli from other nerve cells into nerve impulses. Axons are not known to convert external environmental stimuli into nerve impulses, although many axons, depending on the histological organisation of a tissue, are capable of converting stimuli from another neurons into nerve impulses. The second physiological distinction between axons and dendrites is that dendrites usually carry stimuli towards the cell body to the axon origin, while axons carry impulses from the axon origin to their own distal tips. The axon

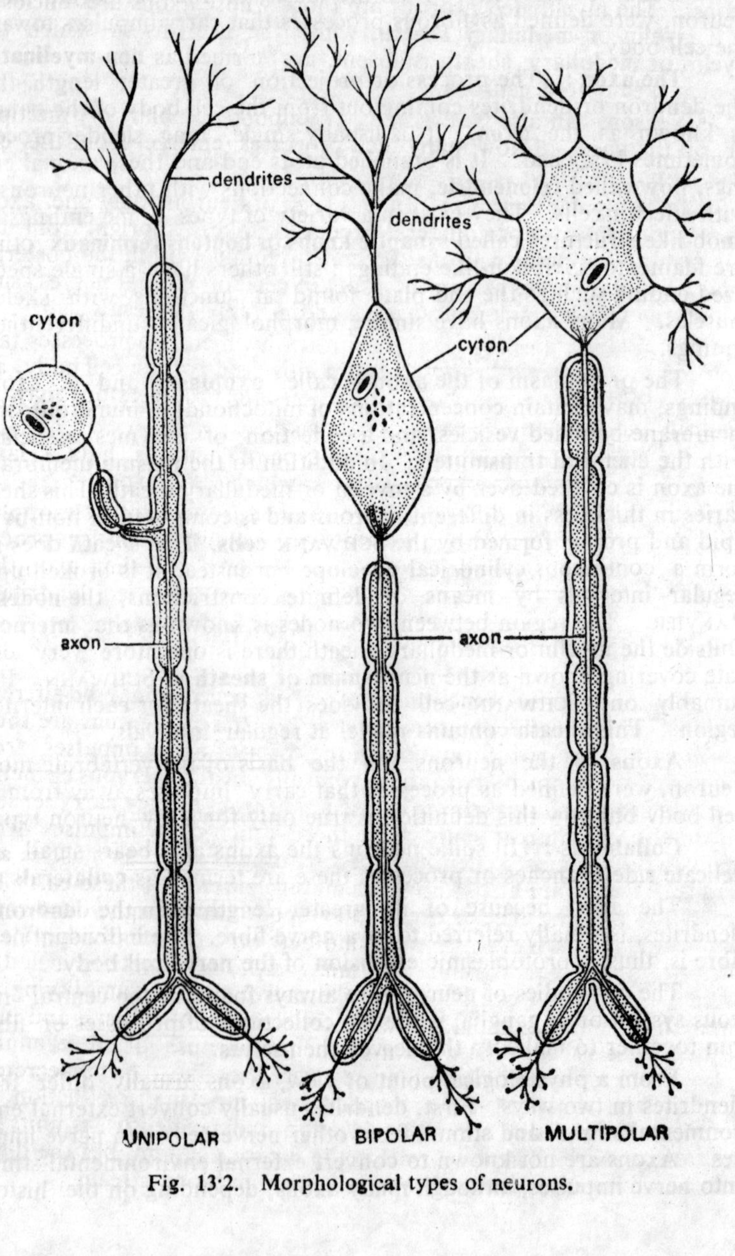

Fig. 13·2. Morphological types of neurons.

origin is by definition the region on a neuron where impulses originate. The region may or may not be near the cell body.

MORPHOLOGICAL TYPES OF NEURONS

The neurons may or may not be myelinated depending upon the presence and absence of the myelin sheath in the neurons around the axons. The myelinated neurons are those whose axons are enclosed in a **myelin** or **medullary sheath,** while the neurons in which the myelin or medullary sheath is absent, are termed as **non-myelinated** or **non- medullated** neurons.

Neurons can also be classified as unipolar, bipolar, or multipolar on the basis of how many cell processes emerge from the cell body (Fig. 13·2).

1. **Unipolar neurons :** In this type of neurons only one axon, often with dendrites or collateral processes, takes its origin from the cell body. Unipolar neurons are mostly found within the posterior roots of spinal nerves and in the roots of certain cranial nerves such as the trigeminal, glossopharyngeal and the vagus nerves.

2. **Bipolar neurons :** The neurons in which two processes take their origin from the cell body, one on either side of the cell body, are known as bipolar neurons. Out of these two processes one may be dendron and one may be axon.

3. **Multipolar neurons :** Such type of neurons have a large number of cell processes, at least one of which is an axon.

Bipolar and multipolar neurons may be isopolar or heteropolar. **Isopolar neurons** usually have two or more somewhat similar processes whereas the **heteropolar neurons** usually have a distinct axon that is long and smooth and one or more irregularly branched dendrites.

PHYSIOLOGICAL OR FUNCTIONAL TYPES OF NEURONS

On the basis of functional properties, neurons may be afferent, efferent, internuncial, or neurosecretory. **Afferent neurons** are those which carry sensory information in the form of nerve impulses from the periphery of the body to more centrally located nervous elements. These neurons are also known as **sensory neurons** as they carry sensory information. **Efferent neurons** are those which carry impulses from the central nervous system to effector organs such as muscles and glands. These are also known as **motor neurons** as these earry the orders towards the effector organs to respond accordingly. **Internuncial neurons** or **interneurons** are found between afferent (sensory) and efferent (motor) neurons and serve many important functions including routing of singals to one of several alternative circuits, amplifying the strength of some signals, reducing the strength of others, and integrating signals from diffuse sources for the purpose of determining appropriate motor activities, learning, and the like. **Neurosecretory neurons** are those which are specialized for the production of neurohormones, they may also function independently as units for the transmission of impulses within the nervous system, but the distal portion of the axon always lies adjacent to the blood stream.

Satellite cells : These cells exhibit a variety of forms and perform various functions. The **satellite** or **neuroglial** cells are usually found invested around the neurons. Some become the myelin sheath of peripheral neurons, while others serve to nourish the axon or to determine what substances will come into contact with the plasma membrane. Satellite cells, therefore, influence nervous activities and play important roles in repairing injured neurons.

The synaptic region or **synapse :** The functional junction between the dendritic region of one neuron or nerve and the axon ending of another neuron is commonly called as the **synaptic region** or **synapse**. SIR CHARLES SHERRINGTON (1861–1954) was first who applied the term **"synapsis"** to the point of contact between two nerve cells or neurons. Its physiological importance for the transmission of nerve impulses was established by MCLENNAN (1963).

According to recent findings of electron microscope there are many different forms and arrangements of synapses, but they nearly all have the same basic structure. The most important feature of synapse is that the membranes of the two nerve cells or neurons or nerves are separated by a space called **synaptic cleft** about 10–50mm wide. It sometimes shows traces of some contained material and in some synapses from the brain it is covered by filaments. The pre-synaptic cell membrane on the side of the cleft at which the nerve impulse arrives, and the post-synaptic or subsynaptic membrane on the side where the new impulse starts, have the usual triple structure. Clusters of synaptic vesicles are found near the pre-synaptic membrane which probably contain the transmitter substance. It is believed that the transmitter substance is produced by the synaptic vesicles at the junction of pre- and post-synaptic membranes of two neurons which induces as change in permeability to certain ions in the post-synaptic membrane. It is thought that an increase in permeability to the inflow of sodium ions promotes the spread of electrical activity in the post-synaptic cell or neuron. An increase in permeability to the inflow of potassium ions or outflow of chloride ions leads to inhibition of the spread of electrical activity.

The various types of synapses found in the nervous system are the following, depending upon the type of contact made between a pre-synaptic and a post-synaptic neuron.

Axo-dendritic synapses : Such type of synapses occur at junctions between an axon and a dendrite.

Axo-axonic synapses : Such type of synapses are found at junction between the axon of one neuron and the axon of a second neuron.

Axo-somatic synapses : Such type of synapses occur where the nerve endings of an axon of one neuron transmit information to the cell body of the another neuron.

Dendro-dendritic synapses : The synapses which occur at the junctions between the dendrites of one neuron and the dendrites of another neuron. In these synapses the electrical excitation from dendrites in one neuron is conveyed synaptically to dendrites of another neuron.

Functions : The characteristic functions of synapses can be summarized as follows :

1. It transmits impulses from one neuron to other neuron or to glands, muscle cells or some other cells.
2. It can transmit impulses only in one direction.
3. It retards the passage of an impulse more so than the nerve fibre does. Consequently, there is a synaptic delay in the passage of an impulse from the pre-synaptic to the post-synaptic neuron.
4. It is a site where repetitive discharge of a neuron may occur.
5. It is very vulnerable to the action of many chemical compounds, some of which increase its responsiveness (strychnine), whereas others decrease it (anesthetics).
6. It has the characteristics of summating and of inhibiting impulses.

One axon may receive the excitatory or inhibitory effects of impulses from several other neurons. Similarly one axon may transmit impulses through its nerve endings to several other neurons.

Nerves : The cell processes of neurons may occur singly within the parenchyma of animal tissues or they may be grouped into bundles to form a nerve. A nerve is thus, an assembly of a large number of cell processes often called the fibres. The nerve is surrounded by a loose connective tissue covering, the **epineurium** due to which it is separated from other body tissues. The bundles of individual nerve fibres within the nerve are covered in a relatively strong sheath of connective tissue known as the **perineurium**. Inside the bundles are individual nerve fibres, each surrounded by the network of connective tissue making up the **endoneurium**. Nerves are supplied with a profusion of blood-vessels. Small arteries and arterioles are present in the epineurium and perineurium and the capillaries in the endoneurium.

Nerves are of different types depending on the functions. Some nerves are **motor** or **efferent nerves** and these carry the messages or impulses from the central nervous system to the effector organs. Other nerves are **sensory or afferent nerves** as these carry the messages or impulses from the different parts of the body to the central nervous system. Motor nerves consist of only motor nerve fibres, while the sensory nerves consist of sensory nerve fibres.

There are also nerves which include both motor and sensory nerve fibres and, hence, these are called mixed nerves.

The motor nerves or fibres terminate in organs (for example—in muscles) with motor endings, while sensory nerves or fibres terminate in organs (for example—in the skin) with sensory endings or receptors. In this way the central nervous system communicates with all organs and tissues through the nerves.

MAIN PROPERTIES OF NERVOUS TISSUE

The main properties of nervous tissue can be summarized under the following heads :

1. Excitability : Excitability is an inherent property of nervous tissue. It possesses the ability to enter into an active state, what is known as a **state of excitation,** in response to stimulation.

The excitation of nervous tissue occurs as a result of stimulation of sense organs or sensory nerve endings known as receptors, by any sort of stimuli acoustic, optic, thermal, gustatory, etc.

2. Conductivity : Nervous tissue is characterized by the fact that excitation does not remain at the site of its origin but is transmitted along nerve fibres. This property of nervous tissue to transmit excitation is called **conductivity.**

The conduction of excitation along nerve fibre is accompained by appearance of bio-electric phenomena, called action protentials (or nerve impulses) in nervous tissue.

Excitation is conducted in the nerves in isolation (separately) along each nerve fibre and is never transmitted from one nerve fibre to an adjacent nerve fibre. It explains that excitation may either be transmitted along all the fibres forming a nerve, or just along some of them. This makes possible the contraction of individual muscle fibres and certain muscles, and not necessarily the entire group of of muscles innervated by a particular nerve.

STIMULUS

A stimulus can be defined as a sudden change in the evironment (external or internal) which is strong enough to excite the nerve or muscle or organism as a whole. If the stimulus is capabie to bring a change or to excite a given tissue, it is called **adequate stimulus or threshold stimulus**. If the stimulus is not capable to excite or fails to elicit any response, it is called **inadequate stimulus**. There are many types of stimuli that may excite the tissues, the most important ones are the following :

1. Mechanical stimuli : When a given tissue is excited by mechanical means, such stimuli are known as mechanical stimuli.

2. Physical stimuli : Heat, cold and humidity serve as physical stimuli for the living tissues.

3. Chemical stimuli : Various chemicals including acids and bases are referred to as chemical stimuli for the living tissues.

4. Electrical stimuli : Electricity functions as electrical stimuli and it is capable to excite the living tissues.

MODE OF ACTION OF NERVES

All the nerve fibres carry messages in the form nerve impulses. The passage of a nerve impulse along a fibre involves a number of physical and chemical changes, none of which is directly visible. The physical changes more correctly the electrical changes are the surest indication of development and propagation of the nerve impulse and are probably essential processes involved in the transmission of the impulse along the fibre, whereas the chemical changes are probably connected with the recovery processes after activity. The energy required for the conduction of impulses is supplied by respiration.

Nerve impulse : PROSSER defined the nerve impulse as the sum total of physico and chemical disturbances created by a stimulus (electrical, chemical or mechanical) in a neuron or nerve fibre and which result in the propagation of a wave of physiological activity along the nerve fibre. GOTHLIN suggested that nerve impulse is a phenomenon which involves both physical and chemical changes.

Properties of nerve impulse : 1. When a nerve is subjected to a threshold stimulus it undergoes excitation or it is stimulated. Thereshold stimulus is roughly constant for a given type of nerve fibre, but is higher in small fibres than in large.

2. Normally nerve impulses pass in one direction only but they may be conducted in either direction too. **Law of forward conduction** of JAMES states that if a nerve is stimulated in the middle, the nerve impulses in the form of electrical changes are conducted to both the ends of the nerve fibre.

3. Generally impulses flowing in a nerve fibre do not spread with the neighbouring fibres because nerves or nerve fibres are generally covered by an insulating nerve sheath called the medullary sheath.

4. After one impulse has passed along the nerve fibre, there is a refractory period during which the nerve fibre is neither excitable nor conductive. During this period a second stimulus, however, of large strength, remains ineffective. In this period the fibre recovers and after recovery it again responds to its maximum. If an inadequate stimulus is followed by another one within a very brief period, say 0·5 ms, there may be a response, so that there can be summation of two inadequate stimuli. The absolute refractory period is followed by a relatively refractory one, during which the threshold is increased.

5. The nerve impulses follow the "all or none law", *i.e.*, if any stimulus is an adequately effective one that is either of threshold strength or more, the nerve fibre responds to a great extent with an impulse of maximum intensity but if it is below threshold strength it is incapable to elicit any response whatsoever. Any stimulus bringing out a response in the nerve, the response cannot be increased further by increasing the strength of the stimulus.

6. The impulse takes a certain time to travel along the nerve fibre. The larger the fibres the more rapidly they conduct, and medullated fibres conduct more rapidly than non-medullated. Thus the speed or the velocity of nerve impulse differs in different animals. In frog it is about 30 metres per second at the room temperature and for the human motor and sensory nerves it is about 120 metres per second. While crossing from one fibre to another there is some resistance at the synapse which breaks the speed of the impulse. Some of the fast-moving animals have single long fibres called the **giant-fibres**—the impulses travel very fast through them.

7. In a nerve fibre the impulse is always conducted at a constant amplitude and velocity.

8. For regular conduction of nerve impulses a continuous supply of oxygen is required in order to liberate the energy and gene-

rally it is believed that nerves do not undergo fatigue, however, the continued stimulation in the absence of oxygen may result in certain disorders or irregularities.

9. Normally the nerve fibres are capable in sending enormous number of impulses.

10. The nerve impulses travel in a nerve in a quick sequence. Each nerve impulse is of the same nature and is similar in all the nerves. Registering of different sensations in the brain is not dependent on the nature of the impulse but on the area in which it is received.

CONDITION AFFECTING THE PASSAGE OF A NERVE IMPULSE

1. Repeated stimulation does not lessen the excitability and conductivity of a nerve fibre or nerve but if the interval between two stimuli is shorter than the relative refractory period, the second and subsequent stimuli produce progressively smaller impulses than the first.

2. Temperature also plays an important role in the passage of a nerve impulse. Below a certain temperature the propagation of the excitatory process in the nerve does not take place.

3. pH value also regulates the conduction and propagation of nerve impulses. Actually pH value does not affect the nerve impulses but alters the excitability and after potentials.

CONDUCTION OF NERVE IMPULSES

1. **Through the nerve fibre :** The nerve fibre or axon, as you have already studied in the beginning of the chapter, is like a cylinder, the interior of which is filled with axoplasm (*i.e.*, the cytoplasm of the nerve cell) and the exterior of which is covered with a thin membrane the axon membrane. The axon membrane is made up of lipo-proteins and is about 100Å in thickness. It has some minute pores or channels of 7 to 10Å in diameter through which water and other substances corresponding to the pores, can pass in and out of the axon membrane, this states that axon membrane is semipermeable in nature. The axon membrane exhibits a **bioelectrical potential** or **membrane potential** which according to BERNSTEIN's membrane-ion theory modified by HODGKIN and HUXLEYS (1952), is caused by unequal concentrations of various ions in and outside the cell due to the semipermeable nature of axon membrane. Membrane permeability to different ions varies and changes regularly depending upon the physiological state of the nerve fibres. The membrane potential is normally about 85 mV. Due to semipermeable nature of the axon membrane and concentration gradient certain ions are transported from the axoplasm (*i.e.*, the protoplasm of the axon) into the interstitial fluid (*i.e.*, the fluid surrounding the axon) and certain other ions from the interstitial fluid into the axoplasm. On account of this, the concentration of ions in the inner axoplasm differs from that of outer interstitial fluid. In the resting state of the nerve, *i.e.*, when the nerve is not being stimulated, the sodium ions are being actively transported from inside

to the outside; with the result the concentration of sodinm ions inside the nerve fibre (*i.e.*, the axoplasm) is made almost negligible and that of potassium ions is more. Normally the concentration of sodium ions in the interstitial fluid is about 100 times more than that in the inner axoplasm. This is simply due to the movements of sodium ions from the axoplasm into the interstitial fluid. The outward transport of sodium ions is known as **sodium-pump**. Steady existence of sodium pump requires a certain amount of energy which is supplied by the ATP. Due to this sodium pump the axon membrane towards its inner side becomes electronegative, while towards outside electropositive. Actually in this state the nerve fibre is said to be in the normal polarized state.

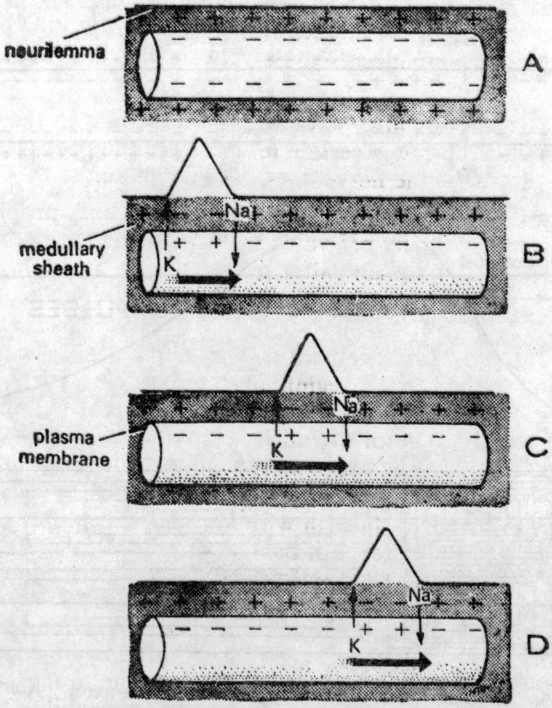

Fig. 13·3. Diagram showing the ionic movement along the axon membrane.

When the axon is excited by an adequate stimulus of any sort (mechanical, chemical or electrical), the axon membrane immediately becomes **depolarized** and the sodium pump stops. Now the permeability of the axon membrane to the sodium ions is suddenly increased (about ten times to that of potassium). This results into the diffusion of sodium ions from the exterior into the axoplasm. The flow of sodium ions from the exterior, *i.e.*, interstitial fluid into the axoplasm, therefore, begins to exceed the out flow of potassium ions considerably. The inward movement of sodium ions is so heavy that the inside of the axon membrane not only gets depolarized, but the inside or the interior of the axon membrane becomes positive to that of outside. This is just the opposite of the resting state of axon-membrane and it is called **reverse potential** or **reverse polarization**. This change in specific permeability and subsequent depolarization starts from the point where the stimulus was received and spreads in both direction along the nerve fibre in the form of **depolarized wave** which constitutes the **nerve impulse**. If the stimulus is too weak then this change stops after transmitting a short distance and normal polarization will be restored, we would then say that the strength of the stimulus was lower than the threshold. When the stimulus is strong depolarization spreads throughout the length of the axon or nerve fibre.

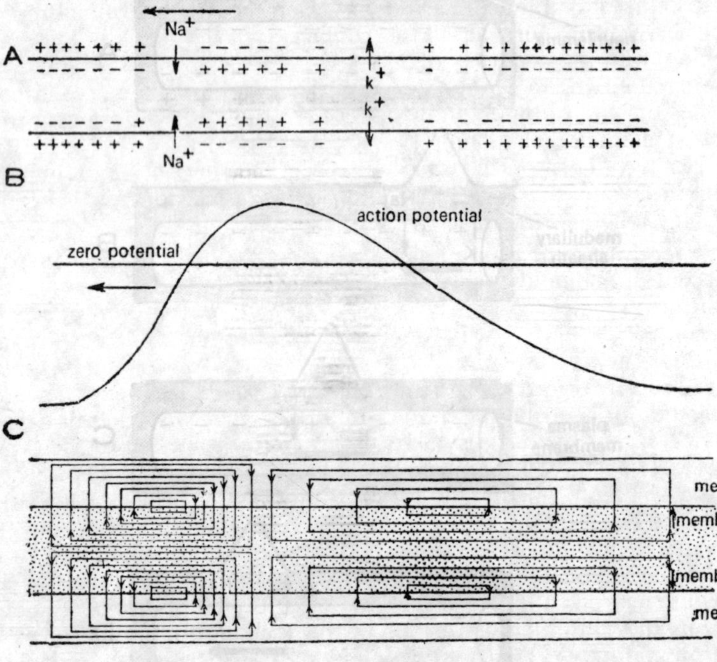

Fig. 13·4. Conduction of a nerve impulse along an axon. A—The distribution of charge and the movement of ions along an axon and across the membrane during conduction of an impulse in the direction of the arrow. The influx of Na+ is initiated by a partial depolarization. As a consequence of Na+ influx the membrane potential is reversed, as in (B). This reversal in turn causes out flux of K+, which returns the membrane to its original state of polarization. The currents that flow across the axon membrane and in the medium and axoplasm are shown in (C).

As soon as the wave of depolarization has travelled the entire length of the fibre, no more sodium ions can enter from outside because, now the axon membrane becomes totally impermeable to sodium ions due to simultaneous increase in potassium permeability. A large quantity of potassium from the interstitial fluid diffuses through the membrane into the axoplasm of the axon and sodium ions begin to diffuse outside. Because of this the reverse potential disappears and the normal resting membrane potential returns. This process is known as **repolarization**. This repolarization starts exactly on the same spot where the depolarization had started and then continues to advance from there in both directions. Now once again sodium pump starts functioning. The entire process of depolarization and repolarization is completed within a fraction of a second and the fibre is ready for a fresh impulse.

2. Through the synapse : You have already studied about the synapse that it is a functional region when the axon-terminations of one nerve fibre form a close contact (but not actually joining or sticking) with the dendrites of the next nerve cell, there is a little gap between the two. When the depolarization wave or the nerve impulse reaches the end of the axon or nerve fibre, how is the next neuron or other structures such as muscle and gland, influenced by it ? Two theories have been put forward to explain the conduction of impulses along the synapse. These are—electrical transmission and chemical transmission theories.

1. Electrical transmission theory : It is believed that in some synapses of mammals, the transmission through the synapse (junction between two neurons) is accomplished by current flow across the **plasma membranes** and between the **presynaptic** and **postsynaptic** neurons. The neuron contributing the axon or axon terminals, that is, the source of the incoming action potential, is the **presynaptic neuron,** while the cell contributing the dendritic sites is the **postsynaptic neuron**. The membrane of the presynaptic neuron that is applied to the synaptic area of postsynaptic neuron is the **presynaptic membrane**. The area of postsynaptic membrane—the membrane of the postsynaptic neuron affected by the presynaptic neuron – that is directly under the presynaptic membrane is the **subsynaptic membrane**. When a wave of depolarization or nerve impulse reaches the synapse along the presynaptic neuron, it serves as a stimulus for the successive postsynaptic neuron and causes depolarization of the dendrites of the postsynaptic neuron and now the wave of depolarization or nerve impulse is triggered in the second neuron.

Electrical synapses are structurally much different from typical chemical synapses ; the membranes of neurons at electrical synapses are generally in much closer proximity, and even in some instances the presynaptic and postsynaptic membranes are fused together. Such anatomical modifications reduce the intercellular resistance and help to assure an adequate flow of current for depolarization of the postsynaptic neuron.

2. Chemical transmission theory : According to this theory when a depolarization wave or nerve impulse, after transmitting down the presynaptic neuron, reaches at the synapse it brings about the release of a chemical compound (**neurohumor**) which is known as a

transmitter ; this substance in the gap of synapse is responsible for the conduction of nerve impulse through synapse. This transmitter is a **acetylcholine**. This substance goes on accumulating in the cleft of synapse to an extent that it causes depolarization of the dendrites of the postsynaptic neuron and now the wave of impulse is triggered in the second neuron. Acetylcholine is secreted only by the axon terminations—this is the reason why the axon is not excited anywhere and the nerve impulse travels in it both direction, it will be continued further only on the synaptic end of the axon and not on the dendritic end. This is the answer to the question why the nerve impulses always travel in a single direction.

For a proper continuous working of the synapse it is necessary that chemical excitation must come to an end after it has acted, so that the synapse may become ready for the passage of the next impulse. So, the act of removing acetylcholine is performed by an enzyme called **cholinesterase**. The time taken for the impulse to travel from one neuron to another through the synapse is 0·6 milli second. It is also believed that single impulse dies out on reaching synapse but several impulses reaching a synapse within a short period "fire" the impulses into the next neuron. The reason for it may be that a single impulse is unable to produce the adequate quantity of the transmitting substance. On the contrary, this also has been found that a certain synapse gets 'fatigued' by working repeatedly, and its capacity to transmit subsequent impulses is reduced. The possible reason for it may be that the secretory reserve of that particular transmitting substance is depleted.

Neuromuscular junction : As the name itself indicates neuromuscular junction is the junction between the nerve and the muscle. When a nerve fibre approaches the muscle, it loses its myelin sheath and it branches which invaginate the sarcolemma of muscle fibres but does not merge with the muscle fibre membranes and thus form special structures—the neuromuscular junctions also called **motor end plates**. The invagination membrane of nerve fibre is called **presynaptic membrane**, while the membrane of muscle which is in contact with the end plate is called **postsynaptic membrane** which is highly folded structure. The gap between the pre-and postsynaptic membrane is called **synaptic cleft** which is always filled with a intracellular transmitter substance.

The endplate structures have numerous acetylcholine vesicles and mitochondria. The mitochondria produce either the acetylcholine—a chemical transmitter or supply energy for its synthesis.

When the motor impulse from the nerve is received on the endplate a local depolarization occurs there and reaches the sarcolemma with the accompanying secretion of acetylcholine resulting in the excitation of the muscle-fibres.

Difference between nerve impulse and electrical flow : Both these processes are entirely different. An electric current flows at the speed of light and there is no chemical action involved in it, only a physical action occurs. But, in the nerves the impulse is an electrochemical action which proceeds from one spot to another and relatively slow. The nerve impulse can be compared with the fire proceeding in the fuse of a fire-work where the fire reaches the other end without diminution, and if the fuse was branched the fire would extend into each branch.

SALTATORY CONDUCTION

Non-myelinated fibres have a lower conductivity than a myelinated one of the same diameter thereby suggesting a different mechanism of conduction of nerve impulse known as the **hypothesis of saltatory conduction**. This is a process by which activity leaks from a node to another node of RANVIER. According to this hypothesis the depolarization wave due to the entry of sodium ions is confined to the nodes but the whole internode length of the fibre is depolarized by local circuit action. It appears that in myelinated nerve fibres the depolarization wave dances (saltare) from node to node in contrast to the continuous, wave-like progression in unmyelinated nerve fibres. This form of impulse propagation is known as **saltatory conduction**.

Chemical and thermal changes during nerve activity : Nerves both at rest and during activity require a steady supply of oxygen which is consumed by the nerves in order to release the energy for the activity of the nerves. The oxygen consumption results in the liberation of certain amount of CO_2 and heat. Heat is essential for the normal temperature of the body.

REFLEX ACTION

A reflex is a stereotyped response to a stimulus. The term implies a **"reflection"** of stimulus excitation by the central nervous system to the effectors. Reflex action was first discovered by MARSHALL HALL in 1833. BEST and TAYLOR defined reflex action as an automatic involuntary and often unconscious action brought about when afferent nerve endings or receptors are stimulated. Reflex action is a elementary function discharged by the nervous system. It is important in protective behaviour such as the withdrawls of the limbs from pain and in locomotion and in standing.

Reflex action involves a reaction which is started by the environment which acts as a stimulus, stimulating one of the receptors. Receptor sends impulses to the central nervous system through the chains of neurons. From the nervous system impulses are passed outward through the motor neuron and reach either a muscle or a gland which acts accordingly. The nervous elements involved in carrying out the reflex action constitute a "reflex arc".

Reflex arc : The reflex arc is the nerve chain between a receptor and an effector organ (muscle or gland). The reflex arc includes the following parts : (1) **A receptor**—it is a sensory structure which receives the stimuli. (2) **A afferent** or **sensory neuron**—it passes into the spinal cord by way of a dorsal root of the spinal cord. Its primary function is to convey the impulses received from the receptor to the central nervous system (spinal cord). (3) **Interneuron**—it is present in the central nervous system (spinal cord). It is generally one but sometimes two interneurons may be present. It simply serves to transmit the impulses from the afferent or sensory neuron to the motor neuron. (4) **A motor neuron**—it is also known as efferent neuron and is situated in the ventral root of spinal cord. It transmits the impulses to the effector organs which may include muscles or glands. (5) **An effector organ**—it may be a muscle or gland which responds to impulse received.

The reflex arc constitutes the functional unit of the central nervous system.

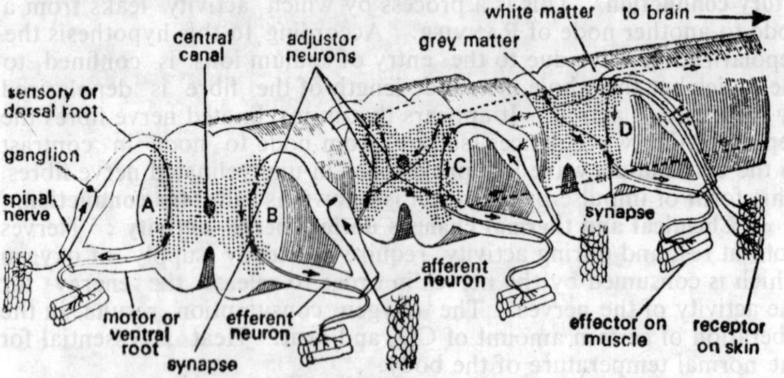

Fig. 13·5. Diagram showing the various components of reflex arc.

Reflexes are of common occurrence in animals. In higher animals a large number of reflex actions are performed in daily life. They are performed unconsciously, therefore, they are of simple nature and termed simple reflexes. Now we shall discuss few examples of reflex action. Suppose that you have place your hand in a bowl of cold water. Receptors in the skin are stimulated by the coldness of the water, nerve impulses pass via the sensory nerve fibres through the dorsal root of the spinal nerve to the spinal cord which is a part of central nervous system where the water temperature is appreciated. Impulses from the brain are carried down to the arm muscles through motor neurons. The arm muscles act accordingly. They contract and withdraw the hand from the water.

Similarly, if you place your hand in a boiling water without knowing that it is hot, immediately withdraw it rapidly and then realise how hot it was. To protect your hand from damage a mechanism has acted in an involuntary manner by making use of a "short circuit". The path of the nerve impulses is as follows : sensory fibres convey impulses from the receptors into the spinal cord as before, but they pass not only to the brain but also by appropriate connector or interneurons to motor neurons which carry them to arm muscles which contract to remove the hand from the water.

Other examples of reflex action are the wrinking of an eye when a particle of dust touches the eyelids, excitment of the salivary glands after seeing the food, etc. Other reflex actions are shown in the following table :

Table 13·1 : Table showing different reflex action.

Reflex	Stimulus	Response
Blinking	Foreign body on surface of eye.	Eyelids close and eye "waters".
Swallowing	Food touches sensitive spot at back of pharynx.	Peristaltic waves passdown oesophagus.
Sneezing and coughing	Foreign particle irritating lining of nose or larynx.	Chest muscles and diaphragm contract and relax violently to produce a gust of air.
Knee-jerk	Sharp tap of tendon below knee-cap of crossed leg.	Leg 'kicks' up.
Earflick of cat	Ear touched lightly.	Ear flicks rapidly.

These reflexes are said to be unconditioned or inborn because they are a natural part of an animal's make up. Another class of reflex produced by the previous experience of an animal is considered below.

Conditioned reflexes : When a reflex which does not naturally exist had become a part of the animal behaviour. Such a reflex is said to be conditioned. Conditional reflexes were first demonstrated by the Russian physiologist, PAVLOV. The cerebrum controls the conditioned reflexes.

BRAIN

Brain constitutes the most essential part of the central nervous system. It is made up of countless number of neurons. It sends out number of cranial nerves to different parts of the body and thus controls the various activities of the body. It is very simple in higher invertebrates, not developed in lower invertebrates such as protozoans and coelenterates but well developed in higher vertebrates.

The brain of cyclostomes, elasmobranchs and fishes is of primitive type. It appears a more or less linear elaboration of the swollen end of the dorsal tubular nerve cord of the embryo. At its anterior end the brain is primarily an olfactory sensory centre, and its removal does not bring any serious effects in behaviour, nor does its stimulation elicit motor responses. In fishes, however, the rhinencephalon does exhibit spontaneous rhythms, and in its absence some instinctive behaviour may be lost. The role of diencephalon in the lower invertebrates has been little explored, and this is unfortunate in view of the importance of that region in the higher vertebrates. The hypothalamus is, however, the site of numerous neurosecretory cell bodies, with axons leading to the neurohypophysis and directly involved in the regulation of osmoregulation and other functions. The midbrain of the lower vertebrates appears to be the highest integrative centre. It consists of the dorsal tectum (terminus of the optic fibres) which exhibits a close correspondence to the retinal receptor pattern, and the ventral tegmentum, including the oculomotor and other motor centres.

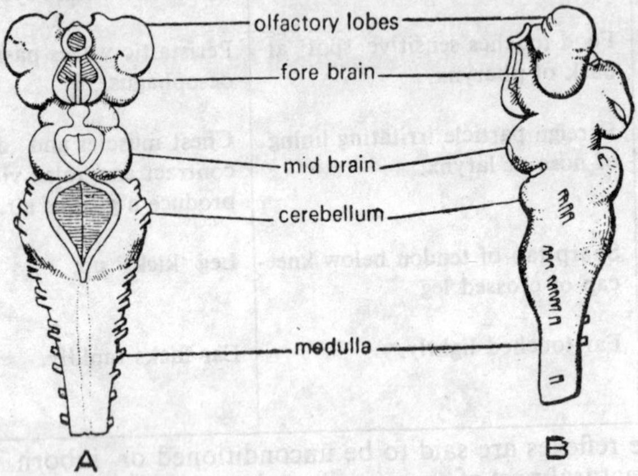

Fig. 13·6. Brain of ammocoete larva of lamprey.

The cerebellum is absent from the cyclostomes and appears merely as a small commissure in elasmobranchs, but in fishes it appears as the archicerebellum, receiving the vestibular input. It has coordinating and locomotor functions. The medulla contains the respiratory centres, generally two in number, coordinating the alternating movements of inspiration and expiration. The circulatory centre

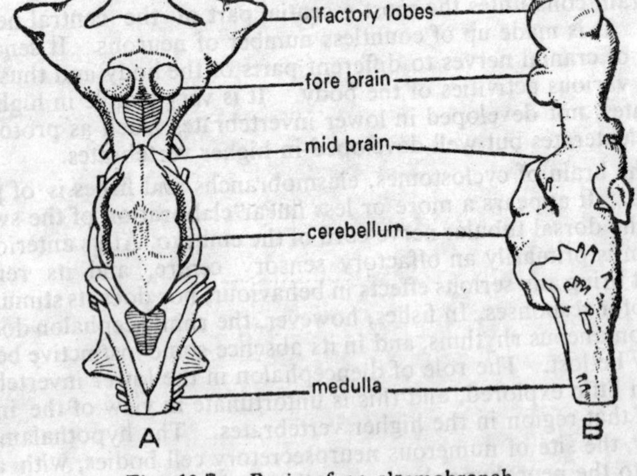

Fig. 13·7. Brain of an elasmobranch.

for cardiac and vasomotor reflexes is also found here. In the lower vertebrates the medulla also contains an important vestibular centre which is the main centre for locomotor integration. In fishes, the nervous centre for regulation of colour change is also found. It appears that the brain of lower vertebrates contains a set of motor centres in the midbrain, sensory centre in the forebrain and an integrative region between.

In the amphibians some notable differences have been observed. Most notable is a movement of integrative functions from midbrain to diencephalon. The latter appears to have some motor control. The

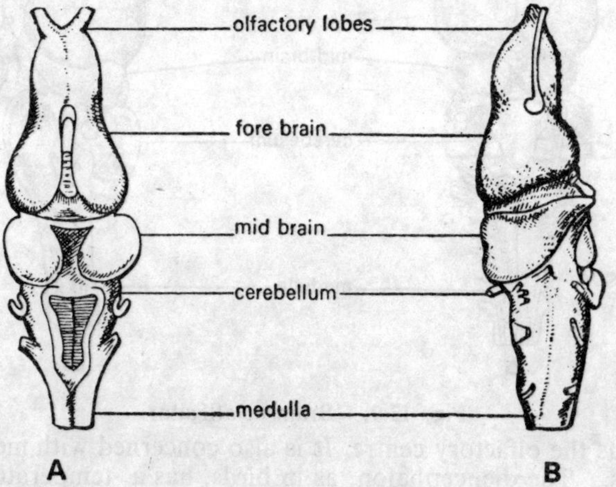

Fig. 13·8. Brain of frog.

optic termini have moved from the optic lobes to the diencephalon. The midbrain retains its function in coordination of proprioceptive and exteroceptive information into motor patterns and has been shown to control aspects of reproductive behaviour.

The brain of reptiles lies between the amphibians and the higher vertebrates from functional point of view. The brain of the bird shows extensive structural differences from brains of the lower vertebrates, especially in the well developed cerebrum and cerebellum. The cerebrum of the birds appears to be primarily a sensory projection area with limited motor function. The cerebellum is highly developed, as the palaeocerebellum, and has extremely important functions in locomotor and postural coordination. The diencephalon of birds has a well-developed centre for control of body temperature. The red nucleus and an extensor function in the vestibular centre appear first in the reptiles and are well-developed in birds as well.

Finally, in the mammals the cerebrum comes to dominate in size and function, the entire brain. It contains projection areas for all the sensory functions, hence is the terminus for all sensory information. The cerebrum contains extensive motor areas dominating and coordinating all the voluntary aspects of the movement. The cerebral cortex contains, in the prefrontal region, association areas not directly involved in either sensory or motor function but concerned with the functions of association, learning and memory. The rhinencephalon

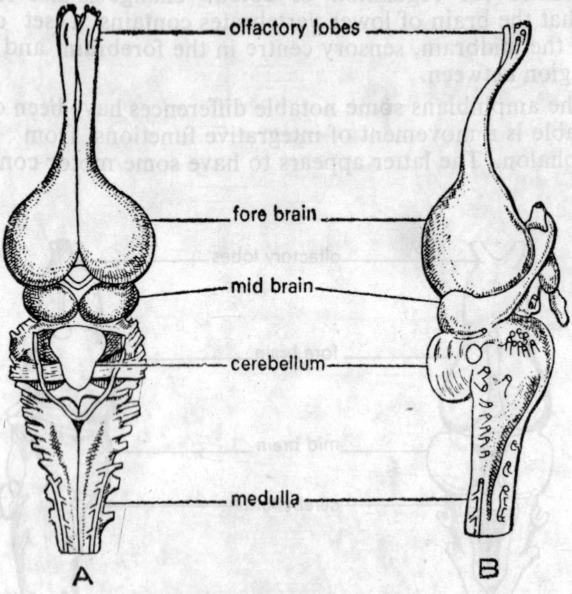

Fig. 13·9. Brain of alligator.

persists as the olfactory centre. It is also concerned with memory and emotions. The diencephalon, as in birds, has a temperature centre

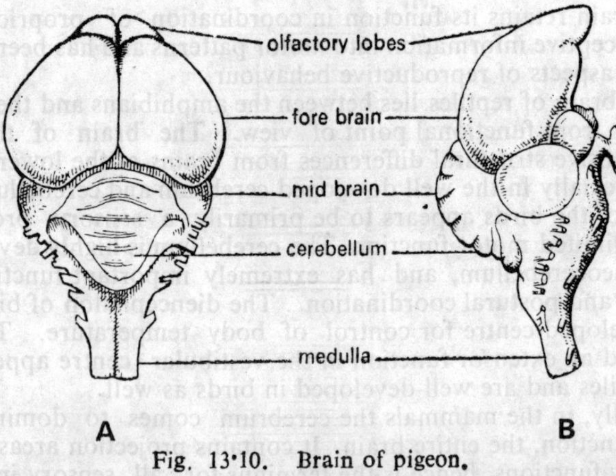

Fig. 13·10. Brain of pigeon.

and, as in the lower vertebrates, centres for communication with the endocrine system through the neurohypophysis. The reticular formation, which extends from the medulla to the rhinencephalon has recently attracted extensive study. Through this region, with its extensive sensory, cortical and motor connections, sweep the regular rhythms which are connected with sleep, arousal, and intellectual

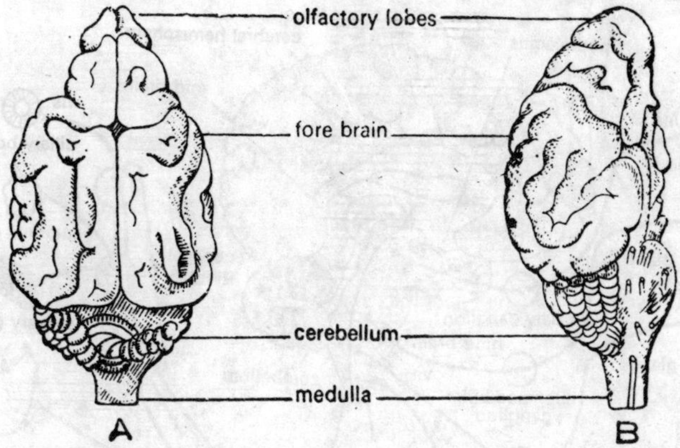

Fig. 13 11. Brain of dog.

activity. By contrast, the midbrain, in lower vertebrates the highest integrative centre, becomes in the mammals only a relay for sensory information, and locomotor centres of the medulla are subordinated to the cortex, diencephalon, and cerebellum. The cerebellum is more elaborate in mammals than in birds, with the lateral hemispheres constituting the neocerebellum, where sensory information is coordinated with information from all parts of the brain related to movement.

AUTONOMIC NERVOUS SYSTEM

The autonomic nervous system is so called because it is partly independent and not under voluntary control, though it is involuntarily controlled by the nerve centres located in the central nervous system, it is also connected to spinal nerves and some cranial nerves.

The concept of an autonomic nervous system was introduced by LANGLEY in 1921. LANGLEY was concerned only with the mammal. The autonomic, as he described it, was the system of nerves controlling the visceral effectors for digestion, circulation, excretion or more or less involuntary functions ; more precisely, it was the motor innervation of smooth muscle, cardiac muscle, and glands. It is an offshoot of the visceral motor division of the central nervous system, it has **visceral motor** neurons which differ from somatic motor neurons of the cranial and spinal nerves in that they are not connected directly with effector organs but through **two neurons**, one from the brain or spinal cord to an autonomic ganglion where it forms synapse with a second neuron which goes to an effector organ (muscle, visceral organ, or gland). Because of these synapses the visceral motor fibres are of two types; the first neurons are **preganglionic** whose cell bodies are located in the gray matter of the brain or spinal cord, their fibres are medullated and go through ventral roots of spinal nerves and white rami communicates to the autonomic ganglia. The second neurons are **postganglionic** which form synapses by their dendrites with the first

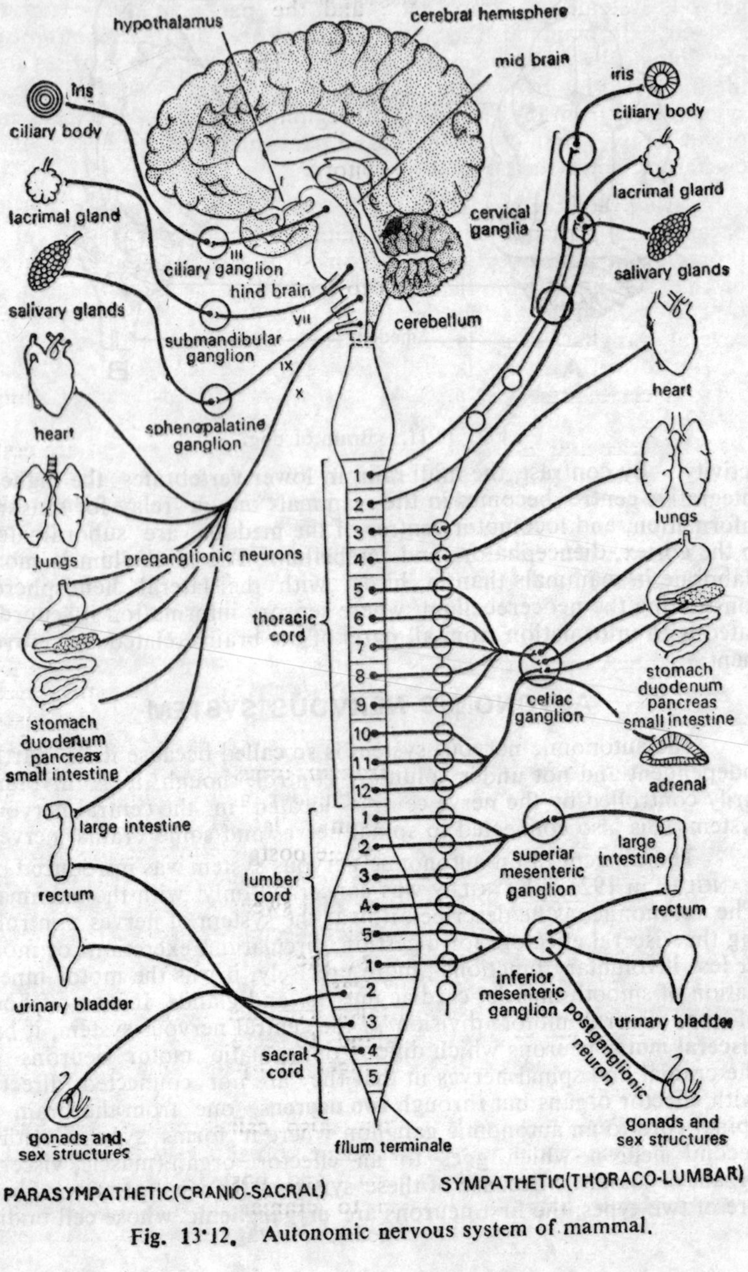

Fig. 13·12. Autonomic nervous system of mammal.

neurons in autonomic ganglia ; they go to the effector organs. The postganglionated neurons have non-medullated axons and go through gray rami communicantes and spinal nerves.

LANGLEY distinguished two divisions of the vertebrate autonomic nervous system, the sympathetic and the parasympathetic nervous systems. In mammals these two divisions are distinct anatomically physiologically and pharmacologically. The sympathetic arises from the thoracico-lumbar region of the central nervous system, the para-sympathetic from the cranio-sacral region. Typically, each visceral organ receives both sympathetic and parasympathetic fibres ; one is excitatory, while the other is inhibitory.

1. **Sympathetic system** is also called **thoraco-lumbar out-flow** because its preganglionic fibres join the spinal cord only in the thoracic and lumbar regions. It consists of two **sympathetic nerves** or chains running from the head to the end of the sacral region, one on each side of the vertebral column. Each sympathetic chain bears several ganglia called **lateral** or **chain ganglia,** some chain ganglia fuse to form three ganglia in the neck known as superior, middle, and inferior **cervical ganglia**, after which there is a linear series of ganglia in the thoracic and lumbar regions. Other sympathetic ganglia (**coeliac, superior** and **inferior mesenteric**) lie in the viscera and are collectively known as **prevertebral ganglia**. All these ganglia are connected to spinal nerve through rami communicantes. **Preganglionic fibres** arise from the spinal cord and pass through ventral roots of spinal nerves and white rami communicantes of all the thoracic spinal nerves and first three lumbar spinal nerves and then go to the chain ganglia. Some preganglionic fibres pass through the sympathetic chain of each side and go to the head to communicate with some cranial nerves. Some other preganglionic fibres arise from lateral ganglia and go to prevertebral ganglia in the viscera. From the lateral or prevertebral ganglia arise non-medullated **post-ganglionic fibres** going to visceral organs under the involuntary control (nose, eye muscles, salivary glands, heart, larynx, trachea, bronchi, lungs, alimentary canal, liver pancreas, adrenal glands, kidneys, bladder, and gonads). The post-ganglionic fibres form plexuses in collateral ganglia, such as **solar plexus** in the coeliac ganglion. Some postganglionic fibres go back through spinal nerves and gray rami communicantes to the skin, small blood vessels, erector hair muscles, and sweat glands. The postganglionic fibres of the sympathetic system secrete **sympathin** which generally stimulates these organs, *e.g.*, the fibres dilate the pupils and bronchi, increase heart beat, decrease secretion of saliva and digestive juices, temporarily reduce peristalsis, contract hair muscles causing hairs to stand up; cause sweat glands to secrete. All these reactions are usually associated with fear, anger, and pain, they cause expenditure of energy.

2. **Parasympathetic system** also called **craniosacral out-flow** because some cranial nerves and sacral spinal nerves are involved. It consists of ganglia, preganglionic, and postganglionic fibres. The **preganglionic fibres** are joined to cranial nerves III (oculomotor), VII (facial), IX (glossopharyngeal), X (vagus), and to the second,

third, and fourth sacral spinal nerves. The preganglionic fibres having their cell bodies in the mid-brain and medulla go to four ganglia in the head, those of the III cranial nerve to a **ciliary** ganglion, those of the VII nerve to **sphenopalatine** and **submaxillary** ganglia, those of the IX nerve to an **otic** ganglion, and those of the X nerve to very small ganglia situated in the organs concerned. The preganglionic fibres having their cell bodies in the gray matter of spinal cord pass through the sacral spinal nerves 2 to 4 and go directly to a **pelvic** ganglion without passing through the sympathetic chain. Thus the preganglionic fibres of cranial and sacral nerves terminate in ganglia located in or close to the organs supplied by them, hence these fibres are rather long.

From the ganglia arise very short **postganglionic fibres**, those from the ciliary ganglion go to eye muscles, those from sphenopalatine ganglion to lacrimal glands and nose, those from submaxillary ganglion to salivary glands, those from otic ganglion to salivary glands, other very short postganglionic fibres arise from small ganglia located in the organs (innervated by the vagus) and go to the heart, larynx, trachea, bronchi, lungs, alimentary canal, liver, pancreas, adrenal glands, kidneys, bladder, and gonads. The postganglionic fibres from the pelvic ganglion go to the colon, kidneys, bladder, and gonads, thus the postganglionic fibres are very short. The preganglionic parasympathetic fibres secrete **acetylcholine**, it has an inhibitory effect on organs which is antagonistic to that of the sympathetic system, *e.g.*, they constrict the pupils and bronchi, decrease the rate of heart beat, increase secretion of saliva and digestive juices, increase peristaltic movements, and stop secretions of glands. All these reactions are associated with comfortable sensations in the body, they also conserve energy.

Another peculiarity of the autonomic system is a double innervation of organs from the sympathetic and parasympathetic systems, the two systems work antagonistically in order to control the functions of all involuntary mechanisms in the body. A fibre of the sympathetic system is generally stimulatory and it starts an action in an organ, then the fibre of the parasympathetic system to the same organ stops that action after a time, hence the sympathetic and parasympathetic systems are antagonistic, *e,g.*, the iris is dilated by stimulation of a sympathetic nerve and is contracted by stimulation of its parasympathetic nerve. The autonomic nervous system regulates the functions of those organs which are not under the control of the will, though its actions are very slow. This system controls many involuntary actions, such as heart beat, respiration, digestion, secretions of glands, and excretion, it controls processes continued over long periods of time. The nerve centres controlling the autonomic nervous system are located in the **hypothalamus** through which the system forms connections with other nervous tissues.

Some neurons of the autonomic nervous system form the medulla of the adrenal glands, these are called **chromaffin** cells. Secretion of the medulla of adrenal glands called **adrenaline** or **epinephrine** stimulates the sympathetic but not the parasympathetic nerves. The nerve endings of sympathetic fibres secrete **sympathin** which is very similar to adrenaline, and reactions caused by sympathin are similar to those produced by adrenaline, either substance can stimulate the organs supplied by sympathetic fibres. Similarly stimulation of the parasympathetic fibres produces **acetylcholine**, the acetylcholine from the vagus slows the heart beat, it can be neutralized by adrenaline. Hence the stimulation of the autonomic nervous system produces its effects by releasing some chemical substance in the effector organs, as also in the case of endocrine glands, thus the nervous and chemical stimulations appear to be the same, though the effects of nervous stimulation are very rapid than to those of hormones.

Functions of the autonomic nervous system : The sympathetic and parasympathetic systems perform many functions which are generally acceleratory and inhibitory in nature. The sympathetic system on stimulation performs the following functions—(1) The constriction of blood vessels particularly of the skin. (2) Contraction of muscles. (3) Secretion of sweat glands. (4) Contraction of the heart. (5) Dilation of the bronchi. (6) Contraction of the muscles of urinary bladder. (7) Sudden increase in the blood pressure. (8) Sudden decrease in the number of red blood corpuscles in the blood. (9) Rapid coagulation of blood.

The parasympathetic system on stimulation bring about the following functions : (1) Dilation of blood vessels. (2) Constriction of the pupil. (3) Contraction of the walls of the digestive tract, and (4) Contraction of muscles of urinary bladder.

Revision Questions

1. Describe the structure and functions of neurons.
2. What is impulse ? Describe the physiology of impulse conduction.
3. What is synapse ? Describe different types of synapsis found in animals.
4. Describe the physiology of impulse conduction over the synapse.
5. What is reflex action ? Give brief account of the physiology of reflex action.
6. Describe the structure and functions of autonomic nervous system.

CHAPTER 14

Endocrine and Neuroendocrine Systems

Endocrine systems are systems which generally control long term activities of target organs as well as physiological processes such as digestion, metabolism, growth, development and reproduction, in contrast to more rapid activities under the control of the nervous system. These are effector organs, in that they are under the control of the nervous system either directly or indirectly. Endocrine systems include certain glands called **endocrine glands**, these glands are found in different regions of the body of the animal. Although, endocrine glands were first recognized only as **ductless glands** as they have no external ducts and which discharge their secretions directly into the blood stream. Particular emphasis was placed on the vascularity and specialized anatomy. Now, however, endocrine glands are known to occur in animals with scanty circulating fluids, while very vascular organs such as the brain and the intestinal wall not only produce hormones but also, at the same time, perform quite different physiological activities. As a matter of fact, the first hormonal action to be successfully demonstrated was that of **secretin** produced by the intestinal wall and released into the intestinal lumen (BAYLISS and STARLING, 1902 ; later GABRIEL and FOGEL, 1955), since 1902 a host of chemicals have been added to the list of hormones.

The activity of endocrine systems involves the release of special chemical substances called **hormones** which are usually (although not always) carried by the circulating blood to the sites where they may act.

Hormones are special chemical substances which are elaborated in restricted areas of the animal and influence the activity of various organs of the body and, thus, bring about a harmonious working of the body. Hormonal actions are usually longer lasting than those produced by nerve impulses. Also hormonal effects are capable of being exerted throughout the body, whereas nervous actions are usually more localized. Hormonal actions take place wherever cells are found that possess the necessary reaction or combining sites with the chemical hormones. Nearly all systems appears to be regulated by a combination of both hormonal and nervous systems. Hormones are sometimes called the "chemical messenger" of the body. They are effective in minute quantities. Their regulatory actions are sometimes one of excitation and sometimes one of inhibition; consequently the word hormone, which comes from the Greek root *hormao* meaning "to excite" is really a misnomer. According to SCHARRER and SCHARRER, 1963, hormones, in general, regulate or control the following types of activities :

1. **Reproductive activities :** Control of gametogenesis, development and maintenance of sex ducts and accessary or secondary sexual characteristics, release of sexual behaviour (including the release of pheromones), initiation of spawning and oviposition.
2. **Growth, maturation,** and **regeneration :** Including growth by addition of segments, for example, annelids; moulting and metamorphosis, for example, in crustaceans and other arthropods ; regenerative activities, and diapause.
3. **Metabolism** and **homeostasis :** Regulation of intermediary metabolism, maintenance of internal environmental factors including temperature regulation, water and ion balance, blood glucose levels, and so forth.
4. Adaptation to external factors including visual adaptation to light intensities, control of physiological colour changes, and so forth.

Chemically hormones do not belong to a special group of chemical compounds because some are protein hormones (*e.g.*, insulin glucagon, parathormone, oxytocin, vasopressin and all the hormones secreted by the intestinal mucosa and the seven or eight so-called tropic hormones of the anterior pituitary), others are steroid hormones (*e.g.*, androgens, estrogens and adrenocorticoids) and some are simple substances, neither protein nor steroid, *e.g.*, adrenaline or epinephrine of the adrenal medulla. They are generally specific in their origin, and to a small extent are also specific in their activity or function.

The proper harmonious functioning of the endocrine systems makes for a healthy, normal individual. Excess functioning or lack of function of an endocrine system may result in a serious physiological disorder.

Most of the endocrine systems of all the more specialized animals, invertebrate as well as vertebrate, are of non-nervous origin. These endocrine systems of non-nervous origin become conspicuous in the arthropods and the vertebrates, although a number of comparable tissues of suspected hormonal activity are known among the lower invertebrates—for example, the internephridial organs of the *Physcosoma* (SCHARRER, 1955) and the salivary glands of the cephalopods (JENKIN, 1962). Among the arthropods intensive research has demonstrated endocrine tissues arising from the non-nervous ectoderm (Y-organs of crustaceans, prothoracic glands and corpora allata of the insects) and from the mesoderm (androgenic glands and the gonads of Crustacea). In the vertebrates endocrine tissues also arise from the endoderm (thyroid, parathyroid, islets of Langerhans, gut epithelia), but none of the invertebrate structures has, as yet, been traced to this germ layer JENKIN, 1962. Thus, it appears that, during phylogeny, hormones first made their appearance in the invertebrates.

Para hormones : Very recently there have been reported such substances which meet the classical definition of a hormone and also have hormone-like actions. Two major classes of these substances are : prostaglandins and pheromones.

Prostaglandins : The prostaglandins were originally identified in human semen and found to be in the prostate glands in high concentration, but are now known to present in the central nervous system, the stomach walls, intestinal walls, spleen, adrenals, lungs, eyes, and adipose tissue, etc. They have a wide range of action, including potent oxytocic-effects. For this reason they have been proposed as contraceptive agents. Over twenty different prostaglandins have been isolated and characterised and are cyclic oxygenated fatty acids that contain 20 carbon atoms. The end results of their actions are varied.

A number of hormones, such as epinephrine, norepinephrine, glucagon, and ACTH, stimulate the break down of lipids by stimulating the synthesis of cAMP. Prostaglandins are known to antagonize the lipolytic effect of cAMP. It is thought that they do so by affecting adenyle cyclase, the enzyme responsible for cAMP synthesis from ATP. Since it is known that a variety of other hormonal events also mediated by cAMP are antagonized by prostaglandins these subtances may serve as general modifiers of cAMP in a number of tissues.

Pheromones : Pheromones (Gk., *pherein* to carry; *hormon*, to excite) are chemical substances secreted by organisms that trigger either behavioural or developmental processes when perceived by other members of the same species. As they resemble hormones in that they have an effect at some distance from the point of release they were originally called **ectohormones ;** the term pheromone was proposed by KARLSON and BUTENANDT in 1959 and this term has virtually replaced the original name.

The pheromones are not true hormones since they are products of exocrine glands; however, release from these glands often is dependent on hormonal stimulation. Endocrinologists consider pheromones for two reasons :

1. Hormonal metabolites being eliminated from the organism may function as pheromones in some species.

2. Pheromones initiate adjustments involving pituitary gland and gonads.

Pheromones may be categorized into two following heads :

1. **Releasing** or **signaling** pheromones that produce rapid and reversible responses through the central nervous system or by neuroendocrine control mechanisms.

2. **Primer** pheromones that activate a slow and long-acting series of neuroendocrine events involving prolonged stimulation.

Neurosecretions : These are special hormone-like substances which are secreted in the specialized nerve cells or neurons of glandular function. These are secreted in nerve cells or neurons and function as hormones, therefore, they are also known as **neurohumors** or **neurohormones** or **neuroendocrine substances**.

During the past few decades, the concept of neurohumors or neurohormones, developed specially by G.H. PARKER, has gained increasing acceptance. PARKER'S concept was that many kinds of nerve cells of glandular function produce their ultimate effects by forming and liberating a secretory product a neurohormone. This product may be liberated in very close proximity to the element on which it is to act, as in a synapse, or it may be liberated at a considerable distance from that element.

This neurohumor concept of PARKER has gained strong support from the histological demonstration by ERNST SCHARRER, BARGMANN, and many others that the central nervous system of animals having such systems contain cells which are, from their structure and connections, nerve cells but which contain distinctive secretory granules. Typically these granules are formed in the cytoplasm of the cell body and move into the axon to be carried to the axon terminus where the granule contents are liberated from the cell.

WELSH (1957, 1959) called the neurosecretions as neuroendocrine substances which are of following types : (1) the neurohumors, and (2) the neurosecretory substances.

The neurohumors are produced in nerve cells relatively unspecialized for secretory purpose. They are released at the ends of the fibres, travel only very short distances before being enzymatically destroyed and act on other neurons, muscles or glands in intimate contact with the nerve endings. They are, then, short-range and short lived materials. Acetylcholine, adrenaline and noradrenaline are well recognized members of this group. In addition, there is circumstantial evidence for a similar action by a few other chemicals, such as 5-hydroxytryptamine and gamma-aminobutyric acid, which are widely distributed in nervous tissues.

The neurosecretory substances, on the other hand, are produced by neurons specialized for secretion. The endings are associated with special storage centres called **neurohaemal organs**. The chemicals are released from the storage centres, they are relatively stable and act at greater distances and over longer periods. These substances may control such phenomena as moulting and chromatophore activity in arthropods, or may be associated with water balance and urine production in the vertebrates.

FLOREY (1962 a) uses the term **transmitter substance** rather than neurohumor for the chemicals which mediate synaptic transmission ; he restricts the usage of neurohormone to the hormones released by the neurosecretory cells. He emphasizes the differences rather than the similarities and draws a sharp line between transmitter substance and neurosecretions.

According to recent findings the neurosecretory substances which are involved in the chemical synaptic transmission, are referred to as **chemical synaptic transmitters** or **neurotransmitters** while those involved in regulating certain bodily activities are referred to as **neurohormones** or **neurohumors**.

According to GILBERT, 1963; KNOWLES, 1963; GABE, *et al.*, 1964, neurosecretory granules are found in almost all the groups of multicellular animals including the coelenterate *Hydra*, several of the polycladian turbellarians and the nemerteans.

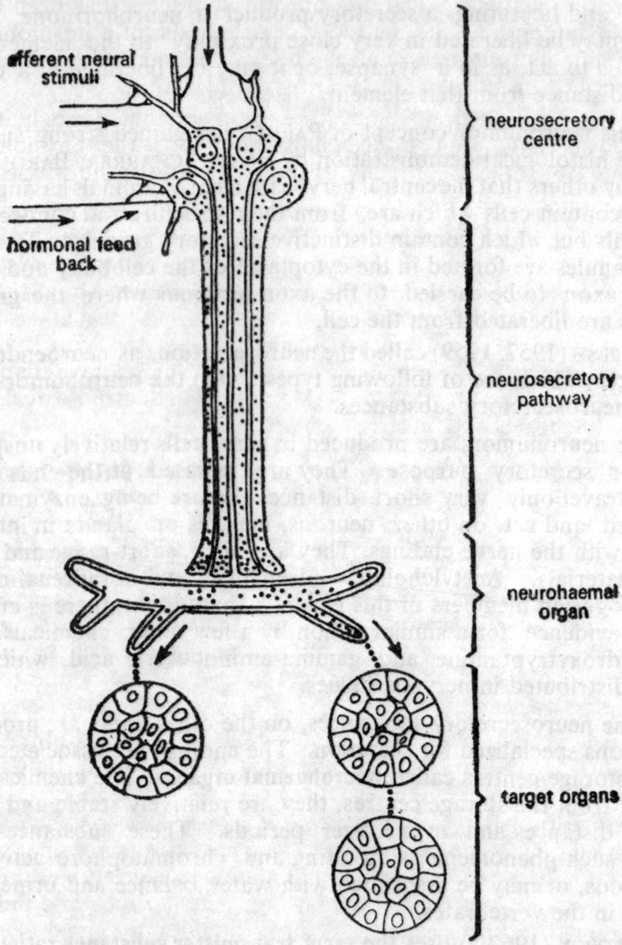

Fig. 14·1. Diagram showing the concept of neurosecretory system applicale both to invertebrate and vertebrate (SCHARRER and SCHARRER 1963).

Methods of investigation : The classical methods of investigation have been : removal of suspected endocrine tissue, either by surgical or by chemical means, the analysis of the resulting physiological or biochemical changes and the restoration of the animal to normal physiological state by replacement of the endocrine tissue or the administration of tissue extracts. Thus, first elimination and then restoration of excised endocrine tissues (sometimes by grafting excised endocrine tissues back into the animal) are basic methods in the investigation of the endocrine functions.

If an extract of excised endocrine tissue is found to restore normal activity or to initiate particular activities, the extract may be analysed in order to isolate and identify chemical substance responsible for the activity. The use of isotopic tracers has been of great value in determining the pathway of synthesis of a hormone or in determining its routes through the body. Endocrinological methods are described by ECKSTEIN and KNOWLES (1963).

Histophysiological studies have also contributed much, and the biochemists have now added substantially to the physiology by purifying or synthesizing many of the better known hormones.

Many endocrine systems or glands of both non-nervous origin and nervous origin, have been reported in the animals which are found at different places in the body. Vertebrate endocrine systems or glands will be discussed first because more is known about them than about those of the invertebrates. TURNER (1960) : WILLIAMS (1962) ; GROBMAN and BERN (1962) ; BARRINGTON (1963) ; VON EULER and HELLER 1963) ; SCHARRER and SCHARRER (1963) ; have contributed most towards the general and comparative endocrinological aspects of the vertebrates and invertebrates as well.

ENDOCRINE AND NEUROENDOCRINE SYSTEMS OF VERTEBRATES

Principal endocrine structures of vertebrates are, now, well studied than those of the invertebrates and include mainly the pituitary, thyroid, parathyroids, pancreas, adrenals, sex glands (ovaries and testes), and gastrointestinal mucosa. Several other glandular tissues are alleged to secrete hormones and these include urophysis, ultimobranchial bodies, juxtaglomerular cells of the kidney, thymus, pineal and placenta.

All of the endocrine glands excluding the parathyroids, are found in all vertebrates, although their structure and functions may differ markedly in different groups.

1. **Pituitary gland** or **system :** It is also known as **hypothalamic system** and is perhaps the single most important endocrine system in the body. Most of the endocrine glands of vertebrates and almost all somatic cells are influenced in their metabolism and function by secretions from this system. In turn, the pituitary gland or system is regulated by nervous or nenrosecretory information stemming from the hypothalamus or by blood-borne agents that are carried through the hypothalamus before entering the pituitary.

The pituitary gland, also called the **hypophysis**, is a small unpaired rounded body about the size of a large pea weighing approximately 5 g. It has been called the **"master" gland** because of the multiplicity of its functions, and because it controls many of the other endocrine glands. It is found at the base of brain, specifically the diencephalon and is lodged in a depression of the sphenoid bone of the skull called **sella turcica**, (*L. sella*, saddle; *turcica*, Turkish) just at the back of the optic chiasma. Thus the pituitary is one of the best protected and inaccessible endocrine glands. It is connected with the brain by means of a short, thin stalk, the **infundibulum**. It is composed of three lobes ; **anterior lobe, intermediate lobe (pars intermedia**) and a **posterior lobe** which differ from one another embryologically, histologically, and functionally. The anterior lobe and the intermediate lobe of the pituitary are known together as the **adenohypophysis**, while the posterior one is known as **neurohypophysis**. It

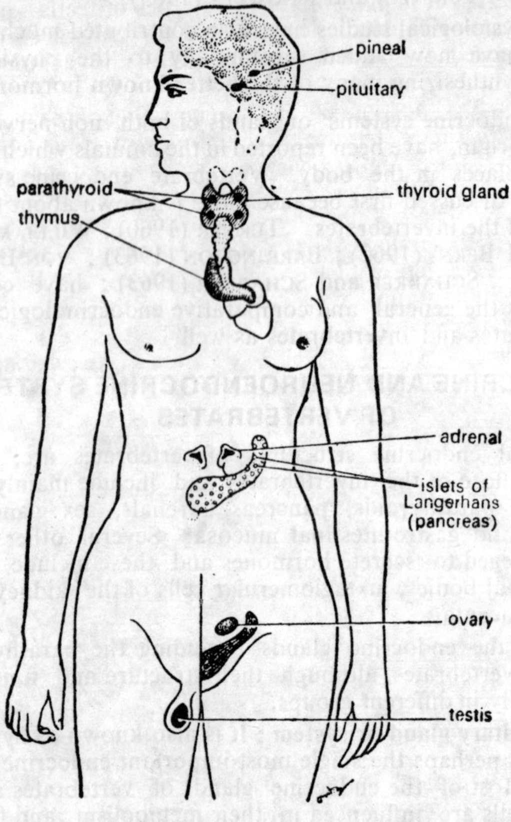

Fig. 14·2. Diagram showing the various endocrine glands with their locations in vertebrates (mammal).

has a dual origin embryologically, the anterior lobe and the intermediate lobe arise from a diverticulum of the roof of the embryonic buccal cavity known as RATHKE'S **pouch**, while the posterior lobe arises from an outgrowth of the embryonic brain, specifically the diencephalon. During embryogeny the two sources of tissue fuse, forming the complele gland. During fusion the anterior lobe along with intermediate lobe becomes pinched off from the roof of the buccal cavity, while the posterior lobe remains permanently attached to the hypothalamus of the brain with a short connection, the infundibulum. Through this connection the endocrine system retains a vital interdependence with the nervous system. In the young human the intermediate lobe is a distinct structure but it gradually fuses with posterior lobe and becomes obscured. However, in most vertebrates the intermediate lobe remains as a distinct structure.

Adenohypophysis : It is the largest part of the pituitary in all vertebrates except cyclostomes and teleosts and is further distinguished into **pars distalis, pars intermedia** (also regarded as intermediate lobe) and **pars tuberalis.** It is made up of three types of cells—peripheral **basophils**, central **acidophils** and scattered throughout **chromophils**, these are supposed to be the precursors of the acidophils and basophils.

No doubt this portion of the pituitary is derived from the different precursor than of neurohypophysis but it does not function in splended isolation from the neurohypophysis, for it is intimately related to the hypothalamus by means of an important vascular connection, the **hypophysial portal system**, which begins as a capillary plexus in the hypothalamus and ends in the sinusoids of the adenohypophysis. The direction of flow in these vessels is from the hypothalamus to the adenohypophysis.

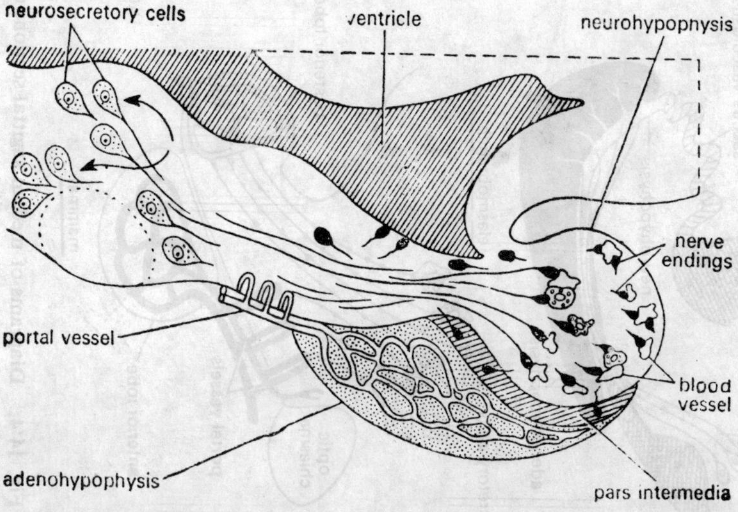

Fig. 14·3. Diagrammatic sagittal section of the hypophysis (pituitary) mammal.

The morphological aspects of the communication between adenohypophysis and neurohypophysis have attracted special attention in recent years. The axon termini of the hypothalamic neurosecretory cells are closely associated in the neurohypophysis with the walls of capillaries in the capillary network of that organ. In addition, in mammals, there are capillaries surrounding the neurosecretory axons in the median eminence of the diencephalon, where the axons often

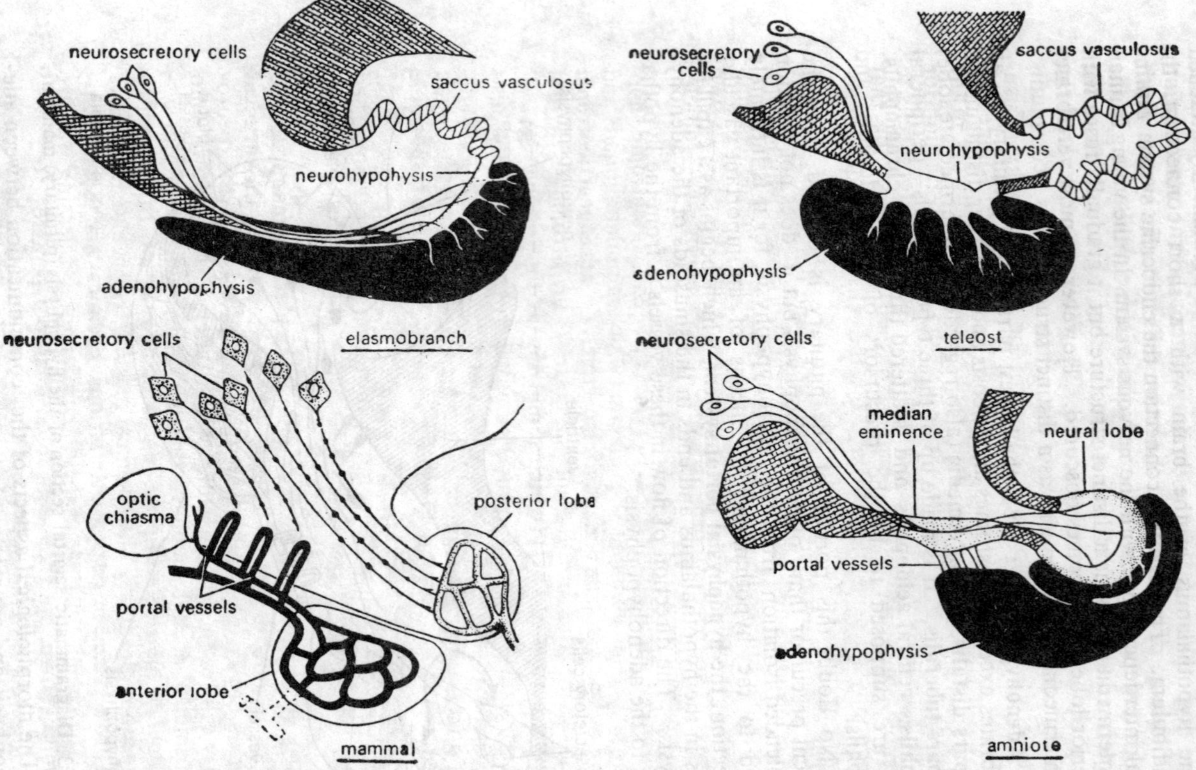

Fig. 14·4. Diagrams of median sagittal sections of the pituitary of an elasmobranchi, *Scylliorhinus*; a teleost, *Anguilla*; an amniote; and the circulatory pattern and distribution of neurosecretory fibres in mammal.

show an accumulation of neurosecretory material. The presumption is that the secretory products of the nerve cells are discharged from the axons and axon termini into the blood of these capillaries. In the lower vertebrates, some of the neurosecretory axons enter into the tissues of the adenohypophysis as well. The capillary network of the median eminence and sometimes of the neurohypophysis, communicates directly with the capillary network of the adenohypophysis, usually through a system of portal vessels; that is, the capillaries unite into larger vessels which then again give rise to a capillary network in the adenohypophysis. Structurally, this provides an ideal situation for chemical transmission from the hypothalamic neurosecretory cells to the adenohypophysis. It has been established experimentally that neurohypophysial principles do act on the cells of the adenohypophysis.

The adenohypophysis of the pituitary gland is characteristically glandular in nature and secretes six different hormones which, with one exception, are the **trophic hormones**. A trophic hormone stimulates some other endocrine gland as its target organ to secrete its own hormones. The trophic hormones are the **gonadotropins—follicle-stimulating hormone** (FSH), **luteinizing hormone** (LH) in the female or **interstitial cell stimulating** (ICSH) in the male and **luteotropic hormone** or **lactogenic hormone** or **prolactin** (which stimulate gonads and lactation by the mammary glands respectively) ; **adrenocorticotropin** (**ACTH**) (which stimulates the adrenal cortex) ; **thyrotropin** or **thyroid stimulating hormone** (**TSH**) (which stimulates the thyroid gland).

The sixth hormone, **growth stimulating hormone** (**GSH**) also called **somatotropin** (**STH**), is not strictly speaking a trophic hormone since it does not require another endocrine gland as its target organ but produces its widespread effects by direct action and is responsible for the growth of the body.

Very recently one more complex of hormones has been reported from the adenohypophysis what are known as **lipotropins** or **adipokinetic** hormones.

The production of these hormones particularly in mammals and birds, as is now believed, is due to secretion of certain **chemical messengers** or **regulation factors or releasing factors** by the neurosecretory cells of the hypothalamus which after passing down into the sinusoids of adenohypophysis through the hypophysial portal system, stimulate the various hormone-producing cells of this gland to discharge their products. The chemical messengers or releasing factors in turn are secreted in response to various chemical and nervous stimuli. The chemical nature of these neurohumoral releasing factors is not fully known, but they seem to be small polypeptides.

Table 14·1 lists the regulatory factors that are released by the hypothalamus and stimulate or inhibit the secretion by the hypophysis of specific adenohypophysial hormones.

1. Growth (somatotropin) hormone : This growth hormone is a protein, a globulin, secreted by special acidophilic cells of the adenohypophysis. Human growth hormone has been isolated in a highly purified state. In 1966, LI, LIU and DIXON succeeded in identifying the α-amino acid sequence in this protein. The molecular weight is 21, 500, and it has the structure of 188 amino acids.

Table 14·1 : Various releasing factors.

Factor	Hormone affected
Adrenocorticotropi chormone releasing factors (CRFs)	Adrenocorticotropic hormone (ACTH) (Adrenocorticotropin)
Follicle-stimulating releasing factor (FRF)	Follicle stimulating hormone (FSH)
Growth hormone regulatory factors : Releasing factor (GRF) Inhibitory factor (GIF)	Growth hormone (somatotropin)
Luteinizing hormone releasing factor (LRF)	Luteinizing hormone (LH)
Melanocyte regulatory factors : Releasing factor (MSHRF) Inhibitory factor (MSHIF)	Melanocyte stimulating hormone (MSH) (melanotropin)
Prolactin regulatory factors : Releasing factor (PRF) Inhibitory factor (PIF)	Prolactin
Thyrotropic hormone releasing factor (TRF)	Thyrotropic hormone

Functions of growth hormone : The basic and the primary function of the growth hormone is to stimulate both body weight and the rate of growth of the body, by growth we mean an increment in the amount of muscle tissue, bone structure, cartilage, etc., of the entire body. It also influences the protein, carbohydrate and fat metabolisms to a great extent.

Hyposecretion of growth hormone : Hyposecretion of the growth hormone during the years of skeletal growth of an individual, results in stunted skeletal development and retarded growth (**dwarfism**). Dwarfism, in contrast to cretinism arising from hypothyroidism, is not accompanied by physical deformity, mental inferiority, or retardation. The hypophysial dwarfs are frequently immature sexually and at adult age may attain a height no taller than 3 or 4 feet. The relative proportions of the different parts of the skeleton do not deviate markedly from normal, although the head is generally large in relation to the body.

Hypersecretion of growth hormone : Hypersecretion of growth hormone prior to puberty, before ossification is complete, results in **gigantism**, a condition typified by a general overgrowth of the skeleton, resulting in individuals of 7 or 8 ft. or more in height. The limbs are generally disproportinately long.

Hypersecretion of the hormone after the usual age of full skeletal growth results in **acromegaly**. In this disease a characteristic enlargement or overgrowth of the bones of the hands, feet, and in particular the jaws, checks, and face takes place. There is also an excessive growth of fibrous tissue, resulting in thickened nose, lips, eyelids, and broadened fingertips. There is bowing of the spine (**kyphosis**) ; the outer posterior diameter of the chest is increased. In the early stages of the disease, increased sexual activity may be evident. Later, there occur atrophy of the gonads and suppression of the sexual functions in both sexes, with importance in men and amenorrhea in women. Acromegaly is sometimes caused by a tumour of the adenohypophysis, and in such a case surgery may be attempted.

2. **Thyrotropin** or **thyroid stimulating hormone (TSH)** : This hormone is produced by basophilic cells of the adenohypophysis. Its structure is not known in detail but it is a glycoprotein of about 10,000 molecular weight. Its primary action is to regulate the growth of the thyroid gland and an increased production of the thyroxine hormone therein. It induces three metabolic effects in the thyroid glandular cells : (i) Increases the rate at which inorganic iodide is absorbed from the blood into the thyroid cells. (ii) Increases the rate of incorporation of iodides into thyroid hormones. (iii) Increases the rate at which thyroid hormones already produced are released from the thyroid gland. A secondary and perhaps unrelated effect is that of causing uptake of water and fat, especially in the eyes of mammals and probably elsewhere. There is evidence suggesting hypothalamic regulation of TSH secretion, possibly as part of a feedback mechanism in which elevated levels of thyroid hormone in the blood act upon the hypothalamus to inhibit secretion of the factor which normally stimulates secretion of TSH. The identity of this hypothalmic factor is unknown.

It is also believed that it increases the rate of metabolism of all the cells of the body, therefore, indirectly thyrotropin is metabolism stimulating hormone. In the absence of thyrotropin, according to SMITH, thyroid gland secretes less than its normal amount of thyroxine or atrophy of thyroid gland results.

3. **Corticotropin** or **adrenocorticotropin hormone (ACTH) :** This hormone is secreted by basophilic cells of the adenohypophysis and is the true physiological stimulus to the adrenal cortex, thus causing the adrenal cortex to grow and to secrete all of its normal hormones at an increased rate. Its presence has been demonstrated from the elasmobranchs up, but some species do not respond to it.

The adrenocorticotropin differs less in its structure than do the growth hormone. The ACTH of beef and human contains 39 and 41 amino-acids respectively and its structure has been fully worked out by C.H.LI, and his collaborators at Berkeley, California and by BELL and his associates. Human ACTH differs in having two additional amino-acids (valine and tyrosine) near the amino end of the chain.

The structure of ACTH is of interest in that it contains a sequence in common with the smaller molecules of the melanocyte stimulating hormones, namely, methionine, glutamine, histidine, phenylalanine, arginine, tryptophan, glycine. In consequence, ACTH has MSH (melanin stimulating hormone) activity as well. The MSH hormone has much smaller molecule than ACTH. The MSH end of the ACTH molecule is apparently not its functional end as regards stimulation of the adrenal cortex is concerned. This stimulation is fully reproduced by a synthetic polypeptide comprising the last 23 amino-acids of the ACTH chain (HOFFMAN).

The adrenal cortex is the main target organ upon which ACTH acts, although a number of extra-adrenal functions have also been reported. The steroid hormonal secretions of adrenal cortex act throughout the body to (i) increase the rate of gluconeogenesis ; (ii) increase the rate of protein catabolism so that blood glucose concentration may greatly be increased (iii) increase the rate of fat-catabolism. Therefore, indirectly corticotropin has profound effects on carbohydrate, protein and fat metabolism.

ACTH is used as a non-disease specific. It is used in the treatment of hypersensitivities and inflammatory reactions and also in the treatment of rheumatoid arthritis and acute rheumatic fever. It is also used in the treatment of chronic and intractable cases of asthma. Its deficiency also causes acute psoriasis and dermatitis. Its oversecretion causes cushing's disease which is characterized by obesity of the trunk (especially of the abdomen), face, and buttocks but not the limbs; cyanosis of the face, hands, and feet, pigmentation of the skin, and excessive growth of hair ; women may grow a mustache or beard ; demineralization of the bones ; loss of sexual functions and hyperglycemia and glucosuria.

It is probable that its production is initiated and increased by the **corticotropin releasing factor** secreted by the neurosecretory cells of the hypothalamus, this releasing factor is transmitted down into the adenohypophysis through the portal system and causes the production of corticotropin hormone. Its production is also increased by adrenaline in the blood and decreased by circulating cortical steroids, so that there is some automatic control of the latter.

4. **Gonadotropins :** The **follicle stimulating hormone, luteinizing hormone** in female or **interstitial cell stimulating** hormone in male and **prolactin** or **luteotropin** hormone are collectively referred to as gonadotropins. The gonadotropin hormones stimulate various activities of the gonads, the testes and ovaries, and also control the sexual cycle.

P. E. SMITH and ASCHHEIM and ZONDEK were the first who discovered that when a piece of anterior pituitary tissue is implanted under the skin of an immature rat, the ovaries develop within a few days. ZONDEK later postulated that the growth of the ovarian follicles is due to a **follicle stimulating hormone (prolan A)** and that the development of lutein tissue is due to a **luteinizing** hormone.

In 1931, HISAW and LEONARD separated the pituitary gonadotropin fractions (hormones) into two components—luteinizing hormone (LH) and follicle stimulating hormone. Recently two other hormones have also been reported and these are interstitial cell stimulating hormone and prolactin or luteotropin hormone, the latter was discovered by RIDDLE.

(i) **Follicle stimulating hormone :** It is not a pure protein but apparently as glycoprotein that contains galactose, mannose, galactosamine, glucosamine, sialic acid, fucose, and uronic acid.

The follicle stimulating hormone stimulates the growth of the ovarian follicles upto the point of ovulation in the female, while in the male it is concerned with the development of the seminiferous (semen-carrying) tubules and the maintenance of spermatogenesis (production of sperm). It is secreted by basophilic cells of adenohypophysis but different from those which secrete TSH.

(ii) **Luteinizing** or **interstitial cell stimulating hormone :** LH is also a glycoprotein of molecular weight about 26,000 (in human). It is secreted apparently by the same basophilic cells of adenohypophysis which secrete FSH. It is identical with the interstitial cell stimulating hormone (ICSH) of males.

In the female, the luteinizing hormone works with FSH and is responsible for the final maturation of the ovarian follicles and ovulation. These two gonadotropins are also responsible for the stimulation of the ovaries to produce estrogens (female sex hormones) and for the stimulation of the testes to produce androgens (male sex hormones).

The luteinizing hormone alone, however, induces luteinizing of the ruptured follicle and stimulates the secretion of progesterone.

In the male the interstitial cell stimulating hormone stimulates the interstitial cells (Leydig cells) of the testes and also induces the secretion of the male hormone, testosterone.

Although the pituitary is the prominent source of gonadotropins, some of the gonad-stimulating hormones are secreted by the placenta and other tissues too. These may be classified as follows : (1) **Human chorionic gonadotropin** (present in blood, urine and tissues of pregnant woman), (2) **human nonchorionic gonadotropin** (present in blood and urine of ovariectomized and post-menopausal woman), and (3) **equine gonadotropin** (present in blood and placental tissue of the pregnant mare).

The placenta is also said to be a site for the production of ACTH. It is known that pregnant woman exhibits great remission in symptoms of asthma and other allergic signs during their pregnancy, but soon after delivery these symptoms return. ACTH has been found to alleviate greatly the asthmatic condition.

5. **Prolactin** or **luteotropin hormone :** RIDDLE was the first who discovered this hormone. This hormone appears to be a pure protein with an isoelectric point of 5·7 and a molecular weight of about 25,000. It is secreted by a special group of acidophilic cells in the adenohypophysis. Its original name (**lactogenic hormone** or **pro-**

lactin) was given because of its role in milk production by the mammary glands.

The functions controlled by this hormone are diverse but all have some connection with reproduction. In mammals the development of mammary glands depends on stimulation by the female sex hormones estrogen and progesterone, and by prolactin. The secretion of milk depends on the thyroid hormone, the steroids of the adrenal

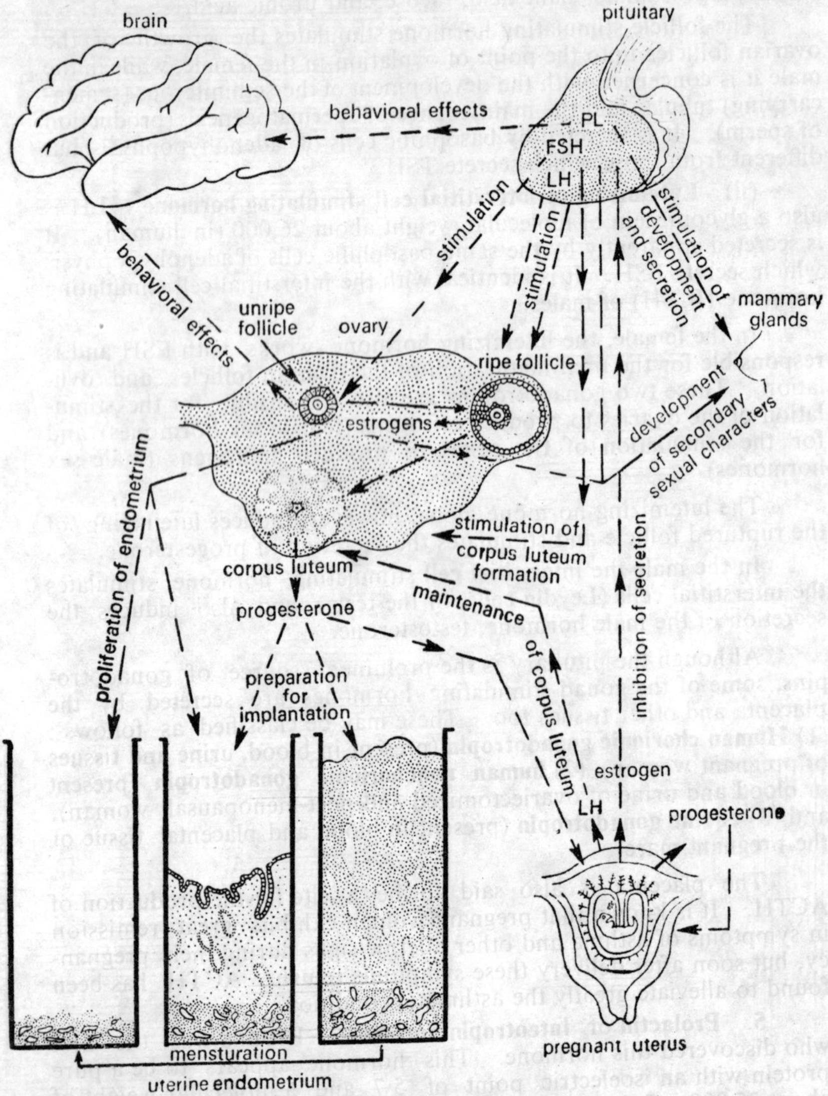

Fig. 14·5. Diagram showing the functions of various gonadotropin hormones secreted by pituitary and sex-glands.

cortex and prolactin, and the actual ejection of milk on oxytocin. Prolactin is clearly not alone in this complex process. The name luteotropin hormone (LTH) refers to the stimulation of secretion of progesterone by the corpus luteum of the mammalian ovary, formed after ovulation under the influence of the luteinizing hormone (LH or ICSH). LTH also has marked effects on behaviour, stimulating the characteri tic maternal behaviour in vertebrates from fishes to mammals. It stimulates the crop glands of the pigeons to form the milk on which the young ones are nourished for a time ; it induces the newt *Triturus viridescens* to return to the water on approach of the breeding season, and it aids in the development of the sexual colouration pattern in certain fishes. There is as yet no known cellular basis for these varied actions, and one wonders if there can be a single common basis. From the biological point af view, LTH offers the first of several examples of the diversity of actions of a single hormone, in this case all directed toward the same biological end of reproduction.

6. Lipotropins or **adipokinetic hormones :** Three lipotropins or adipokinetic hormones have been reported and these are I and II (from porcine tissue) and β (from ovine glands). In the activity, the lipotropins resemble the actions of α- aud β-MSH (melanocyte stimulating hormone) and they have sequences common to ACTH and to the MSHs.

Lipotropins stimulate the liberation of fatty acids from stored body fats (adipose tissue). Some cases of extreme obesity and extreme thinness may be due to hyposecretion and hypersecretion of this hormone.

Pars intermedia or **intermediate lobe :** This lobe connects the posterior and anterior portions of the pituitary gland. Recently it has been referred to as a part of adenohypophysis because it is also derived from the part of alimentary canal of which the anterior lobe is made. This lobe secretes two **melanocyte-stimulating hormones** (peptides) which have been isolated in a pure form by Li and his associates. These hormones, in those vertebrates provided with chromatophores such as amphibians and fishes, cause dispersion of the black pigment in the melanophores or melanocytes. Its role in birds and mammals is uncertain but in 1973 some workers suggested that these hormones (MSH) may be concerned with the maintenance of hair lipid. There is evidence that it stimulates melatin synthesis in mammalian skin by activating tyrosinase.

Posterior lobe or **neurohypophysis :** Among tetrapods the posterior lobe or neurohypophysis, an outgrowth of the primitive brain, forms a distinct neural lobe, the **pars nervosa** of the pituitary gland. It is simply made up of nervous tissue and consists of the terminations of many neurosecretory fibres of the neurosecretory cells located in the preotic or supraoptic and paraventricular nuclei of the hypothalamus. It serves as a storage area (**neurohaemal organ**) for the secretions of neurosecretory cells.

The evolution of the distinct neural lobe or pars nervosa is thought to be related to the water balance demands of the terrestrial living being seven though a number of very different activities (lactation for example) have evidently come under its control. It should be noted that the vascular relationships of the tetrapods pituitary are such that the adenohypophysis receives chemical information from the hypothalmic nuclei via the portal system, even after the establishment of a neural lobe. Thus, in the higher vertebrates, secretions from the hypothalamic nuclei control the adenohypophysis locally and exercise a more remote control over distant organs by neurosecretions which are stored in neural lobe prior to release into the blood. At present there is only scanty information concerning the nature of the biochemical links between hypothalamus and adenohypophysis. The corticotropin releasing factor seems to be closely allied to the neurohypoyphyseal hormones (JORGENSEN and LARSEN, 1963), but this is probably not true for some of the other factors (the thyrotropin-releasing factor, for example).

It is considered that neurosecretory cells of the neurohypophysis secrete many biochemically distinct polypeptide hormones of which several have now been identified : additional ones may be expected when more groups of animals have been examined (HELLER, 1963 ; PERKS and DODD, 1963). Physiologically their most consistent action throughout the vertebrates is on the control of water and salt metabolism. In the primitive groups sodium regulation seems more important, while water regulation assumes greater significance in the terrestrial animals (HELLER, 1963).

The known polypeptides of the neurohypophysis of the various vertebrates are—**arginine vasopressin, lysine vasopressin, arginine vasotocin, oxytocin** (=**pitocin**) and **isotocin,** recently isolated from a fish. Most vertebrates appear to have only two of these; in amphibians, reptiles, birds, and mammals oxytocin (=pitocin) and vasotocin have been identified in posterior pituitary extracts (KAMM and his associates), and in mammals one of the vasopressins replaces vasotocin. Most mammals have arginine vasopressin, but the pig and hippopotamus have lysine vasopressin.

1. **Oxytocin** (=**pitocin**) : The normal action of oxytocin is not fully understood as yet. It has effects on uterine contractions which may be important in parturition in mammals, and it acts as the milk let down hormone in stimulating lactation, also in mammals. Besides uterus, it also causes contraction of other smooth muscles in the body, though usually to a much less extent. It, also seems to have a direct effect on the adenohypophysis to induce production of luteotropic hormone which in turn causes milk secretion.

2. **Vasopressin** (=**pitressin**) : The primary function of vasopressin appears to be that of increased tubular reabsorption by the kidneys, so that it decreases the production of urine. Its fundamental effect seems to be an influence on permeability, which drives potassium out of the cells and sodium and water in. Thus, it exhibits **antidiuretic properties** and is now referred to as **antidiuretic hormone.** Also, like epinephrine, it causes an increase in the cyclic AMP concentration of certain cells.

Large quantities of this hormone cause the arterial pressure to rise, the rise in pressure is due to contractions of peripheral arterioles.

It also causes contraction of almost any smooth muscle tissue in the body, including most of the intestinal musculature, the bile ducts, the uterus and so forth.

The production and liberation of both oxytocin and vasopressin hormones are probably controlled entirely by nerve impulses from the hypothalamus. It has been established that these two hormones are synthesized in the hypothalamus, vasopressin in the supraoptic nuclei and oxytocin in the paraventricular nuclei, and migrate into the neurohypophysis along the non-myelinated nerve fibres of the **hypothalamo-hypophyseal tract**, for storage until released into the blood by neural or nerve impulses carried in the same fibres. In the blood they are transported in loose association with carrier plasma proteins called **neurophysins**. They are destroyed or inactivated primarily in the kidney and liver by enzymatic reactions.

Hypophysectomy : Hypophysectomy (*i.e.*, the removal of entire pituitary or hypophysis) operation was first of all conducted by ASCHNER in 1909, on dogs and later in 1926 SMITH performed such operations on rats and other mammals and noted the following facts :

1. In the adults the gonads degenerate, while in the young ones the gonads fail to mature.
2. The adrenal cortex becomes inactivated that results in the development of fatal symptoms.
3. The thyroid gland shrinks in size, producing a low metabolic rate and other manifestations of thyroid hormone insufficiency.
4. If the hypophysectomy is performed late in pregnancy or after the young ones are born, no lactation occurs.
5. Growth is completely retarted.
6. There is a complete alteration of the carbohydrate, lipid and protein metabolism.
7. In some species, there are resultant pigmentation changes.

In man, a tumor or infraction of the pituitary is called SIMMOND'S disease.

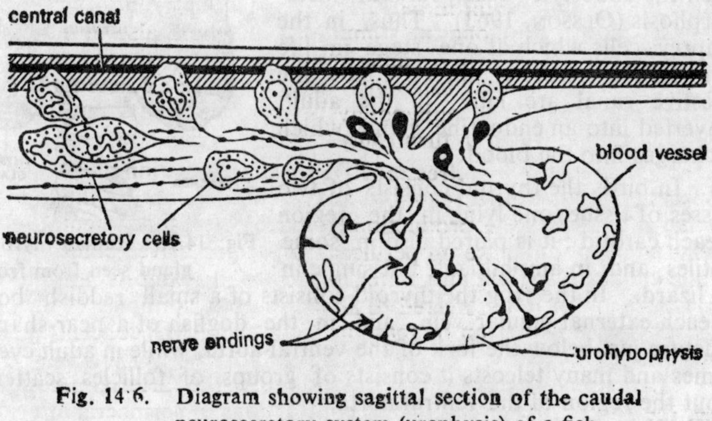

Fig. 14·6. Diagram showing sagittal section of the caudal neurosecretory system (urophysis) of a fish.

Urophysis : In the tails of elasmobranchs, teleosts and chondrosteans there are cells in the posterior region of the spinal cord which, from their staining properties, appear to be secretory or glandular in function. In teleosts these cells are associated with blood vessels to form a neurohaemal organ, the so called **urophysis** or **urohypophysis** similar to the neurohypophysis in the hypothalamic region of the brain. The urohypophysis seems to be concerned with osmoregulation in some fishes, but its hormones and functions have not yet been fully established.

2. Thyroid gland : All vertebrates have thyroid glands. The mammalian (man) thyroid gland, weighing about 25 grams to 40 grams, is composed of two lobes that lie close together on either side of the trachea just behind the thyroid cartilage. The two lobes are joined by a thin connective tissue called **isthmus**. The functional units of the thyroid gland are a large number of small closed follicles, the walls of which are lined with columnar or cuboidal epithelial cells. The follicles are held together by areolar tissue and are surrounded by a profuse network of capillaries. The follicles undergo changes in size depending upon the colloid (synthesized material) that has been produced by epithelial cells. The thyroid gland is larger in females than in males and may enlarge still more in pregnancy. The flow of blood to the thyroid gland is most profuse.

The thyroid gland starts as a single diverticulum from the pharynx, and its position and mode of development show that it is probably homologous with the wall of the endostyle of lampreys and perhaps of *Branchiostoma*—(BARRINGTON, 1959). In the ammocoete larva of lamprey, the endostyle (actually a pouch-shaped sub-pharyngeal gland) contains special iodine metabolizing cells, and these have been shown to form thyroid follicles when the endostyle disappears at metamorphosis (OLSSON, 1963). Thus, in the lamprey, cells which at one stage in life (ammocoete) secrete directly into the digestive canal are later, in the adult, converted into an endocrine gland which discharges into the blood.

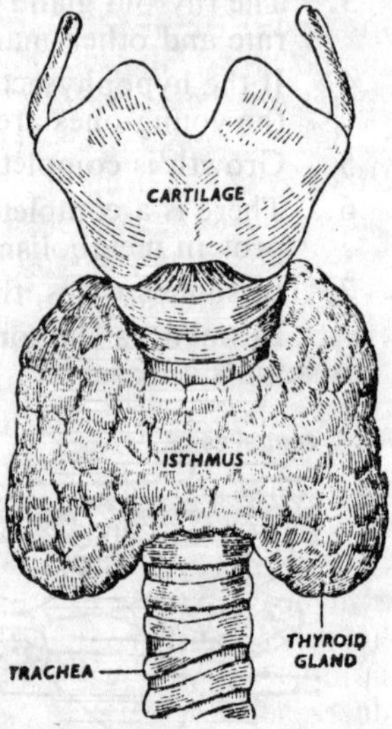

Fig. 14·7. Human thyroid gland seen from front.

In birds the thyroid consists of two masses of tissue, one lying in the region of each carotid ; it is paired also in some reptiles and in amphibians, but single in the lizard. In the frog the thyroid consists of a small reddish body on each external jugular vein, and in the dogfish of a pear-shaped structure just below the fork of the ventral aorta, while in adult cyclostomes and many teleosts it consists of groups of follicles scattered about the region of the ventral aorta.

Function of the thyroid : The function of thyroid gland is to produce mono-, di-, tri-, and tetra-iodothyronine, the last of these commonly known as **thyroxine**, is the most important and active hormone of the thyroid gland.

The cells of the thyroid gland, perhaps through the process of active transport mechanism, absorb the inorganic iodide from the blood and oxidize it to iodine (I_2)—probably enzymatically. This is followed by an iodination of the thyronine, an amino acid derived from tyrosine or thyroglobulin, again presumably by enzyme action, although the precise steps are unknown. This biosynthesis leads to the formation of a series of mono-, di-, tri- and tetra-iodothyronine. Two molecules of diiodothyronine condense to form tetra-iodothyronine. To some degree one molecule of monoiodothyronine condenses with one molecule of diiodothyronine to form triiodothyronine. Tetraiodothyronine or thyroxine and smaller amounts of the other iodothyronines, liberated in the lumen of follicle from the glandular thyroid cells are stored as the glycoprotein thyroglobulin of molecular weight of about 675,000. It is characterized by containing 0.5 to 1% of iodine by weight. KENDLLA showed that thyroglobulin consists of a protein globulin and a simple unit called thyroxine, which contains 65% of iodine by weight. When required by the body these mono-, di-, tri- and tetra iodothyronines are released by action of a protease from the thyroglobulin and secreted into blood, where they combine with the plasma proteins in a loose union so that they may be transported to the tissues of the body. The mono-, di-, and triiodothyronines are bound less strongly to the plasma protein than the tetraiodothyronine, therefore, they are present in low concentration.

The whole process of synthesis and release is stimulated by TSH (thyroid stimulating hormone) from the anterior lobe or adenohypophysis of pituitary. In the absence of this hormone, the thyroid ceases its secretory activity and undergoes atrophy.

Recently it has been found that the thyroid gland also produces **thyrocalcitonin**, an hormone whose action probably is to decrease blood calcium concentrations—an action antagonistic to that of parathyroid hormone.

Functions of the thyroid hormones : The actions of thyroid hormones are many. In mammals the important action is a stimulation of oxygen consumption and heat production. The mechanism of this action is not fully established, but there is an increasing body of evidence in favour of a mechanism by which thyroxine, one of the hormones of thyroid, decreases the efficiency of oxidative phosphorylation. In this view the increased heat production results from a decreased P/O ratio in cellular oxidation, with the result that more of the energy produced by oxidation appears as heat and less as useful work (ATP). At first, this decrease in efficiency appeared as a consequence of a structural effect of thyroxine on the mitochondrion. More recent findings favour the view that thyroxine acts by stimulating the enzyme which transfers hydrogen between nicotinamide-adenine dinucleotide and its phosphate.

$$NADH + NADP^+ \rightleftarrows NAD^+ + NADPH$$

Increased thyroxine results in decreased concentration of NADPH, and this in turn diverts oxidative metabolism from the efficient EMBDEN-MEYERHOF glycolysis pathway to the inefficient pentose shunt pathway. In other vertebrates, particularly poikilothermic animals, in general, oxygen consumption does not increase by thyroxine, though there are numerous reports of increases in individual species or special conditions, even in invertebrates and microorganisms. In amphibians it stimulates metamorphosis in larvae. Thyroid hormones including thyroxine have an effect on growth and development in all vertebrates, although their actions are not clearly understood because other hormones also have this activity. In amphibians, reptiles, and birds the moulting of cornified epidermal cells or of feathers is stimulated by thyroxine. Thyroxine also causes the deposition of melanin in bird feathers and of guanine crystals in the skin of fish, giving them a silvery appearance. In addition, thyroxine affects such diverse processes as schooling behaviour in fish, the threshold sensitivity of sensory receptors, creatine-creatinine conversion, and water diuresis.

Hypofunction of the thyroid gland (hypothyroidism) : Hypothyroidism may be caused by a deficiency of the hormone or a thyroidectomy (*i.e.*, removal of the thyroid gland).

In cases of hypofunction of thyroid gland 20 to 30% of the rate of the basal metabolism is lowered in warm-blooded animals. In the young individuals (mammals) a hypofunction of thyroid leads to **cretinism**, a disease characterized by dwarfism, serious mental deficiencies and a peculiar infantile facial expression. Hypofunction of the thyroid also causes a decrease in the sensitivity to stimulus of the peripheral neuromuscular system, and some changes in the intermediary metabolism are also evident. The serum cholesterol level is high and anemia is quite common. In adults the hypofunction of the thyroid may cause a condition of **myxoedema** which is characterized by a peculiar thickening and puffiness of the skin and subcutaneous tissue, particularly of the face and extremities.

Hyperfunction of the thyroid (hyperthyroidism) : Hyperthyroidism is caused by an excess of the thyroid hormones. This causes an increased metabolic rate, perhaps even double the normal rate, increased sensitivity, sweating, flushing, rapid respiration, palpitation, and increased gastrointestinal activity. The heart is also affected, because the cardiac output is increased.

Over activity of the thyroid or an increase in the size of the gland produces a disease known as **exophthalmic goitre,** characterized by the protrusion of the eyeball. Goiter may be due to an excessive stimulation from the pituitary or more likely, from an indequate supply of iodine in the diet, which causes the gland to overwork. To prevent goiter, the dietary iodine intake should be about 0·1 mg daily.

Thyrocalcitonin : It is secreted by thyroid gland, thefore it is known as thyrocalcitonin. It is a hypocalcemic, hypophosphatemic hormone. It is a polypetide composed of thirty-two amino acids. It acts principally on bone, causing inhibition of bone resorption, therefore, its effects are most striking in young animals. It may be useful in controlling the hypercalcemia resulting from increased bone resorption.

3. **Parathyroid gland :** Parathyroid glands are so named because they are found on the surface of, within, or near the thyroid glands. Generally they are either wholly or partly embedded in the thyroid glands. Their shape, size and number very in different vertebrates. In cyclostomes and fishes so far, it is not certain that whether parathyroid glands are found. In amphibians it appears as a distinct gland, In frog it is a small paired reddish body found near the hyoid cartilage. Parathyroid glands in amphibians are said to control the calcium metabolism. In reptiles it exists in two pairs situated near the neck, located posterior and lateral to the thyroid gland. In birds one or two pairs of parathyroid gland are found on the dorsal of the thyroid gland. In mammals (man) parathyroids are four small glandular bodies, from 2 to 4 mm long and combined weight varies from 0·05 to 0·3 gm. These glands, like the thyroid gland, are derived as epithelial buds, usually from the third and fourth pairs of pharyngeal pouches in the embryo.

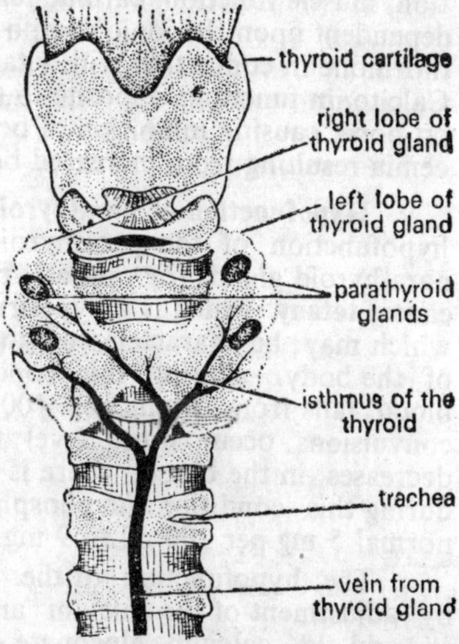

14·8. Human parathyroids.

Function of parathyroid gland : The function of parathyroid glands is to produce a hormone which has been called "**parathormone**" (Collip, 1925). Recently Copp, 1964 has detected one more hormone from the parathyroid extracts and has been given the name **calcitonin** to it.

Parathormone is a large polypeptide of about 9500 molecular weight. It has recently been isolated in the pure state and appears to be a protein of relatively small molecule. It is very much sensitive to changes in pH, and action of acids and bases inactivate the hormone. This indicates a protein structure. Other physical evidence indicates the presence of tyrosine, phenylalanine, and tryptophan.

Calcitonin is also a polypeptide of about molecular weight not less than 3600 and it has opposite effects to those of parathormone. It is composed of thirty-two amino acids.

The release of parathyroid hormones is directly controlled by blood calcium levels. Low concentration of blood calcium causes the secretion of parathyroid hormones.

Functions of parathyroid hormones : The primary function of parathormone is to maintain the metabolism of phosphate and calcium, with particular reference to blood levels and the bone formation. In bone, where calcium and phosphate are mutually involved in the formation of bone substance, parathormone causes movement of calcium and phosphorus from bone to blood. In the kidneys it increases phosphate excretion. While controlling the metabolism of phosphorus and calcium, parathormone plays an important role in homeostasis, for activities such as membrane permeability, nerve function, muscle function, cardiac function and reproduction are strongly dependent upon the ionic composition of the extracellular fluid. Parathormone in conjuction with vitamin D, increases active absorption. Calcitonin functions opposite to the parathormone. It acts principally on bone, causing inhibition of bone resorption. It controls hypercalcemia resulting from increased bone resorption.

Hypofunction of parathyroid gland (hypoparathyroidism) : The hypofunction of the parathyroid glands or the removable of the parathyroid glands (*i.e.*, parathyroidectomy) in both man and animals causes **tetany** which is a hyperirritability of the nervous system and which may be characterized by twitching of the muscles and spasms of the body. During this condition, the calcium ion level in the blood falls from 10 mg per 100 ml (normal) to 7 mg per 100 ml, convulsions occur if the level falls further. As the calcium level decreases in the blood, there is a decrease in the urine. However, during this condition the phosphorus in the blood increases from a normal 5 mg per 100 ml to 9 mg per 100 ml and even higher.

The hypofunction of the parathyroid glands can be checked by adjustment of the calcium and phosphate intake in the diet; that is, add 1% calcium gluconate or lactate to the diet and decrease the phosphate supply.

Hyperfunction of the parathyroid glands (hyperparathyroidism) : Hyperparathyroidism is a condition caused by over activity of the parathyroid glands. Hyperfunction of the parathyroid glands causes extensive decalcification and may lead to bony deformities and fractures. Calcification of soft tissues, especially the kidneys, may occur. If calcium salts precipitate in the kidneys and ureters, kidney stones result and renal insufficiency develops.

Ultimobranchial bodies : In fishes behind the last gill-pouches there occurs a pair of outpushings called **ultimobranchial bodies.** Because these bodies are associated with calcium metabolism in teleosts, they are regarded as the forerunners of, or as homologous with, the parathyroids.

4. Pancreas : In addition to the digestive function of the pancreas, this organ also has certain specialized glandular cells, that seem to be developed as specialized outgrowths of the gut endoderm and differed histologically from the rest of the pancreas tissue. These cells are endocrine in function and these were discovered by LANGERHANS, therefore, they have been known as **islets of** LANGERHANS. It was found that these islets were not connected to the duct system through which the digestive pancreatic juice flows, instead their secretions are freed directly into the blood circulation.

Staining studies showed that the islet cells are of two kinds, the **α-cells** and the **β-cells**. β-cells probably secrete **insulin** hormone and the α-cells seem to be the source of **glucagon** hormone. These hormones play important roles in carbohydrate metabolism. The total islet tissue constitutes 1—3% of the whole pancreatic tissue.

The phylogeny of the islet cells was studied by HOUSSAY, 1959; BARRINGTON, 1962; GORBMAN and BERN, 1962. These cells are found almost in all the vertebrates. In most higher vertebrates these are found in nests—the islets of Langerhans, in the pancreatic tissue and are concerned with carbohydrate metabolism (BARRINGTON, 1962). In the larval lampreys (ammocoetes), distinct follicles of Langerhans are embedded in the submucosa of the anterior intestine instead at the surface of the pancrease as in most vertebrates. Ducts are not present, and the secretions pass directly into the blood. In some teleost fishes and in a few snakes they are grouped in several small but distinct globular mass (principal islets) in the region of the gall bladder. The islets of urodele amphibians are said not to contain α-cells.

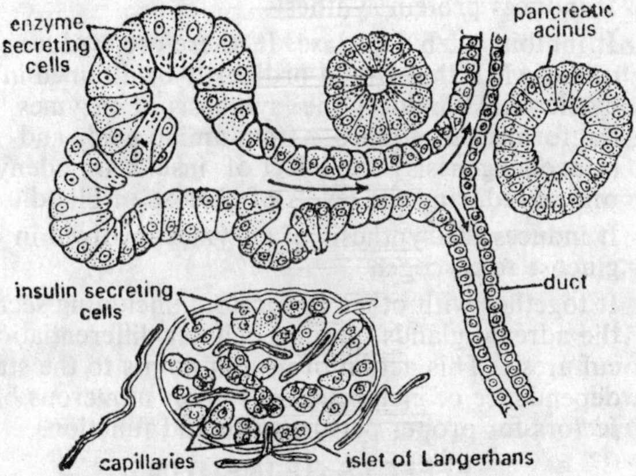

Fig. 14·9. Diagrammatic section of pancreas showing islets of Langerhans.

1. Insulin : Insulin is synthesized in the β-cells. It is a protein of molecular weights 5,734, the basic unit consists of two parallel polypeptide chains of twenty one and thirty amino acids respectively, held together by cross-linkages of two disulfide bonds belonging to the amino acid cystine.

Recent findings showed that insulin is synthesized in the β-cells as **proinsulin**, a molecule with a molecular weight of about 9,000. In the course of producing insulin, proinsulin is hydrolytically cleaved like protrypsin (trypsinogen) into trypsin. Proinsulin is biologically inactive and is present in very small amounts (about 5% of the total insulin content in pancreas). It is, therefore, not to be believed a storage form of insulin. The need of proinsulin is unknown so far.

Functions of insulin : The action of insulin was first demonstrated by BANTING, 1921, and later by BEST, 1959. Two hypothesis have been put forward to explain its action at the cellular level.

1. **Transport hypothesis** : This attributes the major action of insulin to its ability to stimulate the momvents of nutrients (mainly glucose) across cell membranes.

2. **Intracellular enzyme hypothesis :** This attributes the effect of insulin to some as yet unexplained action on the hexose kinase enzyme or the pathways of oxidative phosphorylation.

Although the transport hypothesis has more support at present time, the evidence is still incomplete. Recently J.E. MANERY observed that neither of these hypothesis developed for mammalian tissues is appropriate for some other vertebrates.

The various functions of insulin can be summarized under the following heads :

1. It increases membrane permeability to certain nutrients.
2. It lowers blood glucose levels by stimulating the deposition of glycogen granules.
3. It induces protein synthesis.
4. It inhibits adenyl cyclase. It, therefore, reduces the level of cyclic AMP that would ordinarily be attained in cells. Since cyclic AMP induces the synthesis of enzymes involved in the formation of glucose from amino acids and fatty acids (gluconeogenesis), the effect of insulin on adenyl cyclase is one of reducing the levels of glucose in blood.
5. It induces the synthesis of enzymes involved in converting glucose to glycogen.
6. It together with other endocrines, including secretion from the adrenal glands, permits cellular differentiation in tissue cultures. This action of insulin points to the strong interdependence of metazoan cells upon numerous blood-borne factors for proper performance and function.

EFFECTS OF INSULIN

Effect of insulin on carbohydrate metabolism : Insulin has three basic effects on carbohydrate metabolism—(i) increased rate of carbohydrate (glucose) metabolism (ii) decreased blood glucose concentration and (ii) increased glycogen stores in the tissues. The ability of insulin to increase the rate of glucose metabolism in the tissues is very important to the body. Its complete lack in the body may results in **diabetes mellitus**, a disease characterized by the following symptoms :

1. Hyperglycemia (*i.e.*, high blood sugar level) and glycosuria (*i.e.*, sugar in urine) are common.
2. Diuresis (*i.e.*, increased flow of urine) is common.
3. In liver glycogen levels may be below normal but muscle glycogen in general may be about normal, and heart muscle glycogen level is much above normal.
4. The conversion of carbohydrates into fat is reduced.
5. The formation of glucose from non-carbohydrate sources (mostly protein) is increased to a great extent.

Effect of insulin on blood glucose concentration : In the deficiency of insulin, very little of the glucose absorbed from the gastrointestinal tract is transported into the tissues with the result the blood sugar or glucose concentration is raised from a normal value of 90 mg per 100 ml to as high as 300 to 1200 mg per 100 ml.

On the other hand in the over production of insulin glucose is transported into the tissues so rapidly that its concentration in the blood falls to as low as 20 to 30 mg per 100 ml.

Effect on protein metabolism : The total quantity of proteins stored in the tissues of the body is increased by insulin and greatly decreased by insulin deficiency, thus, insulin promotes protein metabolism.

Effect of insulin on growth : Insulin is essential for growth of animal. In the absence of insulin growth hormone has almost no effect in promoting growth of an animal.

2. Glucagon : The other hormone of the islets is **glucagon** which is secreted by the α-cells. It is secreted in response to hypoglycemia.

Glucagon, like insulin, is a polypeptide of 29 amino acids and has a molecular weight of 3,485.

EFFECTS OF GLUCAGON

1. Glucagon increases blood sugar levels by stimulating the dissolution of glycogen granules and the consequent movement of soluble glucose into the blood.

2. It enhances the uptake of amino acids, the deamination of amino acids within the liver, and the conversion of the carbon skeleton of amino acids and fatty acids into glucose. It is thought that it is a hormone involved in regulation of carbon metabolism.

3. Its essential function is to provide a continuous and immediately available supply of energy fuels (particularly glucose) in accord with bodily needs.

4. It is responsible for lower blood levels of calcium, higher excretory levels of calcium, an increase in heart rate, and an increase in the force of cardiac contraction.

In addition to these, several other physiological actions have also been attributed to glucagon and these are following :

1. Reduction of intestinal motility and gastric secretion.
2. Enhanced excretion of electrolytes by kidneys.

Regulation of insulin and glucagon release : Neither insulin nor glucagon release is under nervous or endocrine control. Rather, the blood glucose level directly stimulates the release of the appropriate hormone. This was first shown by HOUSSAY. Increased levels of glucose in blood stimulate liberation of insulin, while decreased levels of glucose in blood bring about the secretion of glucagon.

5. Adrenal or **suprarenal glands :** The adrenal glands, as implied by the name, are situated near or about the kidneys. In the higher vertebrates these are composed of two distinct parts namely an outer **cortex** and an inner **medulla**. These two parts differ in function and origin, while their association is probably an accident. Though adrenal gland was discovered by EUSTACHIUS but its endocrine function was demonstrated by W. B. CANNON. In the lower category vertebrates, the two parts corresponding to the cortex and medulla occur separately as inter-renal bodies and chromaffin bodies respectively. Chromaffin bodies, because of their affinity for chromate, are termed as chromaffin bodies. Chromaffin bodies are supposed to be homologous with the medulla of higher forms, while the inter-renal bodies are said to be homologous with the cortex.

The adrenals of the mammals (man) are two small yellowish glands, each lying above or near the kidney. They are richly vascularized and they are smaller in the female, while larger in the male. Each adrenal gland weighs about 5 gm. Histologically each gland is composed of two distinct parts : an outer **cortex** and an inner **medulla**. The two parts differ in function and development. The cortex is originated from tissue from which the reproductive glands (gonads) are developed, while the medulla is derived from the embyonic tissue which also gives rise to sympathetic nervous system. We shall discuss the two parts of adrenal gland and their secretions separately.

The two parts of adrenal gland of mammals, though closely associated structurally with each other, are functionally quite distinct, these two parts are found separately in the lower vertebrates. The adrenal cortex of lower vertebrates has its homologue in the inter-

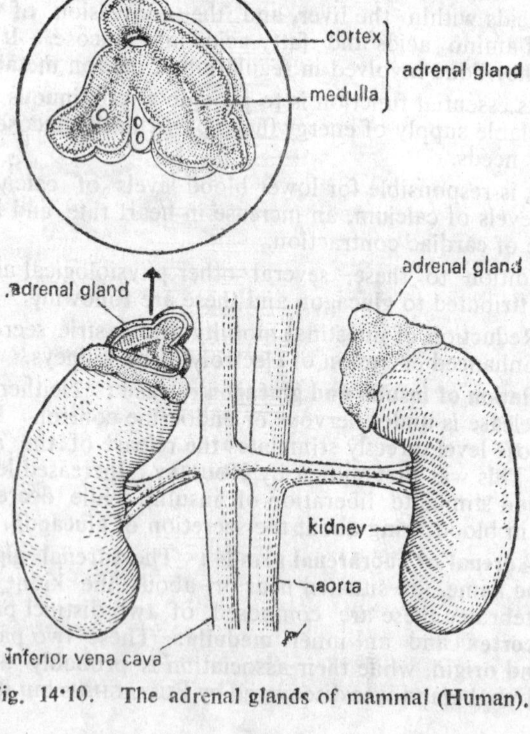

Fig. 14·10. The adrenal glands of mammal (Human).

renal tissue located in more or less discrete clumps surrounding the renal arteries as they enter the kidneys. In the cyclostomes the inter-renal bodies are found as small irregular structures along the blood vessels located above the kidneys. The chromaffin cells are either absent or found as small rounded bodies at the dorsal aorta and its branches. In the hagfish *(Myxine)* the inter-renal bodies are completely lacking. In fishes particularly the dog fishes the inter-renal bodies representing the cortex, lie between the kidneys and the chromaffin bodies or the suprarenals, representing the medulla, consist of small masses on the sympathetic chains. In teleosts, amphibians and reptiles the chromaffin bodies are dispersed widely as islets in or among what are known as inter-renal glands. In amphibian the two elements are intermingled very closely forming a single compound endocrine gland, the adrenal of an orange or yellow colour located on the ventral side of the kidneys. In reptiles the two elements are more intimately connected. In the birds the two elements are intermingled but nevertheless are contained in a single pair of encapsulated glands of yellowish colour found on either side of the post caval vein just anterior to the kidneys and close to the gonads.

Adrenal medulla : The chromaffin cells or bodies which in mammals and some other vertebrates are aggregated to form a central distinct structure of a pair of adrenal glands located just cephalad of the kidneys, are in principle highly specialized nerve endings of the post-ganglionic sympathetic nervous system. Removal of this part from the adrenal gland is not fatal. This part in mammals, birds and reptiles, produces the **catecholamines**, **adrenaline**, (**adrenine** or **epinephrine**) and **noradrenaline** (**noradrenine** or **norepinephrine**). These hormones are secreted when adrenal glands are nervously stimulated and it is thought that separate cells release adrenaline and other cells release noradrenaline (Bard, 1961 ; Gorbman and Bern, 1962). In this regard two types of cells can be distinguished morphologically.

These catechols are amino acid derivatives and probably synthesize in the cell from **tyrosine amino acid** by way of **dihydroxyphenylalanine** and its decarboxylation product **dopamine.** These are rapidly inactivated by the enzyme **O-methyl transferase** and possibly **monoamine oxidase**. Noradrenaline hormone differs from adrenaline in that the methyl group on the nitrogen is lacking. Both of these hormones contain an asymmetric carbon atom, and both naturally occur as D isomers. D (—) adrenaline is about 15 times more potent than the synthetic L (+) isomer ; D (—) noradrenaline is 20 times more potent than its synthetic mirror image.

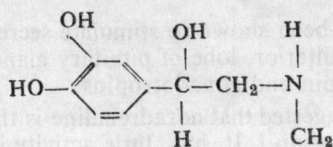

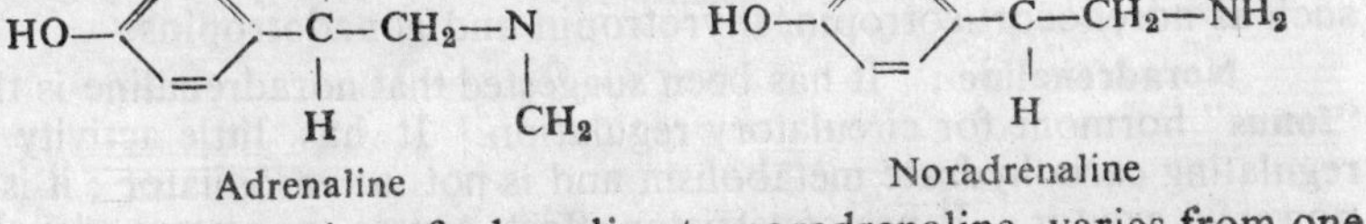

Adrenaline Noradrenaline

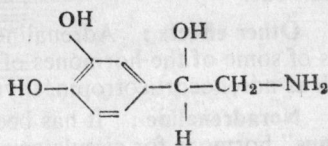

The proportion of adrenaline to noradrenaline varies from one species to another, and even within an individual from one circumstance to another. Normally adrenaline is secreted 5 to 10 times the concentration of noradrenaline in the adrenal medulla but noradrenaline is also secreted by chromaffin tissue of sympathetic nervous system.

Adrenaline : Adrenaline is also known as **emergency hormone** and is secreted in proportional to the stimulus through the central nervous system. Under the conditions of stimulus, such as cold, heat, drugs and emotional excitement, the secretion of adrenaline is greatly increased. In circumstances of great stress, strain, and emotional upheavel, the body requires additional energy in a hurry. Adrenaline accordingly increases the conversion of glycogen to glucose and thus provides quickly available energy. Because it also increases the blood flow, the available glucose may quickly go where it is needed. Adrenaline not only stimulates the formation of glucose from liver glycogen but also stimulates glycolysis in tissues generally.

Another important function of adrenaline is that it dilates the blood vessels in the heart and skeletal muscles, thus decreasing the peripheral resistance and permitting a copious flow of blood to meet the needs of fight or flight (liberated large amounts of catecholamines due to stress or excitation).

EFFECTS OF ADRENALINE

Cardiovascular effects : Adrenaline is a powerful vasoconstrictor ; it affects the cardiac and smooth muscles. It increases both the force and frequency of the contraction of the heart muscles, resulting in rise in blood pressure, rise in pulse rate, and rise in cardiac output. It is used as a heart stimulant in acute emergencies.

Smooth and skeletal muscles : Almost all the smooth muscles of the body as well as the skeletal muscles are affected by adrenaline, some being stimulated to contract while others inhibited.

Effects on metabolism : Adrenaline affects the carbohydrate metabolism by increasing the rate of conversion of glycogen into glucose, which then may be rushed to the muscles to form ATP, which is used to perform work. It, thus, by increasing glucogenolysis in the liver, raises the sugar content of the blood (hyperglycemia).

It also causes increased oxygen consumption and increased heat production. It also accelerates the break down of glycogen.

Bronchodilating effect : Adrenaline shows a strong bronchodilating effect and is used in the treatment of asthma and other allergic conditions.

Other effects : Adrenaline has been shown to stimulate secretions of some of the hormones of the anterior lobe of pituitary gland, such as adrenocorticotropin, thyrotropin and gonadotropins.

Noradrenaline : It has been suggested that noradrenaline is the **"tonus"** hormone for circulatory regulation. It has little activity in regulating carbohydrate metabolism and is not a vasodilator ; it is a powerful excitor. It has constrictor effect, causes a greater rise in blood pressure than adrenaline. It raises both systolic and diastolic pressures. The main function of noradrenaline is the normal control of blood circulation.

Adrenal cortex : This part of the adrenal gland is very essential for the life as its hypofunction or removal causes certain metabolic and other disorders and finally the death of individual. The adrenal cortex of mammalian adrenal gland is more or less distinctly divided into an outer layer, the **zona glomerulosa**, a middle layer, **zona fasciculata** and an inner layer, **zona reticularis**. No comparable differentiation has been described in other vertebrates. The entire cortical part is made of steroidgenic tissue.

The zona glomerulosa is believed to be the site of biosynthesis of **mineralocorticoids**, the principal ones of which are **aldosterone** and **deoxycorticosterone**. These are responsible for electrolyte and water balance. The zona glomerulosa is stimulated by Na^+ deprivation or K^+ administration and is depressed by K^+ deprivation or administration of deoxycorticosterone. The zona fasciculata is associated with the production of **glucocorticoids**, the principal ones of which are **cortisol, cortisone** and **corticosterone**, these are steroïds having an oxygen atom at position 11. This zone is stimulated by the adrenocorticotropic hormone (ACTH) of the anterior lobe of the pituitary gland. The role of the zona reticularis is not clear so far but it is believed to be associated with the production of sex-hormones which are secreted by the gonads. The physiological importance of sex-hormones liberated by the adrenal glands has yet not been established. This zone is also stimulated by the ACTH.

Mineralocorticoids : The most important mineralocorticoids are the **aldosterone** and **deoxycorticosterone**. These hormones are concerned with electrolyte and water balance. They react with the cells of the colon, kidneys and other cells including the red blood cells and the epithelial cells of the salivary glands to bring about (1) reabsorption of both sodium and water, and (2) the excretion of potassium. The hypofunction of cortex or removal results in an imbalance of sodium and potassium that inturn leads to nervous disorders, convulsions, and death (CONN'S **disease**). An overproduction of these hormones also leads to neuromuscular disorders (*e.g.*, hypertension).

Glucocorticoids : The important glucocorticoids are **cortisol, cortisone** and **corticosterone**. Glucocorticoids are initially concerned with the synthesis of enzymes that are involved with **gluconeogenesis**, the synthesis of glucose from amino acids and fatty acids. They are responsible for increased synthesis of glycogen in liver, increased protein break down, increased leucocyte count in the blood, and decreased inflammatory reaction to tissue damage. Hyposecretion of glucocorticoids causes a disease, ADDISON'S disease characterized by shock, reduced blood pressure, high levels of urea in blood and an elevated body temperature. Inadequate levels of glucocorticoids also prevent an individual from responding normally to stress.

An overproduction of glucocorticoids, on the otherhand, causes a disease called CUSHING'S **disease** characterized by (1) hyperglycemia,

(2) marked deposition of glycogen in the liver, (3) increased protein catabolism (4) muscle weakness (5) disorders of the **skin** and bones (6) hypertension (7) diabets mellitus and (8) possible retention of body water and sodium chloride.

Overproduction of these glucocorticoids also causes precocious development of the secondary sexual characters in children. Boys, for example, may appear as miniature Herculus, and girls may exhibit virilism (development of hairs on face and chest, etc.). Adult women also display signs of virilism.

Juxtaglomerular cells of the kidney: Recent research indicates that the juxtaglomerular cells of the kidney produce two hormones, **renin** and **erythropoeitin**. Renin stimulates the release of substances which cause a rise in blood pressure. Erythropoeitin stimulates the release of red blood cells (erythrocytes) from the bone marrow. Anoxia, a deficiency of oxygen in the blood, causes kidneys to produce this hormone.

6. Sex glands : The sex glands are of two types, the testes and the ovaries. The testes are the sex glands of the male, while the ovaries are the sex glands of the female. The shape, size and number of these male and female sex glands may vary in different animals. The characters which serve to distinguish between the male and female individuals are known as secondary sexual characters. In man they include the growth of hairs on the face, chest, axillae and pubis and the masculine voice. In women they include the development of the breasts, widening of the pelvis and sweet voice, no development of hairs on face. If the sex glands do not function properly or they are removed the secondary sexual characters of either sex are supressed. Generally the females, on the removal of sex glands, develop the male charactcters.

Testicular hormones or **male sex-hormones :** Testes in the presence of follicle stimulating hormone (FSH) and interstitial cells stimulating hormone (ISCH) produce two distinct hormones, the **testosterone** and the **androsterone**, collectively called as **androgens.** Actually, some estrogens in addition to testosterone and androsterone are also produced in the testes, due to stimulation from the pituitary. The exact site of estrogen production in testes is not known so far.

1. Testostèrone : Is is a most important and potent androgen secreted by the interstitial cells of Leyding which lie in the interstices between the seminiferous tubles and also from **Sertoli cells** as reported recently. Interstitial cells are not present in the testes of a child but they are present in the testes of a new born infant and also in the testes of the adult male any time after puberty. Its production dwindles rapidly beyond the age of 40 and becomes almost zero by the age of 80.

2. Androsterorone : It is also a fairly strong androgen, second in activity to testosterone, secreted in the testes in the presence of FSH and ISCH.

Action of the androgens (testosterone and androsterone) : The androgens, in general, are not required for the stimulus of spermatogensis but they are responsible for the secondary sexual characteristics. The androgens affect the following :

1. **Genital organs :** The androgens are responsible for the normal growth of the penis, scrotum, prostrate gland, and the other portions of the genital tract.

2. **Secondary sexual characters :** They are responsible for the secondary sexual characters of the male, *i.e.*, the growth and development of hairs at face and chest and the musculine voice.

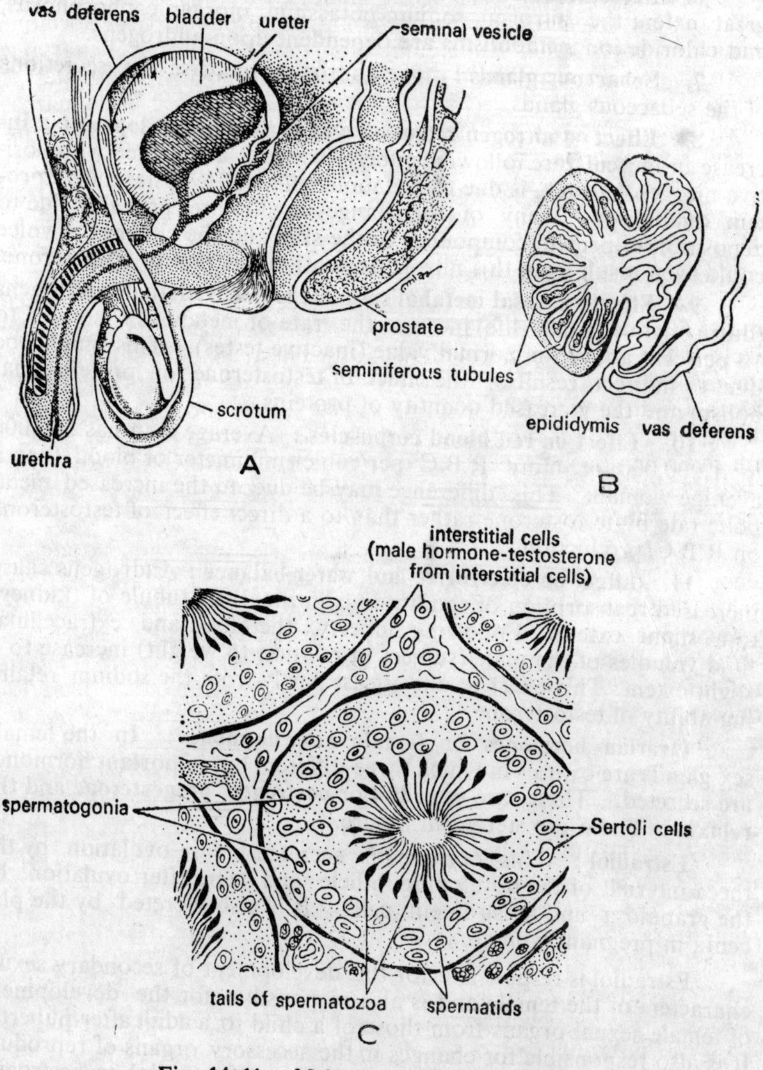

Fig. 14·11. Male sex glands of mammal.

3. Muscles : The androgens are presumably responsible for the muscular strength of males as compared to that of females.

4. Skeleton : The androgens affect the skeletal growth to a reasonable extent.

5. Colour and texture of skin : Androgens are responsible for the normal skin, that is firm, ruddy, and of a darker colour (flesh colour). In the absence of androgens skin becomes furrowed, soft and pale in colour.

6. Metabolism : Recent findings have established that to a great extent the nitrogen, sodium potassium, inorganic phosphorus, and chloride ion metabolisms are dependent upon androgens.

7. Sebaceous glands : The androgens increase the secretions of the sebaceous glands.

8. Effect on nitrogen retention and muscular development : Increase in musculature following puberty which is a ssociated with positive nirogen balance, is due to metabolic effect of testosterone on protein anabolism. Many of the changes in skin are probably due to deposition of protein compounds in the skin and the changes in voice could even result from this nitrogen retention function of testosterone.

9. Effect on basal metabolism : It is believed that androgens (during active sexual life) increase the rate of metabolism some 5 to 10 per cent above the normal value (inactive testes). This might be due to indirect result of the effect of testosterone on protein anabolism and the increased quantity of proteins.

10. Effect on red blood corpuscles : Average man has 500,000 to 1,000,000 or more R.B.C per cubic millimeter of blood than a average woman. This difference may be due to the increased metabolic rate by testosterone rather than to a direct effect of testosterone on R.B.C production.

11. Effect on electrolyte and water balance : Androgens cause increased reabsorption of sodium ions in the distal tubule of kidneys to a slight extent. Following puberty, the blood and extracellular fluid volumes of the male (subject in relation to weight) increase to a slight extent. This effect results atleast partly from the sodium retaining ability of testosterone.

Ovarian hormones or **female sex hormones:** In the female sex glands are ovaries in which three different but important hormones are secreted. These are **estradiol** or **oestrogen, progesterone** and the **relaxin.** These are steroid in nature.

Estradiol : This hormone is secreted before ovulation by the theca interna of the developing follicle and later, after ovulation, by the granulosa and theca-lutein cells. It is also secreted by the placenta in pregnant woman.

Estradiol is responsible for the development of secondary sexual characters of the female and is also responsible for the development of female sexual organs from those of a child to a adult after puberty. It is also responsible for changes in the accessory organs of reproduction during oestrus cycle. Over production of estradiol or oestrogen leads to the disturbance of menstrual cycle and sometimes causes cancer formation. Lack of oestrogen leads to the failure of menstrual cycle and ill developed genital tracts.

Progesterone : This hormone is essential to promote secretory changes in the uterine endometrium, thus preparing the uterus for implantation of the fertilized ovum. It promotes the development of the mammary glands to full maturity during pregnancy and inhibits contraction of the uterus. It causes the breasts to swell. It also regulates the menstrual cycle. In its deficiency the pregnancy does not stand.

Relaxin : This hormone is essential relaxing the pelvic ligament during the pregnancy.

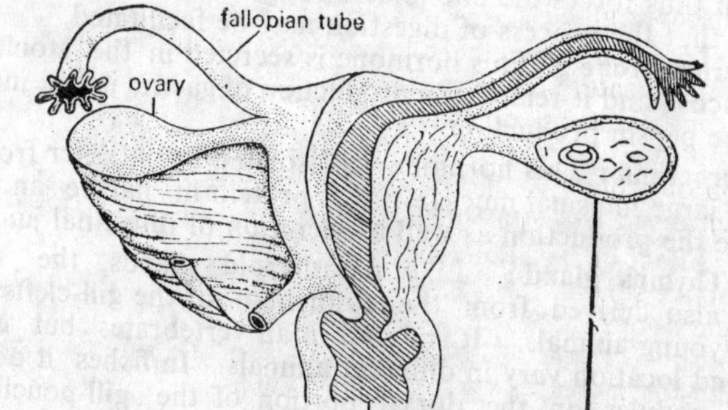

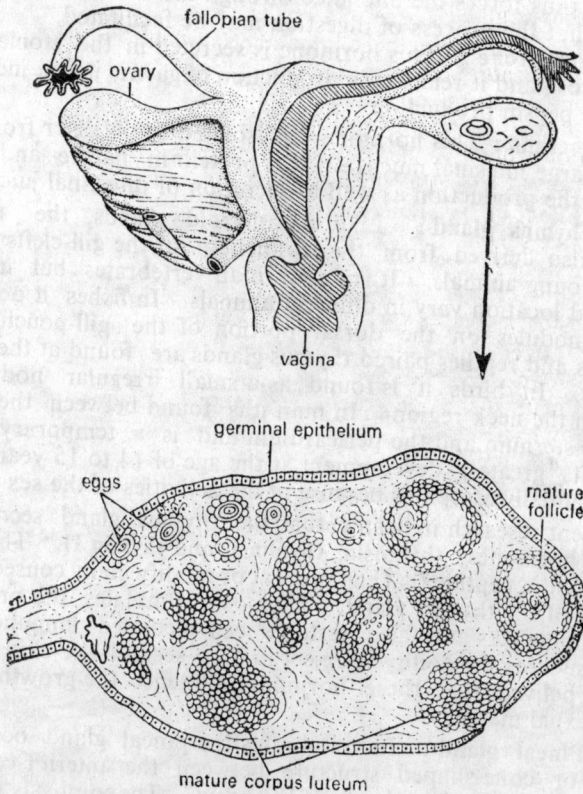

Fig. 14·12. Female sex glands of mammal.

7. Gastrointestinal mucosa : In the gastrointestinal mucosa are found certain cells which are glandular in nature. These cells produce certain hormones which are involved in the digestive processes, and are usually concerned with the flow of the various digestive juices. These hormones are—gastrin, secretin, pancreozymin, cholecystokinin, enterogastrone and enterocrinin.

Gastrin : This hormone is secreted in the pyloric mucosa. It stimulates secretion of the gastric digestive juice and also causes contractions of the stomach wall.

Secretin : This hormone is secreted by the duodenal mucosa, it stimulates the flow of pancreatic juice of pancreas.

Pancreozymin : It is another hormone secreted by the duodenal mucosa and which stimulates the flow of enzymes from the pancreas. The hormones, secretin and pancreozymin work together : one causes the pancreatic juice to be formed, and the other stimulates its flow into the duodenal canal to aid in digestion of food.

Cholecystokinin : It is secreted in the duodenum when food enters the duodenum. It is responsible for the contraction of gall bladder ; it thus forces the bile juice through the ducts into the duodenum so that the process of digestion may be facilitated.

Enterogastrone : This hormone is secreted in the stomach by gastric mucosa and it retards the production of gastric juice, indirectly inhibits the pepsin production.

Enterocrinin : This hormone was isolated by NASSET from both small and large intesinal mucosa. It is protein in nature and seems to regulate the production as well as secretion of intestinal juice.

8. **Thymus gland :** Like parathyroid glands, the thymus gland is also derived from the epithelium of the gill-clefts and is largest in young animals. It is found in all vertebrates but its size, number and location vary in different animals. In fishes it occurs as a pair of nodules on the dorsal portion of the gill-pouches. In amphibians and reptiles paired thymus glands are found at the angels of the jaws. In birds it is found as a small irregular nodule-like structure in the neck region. In man it is found between the upper part of the sternum and the pericardium and is a temporary gland reaching its greatest development at the age of 14 to 15 years, after which it gradually atrophies because of the activities of the sex glands.

Current research indicates that the thymus gland secretes at least three hormones, **thymosin**, **thymin I**, and **thymin II**. Thymosin is a small acidic peptide that can prevent or modify many consequences of removal of the thymus gland. Thymosin stimulates the proliferation of lymphocytes and also restores cell mediated immunological functions such as the ability to reject first or second set skin grafts. It is also belived that these hormoncs influence the growth on the onset of sexual maturity.

9. **Pineal gland :** In vertebrates the pineal gland occurs as stalk-like or cone-shaped structure between the anterior corpora quadrigemina on the dorsal side of the brain. The epiphysis of some reptiles, birds and mammals has been interpreted as a glandular organ, homologous to that of pineal gland.

There is a clinical evidence that the pineal gland has connection with sexual development. It has been found that it secretes a substance called **melatonin**, closely related to tryptophan which has a very strong concentrating action on the chromatophores of frog (but not of urodeles) and also controls matabolism (GORBMAN and BERN 1962).

There are also evidences that the pineal is connected with colour-change in lampreys and salmon.

INVERTEBRATE NEUROENDOCRINE SYSTEMS

Neurosecretions are the main source of hormones in the invertebrates (HAGADORN, 1967). Neurosecretory cells have been found in all metazoan invertebrates including some coelenterates SCHARRER and SCHARRER, 1963 ; LENTZ and BARRNETT, 1965 ; BERN, GORBMAN and HANGADORN, 1965 ; MARTINI and GANONG, 1967 : studied the comparative aspects of neuroendocrine system of invertebrates.

Although neurosecretions are main source of hormones, some invertebrate endocrine structures, derived from non-nervous tissue, are known. For example—the internephridial organs of the *Physcosoma* (SCHARRER, 1955), the salivary glands of the cephalopods (JENKIN, 1962) or the ecdysial glands of insects. Other structures of suspected endocrine function are found in arthropods and these are the Y-organ of crustaceans, corpora allata and prothoracic glands of the insects, these structures are originated from the non-nervous ectoderm. In Crustacea there are some other glands which arise from the mesoderm and these are androgenic and gonadal glands. In the vertebrates the endocrine glands of non-nervous tissue are derived from the endoderm (thyroids. parathyroids, islets of Langerhans, gastro-intestinal mucosa) but no invertebrate endocrine structures have yet been shown to originate from this germ layer (JENKIN, 1962). Thus, it appears that, during phylogeny, endocrine function was primarily associated with nerve cells and the tissues from which they are derived. In the most primitive condition, these may have been used as they were produced, but later their secretions came to be transported down the axon to special receiving or storage structures in the axon terminals, where they might accumulate in order to meet the regular and urgent demands more effectively. Subsequently, with increasing complexity of physiological organisation, endocrine glands developed in several other places.

In lower invertebrates neurosecretions appear to regulate growth, regeneration, metamorphosis and reproductive activities. Little is known about their chemical nature and about role of these substances in, so far as, mechanisms or feed back pathways are concerned.

Neuroendocrine systems in Annelida : In annelids neurosecretory cells are predominent in all three classes. In *Nereis* and *Nephthys*, the brain produces a **juvenile hormone** that influences the reproductive system. In some species there is transformation (epitoky) from an asexual state (atokous) of the worm into a sexually active reproductive form *(Heteronereis)*. This change is also due to this juvenile hormone (BOBIN and DURCHON, 1952 ; CLARKE, (1959). Another hormone has recently been reported which is secreted in the posterior part of the supraoesophageal ganglion and is needed for ripening of eggs and spawning in *Arenicola*. AVEL, 1947 ; SCHARRER and SCHARRER, 1954 ; HUBL, 1959 ; concluded from their studies that neurosecretory cells of the sub-and supra-oesophageal ganglia are involved in the development of somatic sex characteristics, regeneration and physiological colour changes.

It has been reported that in sub-and supra-oesophageal ganglia of some annelids (leeches) are found certain specialized neurosecretory cells which according to PROSSER and BROWN 1962 ; and GORBMAN and BERN 1962, secrcte both **adrenaline** and **noradrenaline** hormones.

Sipunculid worms also have neurosecretory systems and *Sipunculus* possesses a neurohaemal organ with finger organs, or papilliform processes of the cerebral ganglion, extending into the haemocoel. The secretory axons invading the finger organs first loop into a sensory organ that extends into a ciliated canal opening to the outside ; but the significance of this structural arrangement is not known (AKESSON, 1961).

Endocrine systems of Crustacea : In crustaceans neurosecretory cells are found in various parts of the brain ; in all ganglia of the ventral nerve cord ; and in the optic ganglia. Atleast a dozen

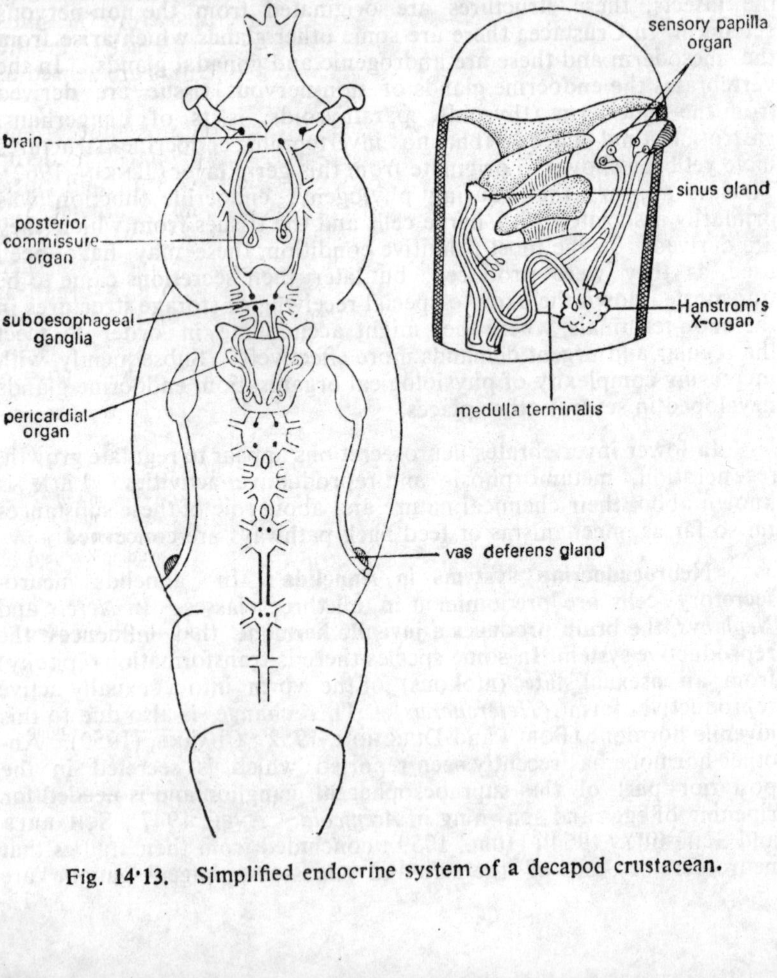

Fig. 14·13. Simplified endocrine system of a decapod crustacean.

diflerent cell types are found with neurosecretory activity. The given diagram (Fig. 14·12) illustrates all the endocrine tissues of the crustaceans.

The **sinus gland** is found in nearly all malacostracans. It is not a gland but merely a neurohaemal organ—a storage and release site for hormones produced in neurosecretory cells. It is first clearly described by HANSTROM and is made up of axon termini of neurosecretory cells with cell bodies in the brain or in the ganglia of the eye-stalk. It also contains axon endings of neurosecretory cells from several sources including the medulla terminalis ganglionic X-organ, optic ganglia, brain, and connective and thoracic ganglia. In its simplest form the sinus gland is a thickened disc separated from a blood sinus by a thin membrane. Upon appropriate stimulation hormones are released into the blood. Experimental studies by B. HANSTROM in Sweden and by F. A. BROWN and others in United States showed that the sinus gland contains **chromatophorotropic hormones** and **molt-inhibiting hormone** as well but recently PASSANO, 1960, observed that the moult-inhibiting hormone arises in neurosecretory cells in the eye-stalk ganglia and is merely liberated from the sinus gland. The hormones of the sinus gland are especially active in the regulation of colour changes although other functions are served as well.

In many crustaceans the eye stalk also contains another end organ, the **X-organ** or the **organ of** BELLONCI. It is primarily made up of axon termini from neurosecretory cells in the brain and optic ganglia. The X-organs are of two types ; **ganglionic X-organs** and **sensory pore** or **pars distalis X-organs** (CARLISLE and PASSANO, 1953). In some species the two types are combined to form a complex. When separate from the medulla terminalis, the sensory pore X-organ is a stucture situated in a blood sinus near the sensory pore and it con-

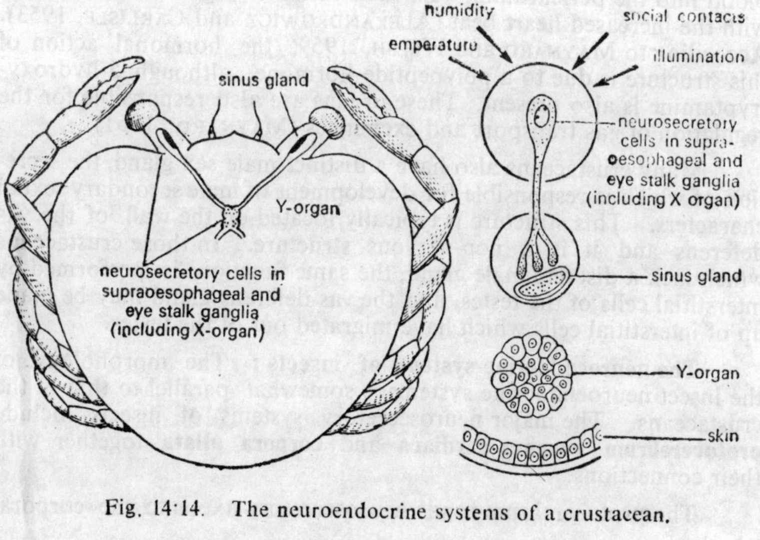

Fig. 14·14. The neuroendocrine systems of a crustacean.

sists of bipolar sensory neurons, elongate cells of epithelial appearance and lamelated concretions (the onion bodies), which are axon endings of neurosecretory cells from the medulla terminalis ganglionic X-organ and from the brain. These axons form the X-organ connective. Ganglionic X-organs are composed of cell bodies of neurosecretory cells.

The X-organ-sinus gland complex is responsible for moulting, pigment changes, distal retinal pigment movements. It produces hormones which inhibit the ovary (all crustaceans) and some which inhibit the testis in crabs. The hyposecretion of these hormones leads to various metabolic changes such as an increased oxygen consumption, changes in calcium, and water levels during moulting and alterations of blood sugar levels.

Y-organs are another neurosecretory complex which produce a hormone that influences moulting. It is under the regulation of the X-organ complex. An eyestalk hormone appears to inhibit production of Y-organ hormone.

In addition to these endocrine structures, many of the higher crustaceans have **postcommissure organs**, these are found just behind the oesophagus as paired extensions of the epineurium of the post-oesophageal commissure and serving as centres for storage and release of secretions arising in the posterior portion of the brain (tritocerebrum). The secretions of these structures affect the colour changes (chromatophorotropic hormone).

Finally, in addition to these neurosecretory systems in the head and thorax, the crustaceans also possess **pericardial organs** consisting neurosecretory cells and axon terminals of neurosecretory cells originating in various ventral ganglia. These structures are found in wall of the pericardium in front of the gill veins that carry oxygenated blood into the pericardium. Extracts of these organs are concerned with the increased heart beat (ALEXANDROWICZ and CARLISLE, 1953). According to MAYNARD and WELSH, 1959, the hormonal action of this structure is due to a polypeptide hormone, although 5-hydroxytryptamine is also present. These organs are also responsible for the regulation of gas transport and exchanges (MAYNARD, 1961).

Many crustaceans also have a distinct male sex gland, the secretion of which is responsible for development of male secondary sexual characters. This structure is typically located on the wall of the vas deferens and it is a non-nervous structure. In those crustaceans which lack a discrete male gland, the same function is performed by interstitial cells of the testes, and the vas deferens gland may be made up of interstitial cells which have migrated out of the testes.

The neuroendocrine systems of insects : The morphology of the insect neuroendocrine systems is somewhat parallel to that of the crustaceans. The major neurosecretory systems of insects include **protocerebrum**, **corpora cardiaca** and **corpora allata** together with their connections.

The protocerebrum sends neurosecretory axons to the corpora-cardiaca that are located just posterior to the brain. The corpora-cardiaça are neurohaemal organs, somewhat comparable to the neurohypophysis of vertebrates except that they also contain neurosecretory cells. Axons from the corpora cardiaca and from the supraoesophageal ganglia are sent to the corpora allata which are also paired structures in most insects. Corpora allata appear to have intrinsic non-nervous secretory cells as well as termini and hence are true endocrine glands of non-nervous tissue origin.

The corpora cardiaca are responsible for the control of the heart, they secrete an **orthodiphenol** which activates the heart indirectly by causing pericardial cells so release a heart accelerating

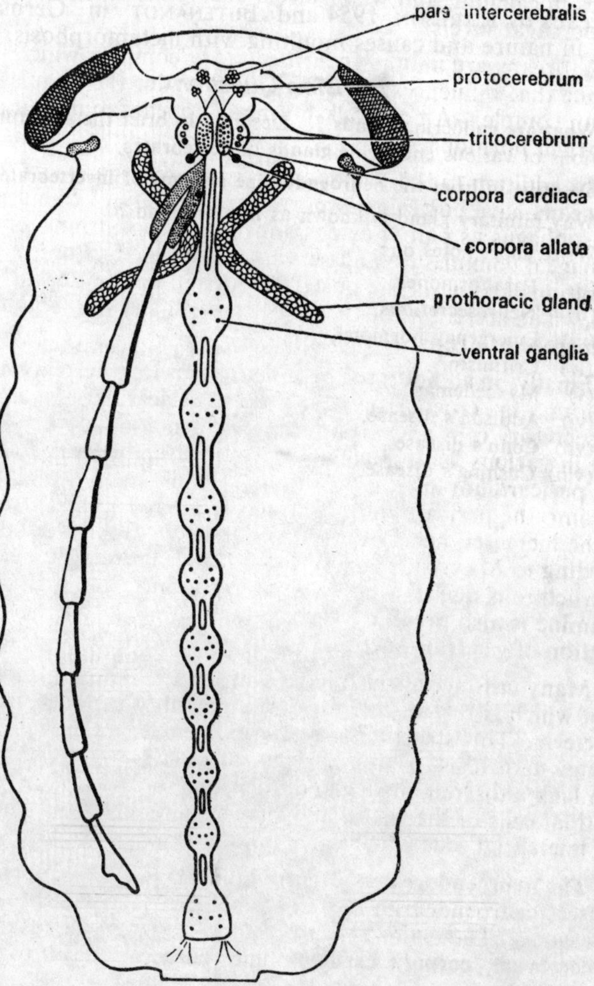

Fig. 14·15. The neuroendocrine systems of an insect.

substance (DAVEY, 1961 ; CAMERON, 1963). Corpora allata, on the other hand form a fat soluble **juvenile hormone** which has not been identified chemically, but which has specific developmental actions in retarding the apperance of adult characteristics. This hormone with ecdysone (secretion of ecdysially gland) also controls the moulting.

The principal endocrine organ of insects, as distinct from neuro-secretory structures, is the **prothoracic** or **thoracic gland**, variously known in arthropods as the **moult gland** and in crustaceans as the Y-organ or ventral gland. It is also known as ecdysial gland. It is a gland of non-nervous origin and is supplied with nerves originating from neurosecretory cells of the brain. It exists in a pair. This gland secretes a hormone called **ecdyson** first isolated from insects and crustaceans by KARLSON. 1954 and BUTENANDT in Germany. It is steroid in nature and causes moulting with metamorphosis.

Revision Questions

1. What are endocrine glands ? Describe in brief the structure and functions of various endocrine glands of vertebrates.
2. Describe in brief the neuroendocrine systems of invertebrates.
3. Why pituitary gland is known as master gland ?
4. Write short notes on :
   (i) Parahormones,
   (ii) Neurosecretions,
   (iii) Emergency hormone,
   (iv) Critinism,
   (v) Myxoedema,
   (vi) Addison's disease,
   (vii) Conn's disease
   (viii) Cushing's disease.

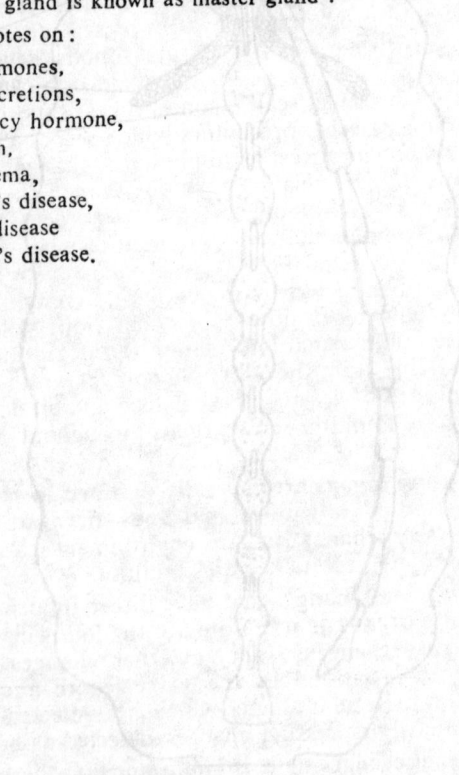

CHAPTER 15

Sense Organs

The **sense organs** or the **receptors** are structures specialized to perceive conditions outside and to some extent inside the body, and these provide the information to the central nervous system upon which action may be taken. In some cases the stimuli act on special cells which may be grouped into organs, while in others the ends of nerve fibres appear to be sensitive. The sense organs are in no way responsible for the sensations because the power of sensation is restricted only in the central nervous system, these serve merely as avenues of approach to central nervous system.

Sense organs are of common occurrence from invertebrates to vertebrates. In protozoans they are represented by pigment spots which are sensitive to light. In arthropods **sensilla** represent the simple types of sense organs, they contain only a few nerve cells and occur in hairs or setae, scales, cones, tubes, slits, or the like. Some sensilla serve as mechanoreceptors which respond to tactile stimuli, to sound, to other vibrations or to pressure changes. Another type of sense organs that are relatively simple, though they are complex in higher arthropods and molluscs, are the **statocysts** of lower invertebrates. The simpler statocysts are multicellular structures that are innervated by the dendritic ends of bipolar sensory neurons. They are located in various regions of the body and contain a mineral concretion. As an organism moves, the concretions follow the course of the movement and signal information to integration centres through the nerve endings that they come in contact with. In vertebrates the sense organs are of very complex nature and all degrees of structural complexity are found in sense organs throughout the vertebrates and invertebrates.

The sense organs are generally destined to receive only one kind of stimulus and not any other and thus they are adequate for the stimulus. Any change in the environment capable of causing an active response is referred to as stimulus.

The various changes that have direct influence on the variously located sense organs or receptors are the following :

External changes : The external changes of biological importance are variations in the wave length and intensity of light, variations in molecular shapes which are detected as smells or tastes, and mechanical or pressure changes detected as hearing or touch.

Internal changes : It is important that the body monitors the changes that are taking place within it so that homeostatic adjustment can be made. Chemical and osmotic changes, temperature fluctuations, mechanical changes and gas tensions are some of the important pieces of information that are fed into the central nervous system by the internal sense organs.

Functioning of the sense organs : All sense organs cause the generation of impulses either from themselves or from the cells immediately in contact with them, *e.g.*, the rods and cones of the eye. It seems that the particular stimulus that the receptor is specialized to detect causes a depolarisation of the cell membrane and the initiation of the action potential.

In many sense cells, *e,g.*, touch receptors, prolonged stimulation leads to adaptation so that impulses are no longer generated (clearly in pain receptors this does not occur). The code of the sense organs or receptors, and of the nervous system as a whole, is a frequency rather than an amplitude code. Intense stimulation thus causes the receptor to produce impulses at high frequency but not to produce higher amplitude impulses.

CLASSIFICATION OF SENSE ORGANS

The sense organs or receptors found in the animals may either be classified in terms of their position or function in the body or else by their sensitivity to various stimuli.

1. The classification of sense organs according to their position: According to their position the sense organs or receptors fall under two heads—exteroceptors and interoceptors.

Exteroceptors : These are receptors found outside the body, in the skin and are responsive to stimuli such as light, temperature and pain, etc., which are impinging directly upon the body from the external environment.

Interoceptors : These are receptors located within the body and are responsive to internal changes including pH, osmotic pressure, oxygen tension and pain.

2. The classification of sense organs according to their functions : With regard to function the sense organs or receptors may be proprioceptors, noiceptors, teleceptors and labyrinthine receptors.

Proprioceptors : These are receptors found in muscles, tendons and joints and sense the movements and positions of muscles and appendages relative to each other and also the spatial position of the body as a whole. These occur in both vertebrates (*e.g.*, muscle spindles and Golgi receptors) and invertebrates (*e.g.*, statocysts in coelenterates and higher phyla, chordotonal organs in antennal and leg joints in insects, and campaniform sensilla in halters of insect flight apparatus and in spiders).

Noiceptors : These are deep pain sensors, responding to injury or any damaging stimulating agent or noxious stimuli.

Teleceptors : These are sense organs which are responsive to changes in the more remote parts of the environment and include such organs as the eyes and ears.

Labyrinthine receptors : These sense organs are sometimes included under the proprioceptors and are very much responsive to the position of the body in space and include statocyst or labyrinth receptors.

This classification of sense organs was put forward by SHERRINGTON in the year of 1906.

3. The classification of sense organs according to the type of energy or **stimulus sensed :** This classification of sense organs was recognized in the early 19th century by JOHANNES MULLER. This classification is based on the stimulating agent. According to this classification the receptors fall under following categories with respect to their sensitivity to various stimuli—mechanoreceptors, thermoreceptors, chemoreceptors and photoreceptors.

Mechanoreceptors : Such type of receptors are responsive to tactile stimuli, to sound, to other vibrations or to pressure changes. The sense organs which are concerned with the pressures of the body fluid or other pressures of the body are referred to as **pressoreceptors** or **basoreceptors.**

Thermoreceptors : These respond to thermal or temperature changes.

Chemoreceptors : These receptors respond to changes in concentration of specific chemical substances or classes of substances in the internal and external environment, for example the cells for taste and smell.

Photoreceptors : These are sense organs responded to changes in intensity or wavelength of light.

In addition to classifying the sense organs or receptors on the basis of location, function and the type of stimuli with which they are attuned, the sense organs can also be classified depending upon the structural organisation and in pattern of nerve impulses produced by the sense organs.

The classification of sense organs according to structural organisation : With regard to structural organisation the receptors may be simple or more complex. The simplest receptors consist of single neurons, usually multipolar, with one axon, called afferent, extending to a junction or synapse with another neuron or neurons, or in a few cases with a muscle cell, at some distance. Such receptors are called **primary sense cells**. The second order of complexity finds the primary sense cells surrounded by or associated with other cells or structures constituting some sort of auxiliary apparatus. In some cases it is possible to infer from the nature of such auxiliary apparatus a possible function, and in some instances the function has been demonstrathe experimentally. The lens of the eye is a familiar example which is known to function in increasing the light-gathering power of the eye and in forming a definite image on the retina, where are located ted primary sense cells. The third order of complexity finds a number of primary sense cells and possibly auxiliary structures as well, associated to form a sense organ. Finally as in the vertebrate eye or ear, the sense organs may have, in addition to primary sense cells and auxiliary structures, **secondary sense cells**—neurons with which the axons of the primary sense cells synapse. In the retina of the vertebrate eye, for example, there are very complex synaptic connections between the primary sense cells (rods or cones) and several layers of secondary sense cells. The axons of the optic nerve arise in these secondary cells rather than the primary sense cells.

The classification of sense organs depending upon the pattern of nerve impulses produced by the sense organs : Such a basis for classifying the sense organs exists but to understand this classification one must know more about the action of receptors in general.

The rate and the extent of adaptation are also important characteristics of receptors and are used as a basis for classification of the receptors. With regards to rate and extent of adaptation the receptors are of two types—(1) **tonic receptors** and (2) **phasic receptors**. In tonic receptors the adaptation is slow and there is a continuing discharge as long as stimulation continues, while in phasic receptors the adaptation is rapid and ultimately complete, with no impulses discharged even though the stimulus continues to act. Tonic receptors include the stretch receptors in the limb of vertebrates and cray fish, the pressure receptors of carotid sinus and statocyst receptors, while the phasic receptors include the Pacinian corpuscle and chemoreceptors.

MECHANORECEPTORS

Mechanoreceptors are responsive to mecchanical forces such as stretch, compression, or applied torque. These forces are produced by factors including movements of body parts relative to one another, gravitational fields changes resulting from alterations of the animal's orientations in space, stretches resulting from the contraction of muscles, or vibrations transmitted through either the internal or external medium.

Primary sense cells : These are very simple type of mechanoreceptors and occur frequently in the vertebrates and invertebrates as simple unicellular structures with only limited auxiliary apparatus. These receptors are responded directly to mechanical forces.

Pressure, pain and touch receptors : Pacinian corpuscles are said to be the pressure receptors (LOEWENSTEIN, 1961). They are found in the deeper layers of the skin, in the connective tissue around the tendons, muscles and joint and in the serous membranes and mesenteries of the viscera. They are very large in size, reaching almost 1 mm in length and 0·6 mm in diameter. These are primary sense cells in which the terminal portion of the nerve fibre that is nerve end which is non-myelinated, is surrounded by a relatively thin granular mass-and covered by concentric layers of connective tissue like the many coats of an onion. The concentric layers or laminae are separated by fiuid which helps in transmitting the superficial pressure changes to the delicate nerve ending within. The concentric layers or laminae constitute the auxiliary apparatus and do not take part in the generation of electric potentials. When the nerve endings of these receptors, either inside or outside, are subjected to increased pressure, the membrane potential in the region of the nerve endings is decreased with the result action potential in the form of nerve impulse is developed which is according to local circuit theory is transmitted down to the axon and ultimatety reaches the central nervous system which acts accordingly.

Pain is a sensory experience initiated by injurious or threatening stimuli. There are both painful sensations and reflex actions in response to painful stimuli. Intensive stimulation of almost any sensory neuron appears capable of producing a painful sensation and only recently has it been shown that specialized receptors for pain exist.

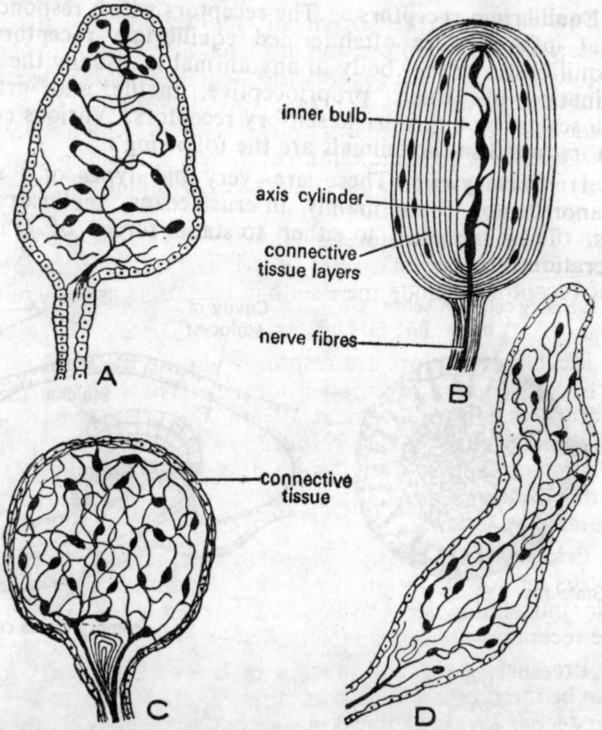

Fig. 15·1. A—Meissner's corpuscle from dermal papilla ; B—Pacinian corpuscle; C—End bulb of Krause; D—Meissner's corpuscle.

Generally the sensation of pain is evoked by either of two sets of nerve fibres. Some C fibres when sufficiently stimulated result in burning pain ; small Δ fibres function in signalling pricking pain. The latter is usually a short term effect, while the former may continue for long periods of time, as long as the stimulus remains.

Touch receptors of various kinds are located in the dermis or in the subepidermal connective tissue, they have a basic structure of nerve endings or tactile cells enclosed in layers of connective tissue. Tactile receptors are found all over the skin in the animals and include wide variety of morphological types. Many have the forms of bulbs (**Krause end bulbs**) or corpuscles (**Meissner's corpuscles**). Meissner's corpuscles are generally found in the outer layer of the skin.

Touch receptors not only convey information a out the presence of tactile stimuli but they also possess discriminatory abilities for both the intensity of the stimulus and for the spatial direction or arrangement of the stimulus. The greatest ability to discriminate between intensities of stimulation or to determine the extent of spatial separation of separate stimuli occurs in areas of greatest concentration of sensory receptors.

Equilibrium receptors : The receptors which respond to gravitational influences are often termed **equilibrium receptors**, though, the equilibrium of the body of any animal is usually the result of a combination of visual, proprioceptive, tactile, and gravitational stimuli sensed by appropriate sensory receptors. Various equilibrium receptors found in the animals are the following :

(i) **Statocysts :** These are very old types of a specialized mechanoreceptor found mainly in crustaceans, molluscs and vertebrates. These respond to either to static forces of gravity or to acceleration forces.

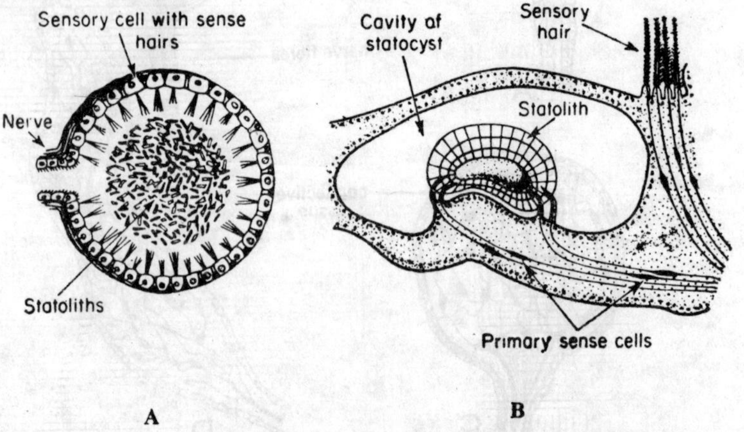

Fig. 15·2. Satatocysts of invertebrate animals. A—*Pecten* with free statolith; B—*Leptomysis* with dendritic hairs attached to the statolith. Free sensory hairs with attachcd dendrites are also shown in (B) (HEIDERMANNS, 1957, after WARMBACK)

Statocysts are generally rounded or cylindrical closed sacs with sensory cells which have sensory hairs. These are filled with a fluid. In crustaceans the sensory hairs of sensory cells are chitinous structures innervated at their bases by the primary sense cells. In most animals such as in *Pecten* and *Leptomysis*, the statocysts contain a harden concretion, the **statolith** which consists either of sand grains picked up from the environment or calcareous concretions produced by the animal. The statolith moves about in the fluid medium of the statocyst and exerts mechanical distortions on the sensory hairs. From the hairs the impulses are transmitted to the nervous system to provide information concerning the position of the animal with respect to gravity. HYMAN, 1940, 1951, reported that such receptors are also present in the coelenterates and flatworms.

(ii) **Tactile and chordotonal sensillae of insects :** These are sufficiently specialized structures or receptors of insects to act as simple **acoustic organs. Sensilla trichodea** are tactile sensory organs that are located in different parts of the body in hollow extensions of the exocuticle. The sensilla (=sensory hairs) are found in abundance at joints and regions between body segments. They consists of a sensory cell in a socket. The shaft of the hair is rigid, and any force applied to it is transmitted to and amplified by the membrane of the sensory neuron. resulting in the production of a nerve impulse (PRINGLE, 1938). Some of these receptors are innervated by only one neuron, others may be innervated by several neurons, indicating a variety of sensory functions including that of chemoreception.

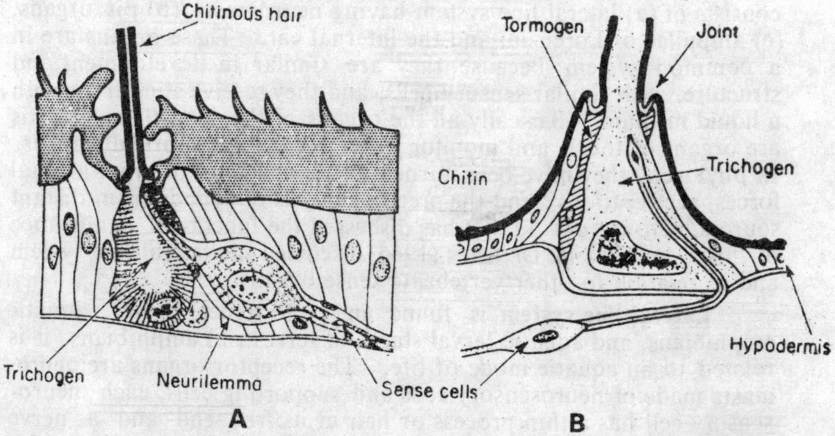

Fig. 15·3. Hair sensilla of insects. A.—From the cercus of the cricket *Liogryllus*; B—From the caterpillar *Pieris*. The trichogen is the hair-forming cell ; the tormogen cell secretes the chitinous joint membrane.

Sensilla chaetae (sensory spines) are also found all over the cuticle of the insect and are mechanoreceptors that function in proprioception as well as other mechanoreceptor activities. **Sensilla campaniformia** (sensory pores) are mechanoreceptors that function in gravity reception and proprioception.

Like sensilla, there are also found some specialized cells over the insect body, these are known as **scolophours sense cells**. A group of scolopidia forms a **chordotonal organ**. These organs may stretch across a fluid-filled space as in the subgenual organs in the legs or „Johnston" organ in the second antennal segment. The former serves as a tactile organ or pickup for low-frequency vibration (200 to 6000 cycles per second) from the ground or substratum where the animal lives ; the latter are probably "statical organ" (AUTRUM. 1959 ; DETHIER, 1963). Chordotonal organs may also stretch across an air sac or tracheal space and form auditory organs of many different patterns and varying complexities.

The chordotonal organs consist one to 400 sensilla (sensory hairs), each with one nerve cell and 2-3 characteristic cells. The distal cell generally called cap cell is attached to an organ or part of the body wall, whereas a ligament is usually attached to some other part of the body. The sensory cell is thus attached to two different parts of the body, and any movements of the body cause impulses to arise in the neuron.

In some insects as many as 1500 or more scolopidia in association with a thin membrane, the tympanum form specialized acoustic organs called **tympanal organs.** The membrane improves the pickup of distal vibrations.

The acoustico-lateralis system : The acoustico-lateralis system consists of (a) lateral line system having neuromasts, (b) pit organs, (c) ampullae of Lorenzini and the internal ear. These organs are in a common system because they are similar in development and structure, with similar sensory cells, and they receive stimuli through a liquid medium. Basically all the organs of the acoustico lateratis are organs of touch and monitor water currents at short distances. In phylogeny they have been turned to the detection of gravitational forces, accelerations, and the precise analysis of sounds from distant sources. DIJKGRAAF (1263) has discussed the functional significance of this system, while DENISON (1966) discussed the lateral line system and its relation to other vertebrate sense organs.

Lateral line system is found in cyclostomes, fishes, aquatic amphibians, and aquatic larval stages of terrestrial amphibians, it is related to an aquatie mode of life. The receptor organs are **neuromasts** made of neurosensory cells and supporting cells, each neurosensory cell has a thin process or hair at its free end and a nerve fibre at the other end. In cyclostomes, a few fishes, larval amphibians, and aquatic amphibians the neuromasts lie on the surface of the skin, especially on the head. In most cartilaginous and bony fishes the neuromasts sink within the skin into depressions, grooves, or canals. The canals contain mucus and open on the surface by many minute pores. The lateral line system is innervated by VII, IX and X cranial nerves. Besides neuromasts the lateral line system has other receptors, they are pit organs and ampullae of Lorenzini in elasmobranchs, and pit organs in other fishes. Lateral line receptors detect movements of water and currents, they furnish information regarding the position of the body in relation to the environment and it is probable that they detect vibrations of water.

In elasmobranchs the lateral line system has two long **lateral line canals** in the dermis, each running along the side of the entire body. In the head region the two lateral canals are joined by a transverse **occipital canal** above the head, then each lateral line canal runs forward as a **postorbital canal** which divides into two branches, a supraorbital canal above the orbit and an infraorbital canal below the orbit, both run upto the snout. Arising from the infraorbital canal is a **jugal canal** below the eye which runs back up to the first gill cleft, it gives off a mandibular canal to the lower jaw.

The canals are lined with epithelium having many mucous gland cells which secrete mucus, the canals open at intervals on the surface by vertical tubes (which pierce the scales in bony fishes). The canals are filled with a fluid and mucus. In the canals are **neuromasts**, each made of a group of sensory receptor cells and supporting cells, each receptor cell has a stiff sense hair at one end and a nerve fibre at the other end, the hairs of receptor cells are tipped with a heavy gelatinous substance. The lateral line neuromasts are current receptors (**rheoreceptors**) detecting any vibrations of water. They are also found in cyclostomes, fishes and aquatic amphibians.

Pit organs are found on the dorsal and lateral surface of the head, they are ectodermal pits, each having a neuromast innervated by the VII cranial nerve. Pit organs are scattered individual neuromasts found in all fishes, they are rheorceptors.

Ampullae of Lorenzini are found only in elasmobranchs, they are highly modified receptors of the lateral line system, though some do not regard them as such. The ampullae of Lorenzini are found in clusters on the dorsal and ventral side of the head embeded below the skin but opening externally on the surface of the skin. Each ampulla has a pore opening on the surface, the pore leads into a duct or canal of Lorenzini filled with mucus, the canal ends below in an ampullary sac having 8 or 9 vertical chambers arranged radially around a central core or centrum. The ampulla has mucous gland cells and sensory cells. The ampullae of Lorenzini are innervated by fibres of the facial (VII) nerve, they detect any changes of temperature (**thermoreceptors**), but the deep position of sensory hair cells show that this is not their prime function, they also detect weak tactile stimulations and hydrostatic pressure at various depths. However, their real function remains uncertain.

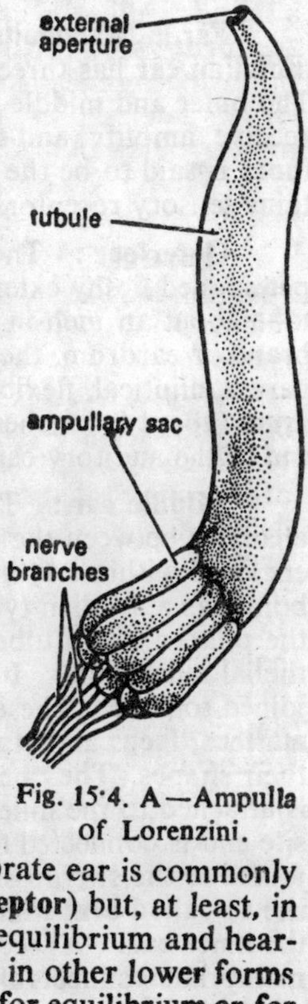

Fig. 15·4. A—Ampulla of Lorenzini.

Vesicles of Savi are like ampullae of Lorenzini but are found on the ventral surface only in the electric ray *Torpedo* as closed sac in the epidermis, their function is unknown.

Ears or statoacoustic organs : The vertebrate ear is commonly regarded only as an organ of hearing (**phonoreceptor**) but, at least, in higher vertebrates it serves the dual function of equilibrium and hearing. In fishes it is only an organ of equilibrium, in other lower forms it is not known definitely whether it serves only for equilibrium or for both equilibrium and hearing. The fundamental ear is a **membranous labyrinth** or internal ear found in all vertebrates with hardly any variations, its cristae and maculae are concerned mainly with equilibrium. The portion of the membranous labyrinth which is concerned with hearing is the lagena which begins to develop in fishes and gradually becomes more complex in the evoluntionary scale of vertebrate series.

In cyclostomes the membranous labyrinth is degenerate with only one (*Myxine*) or two (*Petromyzon*) semicircular canals. In fishes the membranous labyrinth is complete with three semicircular canals, and in many teleosts it is connected to the air bladder by a duct or a chain of bony Weberian ossicles. In tetrapoda the change of environment from water to land necessitated formation of structures to conduct vibrations conveyed through air, thus a **middle ear** or tympanic cavity was formed from the first gill cleft or spiracle of fishes. A tympanic membrane covers the middle ear and small bones come to lie in it for transmitting vibrations. All tetrapods have a middle ear with a Eustachian tube connecting it to the pharynx. In mammals an **external ear** or pinna is formed, though its beginnings are seen in some reptiles and birds. The external ear catches and directs sound waves to the tympanic membrane of the middle ear.

Ear in Mammalia : The ear of man as an example of the mammalian ear has three parts: the **outer ear, middle ear,** and **inner ear**. The outer and middle ears are basically auxiliary structures which receive, amplify, and transmit sound waves while the internal or inner is said to be the actual sense organ for hearing because it contains sensory receptors for the sound waves.

Outer ear : The outer ear consists of three parts : (1) a trumpet shaped fleshy external part, the **pinna** ; (2) a short funnel-shaped tube about an inch in length from the exterior to the tympanic membrane or eardrum, the outer **auditory canal** and (3) a thin, semitransparent elliptical, flexible membrane, the **tympanic membrane** or **eardrum** (about 0·4 inches in diameter), it is stretched across the inner end of the auditory canal and separating it from the middle ear.

Middle ear : The middle ear is a small, hollow, air-filled chamber between the inner ear and the tympanic membrane and is enclosed within a tympanic bone of skull. In many mammals this bone has a swollen tympanic bulla. The middle ear is connected to the pharynx by a tube, the **eustachian tube** and is lined with an epithelial membrane. It has three tiny bones or **ear ossicles** which are joined together. These bones or ear ossicles are commonly called the **malleus, incus** and **stapes** (**hammer, anvil,** and **stirrup**) because of their shapes. The "handle" end of the hammer-shaped malleus bone is attached to the inner surface of the tympanic membrane. Its opposite end is connected to the curved incus (anvil) which in turn is joined to the stirrup-shaped stapes. The foot plate of stapes fits into the so-called **oval window**, a membrane-covered opening leading into the inner ear. A second opening which is also covered by a thin membrane connects the middle and inner ears and is called the **round window**. Three additional openings also lead into the middle ear and these are : one from the external auditory canal covered over by the tympanic membrane ; one from a network of small irregular honeycomb spaces called the **mastoid sinuses** located in one of the surrounding bones of the head ; and one from the so-called eustachian tube.

Inner ear : In its overall structure the inner ear consists of both a bony and membranous labyrinth. In general the bony portion of the labyrinth entirely encloses and protects the similarly shaped membranous labyrinth. The bony and membranous labyrinth remain separated from each other by a layer of lymph-like fluid (perilymph) which serves as a protective cushion or buffer. Of the different sense organs that are present in the inner ear, only one what is known as **cochlea**, is concerned with the sense of hearing. The others namely the **sacculus, utriculus** and its attached **semicircular canals** have to do with the sensation of physical equilibrium and orientation. The anterior and posterior semicircular canals form a crus commune. Utriculus and sacculus are small and the sacculoutricular connection is a narrow duct. From the sacculus arises an endolymphatic duct which ends blindly in the duramater. Also arising from the sacculus is a long, complicated, and spirally coiled cochlear duct which does not have any macula except in monotremes.

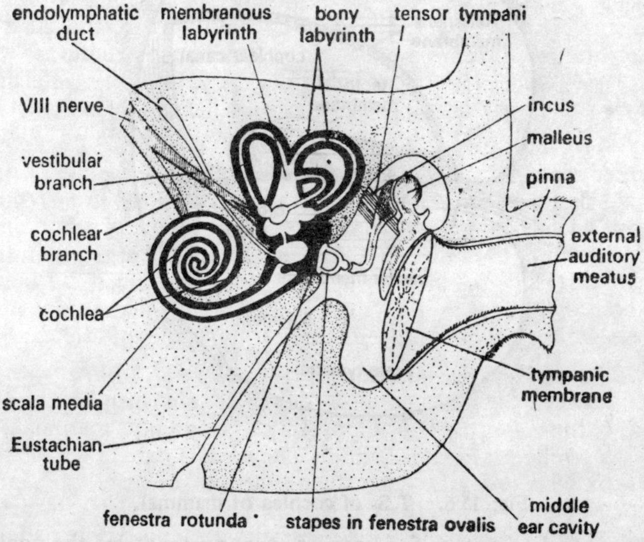

Fig. 15·5. T.S. of the ear region of mammal.

The cochlea : The cochlea is a spirally coiled tube (2½ turns) of about an inch around a central, cone-shaped bony pillar, the **modiolus.** It lies deeply embedded in bone which resembles in shape to a snail's shell. The cochlea consists of three tapering, parallel canals. The uppermost canal known as **vestibular canal (scala vestibuli)** is attached at its end to the membrane-covered **oval window** (fenestra ovalis). This window holds the base of a small middle ear bone, the **stapes**. At its other end located at the apex of the cochlea the vestibular canal has a small opening which communicates with the lowermost canal called the **tympanic canal (scala tympani)**. The tympanic canal also has a membrane-covered **round window (fenestra rotundum)** at its base end, near the middle-ear chamber, but has no bone fitted to its surface. Both the vestibular and tympanic canals are filled with a clear fluid called **perilymph** which comes from the cerebrospinal fluid and remain connected at the tip of the cochlea by a narrow passage called **helicotrema**. The third and the smallest central canal called the **cochlear canal (scala media)** is filled with a clear fluid called **endolymph** and remains separated from the overlying vestibular canal by a membrane called the **vestibular membrane** (REISSNER'S **membrane**) and from underlying tympanic canal by a ledge-like projection of the bony cochlear wall plus a membrane, the **basilar membrane.**

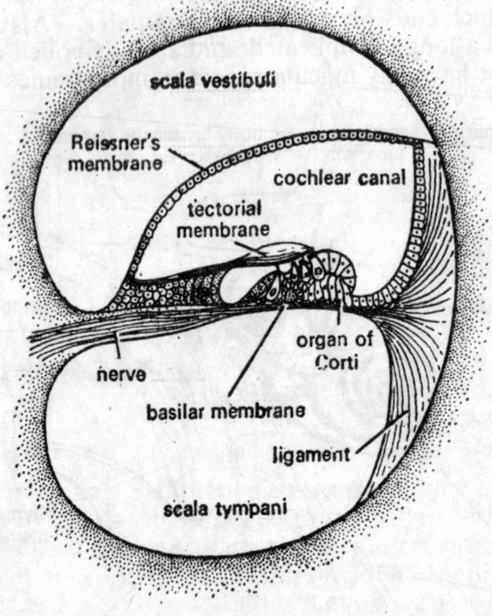

Fig. 15·6. T.S. of cochlea of mammal.

The actual receptors for hearing are present in the cochlear canal as several rows of specialized **"hair cells"**, approximately 24,000 in number. They contain numerous cilia **(stereocilia)** which project into the endolymph from the free end of each cell. The rows of hair cells together with supporting cells and surrounding dendrites constitute the specialized **organ of Corti**. It rests on the basilar membrane within the cochlear canal. The cell bodies from which the dendrites originate make up a ganglion with one of the numerous passages of the inner ear ; their axons form the auditory nerve a branch of one of the cranial nerves.

Sound is transmitted in the ear as follows. Vibrations enter the auditory canal and cause the tympanic membrane to vibrate. Vibrations of the tympanic membrane are transmitted through middle-ear bones or ear ossicles to the membrane of fenestra ovalis or oval window, and then pass to the perilymph and scala vestibuli (vestibular canal) causing the REISSNER'S (or vestibular) membrane to vibrate up and down, these vibrations pass to the endolymph which makes the basilar membrane to vibrate by which sensory hair cells of organ of Corti are stimulated, from here impulses travel through the fibres of auditory nerve to the brain where the temporal lobes of the cerebrum interpret them as sound. Vibrations of the basilar membrane pass to the perilymph in scala tympani (tympanic canal) and the membrane of fenestra rotunda (round window) is bulged in and out to dampen the vibrations. The helicotrema connecting the scala vestibuli (vestibular canal) with scala tympani (tympanic canal) is very narrow and is unable to transmit vibrations. Thus the sense of hearing is confined to the cochlea. The explanation of hearing is based on evidenee but is primarily a theoretical one.

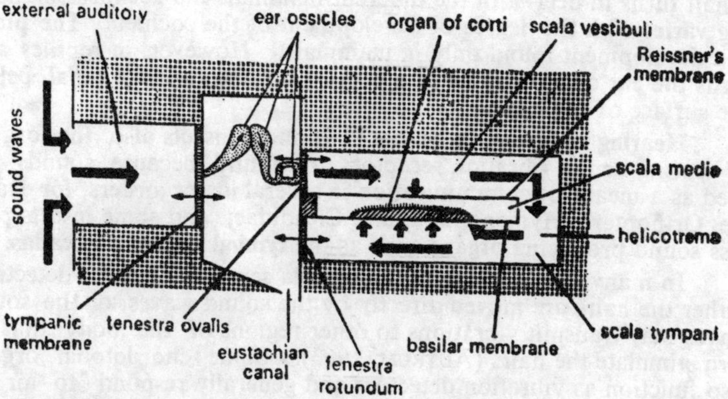

Fig. 15·7. Diagrammatic represeniation of the mechanism of hearing.

The three semicircular canals and their cristae detect changes in acceleration in any of the three planes of space. The cristae also react to rapid changes of position and rotations, this is due to movements caused in the endolymph. The macula of sacculus is concerned with detection of gravitational stimuli.

Ear in other vertebrates : A comparison of the ears of the different classes of vertebrates reveals that the outer and middle ears are relatively recent acquisitions. All vertebrates possess an inner ear whereas the progressive additions of the middle and outer ears, starting with the Amphibia, achieve their most advanced states in the higher animal forms.

Although the outer and middle ears function solely in hearing, certain parts of the inner ear which are not involved at all in hearing, play a central role in physical balance or equilibrium. The inner ear of vertebrates appears to have been originally concerned with balance. Its anatomical and functional development as an organ of hearing is considered to be a more recent evolutionary development.

The ear of fishes is composed of the **utriculus**, the **sacculus**, and the **semicircular canals**. There is neither cochlea nor middle or outer ears ; it is primarily an organ for equilibration. This non-acoustic part of the ear remains essentially the same in all the other vertebrate classes ; the changes are found in the parts of the ear involved with hearing. The amphibian ear possesses a small outgrowth of the sacculus, the **lagena**, that develops into the cochlea in mammals and is present as a very small diverticulum in some fishes. The middle ear cavity is formed from one of the embryonic pharyngeal pouches, the hyomandibular cartilage is incorporated in the middle-ear cavity as the **columella** (in higher forms called the **stapes**). With this, the thin membrane separating the middle ear from the outside becomes the ear drum. There is no external ear in the amphibians. In reptiles the **lagena** becomes an elongated cochlea, and in mammals the cochlea reaches its greatest development, varying from a half turn in the primitive mammal, the echidna to three and a half turns in deer. In the different mammals the keenness of hearing varies with the degree of development of the cochlea. The pinna is a development found only in mammals. However, in reptiles and birds the ear drum may be in a depression; the auditory canal, below the surface of the head.

Hearing receptors in insects : In most insects also the organs of hearing or the vibration receptors are found because sounds are used as a means of communication in several insect orders, for example, Orthoptera, Hymenoptera, and Cicadidae; and some insects possess sound producing organs such as the **tymbal organs** of cicadas.

In many insects some of the sensilla serve as vibration detectors. Either the hairs are moved directly by the sound waves, or the sound waves may transmit vibrations to other regions of the body that in turn stimulate the hairs (AUTRUM, 1963). Some chordotonal organs also function as vibration detectors and generally respond to air or water borne vibrations. JOHNSTON's organ is also said to be responsive to air-borne vibrations. This organ is generally found in the antennae of all insects and resembles to chordotonal organs in structure except that in this organ the cap cells are absent.

Tympanal organs are specialized sound receptors. They are modified tracheal structures with a thin membrane separating the external evironment from a chamber containing sensory hairs. They are present in antennae in Diptera, the legs in Orthoptera; or in thorax or abdomen Lepidoptera and Hemiptera.

TEMPERATURE RECEPTORS OR THERMORECEPTORS

Those receptors which respond to temperature changes are termed **thermoreceptors**. In warm-blooded animals, these are located in the hypothalamus, while in cold-blooded animals, the temperature is regulated through behaviour responses which are triggered by the cutaneous receptors ; blood-sucking arthropods such as mosquitoes and lice often locate their warm-blooded prey by delicate receptors sensitive to a gradient as narrow as 0·5°C ; the rattle-snake, with the most acute temperature receptors known, is able to detect a rat-sized object 10°C warmer than its environment at a distance of 40 cm after only 0·5 sec. In arthropods, thermoreceptors are found on the antennae or legs. In mammals only two sets of sensory fibres or nerve endings are involved in temperature reception. One is concerned with measuring the degree of warmth, and the other measures the degree of cold of the environment. Temperature receptors on being stimulated send impulses to the thalamic region of the brain where their information is integrated and co-ordinated to produce responses. Fore head generally contains very few warm receptors but many cold receptors are present. Ampullae of Lorenzini are also responsive to temperature.

CHEMORECEPTORS

Chemoreceptors are concerned with chemical reception that these are receptors which detect the chemicals in the environment. Detection of chemicals in a gas phase is called **olfaction** (smell) and the detection of chemicals in liquid phase is called **gustation** (taste). In vertebrates the receptors for the said activities differ morphologically and send impulses to different parts of the brain.

In vertebrates and invertebrates generally two types of cells are concerned with chemical reception. One of these is a bipolar nerve cell in which the membrane of the dendrite is specialized and highly excitable in the presence of specific chemicals. The other is a columnar epithelial cell in which the excitable and exposed apex of the cell is extended in a hair-like process, while the base is in contact with a dendritic arborization of an associated neuron. These are often called primary and secondary sense cells respectively.

Chemoreceptors of invertebrates are usually separate single cells, often scattered through the surface epithelium. MONCRIEFF (1951), PROSSER and BROWN (1960), MILNE and MILNE (1962), tried to discuss the chemoreceptors of various animal groups. Various chemoreceptors found in vertebrates are the following :

Olfactory organs and organs of taste, both these organs are stimulated only by substances in solution, thé medium for dissolving substnnces is water for aquatic animals and mucus for land animals. The olfactory organs can respond to a low concentration of the dissolved substance, whereas organs of taste need a higher concentration of the dissolved substance for a response.

Olfactory organs or **olfactoreceptors :** These are invaginations of the ectodermal cells of the skin forming olfactory sacs in fishes or lining of a part of the nasal passages in tetrapoda. The olfactory organs have a lining of olfactory mucous membrane or SCHNEIDERIAN **membrane** made of basal cells, supporting cells and elongated neurosensory cells. Neurosensory cells are bipolar neurons, they are produced at their free ends into hair-like processes or sensory hairs. The structure of sensory hairs is like that of cilia with two central fibrils and nine double peripheral fibrils enclosed in a protoplasmic sheath. Backward growth of neurosensory cells forms fibres of olfactory nerves which join the olfactory lobes. The neurosecretory cells are covered with mucus which is not only protective but also dissolves substances for smelling. The olfactory organs open to the exterior by external nares, the olfactory organ is single in cyclostomes with a single naris but it has two olfactory nerves. In gnathostomes the olfactory organs are paired. In Dipnoi and tetrapoda each olfactory organ communicates with the mouth cavity by an internal naris. In mammals the sense of smell is very acute, they have elongated nasal passages containing folded **turbinal bones** called ethmoturbinals, maxilloturbinals, and nasoturbinals. The turbinals are covered with Schneiderian epithelium which is not olfactory over the maxilloturbinals but only moistens and warms the inhaled air, on the ethmoturbinals it is olfactory, on the nasoturbinals it is only some neurosensory cells.

Organ of Jacobson or **vomeronasal organ** : In many tetrapoda there is a pair of vomeronasal organs which are sac-like chambers lying below the nasal cavities but above the buccal cavity, they have a pigmented epithelial lining like that of the olfactory organs. Each opens by a short duct into the olfactory organ in amphibians, but in others the duct opens into the buccal cavity. The organ of Jacobson receives nerves from the nervous terminalis, a branch from the olfactory nerve, and a branch from the trigeminal nerve. The organ is believed to aid by smelling the recognition of food held in the mouth, and in lizards and snakes it appreciates the scent introduced into it by the tip of the tongue.

The organ of Jacobson first appears in amphibians as an evaginations of the olfactory organ, in the frog it lies in the anterior portion of each nasal cavity, but it is best developed in *Sphenodon*, lizards, and snakes, it is also well formed in monotremes, marsupials, insectivores, and rodents. But in turtles, crocodiles, birds, and many mammals such as Primates and Cetacea, it is found only in the embryo and is absent in the adult.

Organs of taste or **gustatoreceptors :** These consist of **taste buds,** their structure is fairly uniform in vertebrates, each consists of cluster of neurosecretory cells and supporting cells arranged to form a barrel-like structure embedded in stratified epithelium. The supporting cells form the outer covering, while neurosensory cells are in the interior but some supporting cells may be interspersed among neurosensory cells. Each neurosensory cell is long and narrow with a thin **taste hair** at its free tip and a sensory nerve fibre at its base, the taste hairs have a structure like cilia, they project into a depression or **taste pore** in mammals, in others they project above the surface, there being no taste pore. Substances in solution can be tested by taste buds, there are four fundamental sensations of taste, they are sweet, salty, bitter, and sour.

Taste buds are widely distributed in fishes, being found in the mouth, pharynx, branchial cavities, and outer surface of the head, in some fishes they occur on the entire body surface and the sensation of taste is used to detect substances in the environment, the taste buds are innervates by the V, VII, IX and X cranial nerves. In other

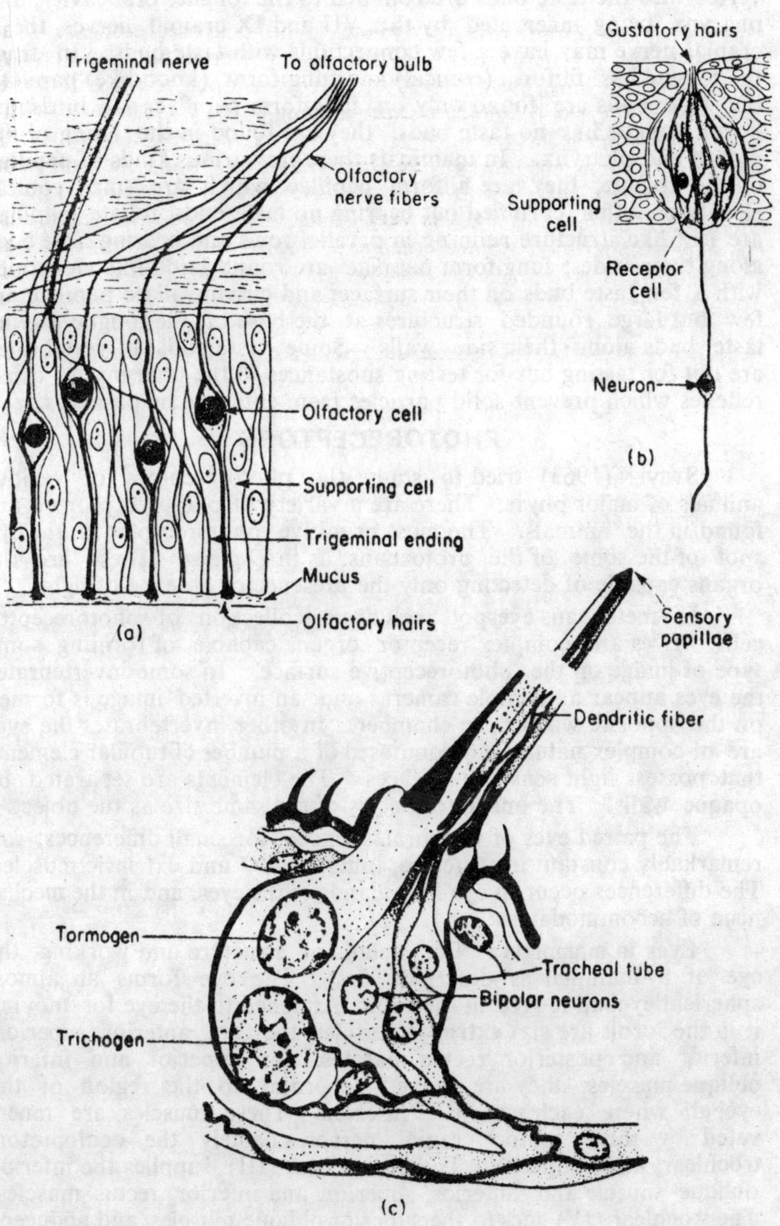

Fig. 15·1. Three types of chemoreceptors. (a) Olfactory epithelium of the mammal to show primary sense cells ; (b) taste bud (secondary sense cells) in the mammal ; (c) chemo-sensory hair of the blowfly *Phormia regina.*

vertebrates the taste buds are confined to the tongue, oral cavity, and pharynx being innervated by the VII and IX cranial nerves, the X cranial nerve may have a few connections with taste buds. In frogs the tongue has filiform (conical) and fungiform (knob-like) papillae, but taste buds are found only on fungiform papillae. In birds the horny tongue has no taste buds, they are found in the lining of the mouth and pharynx. In mammals there are various kinds of **papillae** on the tongue, they are filiform papillae which are small conical elevations, often cornified but bearing no taste buds; foliate papillae are leaf-like structure running in parallel rows and bearing taste buds along their sides; fungiform papillae are round knob-like elevations with a few taste buds on their surface; and circumvallate papillae are few but large rounded structures at the base of the tongue having taste buds along their side walls. Some taste buds of vertebrates are not for tasting but for testing substances in the pharynx to cause reflexes which prevent solid particles from entering air passages.

PHOTORECEPTORS

Steven (1963) tried to study the photoreceptors of various animals of major phyla. There are a variety of photoreceptor organs found in the animals. The most primitive photoreceptor is the eye spot of the some of the protozoans as in *Euglena*. These are the organs capable of detecting only the presence or absence of light,

In metazoans eyespot includes a collection of photoreceptor cells. Eyes are complex receptor organs capable of forming some type of image on the photoreceptive surface. In some invertebrates the eyes appear as pinhole cameras and an inverted image is formed on the opposite wall of the chamber. In other invertebrates the eyes are of complex nature and composed of a number of tubular elements that possess light sensitive surfaces. The elements are separated by opaque walls. The image formed is of the same size as the object.

The paired eyes of vertebrates, except for small differences, are remarkably constant in structure, innervation, and extrinsic muscles. The differences occur in eye lids, glands of the eyes, and in the mechanism of accommodation.

Eyes in mammals : For details of structure and working, the eye of a mammal is described first. The eye forms an almost spherical eyeball lodged in the orbit. Attached to the eye for moving it in the orbit are six **extrinsic muscles,** they are anterior, superior, inferior and posterior **rectus muscles,** and superior and inferior **oblique muscles,** they are named according to the region of the eyeball where each muscle is inserted. These muscles are innervated by three motor cranial nerves, namely the oculomotor, trochlear, and abducens. The oculomotor (III) supplies the inferior oblique muscle and anterior, superior, and inferior rectus muscles. The trochlear (IV) goes to the superior oblique muscles, and abducens (VI) innervates the posterior rectus muscle. The eyeball is made of three coats. The external coat is a **selerotic** made of dense fibrous connective tissue which is opaque, but in front the sclerotic forms a **transparent cornea** of connective tissue. The cornea is covered by a thin transparent but vascular **conjunctiva** which is formed by modified epidermis and continuous with the lining of eyelids, to distinguish the two, the conjunctiva is more specifically termed the **bulbar conjunctiva** and the lining of eyelids is called **palpebral conjunctiva**. The second coat or **choroid** is made of highly vascular connective tissue, pigment cells, and some smooth muscles. In some mammals, especially the nocturnal mammals, the choroid has silvery connective tissue which reflects light rays causing the eyes to shine at night. The choroid closely lines the sclerotic, but it separates in front to form an **iris** which is perforated by a rounded **pupil**. The iris seen through the cornea is pigmented in mammals including man, it may be

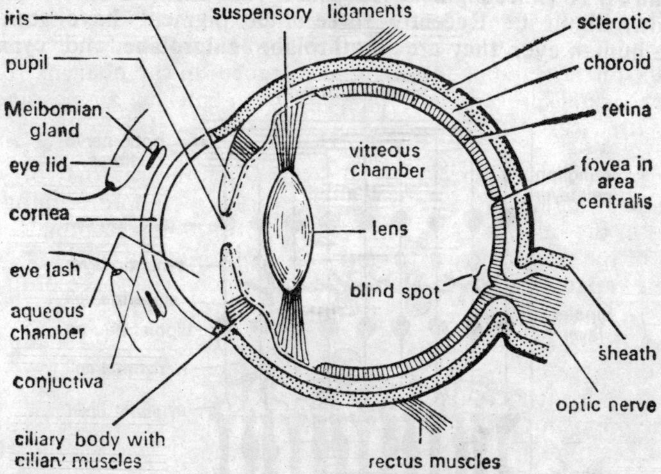

Fig. 15·5. V.S. of the eye of a mammal showing its structure.

blackish, dark brown, light brown, blue, green, gray, or yellow. The iris contains ectodermal intrinsic circular and radial smooth muscles, the circular muscles contract the pupil, while radial muscles dilate it. Along the peripheral margin of the iris the choroid mainly and to some extent the outer part of the retina form a ring-like **ciliary body** which is a thick folded band containing smooth ciliary muscles, radiating ciliary processes to which suspensory ligament fibres are attached, blood vessels, and glands. The circular and radial muscles of the iris and ciliary muscles are under the control of the autonomic nervous system receiving sympathetic and parasympathetic fibres.

The innermost coat of the eyeball is a thin, light-sensitive nervous layer called **retina**. The retina lines the choroid and extends in front over the ciliary body to form a posterior lining of the iris up to the pupil. Only the posterior part of the retina called **pars optica** is sensory, while ends at the ciliary body along an irregular line termed **ora serrata**. The anterior part of the retina over the ciliary body and on the inner surface of the iris is thin, non-sensory, and has a simple structure.

The **retina** has a complicated structure and is made of two layers, an outer **pigment layer** and an inner **nervous layer.** The pigment layer is made of cuboidal cells containing dark brown pigments, the cells have fringe-like protoplasmic processes, it is this layer alone which is continued forward beyond the ora serrata. The nervous layer is confined to the pars optica, it is transparent but is very complex and has an outermost photosensitive layer of visual cells called **rods** and **cones.** Rods respond to low intensities of light but not to colours, while cones are sensitive to bright light and to various colours. The rods bear a long thin cylinder, each of which contains a purple pigment **rhodopsin** made of a protein and vitamin A, it is bleached and destroyed by light but is formed again in darkness by vitamin A. A violet pigment **iodopsin** is found in cones, but very little is known about it. Recently three other pigments have been found in the human eye, they are **erythrolabe chlorolabe,** and **cyanolabe,**

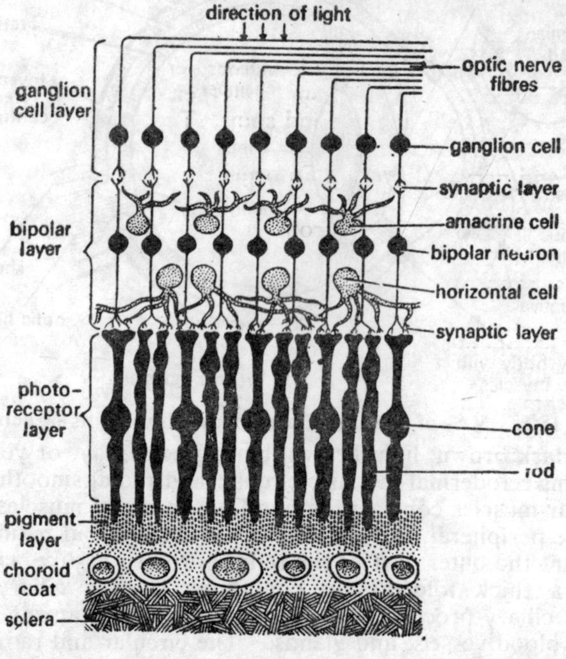

Fig. 15·10. V.S. of retina showing its detailed structure.

they are sensitive respectively to red, green and blue light, these pigments appear to be located in cones and may be responsible for colour vision. Rods detect differences in the intensity of light but they do not discriminate finer details of objects, hence the eyes of nocturnal animals and those living in dark places and deep water (moles, whales) have only rods in the retina. The cones bear a short tapering process each, they are concerned both with discrimination of acute details of structure in bright light and perception of colours. The rods and cones are nucleated cells and their inner ends are continuous with slender nerve fibres. The retina has mixture of rods and cones, towards the periphery there are more rods than cones, but towards the posterior side there are more cones than rods, and in a small depression called **area centralis** (in primates it is called **macula lutea** because it contains yellow pigment) there is a smaller depression in the centre known as **fovea centralis** which contains only cones. The fovea centralis lies at the posterior end of the eyeball on the main visual axis and is an area of most acute vision for discriminating the minutest details of structure and colour. Internal to the rods and cones the retina has a layer of **bipolar neurons** whose dendrites form synapses with nerve fibres of rods and cones. The innermost layer of the retina is a **ganglionic layer** of neurons, the ganglionic neurons by their dendrites form synapses with the axons of bipolar neurons, the axons of ganglionic neurons pass along the inner surface of the retina and then bend sharply to run parallel to each other, they become medullated and form an **optic nerve**. In higher mammals the optic nerves decussate only partially, these forms have binocular and stereoscopic vision. The optic nerve pierces through the retina, choroid, and sclerotic at a **blind spot** and then joins the diencephalon, there are no visual cells at the blind spot. The retina also contains neuroglia cells and MULLER's fibres which fill the spaces between the visual cells and nervous layers. In order to reach the photosensitive rods and cones light must pass through the layers of nerve fibres, ganglionic layer and bipolar neurons, a retina of this type with visual cells away from the source of light is characteristic of the vertebrate eye and is spoken of as an **"inverted retina"**.

Behind the iris is a transparent biconvex **lens** enclosed in a transparent and elastic membrane called **lens capsule**. The lens is attached to the ciliary processes of the ciliary body by very fine fibrous **suspensory ligaments**. The space between the cornea and iris is an **anterior chamber**, and the narrow space between the iris and the lens is a **posterior chamber**, both these chambers are filled with a watery **aqueous humour**. The large cavity of the eyeball behind the lens is a **vitreous chamber** filled with a transparent jelly-like **vitreous humour** or body. A lymphatic **hyaloid canal** passes through the vitreous humour from the lens to the blind spot, in the embryo it contained a hyaloid artery which disappears later and its place is taken by a retinal artery which capillarizes in the choroid.

There are two movable **eyelids** as transverse folds of skin, but in some mammals only the upper eyelid is movable, the eyelids are protective and close the eye. A third eyelid is a transparent nictitating membrane which is movable and can cover the entire cornea. In man the nictitating membrane is reduced to a vestigial **plica semilunaris** lying as a reddish patch in the inner corner of the eye. Below the outer angle of the upper eyelid are **lacrimal glands** which produce a salty secretion or tears continuously to keep the surface of the eye moist and free from dust and to provide nourishment for the cornea, the tears are drained from the eye into a nasal sinus by a lacrimal duct. In emotional stress or injury excess tears are secreted so that they overflow over the lower eyelid and cheeks. In man the edge of eyelids have rows of hairs called eyelashes, modified sweat glands known as **glands of** MOLL open into the follicles of eyelashes. On the margins of both eyelids are openings of tarsal or MEIBOMIAN **glands** regarded as modified sebaceous glands, they produce an oily secretion which forms a film over tears to hold them evenly over the surface of the eyeball. In some aquatic mammals (whales) and mice there is a large HARDERIAN **gland** lying behind the eyeball which secretes an oily fluid for lubricating the nictitating membrane, it is absent in most mammals.

Image formation : In order to see objects at various distances an image must be focussed clearly on the retina so that the photoreceptive rods and cones are stimulated, this is done in different ways by various vertebrates. The cornea, aqueous humour, lens, and vitreous body are transparent, together they form a dioptric apparatus to focus the image of an object on the retina. But the cornea and lens are of primary importance, the cornea places the image on the retina and the lens makes adjustments for sharp focussing. In most animals the eye is focussed for objects beyond 20 or 30 ft., for near objects the lens has to be adjusted to form a clear image, this adjustment is called **accommodation**. In mammals accommodation is brought about by changing the convexity or focal length of the lens, the nearer the object the lens becomes more convex by means of **ciliary muscles** which contract in such a way that they pull towards the lens, thereby releasing the tension on the suspensory ligments and lens capsule, this reduces the radius of curvature of the lens. and the lens bulges out becoming more spherical and displacing the iris anteriorly. At the same time the pupil contracts and a clear inverted image of the object is formed on the retina.

The image on the retina is **inverted**, light first passes through the layers of the retina before it stimulates the rods and cones, impulses thus set up then pass back from the visual cells through the bipolar and ganglion cells to reach the fibres of the optic nerve. The retina is said to be "inverted". The inversion of the retina occurs during development of eye. In the neural plate of the embryo there are patches with an outermost pigment layer, next to these are sensory receptor cells, and inside are conducting nervous cells. When the neural plate folds to form a neural tube then this arrangement of cells is reversed, the photo sensitive receptor cells come to lie towards the inside of the neural tube, *i.e.*, away from the source of light. Hence in the fully-formed vertebrate eye the position of rods and cones is reversed one, and this explains the inversion of the retina and formation of inverted images.

Eyes in Elasmobranchs : The eyes are very large, each is elliptical and supported not only by the recti and oblique muscles but also by a cartilaginous **optic pedicle** which connects the eyeball to the skull. The sclerotic is cartilaginous and cornea is fused with the conjunctiva. Posteriorly is a **suprachoroid layer** formed by the choroid, it lies between the sclerotic and the choroid, this layer is developed in those forms which have an optic pedicle. The choroid is heavily pigmented and lined by a silvery **tapetum lucidum** having crystalline plates of guanin, it reflects light and causes the eyes to shine in the dark, it also reflects additional light on the retinal cells to enable the fish to see in water where light is poor. The retina has elongated photosensitive rods but no cones (except in *Mustelus* and a few others), so that colour vision is lacking. Posteriorily the retina has a non-pigmented **area centralis** having rods where most acute vision occurs. The iris has poorly developed circular and radial muscles which can make the pupil a narrow slit. The lens is very hard and round, it pushes the iris forwards almost touching the cornea. The ciliary body has no intrinsic muscles, and a dorsal suspensory ligament forms a **gelatinous membrane** or **zonule** attached to the ciliary body and the equator of the lens. One the ventral side of the lens the gelatinous membrane has a small **protractor lentis** muscle which can swing the lens forwards in accommodation for near vision. Eyelids are small, stiff folds of the skin, only the lower eyelid is slightly movable, the eye has no glands. The fish is colour blind and the eye cannot discriminate minute details, but the eye is adapted for near vision in dim light and the animal can see its prey or predator.

Eyes in Amphibia : There is more scope for eyes in air than in water because light passes through air with less disturbance, for this eyes must be kept moist and free from dirt. Thus sight has become the dominant sense in terrestrial amphibians which can see distant objects better than the aquatic forms.

The eyeball is spherical, the sclerotic is made of dense connective tissue containing some cartilage, and the cornea is curved outwards and is rounded, so that the anterior chamber is large. The iris has well developed circular and radial muscles which are partly under the control of the autonomic nervous system but are also directly sensitive to light and they can dilate or reduce the pupil. The retina has rods and cones, there are two kinds of rods, red rods with visual purple or rhodopsin, and green rods with visual green, but there is no colour vision. Posteriorly the retina has an **area centralis** with yellow pigment, it contains only cones. The **ciliary body** is attached to the sclerotic and posterior surface of the iris, but it has no ciliary muscles. From the ciliary body arise a complete ring of **suspensory ligaments** which are jointed to the lens capsule. The **lens** is rounded but more flat than in fishes. On the dorsal and ventral side are **protractor lentis muscles**, each running from the cornea to the ciliary body and suspensory ligaments. **Accommodation** for near objects is brought about by the protractor lentis muscles moving the entire lens forward towards the cornea, in doing so the outer surface of the lens may become slightly flattened. Other muscles called **masculus tensor chorioidea** run radially around the lens, they help the protractor lentis muscles in accommodation, they are probably the forerunners of the ciliary muscles of higher forms. Thus the frog can clearly focus not only near objects but also distant ones.

Well developed eyelids appear for the first time in air, the lower eyelid is more movable and its upper border forms a transparent **nictitating membrane** which is probably not homologous with that of higher forms since it arises from one eyelid. The nictitating membrane can be pulled rapidly over the cornea, thus it protects and cleans the eye. **A retractor bulbi** muscle pulls the nictitating membrane upwards and a **levator bulbi** muscle pulls it downwards to fold it below the lower eyelid. Glands of the eye appear for the first time in terrestrial amphibians, below the lower eyelid are **lacrimal glands**, their secretion called tears or lacrimal fluid keeps the eye moist and clean, they nourish the cornea, the tears are drained from the eye into the nose through a lacrimal duct. A HARDERIAN **gland** develops during metamorphosis and lies in the inner angles of the eye in the orbit, its oily secretion lubricates the nicitating membrane.

Eyes in Reptilia : The chief changes in eyes of reptiles are perfecting of the structures which developed in amphibians in transition from water to land. The **sclerotic** is cartilaginous having a ring of small scleral (sclerotic) bones externally. The **retina** has both rods and cones and the number of cones has increased, especially in diurnal forms, so that there is some colour vision in lizards. Posteriorly the retina has an area centralis which in some lizards has a fovea centralis containing only cones. The **iris** does not have circular and radial muscles, hence the shape of the pupil can not be changed. The **ciliary body** has striated intrinsic **ciliary muscles** developed in vertebrate series for the first time in reptiles from the ciliary body, ciliary processes run to the lens capsule having a thickened **annular pad** of the lens. Accommodation for near objects is brought about for the first time in the vertebrate series by ciliary muscles and their processes squeezing the periphery of the lens by which the curvature of the outer surface of the lens and cornea is increased by becoming more rounded. From the blind spot a conical ectodermal **conus** or **pecten** projects into the vitreous chamber, it is made of pigmented neuroglia cells and blood vessels, it nourishes the retina.

The eyelids are well formed but the lower eyelid is still larger and more movable. There is a true nictitating membrane lying between the upper and lower eyelids, it is lubricated by a well developed Harderian gland lying in the inner angle of the eye. A lacrimal gland is present in the outer angle of the lower eyelid and a nasolacrimal duct drains the tears from the eye into the nasal passage.

Eyes in Aves : The eyes of bird are very large in correlation with an aerial life for a precise vision over considerable distances. The eyeball is not round but partly concave in front and biconvex sideways, the cornea projects outwards and the posterior part is expanded so that the eye is broader than deep. The retina lies in the broad posterior portion so that distant objects are sharply focussed on it. Though the eyes are lateral in position there is an overlapping of the two visual fields to some extent, and in birds of prey, such as owls and eagles, the two visual fields overlap to a much greater degree so that there is binocular vision.

The sclerotic has a ring of sclerotic bones surrounding the eyeball in the region of the ciliary body, the cornea bulges outwards. The iris has striated sphincter muscles for intense contraction. The retina has rods and cones, in diurnal birds cones are predominant, but rods outnumber cones in nocturnal birds. Colour vision is highly developed The retina has an **area centralis** with a fovea centralis, in some birds there is more than one area centralis and fovea which explains the great visual acuteness of birds and they can detect the smallest movements. The **lens** is biconvex and soft. The **ciliary body** is well formed with ciliary processes attached to the lens capsule, it has striated ciliary muscles divided into anterior CRAMPTON **muscles** and posterior BRUCKE **muscle**. In accommodation for near objects the Brucke muscles and ciliary processes draw the lens forward increasing the curvature of its anterior surface, and at the same time the Crampton muscles pull the cornea reducing its curvature.

Arising at the blind spot from the choroid is a pleated, highly vascular, serrated, and pigmented **pecten**, it is more developed in diurnal birds than in nocturnal ones. There are many speculations

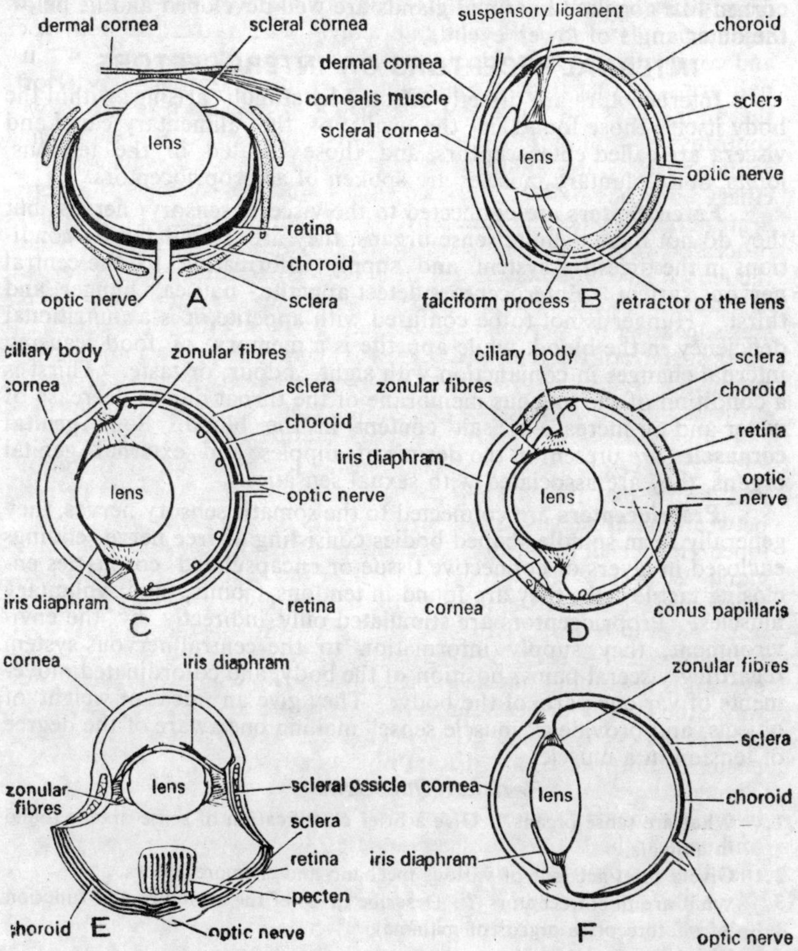

Fig. 15·11. Diagrams of sections through the eyes of the vertebrates showing structural variations. A—Lamprey ; B—Teleost; C—Frog ; D—Snake ; E—Eagle or Owl ; F—Mammal.

about the function of pecten but none is known definitely, it may perform its original function of nourishing the retina by diffusion through the vitreous humour. The other functions attributed to it are, (*a*) it helps in accommodation by pressing the lens forwards, (*b*) it regulates the pressure of fluids in the eye, (*c*) its shadow falls on the retina and aids in the perception of movements, (*d*) it supplies additional nourishment to the retina.

The two eyelids are well-developed which do not blink though they are closed in sleep. The third eyelid or nictitating membrane is attached below the other two eyelids, it keeps the cornea and other eyelids clean, it is lubricated by a Harderian gland in the inner angle of the eye. In flight the nictitating membrane covers and protects the cornea like goggles. Lacrimal glands are well developed and lie below the outer angle of lower eyelid.

INTERNAL RECEPTORS OR INTEROCEPTORS

Interoceptors are directly affected by stimuli arising within the body itself; those found in the walls of the alimentary canal and viscera are called enteroceptors, and those located in the tendons, joints, and voluntary muscles are spoken of as proprioceptors.

Enteroceptors are connected to the visceral sensory nerves but they do not form definite sense organs, they are stimulated by conditions in the digestive system and supply information to the central nervous system. Enteroceptors detect appetite, nausea, hunger, and thirst. Hunger is not to be confused with appetite, it is a nutritional deficiency in the blood, while appetite is a memory of food causing internal changes in conjunction with sight, odour, or taste. Thirst is a condition of the mucous membrane of the throat due to decrease of water and an increase of salt content in the blood. Some **genital corpuscles** are present in the dermis of nipples and external genital organs, they are associated with sexual sensations.

Proprioceptors are connected to the somatic sensory nerves, they generally form spindle-shaped bodies consisting of free nerve endings enclosed in layers of connective tissue or encapsulated corpuscles enclosing tactile cells, they are found in tendons, joints, and voluntary muscles. Proprioceptors are stimulated only indirectly by the environment, they supply information to the central nervous system regarding visceral pains, position of the body, and co-ordinated movements of various parts of the body. They give an idea of weight of objects, and provide a "muscle sense" making one aware of the degree of tension in a muscle.

Revision Questions

1. What are sense organs ? Give a brief classification of sense organs found in animals.
2. Give a brief account of various mechano-and chemoreceptors.
3. What are photoreceptors ? Describe in brief the structure and function of photoreceptive organs of mammal.
4. Write short notes on :
   (i) Pacinian corpuscles, (ii) Equilibrium receptors, (iii) Lateral line system, (iv) Statoacoustic organs, (v) Thermoreceptors, (vi) Interoceptors.

CHAPTER 16

Reproduction

Reproduction is one of the fundamental properties of living organism by which every kind of living organism multiplies to form new individuals of its own kind (reproducing or parental organism). It is indispensable condition for preserving the species and is necessary to make up the losses that occur due to competition, predation, disease, starvation and other causes of death. In most organisms, except few lower ones, reproduction is always necessary to renew the genetic material which appears to have an inborn process of ageing. Although the reproduction is accomplished in several different ways depending on the organism, its most essential feature is the separation from the parent organism of a suitable complement of its DNA (accompanied by a portion of other protoplasmic constituents) to form eventually one or more new individuals.

PATTERNS OF ANIMAL REPRODUCTION

Two basic patterns of reproduction have been observed among animals and these are **asexual** and **sexual**. In asexual reproduction an individual can give rise to daughter individuals by mitotic divisions of a part of its own body; no gametes are required. Asexual reproduction is the characteristic feature of lower organisms such as the protozoans, sponges, coelenterates, certain flatworms, ascidians and others. In sexual reproduction genetically distinct two special sex cells, called gametes, fuse to form one celled structure the zygote which inturn divides repeatedly to grow into a fully developed, new individual. The sex cells or gametes are of two types, the male gamete (also called spermatozoan or sperm) and the female gamete (also called ovum) and are usually produced in different gonads, *i.e.* the male gametes or sperms in the testes and the female gametes or ova in the ovaries, by reduction division, the meiosis. The main differentiating feature between these male and female gametes is the presence of **sex chromosomes** of different nature. The male gamete is usually small in size, motile and contains a very little cytoplasm and stored food, while female gamete on the other hand is large in size, immobile and contains massive cytoplasm and stored food materials (mostly fat, some proteins, minerals and a complete collection of vitamins). Each of the gametes frequently comes from a different parent so that in many instances sexual reproduction requires the participation of two parents. This, however, is not always the case as in the example of certain animals such as parasitic flatworms (tapeworm and flukes), free living ringed worms (earthworm and leech) and certain molluscs, where single organism forms the two types of gametes that undergo fusion. Such a organism that produces both types of gamétes (*i.e.*, seperms and ova) is called **hermaphrodite**. In most hermaphrodites the products of two sexes, *i.e.*, sperms and ova, do not mature at the same time, so that self-fertilization does not usually occur; cross-fertilization is common and for successful cross-fertilization the meeting of two hermaphrodite animals is essential. In brief sexual reproduction is often biparental but may also be uniparental depending on the species.

Why is sexual reproduction so widespread ? Asexual reproduction certainly appears to be an economical way to produce large number of new members of the species; new organism can be produced without the necessity of bringing together two reproductive cells and they are the exact copies of the producing parent. Then why is sexual reproduction so widespread ? Asexual reproduction gives offspring with a gene combination identical to that of the single parent This offers little opportunity for variation and variation in organisms is essential in order to cope with change. The sexual reproduction on the other hand gives such variations and for this reason nearly all living orgnisms have developed two sexes and reproduce sexually.

ASEXUAL REPRODUCTION

Asexual reproduction occurs by several methods. The dominant methods that are found among animals for asexual reproduction may be grouped under the following heads :

Fission : This is the common method of reproduction among most protozoans and which is simple division in which the maternal organism divides into approximately two equal daughter parts, each growing until it has acquired the usual size and structure of the full-grown organism. It, thus, consists in a division of the cytoplasm of the animal preceded by an orderly division of the nucleus which is always mitotic.

When fission produces two daughter individuals, each about half the size of the parent, it is said to be **simple** and **binary**. This type of binary fission is found in *Amboeba* and *Paramecium*. In simplest sporozoans a peculiar type of fission has been observed what is known as **multiple fission**. In this multiple fission the nucleus divides several times to produce a temporary syncytium, and then the cytoplasm falls apart round the nuclei. Multiple fission exhibits a specific condition of the nucleus called **polyenergid** in which several sets of chromosomes appear to be present; when the nucleus divides, these sets of chromosomes separate.

Budding or **gemmation :** A few of the multicellular animals (coelenterates and ascidians) produce small buds which grow gradually, ultimately acquiring the form characteristic of the producing organism. The sponge, such as *Scypha,* is such an animal which develops small buds near the point of attachment and each bud grows into an entire animal. Since the buds tend to remain attached to the parent, large colonies of sponges result. In *Hydra,* however, the bud grows into a small animal which breaks away from the parent and becomes an independent new organism.

In compound ascidians buds develop in quite differenet ways. In *Clavellina*, a hollow median stolon grows out from the ventral side of the abdomen andlon this buds grow. Blood vessels and other structure grow into these buds and remain common to the entire colony. In *Botryllus* the buds grow on the paired outgrowths of the atrium while in *Doliolum* at one point. These buds usually detach from the place of their origin and then migrate to another place where they become attached again.

Spore formation : Spore formation is a common form of asexual reproduction which is widely distribted among plants, but there is one class of Protozoa, the Sporozoa, which produce the small asexual reproductive bodies, the spores. It involves a series of cell divisions giving rise to several small cells called spores which temporarily remain within the confines of the original cell membrane or cell wall of the parent cell. They are eventually liberated by rupture of the parental membrane or wall and under favourable conditions resume growth. In general, spores are able to withstand unfavourable environmental conditions such as dryness, extreme heat or cold. The malarial parasite, *Plasmodium*, after invading the human red blood cells, produces many spores within each invaded cell. Also, spores are produced in the body of a mosquito which has sucked the blood of an infected person. These spores may be injected along with mosquito's saliva when another person is bitten.

Fragmentation : This method of asexual reproduction is common in coelenterates, flatworms and a few oligochaetes. In this method the body of adult organism breaks apart into two or more pieces (fragments), each of which then grows and reforms the parts it lacks to reconstitute a complete animal. Sometimes the fragments are roughly equal in size; more often, one is smaller and removes no vital organ from the parent animal.

Parthenogenesis : The development of an egg cell into a new individual without the participation of sperm cell from the opposite sex, is called the **parthenogenesis**. It is considered to be a form of asexual reproduction since it does not involve the fusion of two gametes. Parthenogenesis is a naturally occuring phenomenon among insects, crustaceans, rotifers and some platyhemininthes.

In the **honeybee**, an example of natural parthenogenesis, the males or drones all have their origin from unfertilized eggs and are, therefore, haploid. The females which are the workers and the queen develop from fertilized eggs and are; therefore. diploid. The queen is apparently only inseminated once but can store the sperm for as long as five years or more using it for fertilizing eggs, apparently at will, during this time. **Aphids**, another example of natural parthenogenesis, lay eggs which hatch without fertilization into other females. In the late summer or early fall, however, some of the eggs hatch into sexual males and females. Mating must take place between these if they are to produce offspring.

Parthenogenesis is rare in the vertebrates. MASTIN (1962); KALLMAN and HARRINGTON (1964) observed parthenogenesis in a few lizards particularly of the genus (*Cnemidophorus*. A breed of white

turkeys were found at the experiment station at Beltsville, Maryland which laid eggs which would hatch without fertilization. When such eggs were incubated a certain small percentage of them would produce embryos and some of them would hatch and grow into adults, turkeys without fathers.

Gynogenesis and androgenesis : Recently PARKES (1960) and KALLMAN (1962a) recognized **gynogenesis** as one of the methods of reproduction in a few non-segmented worms, a ptinid beetle and two species of teleost fishes. In gynogenesis the development of a new individual takes place from the egg which is activated by spermatozoan but spermatozoan does not contribute any genetic material to the egg. The resulting embryo carries only maternal chromosomes. The best examples in which gynogenesis is common are *Poecilia*, a fish and *Ptinus*, a latro beetle.

Androgenesis is the reverse condition of gynogenesis. When chromosome contribution in the developing egg comes exclusively from the male is called androgenesis. Androgenesis in animals is known only experimentally, naturally occurring androgenesis has not reported so far.

SEXUAL REPRODUCTION

Sexual reproduction is the rule in all multicellular animals including sponges and it is the most common way in which new individuals of the same species are produced. In all multicellular animals it consists of the union of two dissimilar gametes an egg nucleus with a sperm nueleus—to produce a single-celled diploid zygote which ultimately develops into a multicellular organism resembling the parents.

The organs which produce the gametes are called the gonads. The gonads which produce sperm cells are called the testes, while which produce egg cells are called the ovaries. In *Hydra* the gonads are temporary structures and appear during period of sexual reproduction as small swellings on the surface of the body. In all multi cellular animals above the coelenterates the gonads are developed within the body.

Sexual reproduction involves two most fundamental events, meiosis and fertilization. **Meiosis** is the means by which gametes from the germinal epithelium of the gonads are formed and reassortment of different genes takes place in the formation of gametes. **Fertilization** involves the fusion of two dissimilar gametes or nuclei and is the means for combining or mixing different genes in the production of offspring. The two dissimilar gametes on fusion or fertilization form a diploid structure called the zygote which gives rise to the new individual. Thus sexual reproduction permits the combining within a single individual of the best genes of its parents.

Almost all species of animals have some means of the sexual mixing of the genes. Let us survey some of the methods of sexual reproduction found in animals.

Conjugation : This is the method of certain protozoans through which the mixing of the genes is achieved. As workedout in certain species of *Paramecium*, the process is as follows. Several defferent mating types exist and any one can conjugate with any of the other mating types, but not with its own. The two animals from two different mating types come together at their oral grooves and form a protoplasmic connection between them. Meanwhile the small micronucleus of each divides meiotically to form four haploid micronuclei. Out of these four micronuclei three become disintegrated and the remaining one undergoes a mitotic division to form two micronuclei. One of these from each animal migrates across the protoplasmic connection and unites with the micronucleus remaining in the opposite conjugant. The animals now separate and the fusion nucleus undergoes divisions accompanied with later cellular divisions.

External fertilization : Most animals produce sperms and eggs, but in many water species these gametes unite in water; there is no union between oppsite sexes. This is called external fertilization. In almost all the aquatic animals the fertilization is external.

Internal fertilization : In all land animals the fertilization is internal. To make the fertilization successful the two opposite sexes of the same species undergo **copulation**, whereby the sperms can be transferred to the body of the female where fertilization takes place. Many water animals also have this process although it is not a rule.

Place of embryonic development : Those animals which lay eggs are said to have **oviparous reproduction**. In such cases the major part of embryonic develomment takes place outside the female body, even though fertilization has been internal. Those animals which give birth directly to the fully developed young ones are said to have **viviparous reproduction**. For example, mammals have a small egg with comparatively little yolk and the embryo develops within the mother's body, deriving nourishment from her body in the process. There is third type of reproduction what is known as **ovoviviparous reproduction** where there is a large eggwhich furnishes food for the developing embryo but there is internal fertilization and the egg remains in the female until it hatches. Thus, the young are born fully developed and active, but they have not derived nourishment from their mother during embryonic development.

SEXUAL REPRODUCTION IN MAMMAL (MAN)

Primary sex characters and sex accessories : The primary sex characters include those structures which actually take part in the process of reproduction. These also make distinction between male and female individuals. These include mainly the **gonods**, *i.e.*, **testes** in the male and **ovaries** in the female. The gonads are concerned not only with the production of gametes or sex cells (spermatozoa and ova) but also secrete hormones that are responsible for the functional state of sex accessories and, to some extent at least, influence the psychobiologic phenomena involved in the mating reaction. Sex accessories, on the other hand, include structures that are involved mainly in the transmission of gametes (sex cells) or developing zygotes from the site of their origin to the exterior of the organism. In male the sex accessories comprise a pair of attached **epididymes** in which sperms are stored, a pair of **vasa deferentia** which serve to transfer sperms from the epididymes away from the testes, a pair of **seminal vesicles** which provide essential nutrients and fructose to the developing sperms especially in men, bulls and other species but not in cats and dogs, a single **prostate gland** that serves to lubricate the passageway to the outside of the body through penis, a pair of COWPER'S **glands** which are also lubricatory in function and a **penis** that may be erected by the circulation to facilitate placement into the vagina of the female for the ejaculation of sperm.

In female the sex accessories include a pair of **fallopian tubes (oviducts)** the distal ends of which (infundibula) serve to receive ova discharged from the ovaries and the proximal ends enter the uterus a single **uterus** which either sloughs off periodically or develops in part into a placenta, depending on whether pregnancy occurs, a **vagina** that serves to receive sperms, one pair (in man) or more pairs of **mammary glands** which produce milk for the new born and **bulbourethral glands** which secrete a fluid similar to that of the glands connected to the urethra of the male.

Secondary sex characters : These include more or less external specializations which are physical differences between the opposite sexes. They have nothing to do with the production and movements of sex cells (gametes), but that may serve to bring the sexes together for mating, to provide for the protection or nutrition of the young.

In mammals the secondary sex characters are less pronounced than in many of the fishes, amphibians, reptiles and birds. The important secondary sex characters of man are the following :

1. The males have high muscular body than the females.
2. The mammary glands are well developed in females and quite rudimentary in males.
3. The pitch of voice is very high in females than males.
4. The males have well developed hairs on chest, face, and all over the body. The females, on the other hand, don't have hairs on the chest and face (usually) and all over the body the hairs are present sparsely.

Male reproductive organs : Sperms are produced in the **testes.** In man and most other mammals the testes are two oval bodies and are suspended in a sac hanging from the lower wall of the abdomen, the **scrotum**. Apparently human sperms cannot develop at the high temperature found within the body cavity, hence the testes are suspended outside the body. Generally a temperature 3°C lower than that of the body is said to be a optimal temperature for the development and maintenance of sperms. This optimal temperature is ensured through the interplay of temperature receptors and cremaster muscles. When the temperature outside the scrotum is suboptimal than the muscles under the skin of scrotum contract, drawing the testes close to the body. In contrast, when the considerable amounts of heat are generated within the body than the muscles of the scrotum loosen so that the testes hang down away from the body. In this way a fairly constant testicular temperature is maintained.

In man testes remain continuously in scrotum because he breeds continuously independent of time of year. In contrast in seasonally breeding mammals such as bats, the testes ascend through inguinal canals and remain in the body cavity during the non-breeding period. During these periods the testes undergo involution. In man the inguinal canal becomes sealed off short after birth. In whales, elephant, seal, and rhinoceros the testes remain permanently in the abdominal cavity.

Each testis is composed of coiled anastomosing **seminiferous tubules** lined with epithelial cells that produce sperm cells ; also **interstitial cells of** LEYDIG around the tubules produce the male sex hormone, testosterone, which promotes the development of the accessory glands and controls male secondary sex characteristics. As sperms are released into the interior of the tubules they are carried by ciliary action to the **epididymis** which lies on the outside of and partially encircling the testis. The epididymis is a convoluted tube of about 20 feet length and is made of a compact mass of small coiled tubules (vasa recta). Testis and epididymis together constitute **testicle.** In the epididymis the sperms are stored so that they become motile. Epididymis connects with the **vas deferens.** Vas defernes is a muscular tube that leaves the scrotum by the **inguinal canal** and empties into the **urethra**, the duct that leads from the bladder. The terminal portion of each vas deferens enlarges to form an **ejaculatory duct**, capable of contraction and expulsion of the sperms which are stored there. A glandular **seminal vesicle** empties into each ejaculatory duct before it connects to the urethra. The seminal vesicles secrete a viscid fluid which is expelled along with sperms. The mixture of this fluid and the sperms is known as **semen.** The secretions of the seminal vesicles contain glucose and fructose which are required by the sperms for their development. In the vas deferens the sperms are inactive but as soon as they mix with this secretion they become quite active.

The urethra is surrounded by a **prostate gland** at the point where the ejaculatory ducts enter. This gland has numerous small ducts emptying into the urethra. Thus prostate gland discharges its secretions directly into the urethra. Its secretions are thin milky in nature containing citric acid, calcium, phosphate, fibrinogenase, fibrinolysin, spermin, etc., and contribute 15—30% of the total volume of the semen.

Another pair of glands, the COWPER's **glands,** are also attached to the urethra about two inches below the prostate gland. Their secretions are also alkaline and serve lubricant for the semen. The secretions of the prostate and COWPER's glands suspend the sperms, motile them, nourish them and neutralize the normally acid environment of the urethra and of the female reproductive tract to a pH more suitable for sperm survival.

Urethra communicates with the exterior of the body through a muscular structure, the **penis.** The penis of man is composed of three columns of spongy tissue, the **corpora cavernosa**, surrounding the urethra, and a layer of skin on the outside. The tip of the penis enlarges slightly to form the **glans**, which is normally covered by a fold of skin, the **prepuce**. The function of penis is to deposit the semen in the genital tract of the female. Penis is found in all the species of mammals, in some reptiles and birds. In most reptiles and birds, one opening, the cloaca serves as the passage for eggs and sperms and also for the elimination of wastes and these animals mate by juxtaposition of their cloaca.

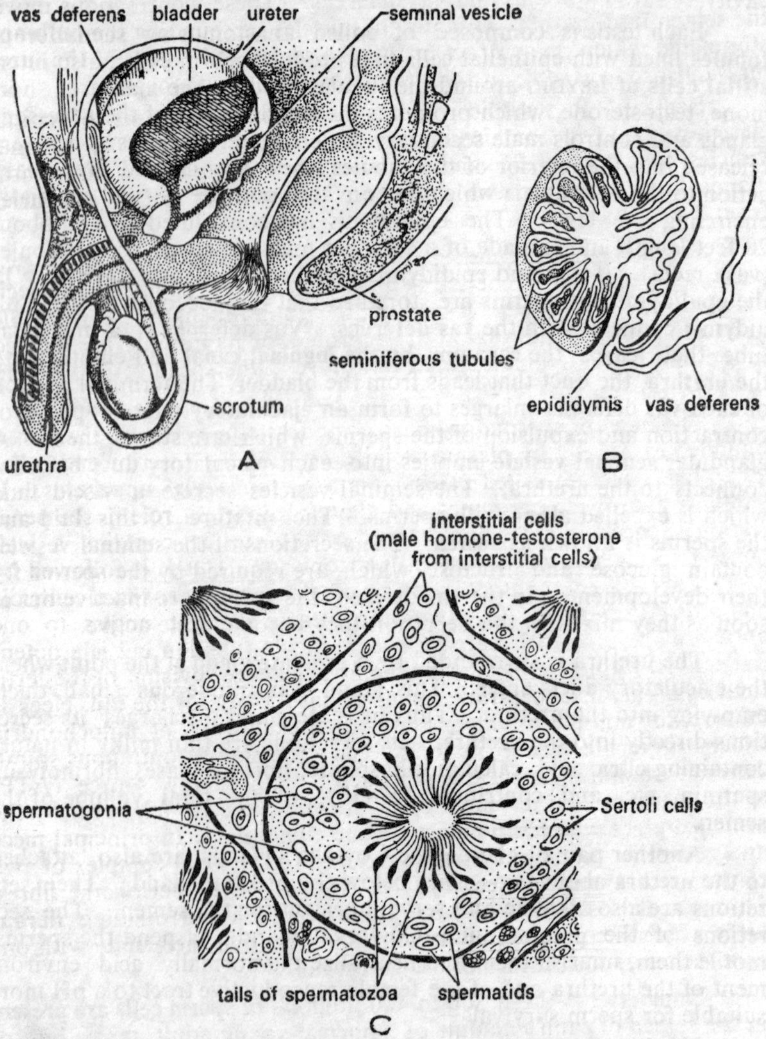

Fig. 16·1. A—Male sex organs ; B—Gross structure of testis ; C—Detailed structure of seminiferous tubule in T.S.

Erection of penis : Erection of penis is associated with sexual stimulation. It is caused by dilation of the blood vessels carrying blood to the spongy tissues, resulting in the collection of blood within these spaces. As the tissues become distended they compress the veins and so inhibit the flow of blood out of the tissues. With continued stimulation the penis and the underlying bulb become hard and enlarged.

Ejaculation : Continued stimulation of the penis leads to contraction of the muscles present in the scrotum, raising the testes close to the body, epididymis and vas deferens. These contractions move the semen into urethra. Finally the muscles surrounding the bulb are stimulated which on contract propel the semen out through the urethra and produce some of sensations associated with orgasm.

Semen : It is a fluid which is ejaculated at the time of coitus (insemination). It contains sperm cells and secretions of seminal vesicles, prostate gland and COWPER'S gland and also of urethral glands. In man the amount of semen discharged per ejaculation varies from 2·5 to 3·5 ml.

Spermatozoon : The spermatozoon of man consists of two parts, **head** and **tail**. The tail is divided into **neck**, **mid-piece**, **principal-piece** and **end-piece**. It is about 0·05 mm in length. It is motile in nature and enzymes that are responsible for its motility are located in the mid-piece. Human sperm can move at the speed of 3 to 8 mm/minute.

The **head** of spermatozoan is a spoon-shaped structure which is bounded externally by plasma membrane. At the anterior end it has a cup-like structure called **acrosome** made up of Golgi apparatus. It contains hydrolyzing enzymes and plays a important role in the penetration of sperm in the ovum. Head in its interior contains a well condensed nucleus and a very little cytoplasm. Head is followed by a short **neck**. Neck consists of a pair of centrioles, a proximal centriole and a distal centriole. The two centrioles lie at right angles to one another. The proximal centriole has no active function but is a potential activist within the egg during the first cleavage division of the fertilized egg. The distal centriole serves as basal body for the tail. Neck is followed by **mid-piece** which is composed exclusively of mitochondria which aggregate about its basal end, forming a continuous spiral. This mitochondrial apparatus provides a ready energy source (ATP) to the sperm tail for its motility. The mid-piece is followed by the **principal-piece** which ultimately ends in a **end-piece**. In principal-piece and end-piece the fibre system is reduced to the axial complex of two central fibres surrounded by the ring of nine peripheral fibres. Throughout the **principal-piece**, there occurs a surrounding **fibrous helix** which is composed of semicircular ribs that articulate with one another on opposite sides of the sperm tail.

Numbers : About 300 to 400 millions of sperm cells are present in the semen of each ejaculate of a normal young adult male but of those only one can fertilize each egg cell. The rest die within three days, retaining their fertilizing ability for only 24 hours. The other sperm cells apparently play an accessory role, however, perhaps by bringing about chemical changes necessary for fertilization. Males that produce less than 35 millions of sperms per millilitre of semen are generally sterile.

MALE SEX-HORMONES AND SEX-LIFE

The principal male sex hormone is **testosterone**. It is secreted principally by the **interstitial cells of** LEYDIG which are located in between the seminiferous tubules in the testes. Testosterone and other chemicals with testosterone like effects are known collectively as **androgens** and render masculine characteristics. The daily testosterone secretion in normal man is 4 to 8 μg/day. In young boys its secretion is very slow but its secretion is suddenly increased with the arrival of sexual maturity (puberty) and continues to rise until 30 years of age, after 30 it falls down but does not stop until death.

Effects of androgens : Androgens produced in the male human foetus are important in the development of external genitalia. When the human male is about 10 years old, renewed androgen production causes the enlargement of the penis and testes and also the prostate, the seminal vesicles and other accessory organs. These hormones also effect the growth of the larynx and an accompanying deepening of the voice, muscle development, skeletal size and distribution of body hair. They also stimulate the biosynthesis of proteins and so of muscle tissue, and also the formation of red blood cells. They stimulate the apocrine sweat glands, whose secretion attracts bacteria and so produces body odours associated with sweat after puberty.

In animals excluding man the androgens are also responsible for the powerful musculature and fiery disposition of the stallion, the cock's comb and spurs, the bright plumage of many adult male bird.

Regulation of male sex hormone production : Androgens or testosterone production is regulated by a gonadotropic (gonad stimulating) hormone called **luteinizing hormone** (LH) secreted by the anterior pituitary. Luteinizing hormone acts on the interstitial cells of LEYDIG so that they may be stimulated to release their androgens. These androgens have profound effects on the male reproductive organs. The androgens, in turn, inhibits the release of L.H. This type of regulating system in which the system is shut off when the products of its operation reach a certain level, is known as negative feed back. According to HOAR (1965 a, b) in hypophysectomised animals both the processes of spermatogenesis and androgen secretion are stopped and the size and functions of male genital organs are considerably reduced.

Another pituitary hormone, the **follicle-stimulating hormone** regulates the activity of the seminiferous tubules, stimulating the development of sperm cells.

Female reproductive organs : The female reproductive organs include a pair of ovaries, a pair of oviducts (Fallopian tubes), the uterus and the vagina.

The human **ovaries** are two small almond-like flattened bodies

They are lying on the sides of vertebral column behind the kidneys in the pelvic cavity. Each ovary is attached to the dorsal abdominal wall through mesovarium and ovarian ligament. Through this attachment numerous blood vessels and lymphatics enter into or emerge from the ovary.

Each ovary is principally composed of **stroma** of fibrous connective tissue and is lined externally at its free surface by a germinal epithelium of cuboid epithelial cells. The germinal epithelium proliferates thousands of primordial follicles during the embryonic life of an individual but less than a tenth per cent mature in the life of the individual and remainder degenerate into **atretic follicles**.

Each ovary is roughly differentiated into an outer **cortex** and an inner **medulla**. Although these two regions appear to contain distinct structures, there is no demarcation line between them in the mature ovary. In mature ovary the cortex contains follicles and corpora lutea in various stages of differentiation and disintegration. The medulla contains only the large blood vessels of the organ. During the growth of the ovary the germinal epithelium at various places dips into the body of ovary, and by the growth of the connective tissue of stroma a mass of epithelial cells becomes separated from the main layer. One cell of this mass gives rise to an **immature ovum** or **oocyte**; the remaining cells form a layer surrounding the ovum or oocyte as sac or follicle called **follicular epithelium** or **granulosa**. Immature ovum or oocyte and surrounding follicular epithelium or granulosa constitute the **primordial follicle**. The stroma of the ovary surrounding the follicular epithelium or granulosa becomes organized into connective tissue layers, the **theca externa** and the **theca interna**. Theca interna in case of man serves as a glandular and vascular oestrogen synthesizing layer.

Ovulation : Ovulation occurs when the follicle bursts liberating the egg. Normally only one ovulation occurs during each monthly reproductive cycle (in case of man), but two or more can occur, resulting in the fraternal type of twins or higher multiples. Ovulation occurs about 14 days before the beginning of menstruation.

Ovulation involves the development of primordial or primary follicle into a GRAAFIAN follicle and release of the ovum or egg from the follicle. The **follicle stimulating hormone** (FSH) from anterior pituitary initiates and thereafter together with environmental, nutritional and other hormonal factors including **luteinizing hormone** from anterior pituitary regulates ovulation.

In the course of becoming a GRAAFIAN follicle an egg produced from a primary oocyte first becomes partially detached from its surrounding epithelial cells beneath the **tunica albuginea**. As the primary follicle matures it starts moving further into the stroma and assumes deeper position there. Side by side it begins to enlarge and its surrounding cells continue to grow in number, forming many layers. A homogeneous covering of mucoprotein, **zona pellucida**, appears between the developing oocyte or ovum and the follicular cells. The follicle cells start differentiating and form a layer of granular cells (**granulosa**) around the zona pellucida and developing ovum or egg. Zona pellucida actually is the layer in which polar bodies produced during the meiotic division of the primary oocyte, are entrapped and this layer remains with the ovum until its implantation or death. Under the influence of pituitary gonadotropins a **follicular fluid** is secreted by the surrounding follicular cells and the cells immediately surrounding the egg become separated from the more remote cells (zona pellucida) and thus a large cavity, the **antrum** is formed. The fully developed follicle with an antrum, well established granulosa and theca is called a GRAAFIAN **follicle.** GRAAFIAN follicle measures about 10 mm in diameter and contains an ovum some 100 to 150μ in diameter.

The GRAAFIAN follicle during final stages of its development, moves to the surface of the ovary and produces a thin blister-like elevation, which eventually bursts releasing the ovum into the abdominal cavity. Now the ovum is swept into one of the oviducts as a result of the movement of the cilia located at the finger-like projections of the opening of the oviducts. From the oviduct the ovum is transported into uterus. In case the ovum is fertilized, it is implanted in the endometrium of the uterus but if it is not fertilized, it liquefies and discarded through the vagina.

The follicles, which enlarge but do not undergo ovulation, finally degenerate and form **atretic follicles**. The follicle which ovulates, is converted into a blood filled cavity called **corpus haemorrhagicum** just after the discharge of the ovum Blood comes from the blood vessels injured during the ovulation of ovum. Soon after this clotted blood is replaced by yellow body of lipids or luteal cells proliferated from the granulosa and theca cells and thus **corpus luteum** is formed. The formation and maintenance of corpus luteum are under the influence of **luteinizing hormone** of anterior pituitary. If the ovum is fertilized, the corpus luteum enlarges and remains during the first seven months of gestation ; if the ovum is not fertilized, the corpus luteum degenerates and becomes **corpus albicans** which appears as disorganized scarred globules.

Oocyte : A human oocyte is about 0·1 mm or 100 micrometer in diameter. It has an unusual large supply of ribosomes, enzymes, amino-acids and all the other cellular machinery that will be used in the early, rapid stages of biosynthesis characteristics of embryonic cells.

The eggs or ova are usually shed in the body cavity from where they are picked by the **ostia of oviducts** or FALLOPIAN **tubes** and come down into the tube. The pair of oviducts lead to a central pear-shaped **uterus** of about 5 by 10 cm (2 by 4 inches). The uterine wall is thick and is the site where development will take place in the event of fertilization. The uterus is connected to **vagina** through **cervix.** Near the opening of the vagina there are found **bulbourethral glands** which secrete a fluid similar to that of the glands connected to the urethra of the male. The secretion serves to lubricate and neutralize any acid which may be present. Vagina leads to exterior through vaginal orifice which is flanked by an inner pair of moist folds, the **labia minora,** enclosed with the fleshier hair-covered outer **labia majora.** These structures enclose the **clitoris** which is a small mass of tissue, just anterior to the urethral opening. The clitoris is homologous to the penis of the male ; it has a glans and prepuce and is capable of engorgement, but it serves no reproductive function. The **urethra** of the female opens just anterior to vaginal orifice.

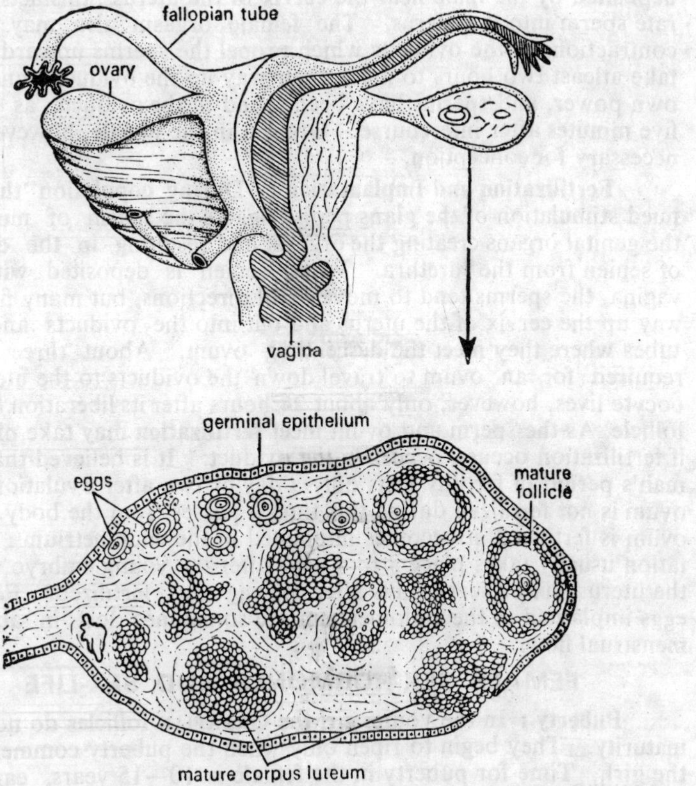

Fig. 16·2. Female sex organs (upper) ; T.S. of ovary showing different developing stages of ovum (lower).

Orgasm in the female : Under the influence of a variety of stimuli, the clitoris, the labia and other tissues in the pelvic region become engorged and distended with blood, as does the penis of the male. The distension of the tissues causes the liberation of secretion from the bulbourethral glands into the vagina that lubricates the walls of vagina and neutralizes its highly acid, and therefore, spermicidal, secretion. The thrusting of the penis into vagina stimulates the lower third of the vagina and also causes the prepuce of the female clitoris to move back and forth over the clitoris, stimulating it. As the female becomes excited the clitoral and genital fold swell and vaginal wall secretes and creates a moist lubricant. Also the breasts usually swell and nipples become erect. On entering the plateau phase, her breathing and heart rates increase. Orgasm in the human female is usually associated with the contractions of the vaginal wall. These may be spasmodic.

At orgasm, the cervix drops down into the upper portion of the vagina. This creates a syringing action on the pool of semen deposited by the male near the cervix of the uterus ; this acts to aspirate sperm into the uterus. The female orgasm also may produce contractions in the oviducts which propel the sperms upward. Sperms take atleast two hours to make their way up the oviducts under their own power, and sperms have been found in the oviducts as soon as five minutes after intercourse. Orgasm in the female, however, is not necessary for conception.

Fertilization and Implantation : During copulation the continued stimulation of the glans penis triggers the spasm of muscles of the genital organs creating the orgasm and resulting in the discharge of semen from the urethra. When semen is deposited within the vagina, the sperms tend to move in all directions, but many find their way up the cervix of the uterus and out into the oviducts and up the tubes where they meet the descending ovum. About three days is required for an ovum to travel down the oviducts to the uterus. An oocyte lives, however, only about 24 hours after its liberation from the follicle. As the sperm and ovum meet **fertilization** may take place and if fertilization occurs, occurs in the oviduct. It is believed that a woman's period of fertility lasts only a day or two after ovulation. If the ovum is not fertilized during this time it passes from the body. If the ovum is fertilized, it becomes implanted in the endometrium ; implantation usually takes place 3 to 4 days after the young embryo reaches the uterus, six or seven days after the egg was fertilized. Fertilized eggs implanted in the endometrium are sometimes lost in abnormal menstrual flow.

FEMALE SEX-HORMONES AND SEX-LIFE

Puberty : In the young girl the GRAAFIAN follicles do not reach maturity. They begin to ripen only when the puberty commences in the girl. Time for puberty in the female is 10—15 years, earlier in tropical countries but somewhat later in temperate or cold countries, indicated by the onset of first menstruation.

Puberty is the period of onset of sexual maturity or sex organ function and beginning of youth. It is achieved under the control of the pituitary-gonad axis. During this period many changes, morphological, psychological, physiological and endocrinological take place in the individual. The accessory sex organs, uterus, vagina and mamma undergo a marked increase in growth. Sex differentiation is complete and secondary sexual characters make their appearance and become fully developed after a short period. These are pubic and axillary hair, the peculiar development of the skeleton (especially the enlargement of the pelvis) and deposition of fat on the hips which causes the body to assume a more famine contour.

All changes undergone by a young girl during puberty are due to pituitary gonadotropins which stimulate the ovaries. Ovaries on stimulation secrete some specific hormones especially the oestrogen and progesterone which regulate the development and maintenance of primary as well as accessory sex characters of the female individual.

Menstrual cycle :. The menstrual cycle is characteristic of primates. It does not occur in other vertebrates. In mammals other than primates the sexual cycle is called **oestrous cycle.** Females of all mammals except primates permit copulation only during a definite period in their sexual cycle, which is known as **fertile period** or **oestrous** or **heat.** In menstruating mammals such periods in their life cycle do not occur. Copulation may be performed at any time during the sexual cycle. In menstruating mammals ovulation always occurs midway between two menstrual periods. In human Female the cycle usually lasts about 28 days. Menstruation begins about 15 days after ovulation and lasts about 2—4 days.

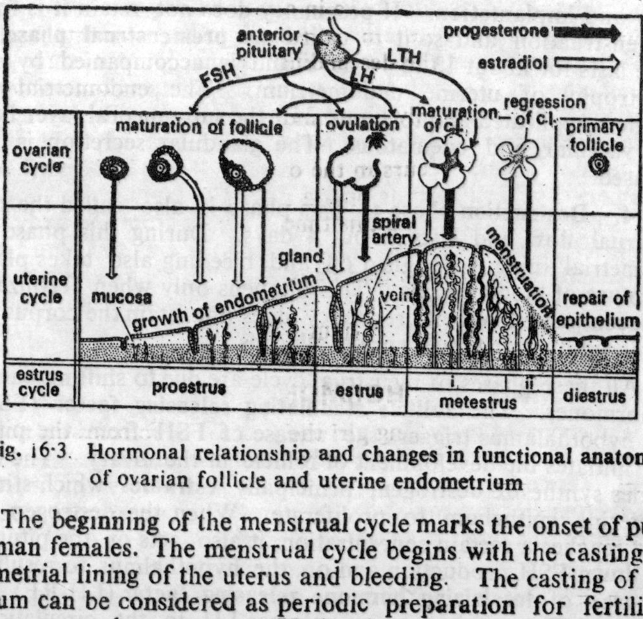

Fig. 16·3. Hormonal relationship and changes in functional anatomy of ovarian follicle and uterine endometrium

The beginning of the menstrual cycle marks the onset of puberty in human females. The menstrual cycle begins with the casting off of endometrial lining of the uterus and bleeding. The casting of endometrium can be considered as periodic preparation for fertilization of the ovum and pregnancy. Under the influence of follicle stimulating hormone (FSH), an ovum and its follicle mature, and the follicle secretes increased amounts of oestrogen which stimulates the regrowth of the endometrium. The increased production of oestrogen triggers the production of luteinizing hormone (an example of positive feedback) Paradoxically, at the end of the cycle, LH and FSH production decline as a result of the increased concentration of progesterone and oestrogen. The menstrual cycle involves the following phases :

1. Proliferative phase : This phase is also called **follicular phase** or stage of repair and proliferation. This is mainly influenced by oestrogens secreted by the follicular cells forming the theca interna. During this phase repairing of endometrial lining of the uterus (which was sloughed off during menstruation) takes place and a functional lining is reappeared in the uterus to receive the fertilized ovum. This phase extends from the end of menstruation to ovulation and generally involves 10 days.

2. **Ovulatory phase :** During this phase no conspicuous changes occur in the uterus endometrium. It occurs midway between two menstrual periods. During this period ovulation takes place and body temperature rises which remains high until the onset of next menstrual period.

3. **Secretory phase :** This phase is under the control of progesterone and oestrogens secreted by the corpus luteum, therefore it is also known as **progestational** or **luteal phase**. It is also known as **progravid phase** as this phase prepares the endometrium for pregnancy and implantation. If pregnancy does not retain it is followed by menstruation and so it is called as **premenstrual phase**. This phase lasts for about 13 to 14 days and is accompanied by marked hypertrophy of uterine endometrium. The endometrial glands become complicated and tortuous and the sub-mucoral layer becomes very vascular and edematous. The glandular secretion is greatly increased.

4. **Destruction phase :** This phase is also called the stage of menstrual flow and lasts about 4 days. During this phase uterus endometrial lining is sloughed off and bleeding also takes place due to rupture of blood vessels. This happens only when fertilization is not accompained. When pregnancy does not retain the corpus luteum is replaced by scar tissue, **corpus albicans.**

All these phases of menstrual cycle are due to shifting balance of the hormones. The **follicle stimulating releasing factor** (FSH-RF) from hypothalamus triggers the release of FSH from the pituitary. FSH initiates the development of follicle in the ovary. The cells of follicles synthesize **oestrogen**, principally **estradiol** which stimulates the uterine epithelium to proliferate. When the oestrogen in the blood reaches a certain concentration, it also acts on the pituitary to shut down FSH production and on the hypothalmus to result in the discharge of **luteinizing hormone releasing factor** (LH-RF) to the pituitary. The pituitary now releases LH to the circulation. In combination with residual FSH, it results in **ovulation**. The LH also causes the cells of follicle to alter, enlarge and fill the old follicular cavity. The structure formed is the **corpus luteum** (yellow body). The cells of corpus luteum soon start secreting the **progesterone** and **oestrogen hormones**. As the levels of these hormones increase, they inhibit the production of the gonadotropic hormones from the pituitary. Production of ovarian hormones than drops, as a result of which the support of the uterine lining fades, leading to menstruation. However, if implantation occurs, the membranes surrounding the developing embryo embedded in the uterine lining contribute to the formation of the placenta and begin to secrete a progesterone-like hormone that finally replaces that produced by the corpus luteum. This prevents the uterine lining from sloughing and supports pregnancy. The hormones oestrogen and progesterone, produced by the developing placenta, also stimulate enlargement of the breasts. **Prolactin** and **oxytocin** hormones regulate the milk secretion and these are released after the inhibitions of the placental harmones are removed at birth. When implantation does not occur than in response to low level of ovarian hormones, the level of pituiary gonadotrophic hormones begins to rise again, followed by development of a new follicle and a rise in oestrogen as the next monthly cycle begins.

Lactation : Human milk is available to feed the baby a few days after the birth. It is made by the mother's **mammary glands** from constituents in the maternal blood and its formation is under control of **prolactin** and **oxytocin hormones** of the pituitary. The milk contains protein, fat and carbohydrate which fullfil the requirements of baby for heat and energy as well as for growth and repair of tissue.

The human **mammary gland** consists of about 15 separate milk producing systems which are arranged radially around the nipple. The milk is produced deep to the surface in the **milk-producing alveoli.** These alveoli lead via series of branching ducts to a main milk duct which opens at the nipple. There are thus 15 openings at the nipple, each leading to its own milk producing alveoli.

The main milk duct has a dilatation just below the surface which is known as the **lactiferous sinus** in which some milk is stored. Baby can take milk from this sinus after taking the nipple and surrounding tissue into its mouth and using a sucking champing action. But the baby can not by this action obtain milk from the deeper milk-producing alveoli.

However, the suckling stimulates the sensory receptors present in the nipple and a nervous reflex via the hypothalamus releases oxytocin from the posterior pituitary gland. The oxytocin contracts the myoepithelium surrounding the alveoli and forces the milk forwards towards the nipple. This process is termed **milk ejection**. Without an adequate level of circulating oxytocin, the breast may be engorged with milk, but the baby can not get an adequate milk supply.

Effects of female sex hormones : Oestrogen and **progesterone** are the female sex hormones. They are produced by the ovarian follicles under the influence of FSH of pituitary gland. Oestrogens, principally **estradiol**, stimulate the development of the breasts, the external genitalia, pubic and axillary hair, and the distribution of body fat. Both oestrogen and progesterone are required to prepare the endometrium for the implantation of embryo; neither can do the job alone. The production of these hormone is regulated by gonadotropic hormones secreted in the pituitary.

Oestrous cycle : In mammals other than menstruating primates the sexual cycle is called **oestrous** cycle. Like menstrual cycle it is also under the control of ovarian hormones-oestrogens and progesterone. The duration of oestrous cycle varies from animal to animal and may be influenced by factors such as light, temperature, nutritional status and social relationships. Like menstrual cycle it also involves four phases : oestrous, metestrous, diestrous and pro-estrous.

Revision Questions

1. What is reproduction ? Describe in brief the various patterns of reproduction found in animals.
2. What is asexual reproduction ? Discuss various methods of asexual reproduction.
3. What is sexual reproduction ? Why is sexual reproduction so wide-spread. Describe in brief the sexual reproduction in mammal.
4. Write short notes on :
   - (i) Hermaphrodite,
   - (ii) Parthenogenesis,
   - (iii) Androgenesis,
   - (iv) Gynogenesis,
   - (v) Secondary sex characters,
   - (vi) Corpus luteum,
   - (vii) Atretic follicle.
   - (viii) Menstrual cycle.

CHAPTER 17

Experimental Physiology

The experimental physiology includes tests of carbohydrates, proteins, lipids (fats), enzymes, estimation of haemoglobin counting of R.B.C., W.B.C., differential count of W.B.C., study of coagulation of blood, bleeding time of blood, clotting time of blood, preparation of haemin crystals, effect of osmolarity of salt solutions and haemolytic agents on R.B.Cs., study of circulation of blood in the capillaries of foot of frog, recording of heart beat in situ, recording of muscular contraction, determination of rate of respiration, reflex action, wiping reflex, knee-jerk reflex, test of urine for urea, proteins, ketones and sugar, and demonstration of osmosis.

CARBOHYDRATES

The carbohydrates are the compounds of the carbon, hydrogen and oxygen. Chemically the carbohydrates are polyhydroxy aldehydes or ketones. Carbohydrates are classified into **monosaccharides (monomers) oligosaccharides (oligomers)** and **polysaccharides (polymers)**. Monosaccharides and oligosaccharides are soluble crystalline substances having a sweet taste and are collectively known as sugars. Polysaccharides are insoluble amorphous substances and are thus called non-sugars.

EXPERIMENT No. 1

Tests for Carbohydrates

1. **Fehling's test :** Take 5·0 ml of Fehling's solution in a test-tube, add 1·0 ml of 0·2% sugar solution and boil. The colour changes with the formation of yellow brick-red precipitate. The precipitate is due to the formation of cuprous oxide. The precipitate indicates that reduction has taken place.

2. **Benedict's test :** Take 5·0 ml of Benedict's reagent in a test-tube, add 0·5 ml of sugar solution and mix it well. Heat to boil for two minutes than cool under tap-water. Green, yellow or red precipitate is formed. The precipitate is due to the formation of cuprous oxide.

3. Take 3·0 ml of given solution of sugar in a test-tube, add 1.0 ml of saturated solution of picric acid and 1.0 ml of 1% or 4% NaOH and heat it. Red colour develops. The red colour is due to the reduction of picric acid to picrylic acid.

4. Take 3.0 ml of given solution of sugar in a test-tube, add

1.0 ml of 4% NaOH and boil for two minutes. Solution becomes yellow and acquires light-brown colour on standing. The yellow colour is due to formation of paramylin.

PROTEINS

Proteins are complex organic compounds. They are composed of carbon, hydrogen, oxygen, nitrogen and sulphur in their molecules. Proteins are the polymers of the amino acids.

EXPERIMENT No. 2

Tests for Proteins

1. **Biuret test :** Take 2·0 to 3·0 ml of protein solution or egg, albumin solution in a test-tube, add an equal volume of 10% NaOH, mix thoroughly and add 0·5% $CuSO_4$ solution drop-by drop. Mix it well until a purple-violet or pinkish-violet colour develops.

2. **Ring Biuret test :** Take 3·0 ml of protein solution or egg-albumin solution or any other solution of protein, add 1·0 ml of 40% NaOH and 1·0 ml of 1% $CuSO_4$ with the help of pipette. The two liquids should not mix. Rotate the test-tube gently. A pink or violet ring is formed at the junction of two liquids.

3. Take 3·0 ml of protein solution or egg-albumin solution in a test-tube, add 5 drops of lead acetate. A precipitate is formed. Add 40% NaOH drop-by-drop until the precipitate dissolves. Boil the solution, brown or black precipitate appears.

4. **Millon's reaction :** Take 5·0 ml of dilute solution of protein or egg-albumin in a test-tube, add 3 to 4 drops of Millon's reagent. Mix and heat the mixture till it boils. White precipitate appears which gradually turns red upon heating. In case of other proteins the solution becomes red only.

LIPIDS (FATS)

The lipids are the organic compounds which are insoluble in the water but soluble in the ether, chloroform, benzene, hot alcohol and petroleum ether. They contain long chains of aliphatic hydrocarbons or benzene ring in their molecules. The lipids are nonpolar and hydrophobic.

EXPERIMENT No. 3

Tests of Lipids

1. **Solubility test :** In five test tubes take 3 ml of water, alcohol, ether, benzene and chloroform. To each test-tube add 2 or 3 drops of lipid. Shake well and note the solubility. The lipid floats on water but it is miscible with ether, benzene and chloroform, while it sinks to the bottom in alcohol.

2. **Formation of a transluscent greasy spot :** Place a drop of olive oil on a piece of ordinary paper. A transparent greasy spot is produced at the point of contact with the oil.

3. **Formation of acrolein :** Take 0·5 gm of powdered potassium

bisulphate ($KHSO_4$) or sodium bisulphate ($NaHSO_4$) in a clean and dry test-tube. Pour 3 or 4 drops of olive oil on the salt. Mix thoroughly with glass rod and heat it. The irritating smell of acrolein is given off. In this reaction the glycerol of the fat is dehydrated and acrolein or acrylic aldehyde is produced.

4. Take 3 ml of olive oil in a test-tube, add 2 drops of oleic acid. Mix them thoroughly. Rancid oil is obtained.

5. Take 2 drops of rancid oil in a test-tube and add 3 ml of sodium hydroxide (NaOH.) The rancid oil dissolves in sodium hydroxide (alkali) forming the soap.

6. Take 10 drops of olive oil in a test-tube, and 2 ml of alcoholic caustic soda and boil. A soft mass of soap separates which dissolves on cooling. This confirm the lipids (fats).

ENZYMES

EXPERIMENT No. 4

Test for Salivary Amylase

The enzyme salivary amylase is present in the saliva. It is also known as **ptyalin**. It changes starch into glucose and maltose. Salivary amylase acts steadily at body temperature but it acts optimally between 40 to 50°C. The optimum pH for activity of amylase is between 6 to 7 and that of salivary amylase of man is 6·6.

For the demonstration of amylase activity of salivary glands, saliva of man and salivary glands of cockroach are the best materials.

Preparation of enzyme extract : For the preparation of extract of salivary glands, 5—10 salivary glands are taken out from the cockroaches. The salivarv glands are ground in a mortar with little thymol. The solution is diluted with 50 per cent glycerol to make 10 per cent solution of salivary gland extract. The extract is stored in a bottle with toluene.

Test for amylase activity : Take 5 drops of extract in an incubation tube and add 5 drops of 1% starch solution, mix it and then cover the mixture with a layer of toluene. In another incubation tube take 5 drops of boiled extract, add 5 drops of starch solution. This will serve as the control experiment. Incubate both the tubes in a bath at 37°C for about one hour. After one hour take a small part of the incubated mixture and add iodine solution. No change in colour of iodine solution will indicate the presence of **amylase enzyme.** Take another part of the extract. add few drops of Benedict's solution and heat it till it boils. A brick-red precipitate will confirm the presence of amylase enzyme in the extract. Repeat the above two tests with the control mixture. In the first test violet colour will confirm the presence of starch as such which has not been digested.

Starch always gives bluish-violet colour with iodine solution. In the experimental solution the enzyme extract is mixed with iodine solution, the amylase present in the enzyme extract hydrolyses the starch into maltose and glucose, therefore, there is no change in the colour of iodine. In the control experiment amylase is destroyed by boiling the enzyme extract, the starch is not hydrolysed into maltose and glucose, therefore, it gives bluish-violet colour with the iodine solution. In the second test with Benedict's solution a red precipitate of cuprous oxide appears. In this glucose is formed. Since glucose is a reducing sugar, it reduces copper sulphate into cuprous oxide.

EXPERIMENT No. 5

Test for Pepsin Enzyme

Pepsin is the most important enzyme of the gastric juice. It hydrolyss proteins into proteoses and peptones. It is secreted by the gastric glands in the gut in an inactive form which is converted into active pepsin by the hydrochloric acid present in the stomach. Pepsin requires a rather definite pH for its optimum activity, The optimum pH for the activity ranges between 1·5 to 2·2.

Preparation of substrate : To test the presence of pepsin acid solution of casein is generally used as substrate. To prepare the substrate take 2·0 gm of vitamin free casein in a beaker, add 5·0 ml of NaOH of 0·1 normality, dilute it upto 50 ml with distilled water and heat it at 60°C. Stirr the mixture well until the casein is completely dissolved. Transfer this mixture into another beaker containing a mixture of 5 ml of HCl of 1 normality and 4·5 ml of distilled water. In this way a 2% solution of casein having 1·5 pH is prepared.

Preparation of enzyme extract : For preparing the enzyme extract, dissect three or four frogs and take out the stomach portion, keep the stomach portion in a petridish, wash it thoroughly with distilled water and add a little thymol to check the bacterial action. Ground the stomach portion in a mortar and add 50% glycerine so as to prepare appoximately 10% solution of stomach tissue. Centrifuge the extract and then filter. Store the filterate in the test-tube with toluene.

Test of enzyme activity : Take 5 drops of extract in a test tube and add equal amount of substrate solution (acid solution of casein) and incubate it at 37°C for 8—12 hours. Perform control experiment by taking 5 drops of substrate with equal amount of distilled water and incubate it at 37°C. To confirm the presence of pepsin take 2 drops of incubated mixture, add 1 drop of 10% sodium acetate solution. Casein is precipitated by addition of sodium acetate but in the presence of pepsin casein is hydrolysed into proteoses and peptones which cannot be precipitated by sodium acetate. Repeat the confirmatory test in the control experimental tube and compare the results of the two tubes.

The presence of pepsin can also be demonstrated by dissolving the **fibres of fibrin** in an acidic medium. Take a few fibres of fibrin, two drops of enzyme extract and two drops of buffer of pH 2·0 on a microslide and cover it by the cover slip. Keep the microslide in the incubation bath at 37°C for 8—12 hours. After incubation for 8—12 hours observe the microslide under the microscope ; dissolved fibrin will indicate the presence of pepsin.

EXPERIMENT No. 6

Test for Trypsin Enzyme

Trypsin is a proteolytic enzyme found in the intestine, pancreas and the liver of animals. In pancreas it is in the inactive condition which is called **trypsinogen** butin in testine it is converted into the active trypsin by the alkaline action of bile. The optimum pH of trypsin is about 8·0 and it is more effective than pepsin. Trypsin hydrolyses the proteins which are not hydrolysed by the gastric juice.

Preparation of substrate solution : Prepare 20% alkaline solution of casein. For this take 1 gm vitamin free casein and dissolve it in 50 ml of buffer of 7·2 pH.

Preparation of enzyme extract : Take a part of pancreas or intestine and prepare the enzyme extract as described in the case of pepsin.

Test of enzyme activity : Take 5 drops of extract, add 5 drops of substrate solution and to this also add 5 drops of buffer of 7·8 pH. Perform control experiment by taking 5 drops of substrate and drops of distilled water in a test-tube Incubate both the tubes in the incubation bath at 37°C for 4—6 hours.

To confirm the presence of trypsin add few drops of 1% acetic acid in the experimental test-tube. Absence of any precipitate will confirm the presence of trypsin. Acetic acid precipitates the casein. Since the casein has been hydrolysed, therefore, it does not form any precipitate and shows the presence of trypsin. In the control experimental tube there is no extract, therefore, it does not contain trypsin and the casein is not hydrolysed and the addition of acetic acid gives the precipitate.

ESTIMATION OF HAEMOGLOBIN BY HAEMOMETER

EXPERIMENT No. 7

Object : *To estimate the amount of haemoglobin in (i) frog's blood, (ii) human blood and (iii) blood of any other vertebrate animal.*

Requirements : (*i*) Haemometer (Haldane's haemoglobin ometer); (*ii*) decinormal (N/10) HCl (1·2 cc. of conc, HCl dissolved in 100 cc of distilled water) ; (*iii*) distilled water ; (*iv*) living frog or any other living vertebrate.

Apparatus : The haemometer consists of two sealed lateral comparison tubes containing a suspension of acid haematin. These are held in a black frame against a white ground glass. Besides, a graduated test-tube of the same diameter is also provided which can fit in the haemometer in between the two side tubes for comparison. A micropipette of 20 cmm is also provided. The other things provided are a small glass rod stirrer, a small bottle brush and a dropper and also a small bottle to contain the decinormal acid solution.

Principle : The method to estimate the haemoglobin content of blood is based on the principle of making an acid haematin solution of blood under experimentation in the graduated tube and then comparing it with the sealed comparison tubes containing the standard acid haematin.

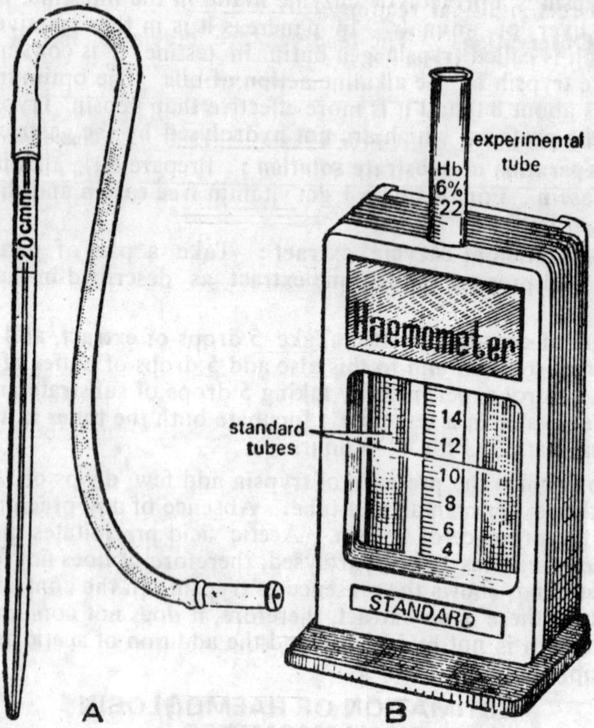

Fig 17·1. A—20cmm micropipette, B—Haldane's haemometer.

Procedure : The graduated tube is first cleaned with distilled water and then with methylated spirit or 90% alcohol. It is thoroughly dried up before being used. Now with the help of a dropper the N/10 HCl solution is filled in the graduated tube up to 2 gms mark. The micropipette is now filled up by sucking fresh blood of the vertebrate under experimentation up to the mark of 20 cmm. The small amount of blood adhering to the outerside of micropipette should be wiped off by sterilized cotton. The blood of micropipette is now added to the N/10 HCl solution in the graduated tube. The pipette should be introduced carefully into the tube and its lower mouth should pass right up to the bottom into HCl solution. When blood has been expelled the pipette is rinsed twice or thrice by distilled water. Every time the contents of micropipette should be expelled into the graduated tube. The acid haematin solution is now thoroughly stirred with the help of a glass rod and then allowed to stand at least for 10 minutes. Afterwards the acid haematin solution is gradually diluted by adding distilled water in a dropwise manner. With the addition of each drop of distilled water the solution should be stirred and its colour matched with that of the standard sealed tubes. This should be continued till the colour of the acid haematin solution just fades away as compared to that of the standard comparison tubes. The reading before the colour just fades is taken as the correct and final reading.

Observations : The readings are tabulated in the following manner :

Readings		Frog's blood	Human blood
I		7·8	12 45
II		7·10	12·60
III		7·9	12·50
Final	=	7 9	12·50

Results : The blood of frog contains 7·9 mgm of haemoglobin in 100 ml of blood.

The blood of man contains 12·50 mgm of haemoglobin in 100 ml of blood.

Precautions : The following precautions are to be taken :

1. In case human blood is being tested, give a light prick to the finger tip of the person providing blood.

2. The finger and the prick (needle) should be disinfected first with methylated spirit or alcohol (90% or absolute).

3. Avoid uncleaned tubes and pipettes, etc.

4. Avoid incorrect filling up of micropipette.

5. Avoid inclusion of blood sticking at the outer surface of the mouth of the pipette.

6. Readings should be taken always of the upper meniscus.

7. Experiment should be performed quickly without waste of time so that fresh blood is not allowed to coagulate before transfer to decinormal HCl.

ENUMERATION OF RED BLOOD CORPUSCLES BY HAEMOCYTOMETER

EXPERIMENT No. 8

Object : *To determine the number of red blood cells (R.B.Cs) present in one cubic mm volume of blood.*

Requirements : Haemocytometer, Hayem's solution, microscope and blood.

Apparatus : The haemocytometer includes two graduated pipettes in which dilution of blood is done. One pipette with a red bead is used for counting R.B.Cs, while the other with a white bead for counting W.B.Cs. Each pipette bears at least 3 graduations those on the pipette for R.B.Cs are 0·5, 1 and 101. The graduations on the pipette for W.B.C. are 0·5, 1 and 11. Besides these pipettes the haemocytometer includes a glass slide; 3 inches long by 1·5 inches wide. It bears two counting chambers with a cover slip. One side of the slide bears two central plateforms bordered by an edged-shaped groove. Beyond this groove on either side is a lateral plateform further demarcated by a lateral groove. Each central platform bears one counting chamber, each counting chamber is formed by several straight perpendicular and horizontal lines enclosing squares of various measurements; the biggest square is 3×3 mm with an area of 9 sq. mm. It is divisible into 9 squares of 1×1 mm each, the central square is again divisible into 25 smaller squares. Each square measures $1/5 \times 1/5 = 1/25$ sq. mm. Each of these squares is further divisible into 16 smaller squares so that there are 400 smallest squares in all. Each with an area of 1/400 sq. mm.

HAYEM'S **solution** has the following composition :

(*i*) Mercuric chloride ($HgCl_2$)$=0·5$ gm
(*ii*) Sodium chloride (NaCl)$=1·0$ gm
(*iii*) Sodium sulphate (Na_2SO_4)$=5·0$ gm
(*iv*) Distilled water (H_2O)$=200$ cc

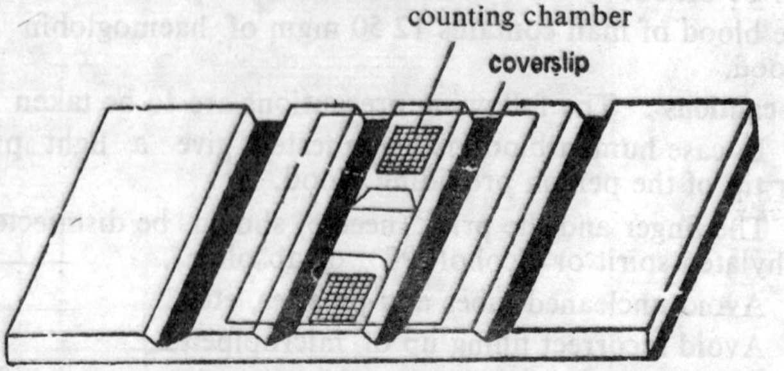

Fig. 17 2. Glass slide with counting chambers.

The mercuric chloride acts as corrosive sublimate and fixes the R.B.Cs present in the blood. The other ingredients act isotonically so that the R.B.Cs may not burst due to haemolysis. The HAYEM'S solutian also serves to dilute the blood.

Procedure : Blood is obtained either directly from the heart of a frog or from some body part externally in case of man. Human blood is generally taken out from a finger (avoid thumb and first finger) which is thoroughly washed and cleaned by spirit or alcohol (absolute or 90%). It is now pricked by an ordinary injection needle which is also sterilized with spirit and dried before use. The finger is pricked quickly and effectively. It should not be pressed hard to let out blood so as to avoid other body juices also oozing out. For this reason also the first one or two drops of oozing blood should be avoided and wiped off.

Now take the pipette meant for R.B.Cs which is already rinsed with alcohol or spirit or ether and thoroughly dried. Suck the blood in the pipette up to 0·5 mark taking care that air bubbles are not included. The excess of blood, if any, may be run out by touching the mouth of pipette to the palm. The blood which is sticking to the outer side of pipette should also be carefully cleaned. The pipette should now be transferred to the container of HAYEM'S solution which is carefully sucked up to 101 mark. The pipette is now held horizontally between the fore-finger and thumb or palm surfaces of the hand and rotated several times so that blood thoroughly mixes with HAYEM'S fluid. The red bead in the pipette also helps in mixing. In this way dilution of blood becomes 200 times.

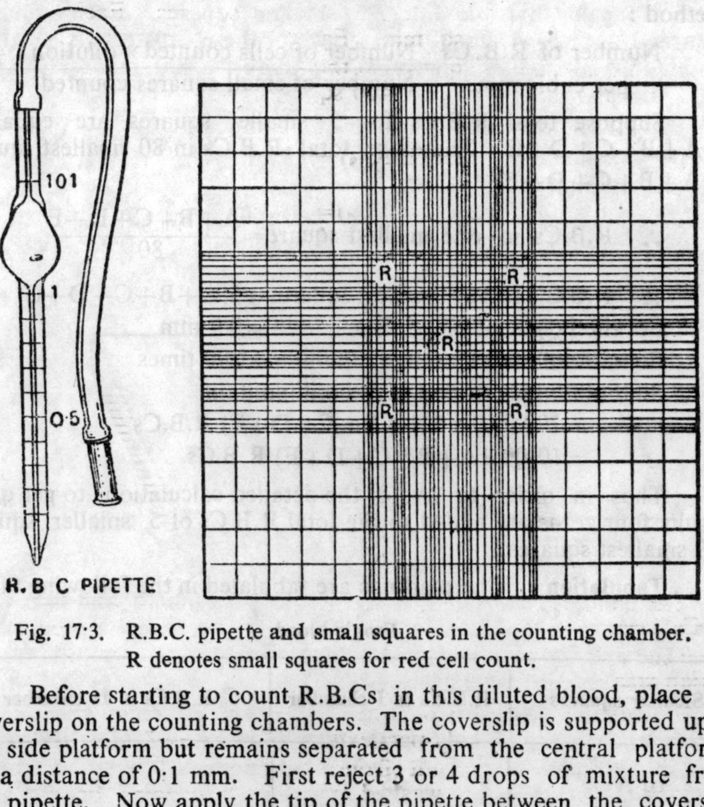

Fig. 17·3. R.B.C. pipette and small squares in the counting chamber. R denotes small squares for red cell count.

Before starting to count R.B.Cs in this diluted blood, place the coverslip on the counting chambers. The coverslip is supported upon the side platform but remains separated from the central platforms by a distance of 0·1 mm. First reject 3 or 4 drops of mixture from the pipette. Now apply the tip of the pipette between the coverslip and the platform and allow few drops of blood mixture to flow in the narrow space between the coverslip and the counting chambers. If necessary both chambers may be filled in this manner. Blood mixture remains filled up between the coverslip and the counting chambers because of capillary action. Air should not be taken into and also pouring excess blood mixture so that the H-shaped groove remains free from it.

When the counting chambers are properly flooded the slide may be kept aside for a few minutes so that the R.B.Cs settled down on the bottom floor of the two counting chambers. Now transfer the slide gently and carefully under the microscope without disturbing the settled R.B.Cs and start counting them.

Counting : It is not necessary to count the R.B.Cs in all the 400 smallest squares or even in the 25 smaller squares, count them only in five smaller squares, *i.e.*, in the 1st, 5th, 13th, 21st and 25th. The R.B.Cs lying on the lower and right sides of a square are to be added in the total, while those lying on the upper and left sides are to be rejected.

Calculations : The calculations may be done by the following method :

$$\text{Number of R.B.Cs per cubic mm} = \frac{\text{Number of cells counted} \times \text{dilution} \times 4000}{\text{Number of small squares counted.}}$$

Suppose total R.B.Cs for 5 smaller squares are equal to $= A+B+C+D+E$. Therefore, total R.B.Cs in 80 smallest squares $= A+B+C+D+E$.

$$\therefore \text{ R.B.Cs in one smallest square} = \frac{A+B+C+D+E}{80}$$

$$\therefore \text{ R.B.C's in 400 smallest squares} = 5(A+B+C+D+E)$$

But the height of blood film $= 0{\cdot}1$ mm

Dilution of blood $= 200$ times

So that 1 cubic mm of blood will contain

$$= 200 \times 10 \times 5(A+B+C+D+E) \text{ R.B.Cs}$$

$$= 10{,}000(A+B+C+D+E) \text{ R.B.Cs.}$$

Thus in order to avoid the detailed calculations to get quick results four zeros are added to the total R.B.Cs of 5 smaller squares (80 smallest squares).

Tabulation : The readings are tabulated in the following table :

Frog's blood

Smaller squares	R.B.Cs in I chamber	R.B.Cs in II chamber
I*st* A	25	15
II*nd* B	30	20
III*rd* C	15	18
IV*th* D	20	25
V*th* E	22	20

Human blood

Smaller squares	R.B.Cs in I chamber	R.B.Cs in II chamber
I*st* A	83	105
II*nd* B	87	97
III*rd* C	74	79
IV*th* D	96	86
V*th* E	98	83

	Frog		Human	
Total R.B Cs=	112	98	438	450
Average=	$\frac{112+98}{2}=105$		$\frac{438+450}{2}=444$	

Results : The result is given as follows :

(*i*) 1 cmm of frog's blood contains 1050000 R.B.Cs.

(*ii*) 1 cmm of human blood contains 4440000 R.B.Cs.

ENUMERATION OF WHITE BLOOD CORPUSCLES (W.B.C.) BY HAEMOCYTOMETER

EXPERIMENT No. 9

Object : *To determine the number of white blood cells (W.B.Cs) per cubic mm in the human blood.*

Requirements : Haemocytometer, glacial acetic acid, solution of gentian violet in water, distilled water. The glacial acetic acid haemolyses the red cells, while the gentian violet slightly stains the nuclei of white blood corpuscles (W.B.Cs), the W.B.Cs may, therefore, be easily recognised.

Procedure : The procedure of counting of W.B.Cs is the same as that of the R.B.Cs. The W.B.Cs are counted in the four corners of 1 square millimetre in the central ruled area on both the sides of the counting chambers of the haemocytometer. The W.B.Cs are recognised by the refractile appearance and by the slight colour given to them by the stain contained in the diluting fluid. The cells touching the boundary lines are not counted.

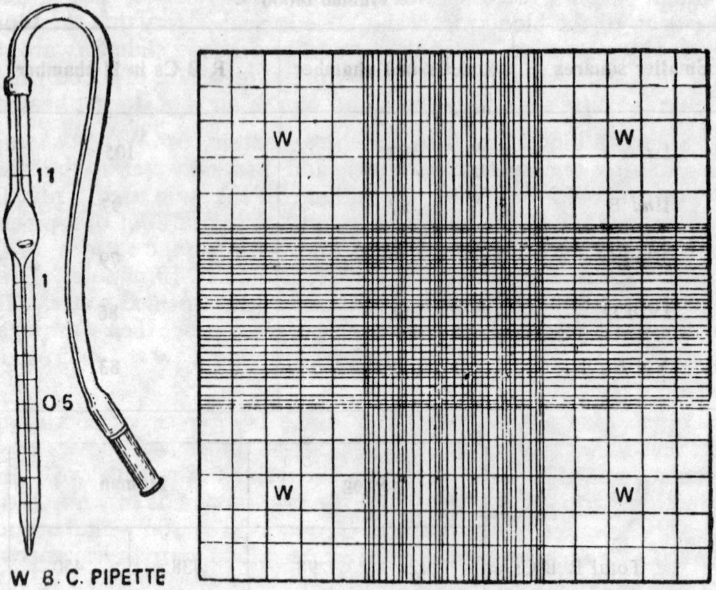

Fig. 17·4. W.B.C. pipette. Big squares in the counting chamber, W denotes big squares for white cell count.

Calculations : The calculations may be done by the following method.

$$\text{Number of W.B.Cs per cubic mm} = \frac{\text{Number of cells counted} \times \text{Dilution} \times 10}{\text{Number of 1 sq mm counted}}$$

Since the dilution is 20 times and thc cubic capacity of the area counted is 1/10 cubic millimetre, the total volume is 1/200 cubic millimetre. Say for instance, the number of W.B.Cs in 4 outer squares is $(10+12+13+10)=45$.. In otherwords number of W.B.Cs in 1/200 cubic millimetre $=45$.

Therefore, the number of W.B.Cs in 1 cubic millimetre $=45\times200\times1/2=4500$ W.B.Cs

Normal healthy man has 4000 to 6000 W.B.Cs per cubic millimetre of blood.

DIFFERENTIAL COUNT OF WHITE BLOOD CORPUSCLES

EXPERIMENT No. 10

Object—*To determine the relative proportions (differential count) of white blood corpuscles in the human blood.*

Requirements : Sterilized needle, clean slides ; Leishman's stain or Wrights stain and microscope.

Procedure : Prick the finger with a sterilized needle, wipe away the first drop of blood. Take the second drop on the end of a clean slide, place the slide on a smooth surface holding it steady with the left hand. Hold a second slide or drawing slide at 45° just in front of the drop of the blood, draw this slide in such a way that the blood spreads along its edge. Push the second or drawing slide forward 45° on the first slide without exciting any pressure, a film of blood will be made on the first slide. Dry the blood film in air, but do not heat it.

Stain the blood film with Leishman's stain or Wright's stain. Wait and after one minute carefully add distilled water to the stain, the water should be twice the volume of the stain used. Mix the water and stain well by sucking the mixture in and out of a pipette. After a time a greenish metallic scum forms on the surface of the mixture. Allow the diluted stain to act from 7 to 10 minutes. Drain of the stain and wash film for 10 seconds with distilled water. The film should be rose pink in colour, if it is too purple, then wash again with distilled water. Dry the slide and examine it by the microscope for white blood corpuscles.

First examine the stained film under low power of microscope to get an idea of leucocytes. Then examine under high power or oil immersion lens after thinly smearing the stained film with oil. In a longitudinal strip of the film count the various kinds of white blood corpuscles from top to the bottom. Count 100 or 200 white corpuscles noting on a paper each type of white blood corpuscles encountered and calculate the percentage.

Calculation : The following formula can be adopted to calculate the percentage of different types of white blood corpuscles :

$$\frac{\text{Number of type cells}}{\text{Total number of white blood corpuscles}} \times 100$$

The normal proportion of white blood corpuscles is as follows :

Neutrophils......50—70 per cent	or	3000—6000	per cu mm
Eosinophils......1·4 ,,	,,	150—300	,, ,,
Basophils......0·5—1 ,,	,,	0—100	,, ,,
Lymphocytes...20—40 ,,	,,	1500—2700	,, ,,
Monocytes......2—8 ,,	,,	300—600	,, ,,

Precautions : It is absolutely essential for differential count of white blood corpuscles that the film of blood is well spread with no clogging of corpuscles.

COAGULATION OF BLOOD

EXPERIMENT No. 11

Object : *To observe the coagulation of blood.*

Requirements : Test-tube and cotton wool.

Principle and procedure : Bleeding takes place if the skin is cut, but after a time blood coagulates to close the cut, thus preventing much loss of blood. Take blood of a mammal in a large test-tube and close the tube with cotton wool. It will be seen that blood forms a soft jelly and a faint yellow liquid appears. In a few hours the liquid or **serum** increases in amount and the solid jelly or **clot**

becomes smaller. The serum consists of blood plasma without its fibrin and almost no blood corpuscles. The clot is scarlet on the surface and bluish-black within. The clot is made of **fibrin,** an insoluble protein derived from the soluble fibrinogen of the blood, and entangled in the fibrin are almost all the blood corpuscles. The fibrin is a mass of fine interlacing fibres.

The blood contains an inactive enzyme **prothrombin** which is kept in check by **heparin** (antiprothrombin) found in blood. Hence the blood flows freely in blood vessels and does not coagulate. If an injury occurs, blood begins to escape from it, but the injured tissue cells and blood platelets on contact with air release a substance called **thromboplastin.** Thromboplastin soon neutralizes heparin, and it combines with calcium ions in the blood to act on prothrombin changing it into **thrombin**. Thrombin is an active enzyme, it acts on the fibrinogen of blood converting it into **fibrin.** The fibrin forms a network of fibres which entangle the corpuscles to form a clot at the site of the injury.

BLEEDING TIME OF BLOOD

EXPERIMENT No. 12

Object : *To determine the bleeding time of blood.*

Requirements : Sterilized needle, stop-watch and blotting or filter paper.

Procedure : Prick the base of the finger nail with a sterilized needle and let a drop of blood appear. Every 30 second dab the blood drop with a piece of blotting or filter paper without touching the skin otherwise the blood may be wiped off. The time from the appearance of the drop of blood to the time when bleeding stops and the filter paper is no longer stained is taken as the bleeding time. Normal bleeding time is 2 to 5 minutes.

CLOTTING TIME OF BLOOD

EXPERIMENT No. 13

Object : *To determine the clotting time of blood.*

Requirements : Sterilized needle ; small test-tubes, stop watch water bath, syringe and glass slides.

Principle and procedure : Itis the time taken for blood to clot in vitro at a standard temperature, or it is the length of time from the moment blood is collected till the appearance of clotting.

Following three methods are used for counting the clotting time of blood.

A. Note the time, then take 1 ml venous blood in a small test-tube, put the tube in a water bath at 37°C. Every 30 seconds shake the tube and tilt it slowly. Note the time when blood stops flowing in the tube and its surface line no longer changes on tilting. The time taken is the clotting time, it is normally 5 to 10 minutes.

B. Take 5 ml venous blood in a syringe and start a stop watch. Put 1 ml of blood in each of two small equal-sized test tubes, put the tubes in a water bath at 37°C. After some time tilt the tubes every half minute. when the tube can be tilted through 90° or even inverted without spilling the blood then the blood has coagulated, note the time.

C. Put a drop of blood on a clean slide and note the time. Every half minute pull a clean needle slowly through the drop of blood. When a fine thread of fibrin can be pulled up by the point of the needle, then coagulation has started, note the time.

PREPARATION OF HAEMIN CRYSTALS

EXPERIMENT No. 14

Object : *To prepare haemin crystals.*

Requirements : Sterilized needle, glass slide, cover glass, acetic acid (glacial) and spirit lamp.

Principle : Haemoglobin imparts a red colour to blood, it is a conjugated protein made of a non-protein pigment **haematin** or haeme, and a colourless protein known as **globin**. The haematin is made of a pigment called **porphyrin** which is combined with iron. A hydrochloride of haematin forms **haemin**.

Procedure : Take some dried blood or a few drops or fresh blood, add a little glacial acetic acid and heat gently, on cooling dark brown haemin crystals (also known as Teichmann's crystals) are formed. Examine haemin crystals under a microscope, they appear as shining rhombic plates prisms and star-shaped clusters. The crystals contain haematin and iron. Haemin crystals test is used in medicolegal work to differentiate fr:sh or dried blood marks from any other red-colouring material.

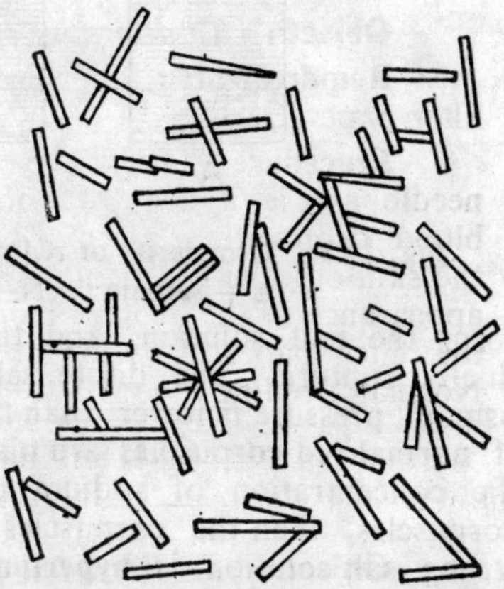

Fig. 17·5. Haemin crystals.

Precaution : In the case of old dried blood it is necessary to boil with glacial acetic acid, and at boiling to add a crystal of sodium chloride.

EFFECT OF OSMOLARITY OF SALT SOLUTION AND HAEMOLYTIC AGENTS IN RED BLOOD CORPUSCLES

EXPERIMENT No. 15

Object : *To study the effect of osmolarily of salt solutions and haemolytie agents on red blood corpuscles*

Requirements : Test-tubes, sodium chloride solutions of different percentage, ether or benzene or chloroform and potassium hydroxide.

Principle and procedure : Erythrocytes are membranous

envelopes filled with haemoglobin. The envelope or plasma membrane is semipermeable, hence substances in solution can pass in and out of the corpuscle. If normal red blood corpuscles are placed in normal sodium chloride solution which has the same concentration as the contents of the corpuscles, then no exchange takes place because the osmotic pressure of corpuscles and of the saline solution is the same. The salt solution is said to be **isotonic** or of equal concentration with the corpuscles. If normal red blood corpuscles are put in a dilute solution of sodium chloride, they absorb water

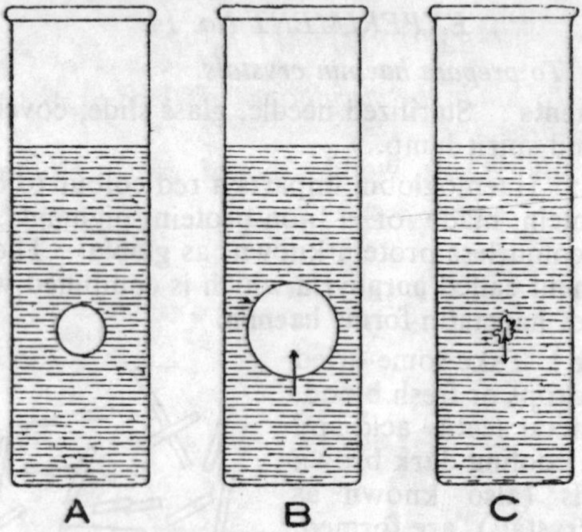

Fig. 17·6. Osmolarity of red corpuscles in different saline solutions A—isotonic ; N—hypotonic ; C—hypertonic.

form the salt solution, and they swell up, become spherical, and finally rupture. The ditute salt solution is **hypotonic** because its osmotic pressure is lower than that of the interior of the corpuscles. If normal red corpuscles are placed in strong salt solution in which the concentration of sodium chloride is higher than in the red corpuscles, then the corpuscles shrink and become wrinkled. The strong salt solution is **hypertonic** because its osmotic pressure is greater than that of the interior of the eorpuscles and the fluid passes from the corpuscles into the salt solution.

A. Put 1 ml of venous blood in each of three small test-tubes. To the first tube add 1 ml of 0·85% sodium chloride solution (isotonic). To the second tube add 1 ml of 0·5% sodium chloride solution (hypotonic). To the third tube add 1 ml of 1·5% sodium chloride solution (hypertonic). Shake the tubes and note the results.

In a hypotonic solution the red corpuscles rupture, in rupturing the haemoglobin passes from the red corpuscles into the plasma, the plasma becomes red in colour. This is called **haemolysis or laking of blood**. The substances inducing haemolysis are called **haemolytic agents**. Some common haemolytic agents are hypotonic solutions, fat solvents, snake venoms, and some toxins.

B. Take a drop of blood in a test-tube, add a fat solvent (such as ether, benzene, chloroform or bile salts), shake the tube, the blood will be haemolysed because the fat component of the plasma membrane of red corpuscles is dissolved and haemoglogin is set free.

C. Take a drop of blood in a test-tube, and a few drops of 0·2% potassium hydroxide, the red corpuscles are dissolved.

CIRCULATION OF BLOOD

EXPERIMEMT No. 16

Object : *To study the circulation of blood in the capillaries of the web of foot of frog.*

Requirements : Soft wooden board, moist cloth, microscope and living frog.

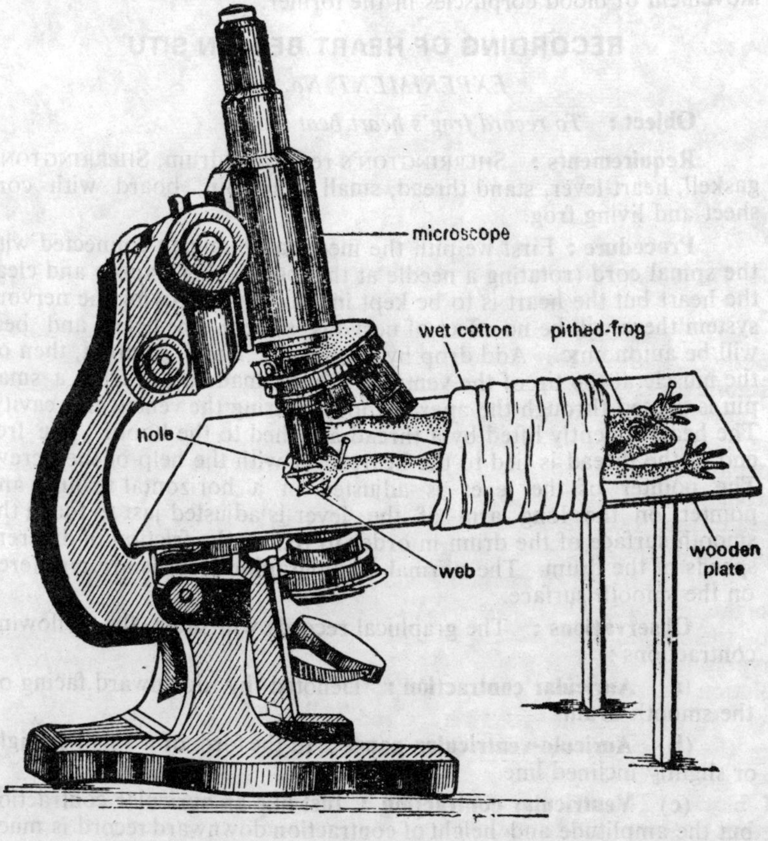

Fig' 17·7. To show blood circulation in the web of foot of frog.

Procedure : Take a soft wooden board and make an aperture of about 2cm, about 4 cm away from one end. Cover a living frog with moist cloth bag and tie it up to the wooden board placing the web between the toes of the foot on the aperture, the foot may be pinned down with web well stretched. Then place the wooden board with the frog on the stage of the microscope so that the aperture in the wood lies on the hole of the microscope stage or platform. Examine the web under the microscope. Blood corpuscles will be seen moving in capillaries, and arterial vessels can be distinguished from veins by the jerky movement of blood corpuscles in the former. Large chromatophores will also be seen in the web.

Alternatively, the frog may first be pithed, in pithing the spinal cord is cut by a sharp pin inserted just behind the head and the brain is crushed. Then place the web of the foot in position on the microscope.

Observations : Blood corpuscles are seen moving in capillaries. The arterial capillaries can be distinguished from veins by the jerky movement of blood corpuscles in the former.

RECORDING OF HEART BEAT IN SITU

EXPERIMENT No. 17

Object : *To record frog's heart beat in situ*

Requirements : SHERRINGTON'S revolving drum, SHERRINGTON'S gaskell, heart lever, stand thread, small bent pin, board with cork sheet and living frog.

Procedure : First we pith the medulla oblongata connected with the spinal cord (rotating a needle at this point), then expose and clear the heart but the heart is to be kept in situ. By destroying the nervous system there will be no effect of nervous system on heart and beat will be autonomic. Add drop by drop 0·6% NaCl solution, then on the muscle at the tip of the ventricle a hook made up of from a small pin is passed through the apex without injuring the ventricular cavity. The heart is gently lifted by a thread attached to the hook. The free end of the thread is tied to the heart lever with the help of the screw. The pointer of the lever is adjusted in a horizontal manner and pointer on the long arm of the lever is adjusted just to touch the smooth surface of the drum in order to reduce the friction of different speeds of the drum. The normal cardiographic records are registered on the smooth surface.

Observations : The graphical records will show the following contractions :

(**a**) **Auricular contraction :** Denoted by downward facing on the smooth drum.

(**b**) **Auriculo-ventricular pause :** In the form of a small straight or slightly inclined line.

(**c**) **Ventricular contraction :** Just like an auricular contraction but the amplitude and height of contraction downward record is much longer than that of the auricular contraction.

Interpretation : The contraction due to sinus venosus is not at all observed only because firstly sinus contraction is very weak and secondly auricular contraction is much more stronger as compared to sinus contraction.

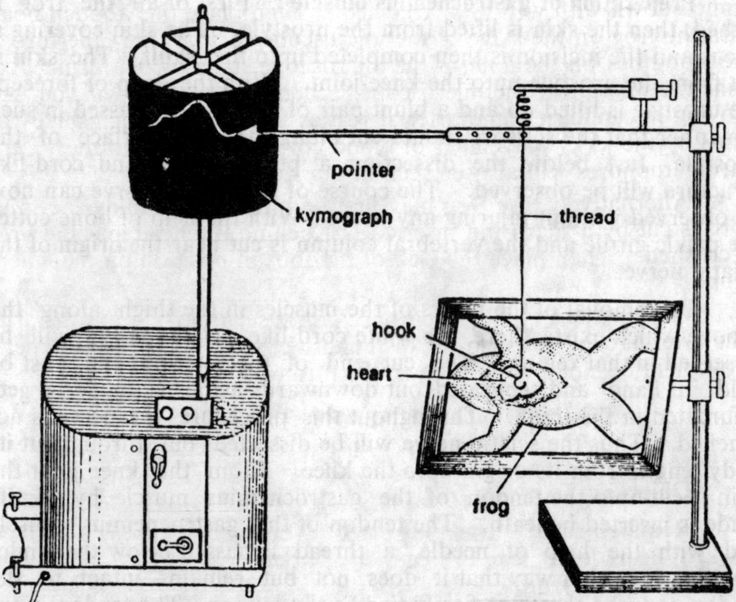

Fig. 17·8. Diagram showing the recording of heart beat of frog.

Inference : Under normal conditions the contraction of sinus venosus is generally followed by auricular contraction and ventricular contraction. The downward tracing is the contraction and upward tracing is the relaxation of the heart, while small short tracings are the pause taken up by heart during series of each changes in each cardiac cycle.

Precautions : The following precautions are to be taken :

1. Rotations of kymograph should not be fast otherwise contractions of the auricles will not be recorded.
2. The paper should be blackened well.
3. Edge of the pointer should be accurate.
4. Current flow should be uniform.
5. Heart should be kept wet in 0·6% saline solution.

RECORDING OF MUSCULAR CONTRACTION

EXPERIMENT No. 18

Object : *Recording of muscle contraction in the gastrocnemius muscle-nerve preparation of frog.*

Requirements : SHERRINGTON's recording drum, Dubois Rejmonds induction coil, muscle chamber, make and break key, lever weight, wire, muscle, 0.65% saline water and battery.

Connections : Drum is primary circuit, movable electrodes form secondary circuit.

Strength of current : Sub-maximal induced break shock should be effective.

Preparation of gastrocnemius muscle : First of all the frog is pithed, then the skin is lifted from the urostyle. The skin covering it is cut and the incision is then completed upto the skull. The skin is cut from the urostyle upto the knee-joint. With the help of foreceps the urostyle is lifted up and a blunt pair of scissors is passed in such a manner that the scissors should cut along the body surface of the urostyle. Just below the dissection a purely white and cord-like structure will be observed. The course of the sciatic nerve can now be observed without injuring any nerve. With the help of bone cutter the pelvic girdle and the vertebral column is cut near the origin of the sciatic nerve.

On removal of the layers of the muscles in the thigh along the groove which exists there, the white cord-like sciatic nerve will be observed in that region. The cut end of the vertebral end must be held in hand and separated out downwards as far as the nerve gets bifurcated in the thigh. Throughout this procedure the nerve is not pinched. Thus the sciatic nerve will be dissected out throughout its body length from its origin upto the knee. From the knee joint the skin is cut upto the tendon of the gastrocnemius muscle by a knife blade is inserted beneath. The tendon of the gastrocnemius muscle and with the help of needle, a thread is passed below the tendo-Achillrs, in such a way that it does not but remains intact to the tendon-end, the thread is firmly tied to the tendon. The tendon is cut away from the tied end by holding the thread the muscle is separated upto the knee joint. With the help of the bone cutter the tibia and femur are cut for about half an inch away from the knee joint. Now the frog's gastrocnemius sciatic nerve muscles preparation is ready. After the dissection is completed the precaution is taken that the muscle and nerve are not allowed to dry due to the exposure to atmosphere. In order to avoid this the nerve muscle preparation is kept moist by normal saline water which in case of frog is generally 0·65% NaCl solution.

Procedure : A nerve-muscle preparation is dissected out and is fixed in the muscle chamber containing 0·65% NaCl solution as it is isotonic with frogs muscle tissue. The apparatus is arranged for a single induction shock and the secondary induction shock will be effective. The drum is included in the primary circuit. A small weight say about 1 gm is attached to the pointer arm which keeps the pointer in a horizontal position. The strength and intensity of the current is adjusted in such a way that the muscle will be in a position to produce a simple twich about $1\frac{1}{2}$ inch amplitude. The primary as well as secondary make and break key is closed and the lever point is bronght just to touch the writing surface in order to have the optimum pressure on the smoked drum. When the contact in the primary circuit is just broken due to revolving base electrodes on the SHERRINGTON's drum the muscle will be stimulated via the sciatic nerve and it will record a single twich. As soon as the writing point has returned to the base line, the short circulating key and secondary make and break key is closed immediately. The drum is dropped with the help of stop key and the writing point is withdrawn immediately from the drum so that the tracing are not in a position to overlap one another. It will be observed that the twitching is in the form of curve only because the drum is rotating with certain speed.

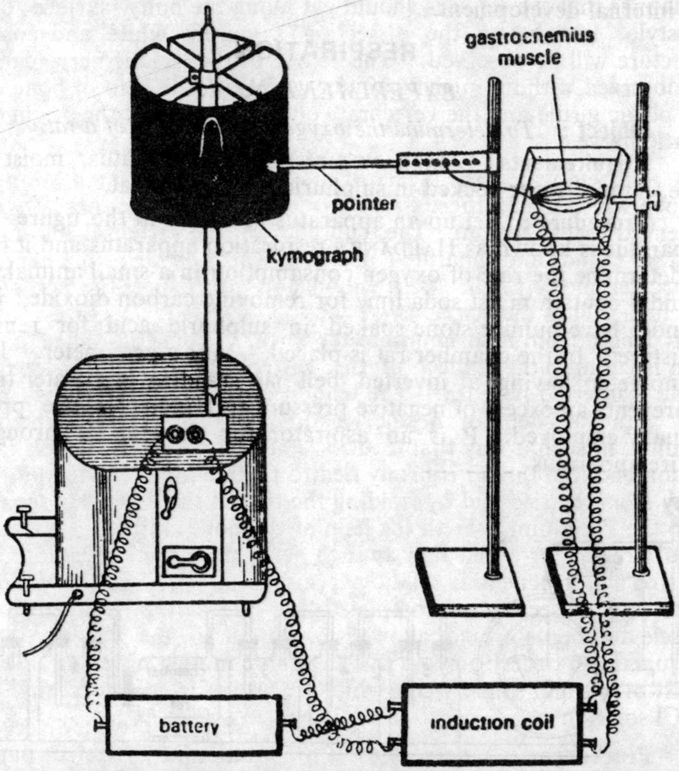

Fig. 17·9. Diagram showing the contraction of gastrocnemius muscle.

In this experiment the tension due to weight on the muscle is merely constant but the muscle fibres are free to shorten in length. Hence the contraction is said to be isotonic (change in length of muscle fibres not in tension).

The latent period is the phase during which no apparatus changes take place in the muscle. The latent period extends from the point at which the contracting muscle begins to raise the lever.

The contraction period extends from the point where the writing arm of the '*l*' shaped lever begins to rise to the point highest above the base line. The contraction period starts at the end of the latent period and is thus in active phase in the contraction.

The relaxation phase or period is a passive phase in the muscular contraction and extends from the highest point of the simple muscle-curve up to the point at which the simple muscle curve twitch retouches the base line.

Inference : An isolated twitch consists of three phases, *viz.*, latent phase, the contraction phase and relaxation phase. Seeondarily a simple muscle twitch represents the response to a single stimulus and a stimulus causes a sudden change in the external as well as in the internal development.

RESPIRATION

EXPERIMENT No. 19

Object : *To determine the oxygen consumption of a rat.*

Requirements : HALDANE's respiration apparatus, moist soda lime, pumice stone socked in sulphuric acid, and a rat.

Procedure : Set up an apparatus as shown in the figure. This apparatus is known as HALDANE's respiration apparatus and it is used to determine the rate of oxygen consumption in a small animal. Jars 1 and 4 contain moist soda lime for removing carbon dioxide, jars 2, 3 and 5 have pumice stone soaked in sulphuric acid for removing moisture. In the chamber rat is placed. M is a gas meter. J is a manometer having a inverted bell jar standing in a water trough, it prevents an excess of negative pressure and indicates the pressure actually employed. P is an aspirator for drawing air through the entire apparatus.

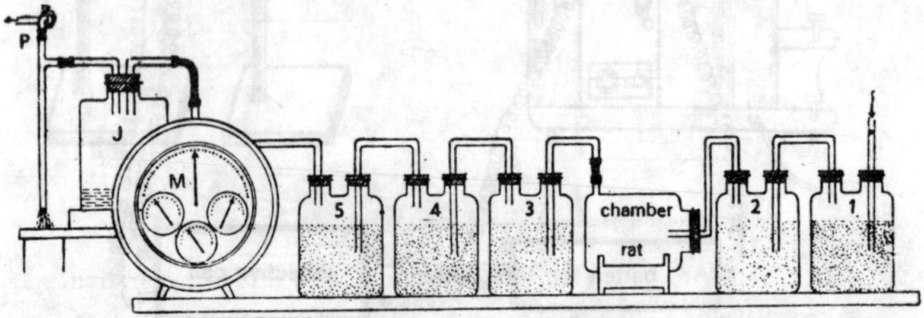

Fig. 17·10. Haldane's apparatus to determine oxygen consumption of a rat. 1 and 4—soda lime ; 2, 3 and 5—pumice stone and sulphuric acid ; rat chamber ; J—manometer ; M—meter ; P—aspirator.

Draw a current of air through the apparatus by means of the aspirator for a specified period of the time. The air entering the rat chamber has been freed of carbon dioxide and moisture by jars 1 and 2 respectively, thus the chamber receives only oxygen. This oxygen is used for respiration needs of the rat. The rat gives out carbon dioxide and moisture into the outgoing air. Jar 3 removes the moisture, jar 4 absorbs the carbon dioxide, jar 5 is to remove any moisture gained from soda lime of jar 4. The quantity of oxygen consumed by the rat is determined indirectly. The oxygen consumed is equal to the carbon dioxide and moisture given out. Thus weigh jars 3, 4 and 5, and the chamber including the rat both before and after the experiment. The increase in weight during a specified period of time gives the quantity of oxygen consumed by the rat during that period. To find the quantity of carbon dioxide given out by the rat weigh jar 4 before and after the experiment.

EXPERIMENT No. 20

Object : *To determine the rate of oxygen consumption of a rat.*

Requirements : One wide-mouth bottle, measured U-tube, cork, a small beaker, potassium hydroxide solution and a living rat.

Procedure : Take a volume of potassium hydroxide solution in a small breaker and keep it at the base of the wide-mouth bottle. Take a U-tube, fix it to the bottle and fill it with a known volume of water. Note the volume of water in the U-tube. Now take a living rat and put it in the bottle. Close the bottle with the cork and seal the apparatus with vaseline to make it airtight.

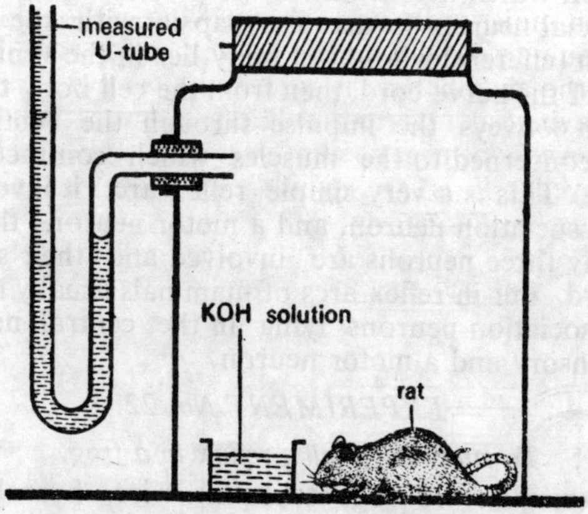

Fig. 17·11. Apparatus for the demonstration of rate of respiration.

Observations : Note the level of water in the U-tube after an hour. It is found to be lowered. This is due to the fact that the rat respires and consumes the oxygen from the air of the bottle. The carbon dioxide released is absorbed by the potassium hydroxide solution. This reduces the pressure in the bottle and hence the level of water in the U-tube falls.

Calculations : The fall in the level of water is measured in a definite period of time that is the rate of oxygen consumption per unit time suppose :

(i) Level of water in the beginning = 25 cc
(ii) Level of water after one hour = 20 cc
(iii) Fall of water level in 1 hour = 5 cc
(iv) Hence the oxygen consumption by rat in 1 hour = 5 cc

Precaution : The apparatus should be air-tight.

NERVOUS SYSTEM

EXPERIMENT NO. 21

Object : *To study reflex action in a frog.*

Requirements : A frog.

Principle and procedure : If a frog is touched, it at once moves away involuntarily as a consequence of the stimulus of touching without involving the will. this type of reaction is called a **reflex action**. The course along which the impulse travelled from the point of touching the skin to the muscles which caused the animal to move away is known as a **reflex arc**. The reflex arc is the structural basis of the reflex action. The impulse of touching the skin goes to the dendrites of a sensory neuron (afferent) and then into its cell body situated in the dorsal root ganglion of the nerve concerned, and then along its axon into the gray matter of the spinal cord where its terminal axon endings form a synapsis with the dendrites of an association or internuncial neuron which lies within the gray matter of the nerve cord. The internuncial neuron forms a synapsis with the dendrites of a motor neuron (efferent) whose cell body lies in the ventral part of the gray matter of the nerve cord, then from the cell body the axon of the motor neuron conveys the impulse through the ventral root of the spinal nerve concerned to the muscles which contract and the frog moves away. This is a very simple reflex arc, it involves a sensory neuron, an association neuron, and a motor neuron ; thus in a simple reflex arc only three neurons are involved and their synapses are in the nerve cord. But in reflex arcs of mammals usually there is a chain of several association neurons lying in the central nervous systems between a sensory and a motor neuron.

EXPERIMENT No. 22

Object : *To study the wiping reflex in a frog.*

Requirements : A frog, stout pins, wire, hyrochloric acid and filter paper.

Procedure : First pith a frog by destroying the brain or by passing a stout pin just behind the head cutting between the brain and spinal cord. In pithing the brain has been severed and destroyed but the spinal cord is intact. Hang the frog with a wire from a stand and fix the other end of the wire to the lower jaw. Prick a pin in one toe, the frog withdraw its foot. A more violent stimulation of one foot will make both legs moves as in jumping. Place a filter paper soaked in hydrochloric acid on the thigh of the frog, the leg will jerk to move the acidified paper, if it fails then the other leg will wipe the paper away by reflex action.

EXPERIMENT No. 23

Object : *To study the knee-jerk reflex in man.*

Requirements : Man and a small hammer or book.

Procedure : Make a man sit on the edge of a chair or table with one knee crossed over the other, and with the legs suspended and relaxed. Tap firmly on the ligament just below the knee cap of the upper leg with the side of the hand or with the edge of a book or with the small hammer. The leg will straighten with a jerk. This is due to sudden contraction of the quadriceps femoris muscle by reflex action whose reflex arc involves the spinal cord. The knee-jerk response will be much more pronounced if this experiment is tried with the fists of the man clenched tightly, because the degree of response depends on the condition of the body.

Explanation : The knee-jerk is an automatic, unlearned and involuntary response to the stimulus which involves contraction of quadriceps femoris muscle stretching the leg forward.

EXCRETORY SYSTEM

EXPERIMENT No. 24

Object : *To test the urine for urea, proteins, ketones and sugar.*

Requitements : Test-tubes, soyabean flour, litmus solution, acetic acid, phenolphthalein, sodium hydrate, copper sulphate, sulphosalicylic acid, ammonium sulphate, ammonia, sodium nitroprusside, BENEDICT's reagent and samples of urine.

Procedure : The following procedure is adopted for the test of urea, protein, ketone and sugar from the samples of urine.

1. **Urea test :** Normal urine contains 2 to 3 per cent urea. Take 2 cc of urine in a test-tube, add a pinch of soyabean flour. a little litmus solution, and a few drops of 1% acetic acid. In a few minutes the mixture becomes alkaline and turns blue because of decomposition of urea into ammonium carbonate. Alternatively, take 5 cc of urine in a test tube, add 2 or 3 drops of phenolphthalein, and 0·1 gm soyabean flour. Mix the contents and place test tube in a water bath at 50°C. The contents will turn pink showing urea.

2. **Protein test : A.** Normal urine is free from proteins, but at times it may contain albumin, globulin, and other proteins. Heat urine to almost dryness by evaporation. Then heat this in a dry test-tube, fumes of ammonia are given off. Remove from flame and cool, add a few drops of sodium hydrate and a drop of copper sulphate solution. A pink colour appears showing the presence of proteins.

B. Filter urine to remove any turbidity and to make it clear. Take this clear urine in a test tube and boil only the upper part for two minutes holding the test tube by the bottom end. If the upper part becomes turbid and cloudy, then it may be due to albumin or phosphates. To find out add 3 to 5 drops of 5% acetic acid, if turbidity appears or persists, then it is due to the presence of albumin in urines.

C. Take some urine in a test tube and boil the upper part only, add several drops of strong nitric acid slowly. If a flocculent precipitate appears and perists in the upper part of the tube, then it shows the presence of albumin.

D. Filter urine to remove any turbidity. Take 5 ml of filtered urine in a test tube, add 6 drops of 20 per cent sulphosalicylic acid, if a cloud is formed it shows the presence of protein.

3. **Ketone test :** In metabolism fats and some amino acids

are broken down into ketone bodies. The ketone bodies comprise acetone, aceto acetic acid, and beta hydroxybutyric acid. In a test-tube take 20 ml of urine and saturate it with ammonium sulphate crystals. Add 2 or 3 drops of ammonia and a few crystals of sodium nitroprusside, shake the tube well. If ketone bodies are present a reddish-purple colour appears in a few minutes; if, however, a brown colour appears, it shows the absence of ketones.

4. Sugar test : A. Normally there is no sugar (glucose) in urine. If there is any albumin in urine, then it is removed by adding a few drops of 5% acetic acid, then boil and filter to make the urine clear and protein free. Take 8 drops of this urine, add 5 ml of BENEDICT'S reagent, boil the mixture and allow to cool. If the solution is a clear greenish-blue then it shows absence of glucose. Boil a bluish-green colour with some yellow precipitate shows 0·2% glucose. A deep yellow to orange precipitate shows 0·5% glucose. A heavy orange brown to red precipitate shows 1 to 3% glucose.

B. Make urine clear and protein-free as before. Take some of this urine in a test-tube. In another test-tube take an equal volume of FEHLING'S solution. But the two test-tubes separately and when boiling mix them together, a yellow to brick-red precipitate shows presence of glucose.

OSMOSIS

EXPERIMENT No. 25

Object : *To demonstrate the process of osmosis.*

Repuirement : One big beaker or flask, one thistle funnel, strong cane sugar solution, pure water, semipermeable membrane (cellophane), one piece of gummed paper, wax, etc.

Principle : "Osmosis is a process in which water or any other solvent flows through a semipermeable membrane from a lower concentrated solute solution to one of higher concentrated solute solution". It is simply a case of diffusion, but the movement of solvent usually causes an obvious volume change, whereas the movement of solute has a negligible effect.

Procedure : Take a thistle funnel and the piece of a semipermeable membrane is tied tightly across the botton of the thistle funnel and the joint is waxed with melted wax to make it water-tight. Now fill up the bowl of thistle funnel with strong solution of sugar and marked the upper level of the sugar solution on the outside of the thistle funnel by a piece of gummed paper stuck at that level. Now this sugar solution contained in the thistle funnel is stood in a beaker containing pure water with the help of some stand. This entire apparatus is put aside for some hours.

Observations : After some hours it is seen that water from the beaker has entered the thistle funnel with the result the solution in the thistle funnel has increased in its volume and rises in the thistle funnel, wellbeyond the original mark. This level of sugar solution in thistle funnel keeps rising till the osmotic equilibrium is attained. On attaining the osmotic eqilibrium the same number of water molecules pass across the membrane in both directions.

Inference : The rise in volume of sugar solution in the thistle funnel is due to passing of water molecules from the beaker containing pure water of low concentration. The sugar molecules do not pass into the beaker because of semipermeable membrane. Thus, this experiment demonstrates the process of osmosis.

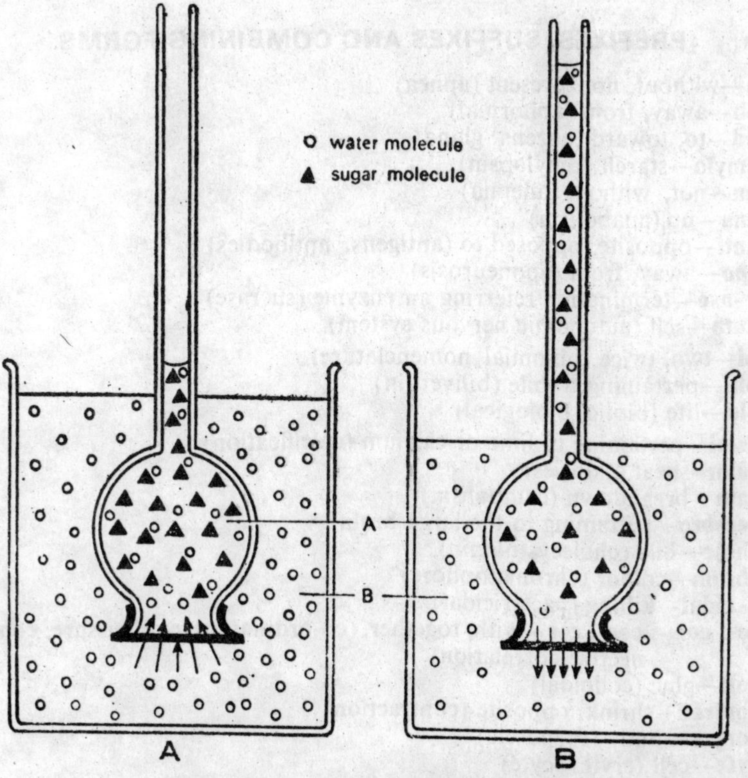

Fig. 17·12. Diagram illustrating the principle of osmosis. A—Beaker and thistle funnel which at the bottom is tied with semipermeable membrane and is filled with strong sugar solution. B—Beaker containing pure water. Arrows show direction of movement of water molecules, which can pass freely across the membrane, while the sugar molecules can not. The right hand side diagram also expresses the state of osmotic equilibrium, at this state the same number of water molecules pass across the membrane in both directions.

Precautions : The membrane to be taken for the experiment should bc semipermeable, *i.e.*, the water passes freely through it in either direction but not the solute such as sugar. If the membrane allows the solute to pass as well as water, then osmosis will not occur.

Glossary

PREFIXES, SUFFIXES AND COMBINING FORMS

a—without, not present (apnea)
ab—away, from (abnormal)
ad—to, toward (adrenal gland)
amylo—starch (amylopsin)
an—not, without (anemia)
ana—up (anabolism)
anti—opposite, opposed to (antigens, antibodies)
apo—away, from (aponeurosis)
—ase—termination referring an enzyme (sucrase)
auto—self (autonomic nervous system)
bi—two, twice (binomial nomenclature)
bili—pertaining to bile (biliverdin)
bio—life (biotic, biological)
calci—pertaining to lime or calcium (calcification)
calor—heat (calories)
cata—breakdown (catabolism)
cerebro—pertaining to the large brain
chole—bile (cholecystokinin)
chrom—colour (chromatophore)
—cidal—killing (bactericidal)
co-, com-, con-, cor—with, together, (co-ordination, commissure, connective correlation)
coll—glue (colloidal)
contrac—shrink, opposite (contraction)
corpus—body (corpuscle)
cyte—cell (erythrocyte)
dermic—dermis-skin (epidermic, epidermis)
di—two, twice (diploblastic, disaccharide)
dia—through, apart (diaphragm)
dis—negative (disinfect)
ectomy—pertaining to operation, *i.e.*, to cutout (hypophysectomy)
em-, en-, endo—in, into (emboly, engulf, endoderm)
—emia—blood (anemia)
entero—intestine (entrogastrone)
epi—on, above, upon (epidermis)
erythro—red (erythrocyte)
ex—out (expiration)
—fer—to carry (efferent)
—fract—break (refraction)
—gastric—stomach (pneumogastric)

—gen=producing (antigen)
—genesis—a process (spermatogenesis, oogenesis)
gluc—glucose, sugar (glucosuria)
—gnosis—knowledge (diagnosis)
haemo—blood pigment (haemoglobin)
hetero—other, different (heterotrophic)
hydro—water (hydrolytic)
hyper—over, above measure (hyperthyroidism, hypertension. hypertonic)
hypo—under, less than (hypothyroidism, hypotonic)
in—in, into (ingestion)
in—not, without (insufficiency)
inter—between, together (interbranchial, intercellular, interstital)
intra—within (intracellular)
—itis—inflammation (tonsillitis)
—ject—to throw (ejection, injection)
kin—to move (kinetic)
—lac=milk (prolactin)
leuco or leuko—white (leucocyte)
—logy—doctrine, science (physiology)
lymphos—pertaining to lymph (lymphocyte)
—lysis—dissolving, destruction (hydrolysis)
—lytic—destruction or dissolving (proteolytic)
macro—large (macrophagus)
micro—small (microphagus, micro-organism)
mole—mass, body (molecule)
mono—one (monosaccharide)
myo—muscle (myoglobin)
nephr—kidneys (nephritis)
neur—pertaining to nerve cell (neuron)
nucleo—pertaining to nucleolus (nucleolus)
—oid—like (amoeboid)
—ole—small (arteriole)
—oma—swelling, tumor (sarcoma)
—osis—a process (phagocytosis)
ovi-or **ovo**—egg (oviduct)
para—near, by, beside, within (parathyroid)
patho—suffering, disease (pathology)
peri—around (peripheral nervous system)
phago—to eat (phagocyte)
philic—pertaining to cell (acidophilic, basophilic)
plasm—form (nucleoplasm)
pneumo—air, lungs (pneumonia)
poly—many (polysaccharide, polypeptide)
post—behind, after (postganglionic)
pre—before (preganglionic)
pro—before, giving rise to (protheca, proenzyme)
proprio—one's own (proprioreceptor)

proto—flrst (Protozoa)
pseudo—false (pseudopodia)
pulmo—lung (pulmonary)
re—back, again (regurgitation)
—**renal**—kidney (adrenal)
—**rrhoea**—flow (diarrhoea)
sarco—flesh, muscle (sarcolemma, sarcoplasm)
—**sclero**—hard (sclerotic)
semi—half (semi-solid, semi-permeable)
sub—under, below (subepidermal)
—**thrombo**—clot, coagulation (thrombin)
—**trophic**—pertaining to target organ (thyrotrophic)
vaso—pertain to blood vessels (vasoconstrictor)

GLOSSARY

Absorption—It is a process of taking up materials by the skin, mucous surfaces or absorbent vessels.

Accommodation—The focussing of images of near objects on the retina, in mammals it is affected by the shortening of the focal length of the lens, while in other animals such as frog, it is effected by moving the lens further from the retina.

Acid-base balance—The proper ratio of H and OH ions in the blood.

Acidosis—It is a condition in which bicarbonate concentration of the blood falls below normal.

Actin—It is a muscle protein which helps actively in twitching of muscles.

Action potential—A change in the resting or membrane potential of an active cell or tissue. It is generally referred to as nerve impulse.

Active state—It is a condition in which a muscle, immediately after its excitation, shows an increased resistance to stretch and capability of shortening or creating tension.

Active transport—It is a process in which materials are transported against gradients by the expenditure of metabolic energy.

Adenohypophysis—It is an anterior lobe of pituitary gland and is derived from the alimentary canal.

Adequate stimulus—That stimulus which is strong enough to excite a given tissue.

Adipose tissue—Tissues in which considerable amounts of fat are stored.

Adrenal glands—These are endocrine glands found just above the kidneys.

Adrenaline—It is a hormone secreted by the medulla of adrenal gland.

Adrenocorticotropin—It is a hormone secreted by posterior pituitary and which regulates the activity of the cortical part of the adrenal gland.

Aerobic respiration—Respiration involving the use of oxygen.

Afferent blood vessels—Blood vessels which carry deoxygenated blood to the respiratory surfaces, *i.e.*, the gills (as in fishes).

Afferent nerve—A nerve which carries impulses from the receptor organ to the central nervous system.

Aldosterone—It is a hormone secreted by the cortical part (zona-glomerulosa) of the adrenal gland.

Alveolar air—The air which is found in the alveoli of the lungs.

Alveolus—It is a unit pocket of lung.

Amino acid—An acid containing an NH_2 and COOH group and having both acidic and basic properties.

Amphoteric—Any substance which is capable to act as acid and base at a time.

Anabolism—It is a building up process of metabolism by which protoplasm is produced from the simple food molecules.

Anaerobic respiration—Respiration not involving the use of oxygen.

Androgen—It is male sex hormone secreted in the testes.

Anemia—A lack of the proper number of erythrocytes per cubic millimeter of blood.

Angstrom (=Å)—A unit of measure equal to 10^{-8} cm.

Antibody—A substance produced in the blood to combat bacteria and virus.

Antidiuretie hormone—This hormone is secreted by the neurohypophysis of the pituitary and which influences the kidney and promotes the conservation of body water.

Anus—It is a terminal opening of the alimentary canal to the exterior.

Aorta—The main artery springing from the heart and leading blood away from the heart. In mammals it heads from the left ventricle.

Apoenzyme—It is a protein part of an enzyme.

Arteriole—It is a fine artery formed after repeated branching of main artery.

Artery—The blood vessel which carries the blood, mostly oxygenated, away from the heart to the tissues.

Assimilation—The incorporation of food substance into the protoplasm of an organism.

Atrioventricular bundle—It is a bundle of special tissue which is responsible for the conduction of nerve impulses of contraction from the auricle to the ventricle.

Atrium (=auricle)—A thin-walled chamber of heart which receives the blood from the veins.

Atrophy—The destruction of the tissues or organ as a whole.

Autolysis—It is a process in which a cell is disintegrated by its own enzymes.

Avitaminosis—A disease due to lack of vitamins.

Axon—It is one of the protoplasmic processes of a nerve cell or neuron which conducts impulses away from the cell body.

Basal ganglia—Group of neurons at the base of the cerebrum consisting of caudate nucleus, putamen, globus pallidus, subthalmic nucleus, substantia nigra, and red nuclei.

Basal metabolism—It refers to the minimum energy required to maintain the normal activities of the body.

Bilirubin—It is a red pigment of bile juice.

Biliverdin—It is a green pigment of bile juice and is derived from the bilirubin.

Biuret test—It is a test for native proteins and peptones.

Blind spot—The spot where the optic nerve is connected with the retina and which is lacking in rods and cones and hence insensitive to light.

Blood plasma—It is a fluid part of the blood which is distinct from the blood cells (corpuscles).

Blood pressure—The force exerted by blood against the blood vessel.

Bouton—It is terminal enlargement of an nerve fibre or axon.

Bowman's capsule—It is a cup-shaped structure found at one end of the uriniferous tubules of ths kidneys.

Bronchial tubes—Tubes which formed by the division of the trachea and go into the lungs.

Brunner's glands—These are glands found in the submucosa of the upper division of duodenum and help in digestion of food.

Bundle of His—Tissue which is capable enough to couduct the impulses from the auricle to the ventricle.

Calorie—It is a unit of heat energy.

Carbhaemoglobin—It is a complex of haemoglobin and CO_2.

Capillary—A tube of very small diameter. In the circulatory system, capillaries connect the ends of the smallest arteries (arterioles) to those of the smallest veins (venules).

Carbohydrates—A compound of carbon, hydrogen and oxygen in which the ratio of the hydrogen and carbon atoms in molecule is 2 : 1.

Cardiac stomach—It is an anterior part of the stomach.

Carnivore—Animal which feeds on animal flesh.

Casein—It is a milk protein.

Catabolism—It is the destructivephase of metabolism ; the disintegration af protoplasm.

Catalysis—It is a change in the rate of chemical reaction brought about by a catalyst.

Catalyst—The substance capable to accelerate a chemical reaction without itself undergoing any permanent change.

Cell—It is the unit of the body of animal or plant consisting of a nucleus and cytoplasm.

Cellulose—It is a polysaccharide.

Cerebellum—It is a part of hind-brain. It is concerned with the co-ordination of muscular movement and with the maintenance of the equilibrium of the body.

Cerebrum—It is a part of fore-brain concerned with intelligence, etc. It is highly developed in mammals.

Cholesterol—It is a fat-like substance present in animal tissues and fluids such as bile juice.

Chorionic gonadotropin—It is a hormone secreted by the placenta responsible for the release of estrogens and progesterone from the corpus luteum.

Choroid—It is a pigmented middle coat of the eyeball.

Chromoprotein—Proteins which contain amino acids and coloured pigments in their molecules.

Chyle—The contents of the intestinal lymph vessels.

Chyme—It is semi-digested food found in the stomach.

Coagulation—The formation of clot of blood.

Cochlea—It is a part of internal ear responsible for the reception of sound waves and the generation of auditory impulses.

Coenzymes—They are organic substances that constitute the part of the catalytic system, responsible for the transfer function of enzyme action.

Colon—It is a large intestine except for the caecum and rectum.

Conduction—Passage of the nerve impuse along the nerve fibre and other protoplasmic tissues.

Conjugated protein—Protein which is made up of amino acids and other chemical compounds, *e.g.*, chromoprotein, glycoprotein, lipoprotein.

Conjunctiva—The thin transparent skin overlying the front of the eyeball and continuous with that of the eyelids.

Connector neuron (interneuron)—A neuron linking a sensory neuron to a motor neuron and is always found in the central nervous system (spinal cord).

Convoluted tubule—A coiled tube-like structare of a uriniferous tubule of the kidney.

Cornea—It is a transparent part of the sclerotic coat of the eye at the anterior region through which light penetrates into the eye.

Coronary arteries—Blood vessels which supply the blood to the heart muscles.

Corpora quadrigemina—Four small almost rounded elevations that are found dorsally in the mid-brain.

Corpus—It denotes body.

Cortex—The outer part of a kidney or brain or adrenal gland.

Corticotrophin releasing factor—It is a chemical agent secreted by the neurosecretory cells of the hypothalamus which on reaching the adenohypophysis of the pituitary gland activates its cells, to produce adrenocorticotropin hormone.

Cortisole—It is a hormone of cortex of the adrenal gland.

Cortisone—It is a hormone secreted in the cortex of the adrenal gland.

Cranial nerves—Those which arise directly from the brain.

Creatine phosphate—It is a chemical substance found in abundance in muscle to provide energy to muscles for work.

Creatinine—It is a nitrogenous waste product.

Cretinism—It is a disease caused due to hypofunction of the thyroid gland which is characterized by dwarfed and malformed condition of the body.

Crypts of Lieberkuhn—These are also known as intestinal glands and are found throughout the small intestine.

Cuticle or cutis—It is a non-living covering generally present outside the epidermis, transparent in nature, *e.g.*, earthworms and insects.

Cytoplasm—The protoplasm of the cell excluding the nucleus.

Deamination—It is a chemical process in which amino group (NH_2) is removed from an amino-acid.

Defaecation—The removal of the faeces from the body or alimentary canal.

Deficiency disease—A diease which is caused due to lack of vitamin or hormone.

Deglutition—Passage of food from the mouth to the stomach.

Dendrite—A branching process of a neuron.

Deoxyribonucleic acid (DNA)—It is a high molecular weight compound generally found in the nuclei of cells.

Deoxyribose—It is a sugar of 5-carbon atom generally occur in DNA.

Depolarization—When polarized state of a neuron is reduced.

Dermis—It is a inner layer of the vertebrate skin.

Diabetes mellitus—It is a diseasee caused in the lack of insulin hormone in which carbohydrates not used by the body are excreted in the urine.

Diaphragm—It is a muscular sheath separating the thorax from the abdomen and helps in respiration.

Diastole—It is the state in which heart muscles relax, while heart chambers dilate.

Diastolic pressure—The lowest blood pressure in a blood artery during the diastolic state of the heart.

Diffusion—It is a process in which molecules of gas or substance are transported from high concentration to low concentration.

Disaccharides—Carbohydrates having two monosaccharide units in their molecule, *e.g.*, sucrose.

Digestion—It is a mechano-chemical process by which insoluble, indiffusible food substances are converted into small molecules of soluble and diffusible substances that can be absorbed easily by the intestinal cells. In this process digestive juices secreted by digestive glands take important part.

Duodenum—It is a anteriormost part of the small intestine generally of U-shape.

Dystrophy—A metabolic disease of muscle caused due to faulty nutrition.

Effectors—The responding organs, *e.g.*, muscles and glands.

Efferent neuron (motor neurons)—A neuron connected with an effector organ, *i.e.*, muscle or gland and sends impulses from the central nerous system to these effector organs.

Emesis—Vomiting.

Embryo—The young state of an animal or plant before it is capable of leading a self supporting existence.

Endocrine gland—Gland which produces endocrine secretions, *i.e.*, hormones and generally has no duct for the distribution of its secretions and the secretions are diffused directly into the blood stream.

Endolymph—It is a liquid present in the membranous labyrinths of the ear.

Endoplasmic reticulum—It is a fine network of tubules and vessels in the cytoplasm of the cell.

Endothelium—The epithelial covering of the heart, blood vessels and lymphatics.

Enteropeptidase—It is an enzyme secreted by the duodenal mucosa, in function it transforms trypsinogen into trypsin.

Enzyme—A chemical agent serves as catalyst and which is generally produced in living cells.

Epidermis—It is the outermost covering of the skin.

Epiglottis—It is a structure which closes the glottis when swallowing of food occurs.

Epinephrine—It is the hormone of adrenal medulla.

Ergosterol—A sterol which, upon irradiation with ultra-violet light, acquires the properties of vitamin D.

Erythrocyte—Red blood corpuscle having haemoglobin or other red pigment.

Estrogens—These are female sex hormones secreted in the ovaries generally.

Estrus—Sexual desire in excited or heat period.

Eustachian tube—It is a tube connecting the middle ear with the pharynx.

Excitibility—The property of responding to a stimulus.

Excitation—When a stimulation of any sort (physical, electrical or mechanical) is followed by increased protoplasmic activity.

Excretion—The removal of the nitrogenous wastes produced during metabolism from the body.

Exteroeceptors—Receptors of sense-organs stimulated by changes in the external environment.

Faeces—Undigested food residues egested via the anus.

Fascia—Connective tissue enveloping and connecting muscles.

Fasciculus—A small bundle of muscles.

Fat (or **oil**)—A compound of carbon, hydrogen and oxygen formed from glycerol and fatty acids.

Fatigue—A state during which living cell does not perform normal function.

Fibrin—It is an insoluble protein which is formed from fibrinogen while the blood is co-agulated.

Fibrinogen—It is a plasma protein which takes part in the co-agulation of blood, forming fibrin.

Follicle—Small sac-like structure.

Follicle stimulating hormone—It is one of the gonadotropin hormones secreted in the anterior lobe of pituitary gland, it stimulates the follicular development of ovaries.

Food stuffs—Substances that, when absorbed into the body tissues yield materials for the production of energy, the growth and repair of tissue and the regulation of life processes, without harming the organisms.

Ganglion—A collection of nerve cell bodies.

Gastrin—A hormone responsible for the production of gastric juice in the stomach secreted by the specialized glandular cells of pyloric stomatch.

Gestation—Pregnancy.

Gland—Any structure producing a secretion of any sort.

Glomerulus—The tuft of fine blood capillaries in BOWMAN'S capsule of the uriniferous tubule.

Glottis—The entrance to the larynx.

Glucagon—A hormone secreted by the α-cells of islets of LANGERHANS which regulates the sugar levels of the blood.

Glucokinase—This is an enzyme which causes the addition of a phosphoric acid radical to glucose.

Glucose—It is a monosaccharide sugar.

Glucosuria—The presence of glucose in the urine.

Glucogenesis—The process in which glycogen is formed from the glucose.

Glycogenolysis—The process in which glycogen is converted back into glucose.

Glycolysis—It is an anaerobic process of breaking down of glycogen into lactic acid.

Glyconeogenesis—The process in which glycogen is formed from amino acids.

Golgi bodies—Bodies which are secretory in function and always present near the cell nucleus.

Gonad—An organ producing gametes.

Gustatory—Pertaining to taste.

Habitat—The environment in which a particular type of animal or plant lives.

Haemoglobin—An iron containing respiratory pigment concerned with the transport of oxygen by the blood.

Haemophilia—It is a disease in which blood does not clot, that results in excessive loss of blood during injury.

Haemopoiesis—The formation of blood corpuscles.

Haemorrhage—The out flow of blood from the blood vessels.

Heparin—It is a chemical substance secreted in the liver cell, generally prevents the co-agulation of blood.

Herbivore—An animal whose main diet is vegetable.

Hermaphrodite—Any organism producing both eggs and sperms (female and male gametes).

Homeostasis—The maintenance of a constant or a normal state (with special reference to the internal environment).

Homoiothermic animals—Animals having a constant body temperature.

Hormone—A chemical substance produced by the endocrine structure which influences the growth, develpment and functien of some other part of the body. It is effective in small quantity.

Hydrolysis—It is a process in which a larger molecule is splitted into a number of smaller molecules through the addition of one or more water molecules.

Hyperglycemia—Presence of excessive amount of glucose in the blood.

Hypertension—Sudden increase in the normal blood pressure.

Hypoglyeemia—Presence of small amount of glucose in the blood.

Hypophyseal-portal system—It is the system of blood vessels supplying the adenohypophysis of the pituitary gland.

Hypophysectomy—Removal of the pituitary gland.

Hypophysis—Pituitary gland.

Impulse, nerve—Any change caused due to stimulation and conducted along the nerve fibre.

Ingestion—The taking of food into the alimentary canal of an animal.

Insulin—It is a hormone secreted by the β-cells of the islets of LANGERHANS, it regulates carbohydrate metabolism.

Interdigitation—An interlocking of part.

Interstitial cell stimulating hormone—It is a gonadotropic hormone secreted in the adenohypophysis of the pituitary gland which stimulates the activity of the LEYDIG cells.

Iris—The portion of the choroid coat of the eye visible through the cornea-conjunctiva.

Islets of LANGERHANS—Groups of α and β present in the pancreas concerned with the production of insulin and glucagon hormones.

Isotonic—Having the same tension or pressure (*e.g.*, osmotic pressure).

Jaundice—The pressure of bile in the blood.

Keratin—It is a scleroprotein neither soluble nor digestible.

Kinase—A proenzyme activator.

Lactase—The enzyme which digests the lactose into glucose and galactose.

Lactation—Production of milk.

Lacteals—Lymphatic vessels present in the intestine.

Larynx—The 'voice box' or upper portion of the main air passage leading to the lungs from the glottis.

Latent period—The length of the time elapsing between the stimulus and the response.

Leucocyte—White blood corpuscle.

Liminal—Minimum stimulus which brings about excitation.

Lipase—A enzyme which splits fats into fatty acid and glycerols.

Luteinizing hormone—It is a gonadotropic hormone secreted by pituitary adenohypophysis, responsible for the final maturation of the follicles of ovary, ovulation, and corpus luteum formation.

Lymph—The fluid of lymph vessels.

Lymphocytes—A certain type of leucocytes.

Malpighian body—The upper end of a kidney tubule acting as a filter.

Maltase—An enzyme that splits maltose into glucose.

Mammary gland—The milk producing glands of the female mammal.

Mastication—Break down of larger molecules into smaller ones by mechanical means.

Medulla—The central portion of the kidney, spinal cord or adrenal gland.

Medullated nerve fibres—Fibres having medullary sheath.

MEISSNER's **plexus**—A network of nerves in the submucosa of the small intestine.

Melanophore-stimulating hormone—It is a hormone of anterior pituitary responsible for the colour of skin in most animals including man.

Menstruation—The periodic changes in the uterus during the sexual life of the female, characterized by bleeding.

Metabolism—The sum total of the chemical changes occuring in the body.

Mucin—A glycoprotein of saliva.

Myelin sheath—The inner covering of a medullated nerve fibre.

Myofibril—A sub-division of the muscle fibre found in the sarcoplasm.

Myogenic—Originating in muscle tissue.

Myosin—It is a muscle protein helping in muscle contraction.

Myxoedema—It is a disease caused due to lack of thyroxine hormone of thyroid gland, in which subcutaneous tissue becomes filled with a mucin-like material and mental disorders are also evident.

Nerve—A bundle of nerve fibres.

Nerve cell (Neuron)- It is a structural and functional unit of nervous tissue.

Nerve fibre—It is an elongated protoplasmic process of nerve cell body which is usually covered with one or more sheaths.

Neurilemma—It is the outermost covering of the nerve fibre.

Neurohumor—It is a chemical agent responsible for the transmission and excitation.

Neurohypophysis—It is the posterior lobe of pituitary gland made up of nervous tissue only.

Norepinephrine—It is a hormone of adrenal medulla.

Oesophagus—The portion of the alimentary canal leading from the pharynx to the stomach.

Omnivore—Animal whose diet is derived from both plants and animal.

Ornithine cycle—It is a cyclic chain of reactions leading to the formation of urea from the ammonia.

Osmosis—The diffusion through the semi-permeable membrane of the solvent from a lower to a more concentrated solution.

Ovary—Female sex gland in which ovum is produced.

Oviduct—Tube leading the eggs of an animal from the ovary to the exterior, *i.e.*, outside the body.

Ovulation—The setting free of the ovum from the ovary.

Oxygen—Respiratory gas.

Oxyhaemoglobin—A combination of oxygen with respiratory pigment, haemoglobin.

Oxyntic cells (parietal cells)—Cells which secrete hydrochloric acid and these are found in the stomach wall.

Oxytocin—It is a hormone of posterior pituitary (neurohypophysis) whose principal action is to contract smooth muscles and ejection of milk in the mammary glands of female.

Pancreas—It is an exocrine gland found in the loop of alimentary canal in the abdomen, it secretes pancrcatic juice which by its enzymes digests the food substances.

Parathormone—It is the hormone of parathyroid gland which regulates the calcium and posphorous metabolism.

Parathyroids—These are endocrine glands found embedded in the thyroid gland and secrete parathormone.

Paraventricular nnclei—The specialized groups of neurons in the hypothalamus involved in the regulation of neuro hypophyseal secretion of oxytocin.

Pellagra—It is a disease caused due to deficiency of nicotinic acid and is characterized by changes in the structure of skin (keratinization), gradual impairment of digestive system and changes in the nervous system (mental disordcrs).

Pepsin—A proteolytic enzyme secreted by the gastric glands of stomach.

Pericardium—A thin membrane enclosing a space (the pericardial space) in which heart is located.

Peristalsis—The wave-like muscular movements of the walls of the alimentary canal by which food is made to pass along it. Faeces are removed by peristalsis of the rectum.

pH—The negative logarithm of the hydrogen ion concentration.

Phagocytosis—The process by which a cell engulfs the food particles.

Pharynx—It is a part of alimentary canal between the buccal cavity and the oesophagus and forming a common passage for food and air in land vertebrates. In fishes its walls are pierced by gill slits.

Phosphoglucomutase—It is an enzyme responsible for the conversion of glucose-1-phosphate the initial product of glycogen break down, to glucose-6-phosphate.

Phosphoprotein—Protein made up of amino acids and phosphorus, *e.g.*, casein.

Pineal gland—It is a gland located dorsally on the brain and is regarded as an endocrine gland concerned with sexual development and colour changes.

Pinocytosis—A process of ingestion of molecular particles similar to that of phagocytosis but on a micro-scale, cell drinking.

Pituitary gland (hypophysis)—It is commonly known as master gland of endocrine function and usually found at the ventral side of the diencephalon—a part of fore brain.

Plasma—The liquid constituent of the blood.

Plasma-membrane—The outer layer of cytoplasm of selective permeability.

Plasmolysis—The skrinkage of protoplasm due to withdrawl of liquid from the cell by a hypertonic solution.

Pleura—The membranes surrounding the lungs.

Plexus—A network either of nerves or veins.

Portal vein—The veins which carries the deoxygenated blood (venous blood) from the small and large intestines, spleen and stomach to the liver.

Proenzyme—Precursor of the enzyme.

Progesterone—It is female sex hormone secreted by the corpus luteum and placenta and is responsible for proliferative changes in endometrium and breats.

Prolactin—It is a hormone of anterior pituitary which is responsible for the production of milk in coordination with thyroid and adrenal cortex hormones. In mammal it is also responsible for the development of the mammary glands.

Prosthetic group—A chemical compound which is attached to enzyme protein for its activity.

Protein—A compound of carbon, hydrogen, oxygen and nitrogen (and sometimes sulphur and phosphorus as well) built of units called amino acid.

Prothrombin—Substance from which thrombin formed.

Ptyalin—It is a starch-splitting enzyme of saline secreted by salivary glands.

Puberty—The period of life while the young of either sex become capable of reproduction, *i.e.*, onset of sex organ function.

Pulse—A rhythmical dilation of the artery caused by the systolic output.

Pulse pressure—Difference between the systolic and diastolic pressures.

Pupil—The circular aperture of variable diameter in the centre of the iris of the eye.

Pylorus—The part where stomach and small intestine meet.

Rachitis—It is a disease caused due to deficiency of vitamin D (calciferol) in which bones show incomplete calcification due to which disturbance in bone formation is evident.

Receptor—Structure specialized to perceive condition outside and to some extent inside the body.

Rectum—Posteriormost part of the alimentary canal where undigested food is stored for removal from the body when it is required.

Reflex action—An automatic response to a stimulus made by an animal possessing a nervous system.

Reflex arc—The constituents involved in bringing out reflex action, consisting a receptor, a sensory neuron, a connector or inter-neuron, a motor neuron and a effector organ.

Refractory period—Period during which protoplasmic structures do not respond to stimulus.

Rennin—Milk protein splitting enzyme secreted by the gastric glands.

Respiration—A chemical activity taking place within the protoplasm of the cell and results in the liberation of energy.

Retina—Innermost layer of the wall of the eyeball containing the light sensitive cells the rods and cones.

Ribonucleic acid (RNA)—It is a nucleic acid having ribose sugar and helps in protein synthesis in the cells.

Rigor mortis—The stiffening or hardening of the muscles soon after death.

Sarcolemma—It is a sheath found enclosing the muscle fibre.

Sarcoplasm—Cytoplasm of the muscle fibre.

Sarcoplasmic reticulum—A network of fine tubules and vesicles found in the sarcoplasm of the muscle fibres which is involved in the transmission of excitation from the surface of the cell to the contractile proteins.

Scurvy—It is a disease caused due to lack of vitamin C and is characterized by bleeding gums, loosening and falling out of teeth and subcutaneous and intramuscular haemorrhages.

Secretin—It is a hormone responsible for the production and secretion of pancreatic juice, secreted in the wall of intestine.

Semi-permeable membrane—The membrane which allows the free passage of water through it but not certain substances in solution in the water.

Somatotropin hormone—It is a growth hormone secreted in the adenohypophysis of the pituitary gland.

Spermatogenesis—The process of sperm production from the primitive germ cells.

Sphincter—It is a band of muscles located at the opening to close it, *e.g.*, pyloric sphincter.

Stimulus—An influence of the environment evoking a response by a plant or animal.

Supra optic nuclei—These are specialized neurons found in hypothalamus responsible for the neurohypophyseal secretion of vasopressin (antidiuretic hormone).

Supra-renal glands—The adrenal glands.

Synapse—A functional junction between the axon branches of one neuron and the dendrites of cell body of another neuron.

Systole—The contraction of the heart muscle.

Systolic pressure—The arterial pressure during the systole of the heart.

Testis—Male reproductive gland.

Tetanus—A sustained contraction of muscle produced by the fusion of twitches.

Threshold stimulus—Minimum frequency of stimulus required to bring any change in the protoplasmic structures.

Thrombin—It is a factor which takes part in blood coagulation.

Thromboplastin—It is an enzyme responsible in the formation of thrombin.

Thrombosis—Formation of clot.

Thymus—It is a gland referred to as endocrine gland which is responsible for the secondary sexual characetrs.

Thyroid gland—A gland of internal seeretion generally found in the neck region above and either side of the trachea.

Thyroxine—It is a hormone of thyroid gland.

Trachea—The main passage between the larynx and the branchial tubes.

Transamination—It is a process which brings about the transfer of the amino group from the donor amino acid to a recipient keto acid, under the influence of transaminase or amino transferase enzyme.

Troponin—It is a muscle protein taking part in the contraction.

Trypsin—Proteolytic enzyme found in the intestine

Trypsinogen—It is an inactive form of trypsin, it in the influence of enteropeptidase enzyme, is converted into active trypsin.

Twitch—A single contraction of a muscle.

Urea—It is nitrogenous waste produced at the end of ornithine cycle and is eliminated from the body in the form of urine.

Ureter—The duct which carries the urine from the kidney to the urinary bladder.

Urethra—The tube leading from the bladder to the exterior. In the male mammal it terminates at the tip of the penis.

Uric acid—It is a nitrogenous waste product of birds.

Vein—The blood vessel which conveys blood towards the heart from the capillaries.

Vasopressin—It is a hormone of neurohypophysis responsible for the permeability of the membranes of the uriniferous tubules of the kidney, thus, antidiuretic in nature.

Venous blood—Deoxygenated blood having more CO_2 and less oxygen.

Villus—A minute finger-like structure of the intestinal mucosa projecting into the lumen of the intestine. It is an organ of absorption.

Vitamin—It is an organic compound which does not furnish energy but is responsible for the transformation of energy and for the regulation of the body metabolism.

Zymogen—It is a precursor of the enzyme.